AF709485

HANDBOOK OF FINANCIAL DATA AND RISK INFORMATION

Volume I: Principles and Context

Risk has always been central to finance, and managing risk depends critically on information. In recent decades, derivative markets for parceling and shifting risks have burgeoned, while computational advances have increased the pace of trading. As evidenced by recent events, the need has never been greater for skills, systems, and methodologies to manage risk information in financial markets. Authored by leading figures in risk management and analysis, this handbook serves as a unique and comprehensive reference for the technical, operational, regulatory, and political issues in collecting, measuring, and managing financial data. It will appeal to a wide range of audiences, from financial industry practitioners and regulators responsible for implementing risk management systems, to system integrators and software firms helping to improve such systems.

Volume I examines the business and regulatory context that makes risk information so important. A vast set of quantitative techniques, internal risk measurement and governance processes, and supervisory reporting rules have grown up over time, all with important implications for modeling and managing risk information. Without an understanding of the broader forces at work, it is all too easy to get lost in the details.

Volume II sets out a detailed structural and operational framework for managing a financial risk data repository. As experience accumulates on managing modern risk systems, the knowledge base of practical lessons grows. Understanding these issues and leading practices may mean the difference between failed and successful implementations of risk systems.

Margarita S. Brose, a former Senior Counsel in the Division of Enforcement of the US Securities and Exchange Commission, has almost two decades of experience in the financial markets. After leaving the SEC and earning an MBA from The Wharton School, she spent ten years consulting to financial services companies, with a focus on risk and compliance, with IBM Business Consulting Services (and its predecessor, PricewaterhouseCoopers Consulting). In that role, her clients included leading global investment banks and money center banks, as well as GSEs. Brose has since served in Director positions at Fannie Mae and Barclays investment bank, in their Finance and Risk Management groups, respectively.

Mark D. Flood did his undergraduate work at Indiana University in Bloomington, where he majored in finance (B.S., 1982), and German and economics (B.A., 1983).

In 1990, he received his Ph.D. in finance from the Graduate School of Business at the University of North Carolina at Chapel Hill. He has worked as Visiting Scholar and Economist in the Research Department of the Federal Reserve Bank of St. Louis, an Assistant Professor of finance at Concordia University in Montreal, a Visiting Assistant Professor of Finance at the University of North Carolina at Charlotte, a Senior Financial Economist in the Division of Risk Management at the Office of Thrift Supervision, a Senior Financial Economist with the Federal Housing Finance Agency, and most recently as a Research Principal with the US Office of Financial Research in Washington, DC. His research interests include financial markets and institutions, systemic financial risk, financial data management, securities market microstructure, and bank market structure and regulatory policy. His research has appeared in a number of publications, including the *Review of Financial Studies*, the *Annual Review of Financial Economics*, the *Journal of International Money and Finance*, *Quantitative Finance*, and the St. Louis Fed's *Review*.

DILIP KRISHNA is a Director with the Governance, Risk and Regulatory Consulting practice with a focus on risk architecture and information. He is involved in several regulatory initiatives such as stress testing and capital management implementations. He has 17 years of experience across a range of wholesale and retail banking products in banks across North America. Previously, Krishna was a partner with Teradata Corporation, leading its North-East Banking and Capital Markets consulting team. He worked with many large banks in the North-East to implement some of their largest data warehouse and data management programs. Krishna was also Director of Teradata's Risk Management Center of Expertise in the Americas, responsible for working with many large financial institutions across the USA and Canada implementing Basel II and risk management solutions. In this role he was also active with Teradata's Government Services team, helping to articulate the role of analytics in regulation and transparency, including being called upon to offer expert testimony to the US Congress. He has served as chief architect for Basel II at a major Canadian bank, and has implemented front-offices trading solutions in Canadian capital markets operations. Krishna is widely published in the area of risk information and risk architecture.

BILL NICHOLS has been at the intersection of technology and finance for 25 years. He co-founded and served as CEO of a research firm acquired by Thomson Financial in 1995, and spent the following seven years with Thomson. Subsequently, he spent three years in venture capital-related organizations, followed by consulting stints for major banks, asset management firms, and securities exchanges. He has participated as an expert in multiple ISO standards groups and is the former Convenor of the ISIN Standard for financial instruments. With Mark Flood, he co-chaired the Data Committee of the Committee to Establish the National Institute of Finance. Currently a Senior Advisor for Information Architecture and Innovation at the Office of Financial Research, he was previously Data Architect at Bridgewater Associates in their Data Research group.

HANDBOOK OF FINANCIAL DATA AND RISK INFORMATION

VOLUME I

Principles and Context

Edited by

MARGARITA S. BROSE
MARK D. FLOOD
DILIP KRISHNA

and

BILL NICHOLS

CAMBRIDGE UNIVERSITY PRESS

CAMBRIDGE
UNIVERSITY PRESS

University Printing House, Cambridge CB2 8BS, United Kingdom

Cambridge University Press is part of the University of Cambridge.

It furthers the University's mission by disseminating knowledge in the pursuit of education, learning, and research at the highest international levels of excellence.

www.cambridge.org
Information on this title: www.cambridge.org/9781107012011

© Cambridge University Press 2014

This publication is in copyright. Subject to statutory exception and to the provisions of relevant collective licensing agreements, no reproduction of any part may take place without the written permission of Cambridge University Press.

First published 2014

Printed in Spain by Grafos SA, Arte sobre papel

A catalog record for this publication is available from the British Library

ISBN 978-1-107-01201-1 Hardback

Cambridge University Press has no responsibility for the persistence or accuracy of URLs for external or third-party internet websites referred to in this publication, and does not guarantee that any content on such websites is, or will remain, accurate or appropriate.

This work contains general information only and is not rendering accounting, business, financial, investment, legal, tax, or other professional advice or services. Before making any decision or taking any action that may affect you or your business, you should consult a qualified professional advisor. This work is not a substitute for such professional advice or services. No party involved in the production and publication of this work shall be liable for any loss sustained by any person who relies on information contained within this book to their detriment.

Contents

Contributors

Marc Alvarez is a Senior Director with Interactive Data in New York responsible for Reference Data service infrastructure and enterprise data architecture. He is the author of *Market Data Explained: A Practical Guide to Global Capital Markets Information* (Elsevier, 2007) and an active industry commentator focusing on capital markets data content and management.

Nicola Antoniou is a Bank Expert at the European Banking Authority, where she specializes in policy and regulation for European cross-border banking supervision, in particular on the supervisory review and evaluation of bank risks. Nicola has previously worked at the UK Financial Services Authority focusing on international strategy and policy coordination.

Jim Blair is a Senior Consultant with Teradata, Inc. He is recognized globally for his experience in data warehouse development, implementation, and performance. He is dedicated to driving business value from analytical processing. Jim is an international speaker on data warehousing and a standing member of TDWI and the Society of Information Management.

Jefferson Braswell is the Founding Partner of Tahoe Blue Ltd. and has been active in the development of financial technology for over 30 years. He has delivered comprehensive risk management solutions to many of the world's largest financial institutions. Recently, he has served as a lead coordinator of the G20 LEI Private Sector Preparatory Group.

Margarita S. Brose has 20 years of experience in financial services. She began her career as an attorney at the US Securities and Exchange Commission. After earning an MBA, she specialized in risk and compliance consulting with PwC and IBM. She has degrees from Barnard College, GW Law School and The Wharton School.

Dianne Buonincontri is a data manager with the Federal Reserve Bank of New York. She is responsible for a team of business and data analysts performing functional and data analysis and creating data quality policies, standards, and processes. Dianne holds an executive MBA from Baruch College and has held previous roles in data management.

John Carroll is a well-known industry expert and experienced financial services operations and data manager currently working as a Managing Consultant at element22, a boutique advisory firm that specializes in data management within the financial services industry. John spent 22 years at Merrill Lynch, where he was responsible for Data Solutions and Integrations.

Thomas Dang is a software/database developer working for the University of British Columbia. He holds a Masters in computer science, with a subspecialization in human computer interaction and has been actively involved with research on visual analytics. Thomas's other active interests include data integration, high-density information displays, and interactive video.

Brian Fisher is Associate Professor of Interactive Arts and Technology and Cognitive Science at Simon Fraser University and Associate Director of the Media and Graphics Interdisciplinary Centre at the University of British Columbia. His research explores the cognitive science of interaction with visual information systems (visual analytics).

Mark D. Flood is a graduate of Indiana University in Bloomington (B.S. in finance, and B.A. in German and economics). He earned his Ph.D. in finance from the University of North Carolina at Chapel Hill. He has taught finance and business at universities in the USA and Canada, and has been active in policy development and research publication as an economist in the risk management and research departments for a number of regulatory bodies.

Janine Forsythe is a Vice President at Goldman Sachs leading the technology teams that support the firm's global insurance businesses. Janine has over 25 years in the financial services industry in line management roles and consulting services focused on a broad spectrum of clients and business functions including insurance, capital markets, and commercial banking.

Martijn Groot is an expert in financial data management and has worked in this space in product management roles from the investment banking, financial technology, and services perspectives. His specialties include instrument, entity and analytics data services. Martijn is the author of *Managing Financial Information in the Trade Lifecycle* (Elsevier, 2008).

John Patrick Hunt is Acting Professor of Law at UC Davis School of Law (King Hall). His scholarly interests include credit rating agencies and securitization.

Alan King is a Research Staff Member at the IBM Thomas J. Watson Research Center in Yorktown Heights, New York, in the Industry Solutions Research Department with a focus on the Financial Services Sector. Alan's research focus is on analytics middleware technology in risk management processes and applications of stochastic programming.

Laura Kodres, an Assistant Director in the IMF's Monetary and Capital Markets Department, oversees the analytical chapters of its Global Financial Stability Report, which frequently focuses on systemic risk monitoring and mitigation techniques. Her research centers on financial contagion and interactions of macroeconomic and financial policy issues.

Alejandro Komai is a Ph.D. candidate at the University of California, Irvine. His research interests include economic history and monetary theory.

Dilip Krishna is a Director with Deloitte & Touche LLP, focusing on risk and regulatory analytics, data and architecture in financial services firms. He has previously held roles implementing large data warehouse and data management programs for initiatives such as Basel II.

Jarl Kure has worked as a supervisor for over 20 years; he is currently with the European insurance authority, EIOPA, where his responsibilities include project lead on the Solvency II Preparatory Guidelines. Previously, Jarl coordinated the EIOPA Financial Stability area, including EIOPA Risk Dashboard, stress testing and relations with the European Systemic Risk Board.

Simon Kwan is Vice President and Head of Financial Research at the Federal Reserve Bank of San Francisco. He conducts research and policy analysis in the areas of financial institutions and financial markets. Simon has published research articles in leading finance journals and book chapters.

Victoria L. Lemieux is an Assistant Professor at the University of British Columbia's iSchool with a research focus on financial informatics. She founded and directs the Centre for the Investigation of Financial Electronic Records and is also Acting Director of the University of British Columbia's Media and Graphics Interdisciplinary Centre.

Irina S. Leonova is a member of the Financial Stability Board Secretariat, where she currently specializes in market structure questions related to financial data infrastructures, LEI, and OTC and exchange traded derivatives among others. Previously she held positions at the US Department of the Treasury, CFTC, and Bureau of Economic Analysis.

John C. Liechty is a Professor at the Smeal College of Business at Penn State University. He is an expert in derivative pricing and asset allocation, computational statistics and high performance computing, and marketing research. He has a Ph.D. from the Statistical Laboratory at Cambridge University.

Robert Mark is the Founding Partner of Black Diamond Risk, serves on several boards, has led Treasury/Trading activities, and was a CRO at Tier 1 banks. He was the Founding Executive Director of the MFE Program at UCLA, has co-authored two books on risk management, and holds a Ph.D. in applied mathematics.

Michael J. McCabe is a Director with Deloitte Financial Advisory Services LLP, specializing in the application of analytic and forensic technology in the financial services. He has held previous roles consulting on regulatory and compliance issues, business processes and controls, data collection and retention, as well as transaction monitoring in financial firms.

Karla McKenna is Chair of the International Organization for Standardization Technical Committee 68 (ISO TC68), Financial Services, a position she has held since 2006. When not handling ISO assignments, Karla manages Market Practice and Standards for Citibank's Global Transaction Services, Securities, and Fund Services businesses. She sits on numerous other standards committees worldwide.

Bill Nichols is Senior Advisor at the US Office of Financial Research. Prior to joining the OFR, he was at Bridgewater Associates. He has worked at the intersection of finance and technology for the past 25 years and been active in the global financial standards community for ten years.

Jim Northey (B.S. University of Indianapolis, M.S. Illinois Institute of Technology) is a co-founder and senior partner in The LaSalle Technology Group, LLC. He is co-chair of the FIX Americas Regional Committee, representing FIX within ASC X.9. Jim is a member of the US Treasury OFR Advisory Committee.

Per Nymand-Andersen is an adviser to Senior Management at the European Central Bank. He specializes in European banking, financial markets, statistics, securities settlement systems, management and communication. Per is a key editor of the Statistics Paper Series and lectures at the Goethe University. He is a former international consultant and has an MBA in Economics and Management Science from Copenhagen Business School, Denmark.

John C. Pattison is a financial consultant specializing in regulatory and risk governance matters. He is a former banker and has taught at the Ivey School of Business, University of Western Ontario. His publications are mainly in the field of international finance and economics including the regulation of financial institutions.

Gary Richardson is a professor at the University of California, Irvine. He has recently been named the Federal Reserve System Historian. He is also a research associate at the National Bureau of Economic Research. His research interests include economic history, macroeconomics, the Great Depression, the Industrial Revolution, property rights, and economic development.

Nicholas Robson is an IT consultant with 18 years of experience across Canada, the USA, UK and Italy. He has a depth of experience in Enterprise Data Management, including database design, administration, architecture, and project management of EDM implementations. Nick holds an M.S., Computer Science, from the University of Hertfordshire, UK.

Clifford V. Rossi is an Executive-in-Residence and Tyser Teaching Fellow at the Robert H. Smith School of Business, University of Maryland. Prior to entering academia, he had nearly 25 years of experience in banking and government. He has also held senior risk management positions at Freddie Mac and Fannie Mae. He received his Ph.D. from Cornell University.

David M. Rowe is founder and president of David M. Rowe Risk Advisory, a risk management consulting firm. David has spent almost 40 years in the risk management technology, banking, and economic forecasting industries. He has written the monthly Risk Analysis column in Risk magazine since late 1999.

Jonathan Sparks is an attorney at King & Spalding in Atlanta, Georgia. He recently graduated from the George Washington University Law School in Washington, DC, where he specialized in corporate, regulatory, and securities law and policy.

Richard Stanton is Professor of Finance and Real Estate and holds the Kingsford Capital Management Chair in Business at the Haas School of Business, University of California at Berkeley. He is best known for his research on mortgage markets, term-structure modeling, mutual funds, and employee stock options.

Deborah Stockdale is an Engagement Partner at Teradata focusing on Enterprise Data Management, Data Governance, and Data Architecture. She has held previous roles as Chief Data Architect designing information architectures for satellite constellations, designing and implementing large scale data warehouses for Fortune 100 firms, and teaching data strategy and data management courses.

Charles Taylor is Deputy Comptroller of the Currency for Capital and Regulatory Policy. Previously, he was Director of the Financial Reform Project at Pew Charitable Trusts and Executive Director of the Group of Thirty.

Nancy Wallace is a Real Estate and Finance Professor, Haas School of Business, University of California Berkeley and the Lisle and Roslyn Payne Chair in Real Estate and Capital Markets. She co-chairs the Fisher Center for Real Estate and Urban Economics and co-faculty directs the Berkeley Center for Law, Business and the Economy.

Jesse T. Weintraub is a broker-dealer and regulatory enforcement attorney at Bracewell & Guiliani in New York City, specializing in the representation of various financial institutions in matters arising under federal securities laws and self-regulatory organization rules. Jesse is a graduate of the George Washington University Law School.

Preface

On August 9, 2007, a large European bank announced unexpectedly that it could not obtain valuations on subprime securities underlying three of its investment funds. Accumulated anxiety about the quality of the collateral in repurchase agreements (repos) turned into a full-blown panic attack. Lenders withdrew from this market in large numbers, sending dealers on an urgent hunt for new funding. The world was witnessing a full-scale run on the shadow banking system. At that time, the shadow banking system was not well understood by practitioners outside its arcane confines. Yet this episode demonstrated its ability to quickly spread contagion not only across the financial system, but also to the real economy, culminating in the events of the fall of 2008, where extreme measures had to be taken by the US and European governments to stabilize the economy. All told, the failures represented trillions of dollars in assets and hundreds of billions of dollars in losses, with central banks, finance ministries, and the US Treasury piling resources into the breach.

This was all a prelude to the full catastrophe.

In September 2008, a second intense wave of failures and near-failures claimed a number of victims, mostly in the USA, and many supposedly "too big to fail." The episode sent financial markets into a free fall that lasted for six months. The ensuing failures, interventions, conversions, and mergers have permanently and fundamentally altered the financial industry landscape. Much time, effort, and discussion has been devoted to analyzing these events; yet even today, years later, there are several conflicting theories about what exactly happened and who was responsible. It is facile to lay the blame wholly on first-time homebuyers in subprime and Alt-A mortgage markets, as these borrowers represented only one half of a transaction at the front end of a much longer securitization chain. How did loanable funds make their way into these deals? Observers of various persuasions have laid responsibility at the feet of such varied parties as the US and Chinese governments, Government Sponsored Enterprises, investment banks, and credit ratings agencies. In hindsight, one of the surprising things was how unprepared the world economic community was for the scale and speed of these events.

Anyone who has sat through a college investments course knows that information plays a central role in finance. Markets can be seen as an information-processing system. The Efficient Markets Hypothesis asserts that markets should reliably react and adjust to available information. They may make mistakes for a time, but they should not be wildly or consistently biased. Yet what the events leading up to 2007 showed us was that the markets did not, in fact, properly react to information that was available, and continued to build up stresses. How could sophisticated institutional investors have been so wrong for so long about the values of mortgage-backed securities?

Starting in February 2009 – as the economic situation continued to deteriorate despite extraordinary measures – a small group of academics, supervisors, and financial industry participants began a series of conference calls to discuss the "what happened" question. The initial core group of Professors John Liechty and Arthur Small of Penn State University and Mark Reesor of the University of Western Ontario, Joe Langsam at Morgan Stanley, and Allan Mendelowitz and Mark Flood at the Federal Housing Finance Agency, began soliciting further expertise among their professional networks. Many of the authors and editors of the Handbook you are reading participated in those discussions. Like a garage band, we settled on a name before we learned how to play our instruments: *The Committee to Establish a National Institute of Finance* (CE-NIF, `www.ce-nif.org`). We agreed from the start that a basic problem – although surely not the only challenge – was the supervisors' lack of wherewithal: they simply did not have data of sufficient quantity and quality, nor the analytic capacity required for the scale, pace, and sophistication of modern financial markets. We had the hubris to propose a new federal agency to rectify this shortcoming. The group grew quickly through professional connections. There was never a formal charter, outside funding, or well-defined authority within the CE-NIF. Instead, the incentive structure ensured that the most dedicated participants would take the lead: significant effort was required, with zero remuneration guaranteed and little hope of success or glory.

The idea for a Handbook emerged from the conversations of the Data Subcommittee of the CE-NIF, which assembled a wealth of technical and professional expertise and wisdom. This "data committee" included chief data and technology officers, senior risk managers, and economists, researchers, and engineers from government, industry, and academia. Although the original impetus for the CE-NIF and its data committee was a systemic financial crisis, the discussion quickly moved beyond systemic risk to the broader risk and data management issues in finance. Where economic theory speaks in abstractions about public information sets and conditional distributions, the data committee discussions focused on more mundane topics and concrete realities such as end-user computing, metadata management, messaging formats, data validation, ISO standards, and semantic modeling.

In sharing ideas and swapping war stories, two themes emerged. The first was the extent of disarray in financial data, not just for supervisors but at financial firms themselves. While it is common in policy and management circles to proclaim that it is the "What" that matters, not the "How," it became abundantly clear that the converse was often true: that the How had in fact influenced the What in a profound way.

The second theme was that we all had a lot to learn from one another. Too many facts and lessons were passed along as oral traditions within subspecialties, or re-learned the hard way by individuals confronting a problem that others had seen before. It was observed repeatedly during these sessions that, "this stuff just isn't written down anywhere."

This Handbook is our attempt to write some of those facts and lessons down. We have in mind the financial economist who needs to know about the main industry messaging standards, the database administrator who wants to learn what "volatility clustering" is and why it is important, the risk manager working with the mortgage supply chain for the first time, and the systems engineer trying to understand bank regulatory reporting. In short, we envisioned this Handbook as a practical companion and reference sitting on desks of the wide range of people who need to come together to solve the problems of information infrastructures in large financial institutions and regulatory organizations

The ambition to collect this much information in such a wide-ranging and fast-moving area in one place creates an inevitable tension between comprehensiveness and timeliness. We hope that the resulting compromises and omissions are at least tolerable, and that the Handbook will achieve its desired status as a dog-eared companion for the reader. We encourage the reader to make liberal use of references and sources provided to extend the lessons provided herein.

The Handbook is presented in two volumes. This satisfies the needs of practicality; two 600-page books are undoubtedly less daunting and more tractable than one 1,200-page book, but there is logic to this as well. Any attempt to address information infrastructure in a large organization (be it commercial or regulatory) must stand on a foundation of requirements. Volume I provides this basis in three parts. Part I provides the context and rationale for the need for financial data, describing the various data needs of risk and regulatory processes. Part II describes the supply-chain of data in the financial markets and the real economy, as well as the intersection and transmission of data across their common boundaries. Part III covers a broad range of topics relating to the use of data by regulators.

Volume II addresses topics that a financial institution or regulator would find necessary to implement change in their infrastructure. Part IV addresses a critical component of implementation by describing how data emanate and are modified within an organization. Part V addresses various topics relating to the rapid changes in technology for data processing that makes these problems much more tractable. Part VI covers topics relating to the management of data remediation programs and environments.

The Handbook brings together an eclectic range of experts from a wide range of fields, as befits this expansive topic. Early on in its development we felt it would better serve our readers to let the voices of these experts shine through unadulterated by attempts to standardize the presentation across the Handbook. Accordingly, each chapter employs its own presentation style, with the attempts at normalization restricted to the superficial "look-and-feel" of the chapter.

Bringing this Handbook together has been at once a challenging and rewarding task. We hope that you find as much value in reading and referencing the book as we did in putting it together. Nothing would please us more than if this Handbook were to render itself obsolete over time by playing a small part in fixing the present unwieldy information infrastructures in the financial industry.

Margarita Brose
Mark Flood
Dilip Krishna
Bill Nichols

VOLUME I

PRINCIPLES AND CONTEXT

PART I

RISK MANAGEMENT CONTEXT FOR FINANCIAL DATA

EDITORS' INTRODUCTION TO PART I

Margarita S. Brose and Mark D. Flood

The events of 2007–2009 should have dispelled any doubts about the importance of data and information for the management of risks in the financial system. The litany of information problems is by now painfully familiar: banks too big and too interconnected to fail, structured securitizations too complex to evaluate, shadow banking, inaccurate third-party credit ratings, mis-specified value-at-risk models and six-standard-deviation surprises, robo-signing of legal documents, and on and on.

The consequences of bad information can be enormous. Yet the underlying management issues here would be important even in the absence of a crisis. Many of these challenges are not a one-time fluke, but a basic characteristic of the Information Age. As data proliferate in financial markets and elsewhere, the need to manage those data becomes more urgent. Overall, data volumes are expanding at an exponential rate. Those organizations that successfully tame and harness this data flow will wield a powerful tool. Organizations that fail at this risk being overwhelmed – drowning in data and overmatched by better informed competitors.

The opening section of the *Handbook of Financial Data and Risk Information* provides a big picture view of the institutional and organizational context within which financial data and risk management occurs. Much of this activity occurs at the level of the individual firm, and a vast body of technical expertise, legal and regulatory constraints, and institutional practice have accumulated to guide it. This section of the Handbook examines the historical legacy that brought us to this point as well as the state of the art of risk management, reporting, and governance within financial firms. Risk and information are also issues beyond the boundaries of any one firm, however. The various tasks associated with systemic risk fall largely to regulators, with supervisory risk monitoring and policy making replacing firm-level risk management and governance as the foci. The section concludes with some reflections on the meaning and implications of the brave new data-centric world that dominates modern finance and risk management.

There are economic, intellectual, and historical reasons why we operate within the current system. Our immersion in a financial world dominated by computational infrastructure, large databases, and complicated risk systems has been sufficiently gradual that it can be difficult to imagine that it was ever otherwise. Proper understanding of the high-level forces and constraints that got us here is important for building data and information systems that work. With that in mind, Chapter 1, by Mark Flood, examines how we arrived at the current context of institutions, technologies, and risk practices, starting with the origins of risk management. The overlapping themes of institutionalization, technology, globalization, and complexity dominate this history, and emblematic technologies and events such as wars and financial crises illustrate the evolution of institutions and technologies through time and space. In the end, the more things change, the more they stay the same.

Chapter 2, by Robert Mark and Dilip Krishna, covers the landscape of risk management today and the tools used to carry out the discipline. The emphasis is on the practical realities of implementing risk systems. The authors survey the main categories of financial risks (market, credit, and liquidity) and "non-financial" risks (operational, strategic, reputational, and business). Risk systems are an enabling technology that has supported and encouraged many of the recent innovations in financial products, markets, and regulation in recent years.

Risk modeling, however, is of limited use if the information and insights derived from risk systems are unavailable to decision makers. In Chapter 3, Cliff Rossi looks at risk governance and reporting structures, especially the presentation of risk information to the board of directors and senior management. The running example in this chapter is a mortgage portfolio with its attendant assets, liabilities, business lines, and risk types, but the lessons for risk monitoring generalize to any financial firm. Key principles for effective risk monitoring are that it should be transparent, operationally tractable, forward-looking, integrated, and action-oriented. Building risk systems to achieve these goals requires a clear commitment from the leadership of the organization.

The challenges faced within the firm are only part of the landscape. Individual financial firms may have the tools and skills for managing their specific business risks, but this alone is insufficient. The financial crisis was a sharp reminder of the importance of monitoring and managing risks across the financial system as a whole. Data collection and analysis for systemic risk are the subject of Chapter 4, by Alan King, John Liechty, Cliff Rossi and Charles Taylor. Because systemic risk, by definition, breaches the boundaries of individual firms, the task of monitoring these risks falls to the regulatory community. The authors examine the definition(s) of "systemic risk" and highlight the various policy options available for monitoring and addressing it. We are still just beginning a long journey to explore the implications.

In Chapter 5, John Pattison reviews regulators' increasing reliance on formulas and models in the years leading up to the financial crisis. *Data-driven regulation* means that the information set guiding official supervision is guided and dominated by formal collections of well-defined, primarily numeric data. Alongside the obvious operational efficiencies, this also introduces layers of abstraction and quantification that have the potential to distract supervisors' attention from the underlying financial reality. Pattison argues that data-driven regulation has permanently changed the relationships under which supervision occurs, with important conversations mediated by data and analysis constrained by formal models. The net result is a relative shift in power to an expert caste of regulators with the technical skills to use the new tools.

1

A brief history of financial risk and information

Mark D. Flood

This chapter presents the historical context for the current state of financial information and risk management. At any point in time, the institutions and practices that constrain financial behavior are just the latest point in an evolutionary process along a single historical path. Our basic understanding of events is subject to this path dependence too. Keynes (1936, p. 383) notes that, "Practical men, who believe themselves to be quite exempt from any intellectual influences, are usually the slaves of some defunct economist." In tracing events over the longer term, the rhymes of history begin to emerge. There are recurring dynamics and episodic patterns that make history something more than mere chronology. Reinhart and Rogoff (2009), for example, mock the perennial speculator's incantation that "this time is different" with a sobering litany of financial over-enthusiasms that have ended in tears.

On the other hand, the world has indeed changed over the long run. As we write this, the global financial system is still emerging from the catastrophic events of 2007–2009. While one of this chapter's central goals is to demonstrate that many of the features of the recent crisis have historical antecedents, events that may appear to be defining characteristics of the latest trauma – such as interconnectedness, technological sophistication, and world-wide impact – are in fact the manifestations of broader historical trends that have been at work for decades or centuries. This is clearly true of the four thematic dimensions I emphasize below: institutionalization, globalization, technology, and complexity.

It is impossible to do justice to this rich history in such a brief space. Indeed, entire books have been written on subsets of the topic.[1] In lieu of a comprehensive history, the discussion ranges widely here, illustrating the broad historical themes by

[1] Examples include Goetzmann and Rouwenhorst's (2005) compendium on the history of financial innovation, Bernstein's (1998) history of the intellectual enterprise of modeling risk, Gleick's (2011) history of information theory, or the one-volume histories of various aspects of economic and financial development in the Atlantic sphere, by Kindleberger (1993), Neal (1990), Ferguson (2008), Grossman (2010), or O'Rourke and Williamson (2001).

Handbook of Financial Data and Risk Information, ed. Margarita S. Brose, Mark D. Flood, Dilip Krishna and Bill Nichols. Published by Cambridge University Press. © Cambridge University Press 2014.

identifying emblematic events while providing pointers to additional resources for deeper research on specific topics. I compensate for space constraints by decorating the discussion with recommendations for further reading. This makes for a lengthy bibliography, which the reader will hopefully find useful.

1.1 The institutionalization of financial services and regulation

History documents the gradual and piecemeal institutionalization of financial markets. Starting with the Age of Exploration, new organizations begin to emerge that are still familiar today. Developing and maintaining supply routes that extended to the Far East and the Americas required more capital and greater degrees of coordination than before. Hence, we see the introduction of new forms of ownership and control, such as the limited liability corporation, exemplified by the Dutch East India and British East India companies, described below. In turn, these new firms required access to deeper capital pools to fund their infrastructure; hence the creation of stock exchanges in London and Amsterdam. Also important were enhancements in risk management to spread the losses due to maritime hazards; hence Lloyd's of London, also described below.

The institutionalization of commerce is mirrored by institutionalization of the accompanying supervisory and regulatory superstructure.[2] As noted, there is a powerful historical path dependence that describes our current institutional status quo. Institutional change typically accretes gradually, limited by legal and bureaucratic inertia and frequently in concert with technological advances. Major events occasionally disrupt this process. Calomiris and Gorton (1991) note that the history of financial regulation can be written as a chronology of traumatic events, linking each to the financial institutions created in its wake. Table 1.1 provides a few examples. Even this brief sampling should firmly establish the "crisis mechanism" as a means of generating new institutions. The most recent such examples in the USA are the Financial Stability Oversight Council (FSOC) and Office of Financial Research (OFR), created through the Dodd–Frank Act, and which stand as memorials to the market collapse of 2008. Whether created in reaction to a crisis or to serve another financial purpose, such institutions tend to have remarkable staying power. A few, such as the First and Second Banks of the United States, have been dissolved, but many others, such as the Federal Reserve (1913) or the US National Banking System (1863) are still with us long after the memories of the Panic of 1907 or Civil War financing have

[2] A more detailed chronology of the establishment of US regulatory institutions appears in Chapter 12. A discussion of the current evolution of regulatory institutions in the European Union is found in Chapter 13. The crisis of 2007 and its aftermath is discussed in Chapter 5.

Table 1.1 *Some crises and their institutions*

Crisis	Institution(s) created	Reference
US Civil War (1861–1865)	National Banking System and Comptroller of the Currency	Hammond (1991)
Panic of 1907	Federal Reserve	Johnson (2010)
Great Depression (1930s)	Federal Deposit Insurance Corporation (FDIC)	Flood (1992)
	Securities and Exchange Commission (SEC)	Seligman (2003)
	Federal Housing Administration (FHA), Federal National Mortgage Association (Fannie Mae), and Federal Home Loan Bank system	Wheelock (2008)
	Bank of Canada	Bordo and Redish (1987)
Paperwork Crisis (late 1960s)	Securities Investor Protection Corporation (SIPC)	Markham (2002b)
Bank Herstatt failure (1974)	Basel Committee for Banking Supervision (BCBS)	Goodhart (2011)
Savings and Loan Crisis (1980s)	Office of Thrift Supervision (OTS), and Federal Housing Finance Board (FHFB)	FDIC (1997)
Global Financial Crisis (2007–2009)	Office of Financial Research (OFR), Financial Stability Oversight Council (FSOC), Federal Insurance Office (FIO), and Consumer Financial Protection Bureau (CFPB)	FCIC (2011)
	Financial Policy Committee (proposed) of the Bank of England	BoE/FSA (2011)
	European Systemic Risk Board (ESRB)	Nymand (Chapter 14, this Handbook)

faded.[3] The result, at any point in time, is a patchwork of official institutions, each originated in response to a specific historical event or need.

A second theme that appears repeatedly over time is financial innovation as a response to the inflexibility of official institutions. Kane (1977) posits a "regulatory dialectic" of alternating innovations and countermanding rules succeeding each other over time. For example, the three estates of church, nobility, and commoners dominated the institutional structure in Europe throughout the Middle Ages and up to the Industrial Revolution. In finance, an important practical issue was the need to accommodate the Church's ban on usury, taken from Deuteronomy (23:19–20).

[3] Bruner and Carr (2007) provide a readable and informative account of the Panic of 1907. See Johnson (2010), Friedman and Schwartz (1971), and Markham (2002b) on the founding of the Federal Reserve. Regarding the Bank of the United States and the National Bank Act, see Hammond (1991) or Markham (2002a).

The ban was vigorously debated and enforced for centuries, and it hampered a wide range of economic activity, most notably the financing of military adventures. One workaround was to delegate the business of money lending to Jews, who by definition were not subject to excommunication. The economic segregation of money lending facilitated economic activity, but unfortunately tended to reinforce anti-Semitic attitudes by adding a profit motive for the recurring pogroms that plagued the Jewish community.[4] Munro (2003) documents a number of more intricate circumventions of the rule, many of them variants of perpetual annuities, known as *rentes*, that sufficiently disguised the accrual of interest. These innovations expanded the range of financial techniques, forming a foundation for a subsequent "financial revolution."[5]

Few endeavors are riskier than war, and historically much of finance was devoted to funding the imperial exploits of the European powers in their high-stakes games. The voyage of Columbus was a precursor to several centuries of conquest and settlement, with the potential for violent conflict with indigenous locals or open warfare with national rivals. Large-scale conflict is a speculative undertaking for both the participants and their financiers. Historians sometimes refer to the "fiscal-military state" to describe militaristic societies where much of economic life exists in the service of financing a large standing army and navy (see Knights, 2012). The European powers in the fifteenth to eighteenth centuries fit this description. Frederick the Great, who had to feed a famously effective standing army, advised, "A financial system, handed over by father to son and constantly improved, can change a government's position" (Scott, 2009, p. 23). It was especially in Britain, however, that the needs of military funding supported the development of financial markets. In contrast to absolutist monarchies, which might renege on their debts, the invigorated powers of the British Parliament after the Glorious Revolution of 1688 created a public counterweight that could hold the crown to task for the repayment of debts (North and Weingast, 1989). This credible threat brought lenders into the market, making large-scale public borrowing possible and spurring the development of private financial markets as well.

The stakes for fiscal-military states were indeed high: the arrival of bullion (primarily silver) from Spain's New World colonies was a major contributor to the so-called "price revolution" in Europe, a two-century inflationary episode driven largely by mining discoveries and processing advances (see Braudel and Spooner,

[4] Usury was the subject of a number of ecclesiastical rulings over the centuries, of which the Third Lateran Council in 1179 is one of the more significant. Homer and Sylla (1986) provide a useful and concise chronology (although they appear to neglect Lateran III). Modern Christian teaching has largely relaxed prohibitions on usury. It remains an important feature of Islamic finance: El-Gamal (2008) describes the Shari'a prohibitions on usury (*riba*) and gambling (*gharar*), as well as techniques and ramifications of Shari'a arbitrage. Most US states still impose an explicit legal usury ceiling on interest rates; the cap varies by jurisdiction.

[5] On *rentes*, see also Kindleberger (1993, ch. 12–14). The term "financial revolution" refers to a nexus of activities, with a large increase in public borrowing being a notable characteristic. The financial revolution occurred at different times in different places. See, for example, Tracy (1985) and Wennerlind (2011).

1967). Naturally, the producer of the precious metals was the immediate beneficiary of this increase in purchasing power, at least when it escaped the grasp of Dutch and English pirates. Spain ultimately dissipated most of this windfall on the European wars of the Counter-Reformation (the defeat of the Spanish Armada in 1588 is the most famous example) or squandered it on royal extravagances like the Buen Retiro palace in Madrid. Reinhart and Rogoff (2009) note that Spain was a "serial defaulter" on its sovereign debt during the reigns of Philip II and his successors (roughly 1550 to 1650).

After the political revolutions of the eighteenth century, the fiscal-military states gradually reoriented themselves. Industrialization created less painful and less risky paths to enrichment for the ambitious and enterprising. Not coincidentally, the Industrial Revolution, with its focus on capital-intensive industries, coincided with the formalization of markets for equity shares: the London Stock Exchange was formally organized in 1773, while the Buttonwood Agreement creating the New York Stock Exchange (NYSE) was signed in 1792.[6] It is in the relatively peaceful nineteenth century that the institutionalization of financial markets becomes pervasive.

The relatively peaceful nineteenth century was eventful in other ways. The Panic of 1825 in London saw the first experiments with a lender of last resort (a cornerstone of central banking), which led to a gradual restructuring of the British financial system.[7] In that episode, the Bank of England, then a shareholder-owned public bank without lender-of-last-resort responsibilities, stepped into the breach of a major liquidity crisis. By lending to other banks at a discount (against good collateral), they enabled institutions to satisfy withdrawals, thus stemming the panic. Posting collateral in a crisis to liquidate assets without selling them, addresses what is at root an information problem, namely the inability of creditors to observe directly the quality of the debtor firm's assets. The lesson learned by central bankers from the 1825 experience has become known as Bagehot's Rule: lend freely in a crisis, on the basis of any acceptable collateral ("any good banking securities"), but at a "very high rate of interest" (see Bagehot, 1873, p. 197). The USA would suffer nearly a century of similar clearing crises before founding the Federal Reserve in 1914. Unfortunately, while the existence of a central bank should facilitate crisis response, it does not guarantee good monetary policy, a point driven home after the 1929 crisis.[8]

[6] See Neal (1990) on the creation and expansion of exchanges in Europe and elsewhere. Sylla (2005) focuses on the founding of the NYSE. The Amsterdam Stock Exchange is an exception that proves the rule – it was founded in 1602 to support trading in shares of the Dutch East India Company.

[7] See Bagehot (1873), Neal (1998), and Bordo (1998).

[8] On the 1929 crisis and Federal Reserve response, see Friedman and Schwartz (1971), Galbraith (1955), and Ahamed (2009). There are more extreme examples as well, including the German hyperinflation of 1922–1923; see Kindleberger (1993, ch. 17) and Eichengreen (1992, ch. 5).

The impact of new regulatory institutions is also important to the development of financial markets. The newly regulated entities respond with an institutionalization of their own as they comply with the new rule set. In the 1950s, for example, US banking in the aftermath of Depression-era regulations and the newly revamped system of fixed exchange rates was famously boring, inspiring the so-called 3–6–3 rule (i.e., borrow at 3%, lend at 6%, and be on the golf course by 3 pm). Equally important, regulated entities can push back, lobbying for repeal of the rules. To the extent that the regulations constrain behavior and innovation (there is little point to rules that never bind), financial activity will tend to migrate away from supervised and regulated institutions to less supervised and encumbered markets. Kaufman and Mote (1994), for example, document the decades-long migration of financial assets away from traditional banking. They emphasize that assets are a very inexact measure of financial activity, and that other measures, such as employment and economic value added, paint a more optimistic picture for banking. Nonetheless, by the 1990s, the share of assets held at banks and thrifts had fallen to less than half of what it had been in 1900.[9]

1.2 Globalization

The essence of the "invisible hand" is the sufficiency of price information to achieve the coordination of production and consumption activities across all participants in the economy. However, for such a signaling mechanism to work in practice, price information must be broadly observable. In other words, an adequate communication technology is a prerequisite for economic efficiency. Historically, this fundamental economic force has created a strong tendency toward increasing availability and consistency of prices. In international financial markets, this historical process is often referred to as "capital market integration."[10] While timely knowledge of foreign prices is in itself a valuable information source, full access to foreign financial markets allows for actual diversification of investments and sharing of risks internationally. Neal (1990, ch. 1–3) describes the emergence in the late seventeenth century of a financial information network connecting the equity markets of London and Amsterdam with each other and the rest of Europe. Entrepreneurs in both cities began to publish weekly (or more frequent) stock price lists reporting the "course of the exchange" locally, as well as in ports as

[9] A more recent update by Feldman and Lueck (2007), reinforces this impression. After adjusting for off-balance-sheet activities, they reveal a surge in the asset share held by "other financial intermediaries" in the most recent two decades. Regulatory burdens are only one example of transaction costs driving the long-term migration from traditional financial intermediaries to securities markets and contracting – i.e., away from intra-organization activities and toward inter-organization transactions. Coase (1990) introduced transaction cost economics; Holmström and Roberts (1998) offer a survey.

[10] See, for example, Obstfeld (1998), Lothian (2000) or Obstfeld and Taylor (2004).

far-flung as Cadiz and Leghorn (i.e., Livorno). Prices were transmitted via packet boat and stage coach.

It is unsurprising that these stock price lists first developed prominently in the capitals of the two seafaring empires, since shipping requires both financial capital to fund the cargoes and well-developed insurance markets to spread the risks of loss due to sinkings. Of course, international trade has existed as long as there were nations. Truly global trade goes back at least to the Age of Exploration, highlighted by the commencement of the spice trade, the exploitation of American gold and silver resources, and the development of the slave trade. From the start, trade and risk commingled. Motivated by international trade and the risks of venture capital, the Dutch United East India Company pioneered two of the most important risk-management innovations known, the implicit limited liability of shareholders and a secondary market for equity shares on the Amsterdam Stock Exchange.[11] British Parliament similarly granted limited liability status to the rival British East India Company in 1662. The limited liability provision implied that owners could not lose anything beyond a fixed initial stake in the company. This encouraged the investment of capital in ventures of inherently high risk but potentially very high reward. Unfortunately, it also encouraged leverage, which magnified the positive upside without increasing the (zero) downside; the subsequent British South Sea Company, chartered in 1710, was at the center of one the earliest systemic financial crises, the South Sea Bubble of 1720.[12]

Until relatively recently, international trade overwhelmingly dominated international investment. For centuries, the latter was dominated by various forms of sovereign debt to finance wars, or to pay indemnities to the victors afterward.[13] Also prominent was financing for overseas investment in infrastructure projects, such as railroads and canals. However, even today, trade in goods and services is several multiples the level of investment income as measured in current account statistics (see OECD, 2010). Given the preeminence of trade, the challenges of

[11] The Dutch East India Company is often known by its initials, VOC, an abbreviation for the official name, *Vereenigde Nederlandsche Geoctroyeerde Oostindische Compagnie*. See Neal (2005) for a good history of the early days of the VOC, which initially expanded much more rapidly than the British East India Company; both firms were chartered in 1602. See Ferguson (2008, ch. 3) for a discussion of the limited liability provisions. Neal (1990, ch. 3) traces the limited liability tradition to the earlier medieval shipping ventures known as *societas maris* and *commenda*. The same chapter offers a good treatment of the early days of financial trading in both London and Amsterdam.

[12] Bordo et al. (1999) provide a good summary of commercial and financial globalization since the late nineteenth century. On the broader history of limited liability corporations, see Baskin and Miranti (1997). Reinhart and Rogoff (2009), Kindleberger (1993, ch. 15), and Kindleberger and Aliber (2005) discuss a range of speculative manias and financial crises.

[13] The borrowers and their underwriters were frequently very inventive in structuring these deals, experimenting for example with the sale of a variety of retail life annuities known as *rentes*, described above. Reinhart and Rogoff (2009) provide an authoritative overview of the problem of default on sovereign debt (both external and internal).

payment and settlement were the main focus of international finance. For much of recent history, this meant the gold standard.

The gold standard grew in tandem with the globalization of trade, especially in the Atlantic sphere (see O'Rourke and Williamson, 2001; Lothian, 2000). By rigidly fixing certain key prices, the gold standard eliminated exchange-rate uncertainty, thus playing an important informational role. By propagating inflation and deflation across borders, it was central to economic risks in the system. The gold standard did not spring forth fully formed, but evolved gradually over several centuries.[14] Silver coinage dominated in Europe until the late Middle Ages. Following the introduction of gold coins in thirteenth-century Italy, most of Europe subsequently evolved to a complicated hodge-podge of gold, silver, and bimetallic standards.[15] As industrializing economies were discovering and exploiting enormous gains from trade throughout the nineteenth century, the international gold standard alleviated uncertainty about the value of payments exporters could expect from importers. The chronology of the gold standard is punctuated by three major wars, with the international monetary framework fundamentally reorganized under new institutions after each. The first was the Franco-Prussian War of 1870–1871, during which most of Europe, except for Britain, suspended convertibility.[16] When Germany opted for a gold standard in 1873, using the French indemnity from the war to fund their new currency, the rest of Europe followed, either officially or by committing to fix their currency's exchange rate against the new German mark or the British pound.

The most famous of the postwar indemnities is the German reparations debt under the terms of the Treaty of Versailles after World War I. The unsustainable burden imposed on Germany by the Treaty contributed significantly to a decade of political and economic turmoil there in the first decade after the war, including the hyperinflation of 1923–1924 (see Eichengreen, 1992, ch. 5). In 1929, the Young

[14] The gold standard is a commodity monetary standard. In an international context, if all jurisdictions convert their circulating currencies to a common commodity base, then international payment becomes straightforward. In a world with restricted and unreliable communication networks, such a coordinated valuation system solves an important problem of distributed information by removing exchange rate risk from the calculations, as exchange rates are effectively fixed by the ratio of the relevant mint prices of bullion in the local currencies. It is noteworthy that – in theory – any country can join the international regime by committing *domestically* (i.e., unilaterally) to convert its own currency to bullion at a fixed price. In so doing, it locks into a system of fixed exchange rates as a matter of simple arithmetic, enforced by arbitrage. There is a degree of arbitrariness in the choice of gold as the basis for a monetary system. Gold, silver and copper all had long legacies, along with the practical advantages of being durable, visually distinctive, and sufficiently rare that coins of useful denominations have a portable size.

[15] Kindleberger (1993, ch. 4) highlights the operational burdens inherent in keeping a bimetallic standard balanced. For example, Britain in 1717 unintentionally became the first major economy to abandon bimetallism. Sir Isaac Newton, as Master of the Royal Mint, undervalued silver (relative to gold) in the official *mint exchange ratio* – the rate at which it would trade gold for silver. With ongoing mineral discoveries in the New World, the correct ratio was uncertain. In typical Newtonian fashion, his intention was to monitor market prices and adjust the mint ratio adaptively over time. However, the mint ratio went untouched after his retirement, and Gresham's Law drove silver out of circulation.

[16] It is standard practice to suspend convertibility during wartime, to protect bullion supplies and facilitate potentially inflationary military spending needs.

Plan to reduce the reparations burden created the Bank for International Settlements (BIS) in Basel to manage the payment flows (see Kindleberger, 1993, ch. 16). Although the payments plan quickly fell apart in the aftermath of the 1929 stock market crash, the BIS is still with us as a key policy body for coordinating international banking regulations.

The trauma of World War II provided an opportunity to restart international monetary coordination. The crisis mechanism for creating new institutions worked once again. Before the war was even over, the allies (44 nations in all) met at the United Nations Monetary and Financial Conference in Bretton Woods, New Hampshire, to negotiate creation of the International Monetary Fund (IMF) to manage a reinstated system of fixed exchange rates, and the International Bank for Reconstruction and Development (IBRD, later part of the World Bank) to manage the rebuilding of Europe and Japan (see Eichengreen, 1996, ch. 4). The United Nations (UN), General Agreement on Tariffs and Trade (GATT), European Community (EC), and Organization for European Economic Co-operation (OEEC) rounded out the complement of major international economic institutions created in the wake of the war.[17] The Bretton Woods framework was significantly more flexible than either the interwar arrangements or the classical gold standard. The US dollar was pegged to gold, and other currencies pegged to the dollar, with provisions for orderly (re-) devaluation when necessary.

US deficit spending to support the Vietnam War and Great Society ultimately created inflationary pressures, which the fixed exchange rates tended to export to other countries. At the official parity of \$35/oz, gold was undervalued and flowed unrelentingly out of the USA. In August 1971, acknowledging that the USA had exhausted its reserves, the Nixon administration suspended convertibility, effectively ending the gold standard (see Eichengreen, 1996, ch. 4). The network of pegged exchange rates was relaxed and finally abandoned in 1973.

As exchange rates began to float, trading in currencies expanded rapidly. The need for coordinated international regulation was revealed by the failure of Bank Herstatt in 1974. West German regulators closed Herstatt at the close of business in Frankfurt on June 24, 1974, after it had received Deutsche marks, but before paying the corresponding US dollars in New York (see Becker, 1976). The resulting settlement crisis motivated the creation of the Basel Committee on Banking Supervision (BCBS) within the BIS. The BCBS provides a forum for regular cooperation on banking supervision among its 27 member countries, and is best known for its international coordinating role on issues of bank supervision,

[17] The OEEC received much of its funding through the Marshall Plan. It evolved into the Organization for Economic Cooperation and Development (OECD).

especially its standards on capital adequacy.[18] Another example of global coordination is the Continuous Linked Settlement Bank (CLS), created in the late 1990s by several large banks to settle their currency trades. A key feature of CLS is multilateral netting to reduce settlement risk.[19]

The expansion of credit that accompanied the breakdown of Bretton Woods was the leading edge of a decades-long upswing in international financial activity. Facilitated by advances in trading technology, and augmented by "petrodollar recycling" of exporter revenues following the oil embargo of 1973–1974, cross-border holdings of financial securities, especially bonds, grew at double-digit rates through the 1980s (see Abken, 1991). An unhappy side-effect was an overextension of credit, culminating in the Latin American debt crisis of the 1980s.

The most recent episode in international monetary coordination is the Euro Area (or Eurozone), the monetary union of the 17 countries sharing the Euro as their legal tender currency within the European Union. Narrowly construed, the Euro Area is an economic policy initiative, but because (like the gold standard) it necessitates the coordination of fiscal and monetary policies across participating countries, it is an important political commitment. The ongoing negotiations (as of early 2012) around the possibility of a Greek sovereign default forcefully make this point. Indeed, the 1992 Maastricht Treaty that launched the Euro project was itself a powerful commitment to political unification, coming not coincidentally at the end of the Cold War.

Global financial activity has continued to grow undeterred, facilitated by a gradual transition from bank-based to market-based finance (see Lane and Milesi-Ferretti, 2007). From the 1990s, the "Washington consensus" on financial liberalization has encouraged the managed extension of credit to under-capitalized and funding-constrained emerging economies.[20] Once again, there were some unhappy side-effects, including the Asian financial crisis, precipitated by a speculative attack on the Thai baht in May 1997 (see Woo et al., 2000, or Blustein, 2003).

Following upon the heels of the Asian crisis, the Russian sovereign debt default in the fall of 1998 coincided with the expanding role of hedge funds to produce another systemic event. Long Term Capital Management (LTCM) was founded in 1994 to implement trading strategies in the currency and bond markets. Although LTCM was not heavily exposed to Russian debt directly, it

[18] See BCBS (2009) and Goodhart (2011) on the history of the Basel Committee. Goodhart clarifies that international supervisory coordination had begun informally within the EC as early as 1972, well before Herstatt. This so-called *Groupe de Contact* was motivated primarily by concerns around the burgeoning Eurodollar markets. The BCBS is discussed in greater detail in Chapter 10 of this Handbook. For the statement of supervisory principles, see BCBS (2011a). For various versions of the Basel capital standards, see BCBS (1988, 1996, and 2011b).

[19] CLS handles 94% of trading volume for 17 currencies; see Groenfeldt (2011). Aggregate daily trading volume in the global foreign exchange market is on the order of \$4 trillion; see BIS (2010).

[20] For a history of the Washington consensus, see Williamson (2004). For a reconsideration, see Kose et al. (2009).

was highly leveraged and very sensitive to the drop in liquidity that pervaded all markets after Russia unexpectedly defaulted in September 1998 (see Lowenstein, 2001). Fearing a possible chain reaction if LTCM were to unwind its positions in an already fragile market, the Federal Reserve engineered a bailout, with 14 US and European banks contributing a total of $3.5 billion in exchange for an equity stake to refloat the firm.

The latest episode in this litany of global crises is the financial crisis of 2007–2008, the ramifications of which are still underway. What appeared to be a "Great Moderation" in the economy turned out instead to be market euphoria, as a housing bubble in the USA (and elsewhere) was financed largely through a complex web of securitization vehicles. Related structures, such as credit default swaps (CDSs), provided third-party credit enhancement. Because the USA does not save enough to fund its investments (including housing expenditures), much of the securitization debt was placed abroad, ensnaring much of the developed world in a complex web of now troubled financial claims. As Reinhart and Rogoff (2009) emphasize in their sobering historical review, the surprising fact about financial crises, whether domestic or global, is not that they occur, but that they unfold so similarly from one instance to the next.

1.3 Technology

From the start, innovations in financial risk management have been intricately linked with innovations in information technology. It is difficult to distinguish cause and effect among emerging risks and new technologies to manage them. Invention evolves over time, with episodes of mimicry, inspiration, and unintended consequences tumbling over one another.[21]

One humble institution of modern deposit banking – the checking account – had its origins among innovations in risk management and information technology in medieval Europe.[22] The main risk addressed was quite immediate, namely the potential loss of valuable cargo, especially coin or bullion, to shipwreck, piracy, or highway robbery.[23] The innovation was the bill of exchange, a remittance service provided by a "sedentary" merchant who maintained a local presence in major trading centers, with a network of couriers to deliver instructions and reports.

[21] For an instructive historical example involving the invention of double-entry bookkeeping, see Heeffer (2010).

[22] Usher (1934) is a classic history of the early development of banking in Europe. Grossman (2010, ch. 2) offers a more recent account of the same epoch. Kohn (1999a) focuses specifically on the details of the process of issuing bills of exchange, along with related payment innovations. Ferguson (2008) highlights their importance to the rise of the Medici dynasty in Italy.

[23] We would classify these today as operational risks. Kohn (1999b) also describes commonplace medieval market risk – namely the risk that a shipment would arrive in port to discover that local prices for the cargo were lower than anticipated.

Instead of accepting cash, an itinerant trader would deliver goods to the sedentary merchant's local office in exchange for a *cambium per literas* (bill of exchange). This was a written payment instruction that the trader could transport safely and easily, and that he could redeem on a specified future date at a distant market center. The dates were calibrated against estimated travel times and a well-known calendar of major trading fairs. For our purposes, the most interesting aspects of these arrangements are the early information technologies supporting the risk management innovation. First, the sedentary merchant maintained a transactional database journaling the bills of exchange issued and the payments and deliveries made.[24] The sedentary merchant also maintained a communications network of couriers and agents for messaging. Simultaneously and separately, the local sedentary merchant would send along instructions (so-called *avisas*) to the destination that the bill of exchange should be honored upon presentment. Message formats were roughly standardized, and settlement procedures were well established and enforceable in the courts. To ensure the resilience of the communications network, the sedentary merchant would typically transmit multiple redundant copies of the *avisas* (but not the bill of exchange) via different channels.

A subsequent innovation, related to the bill of exchange, was the *cambium nauticum* (maritime exchange contract), which first appeared in thirteenth-century Europe, and ultimately evolved into modern marine insurance. The *cambium nauticum* was very similar to a bill of exchange, with the important difference that payment was *contingent* upon safe delivery of the collateral cargo.[25] Of course, proper insurance, as a business model, involves more than a willingness to take a gamble: it requires that the insurer be able to calculate odds and select risks. Merchants' record-keeping systems provided a natural statistical data sample of ladings, sailings and sinkings, as well as other partial losses of cargo. The obvious fact that most shipments arrived intact made sea-borne trade enticing for both the shippers and their captains. More interesting is that insurers were able, at least approximately, to select risks. Insurance premiums varied according to the distance shipped, the time of year, state of war or peace, the presence of pirates, and the type of vessel.

Benefiting from a general interest in measurement, more sophisticated insurance markets developed in tandem with the foundations of probability theory during the sixteenth to eighteenth centuries.[26] Daston (1988) argues that the raw materials for probability theory – for example, experience with games of chance, set theory, and

[24] This was, of course, written out longhand; early on, the records typically used Roman numerals. Roman numerals were also an information security measure, as the alternative Arabic representation was seen as more susceptible to forgery; see Arlinghaus (2003).

[25] See de Roover (1945) for a detailed discussion of this and related legal developments.

[26] Bernstein (1998), Daston (1988), Hald (2003), Hacking (2006), and Devlin (2008) recount the history of this period.

insurance markets – remained essentially unused for centuries before the path-breaking correspondence of Pierre de Fermat and Blaise Pascal in 1654. As old cosmological certainties collapsed in the wake of Columbus's explorations, the Renaissance witnessed an invigorated focus on empirical measurement. This was epitomized by the astronomical observations of Copernicus, Galileo and Kepler in the sixteenth and early seventeenth centuries (see Hald, 2003). Christiaan Huygens formalized Fermat's and Pascal's ideas in 1657. Royal societies for science were founded London and Paris in the 1660s, and both worked on the measurement of longitude, a matter of vital strategic importance for the new transoceanic empires. Meanwhile, Lloyd's of London, the maritime insurance society, was growing steadily from its beginnings in the 1630s.

In 1662, essentially contemporaneous with the aforementioned seminal work on probability theory, John Graunt published the first mortality table, compiled from official records of births and deaths (the latter attributed to various causes) in London over the period 1604–1661.[27] Within a decade of Graunt's publication, governments were using the data to refine the pricing of life annuities they sold to raise revenue. By the end of the century, the first modern life insurance companies were beginning to form in London. The new methodologies for compiling high-quality data enabled more accurate risk assessment. An important refinement in data collection was the inclusion, starting in 1728, of age at death in the official London Bills of Mortality. A boom in insurance markets ensued throughout the eighteenth century, alongside intensive development in probability theory. This period also saw Abraham de Moivre's approximation formula for the binomial coefficients, first published in 1738. A century later, Carl F. Gauss would refine and formalize this approximation as the normal distribution, applying it to the Law of Errors to describe the regular patterns in measurement errors and manufacturing variances.

The nineteenth century also saw dramatic improvements in communication technologies. This is important for the history of finance, as contracting and payment (or physical delivery) processes rely on effective messaging to initiate and complete transactions. The first commercial deployments of the telegraph occurred in the 1830s. Prior to that, most long-range communication required couriers or ordinary post, although the Rothschild banking house famously used carrier pigeons to gain an advantage over competitors.[28] By the late 1840s, the telegraph was reducing contemporaneous price discrepancies between New York and markets further west (see Garbade and Silber, 1978). The first successful transatlantic

[27] See Bernstein (1998, ch. 5), for a discussion of Graunt's contribution to insurance.

[28] On Rothschild's use of pigeons, see Stephen (1885). Contrary to popular myth, the pigeons were not a factor in the Rothschilds' profiting from the Battle of Waterloo; Ferguson (2008) details the true story, which was more about financial fortitude and much less about cleverness. Another inventive technology of the same era was Claude Chappe's semaphore telegraph, implemented as a network of visual signal towers throughout France in the eighteenth century; see Gleick (2011).

cable was laid in 1866.[29] The initially high cost of messaging was justified by reductions in the risk of pricing surprises caused by the timing delay of shipments to a distant port or investments in overseas securities. On the other hand, the cable also helped speculators in London to participate actively in US banking panics.

More recently, modern investment strategies and risk management methods have emerged contemporaneously with advances in information technologies. Investors have understood for centuries the benefits of diversification for risky investments. Ancient traders dispersed their most valuable cargoes across many ships in a fleet. Computation expands these possibilities. The cornerstone of modern portfolio theory, namely Markowitz's (1959) formal optimization methods for portfolio selection based on the covariance matrix of observed returns, requires computers for practical implementation. Similarly, the option-pricing models of Black and Scholes (1973) and Merton (1973) are able to assign plausible and well understood values to simple put and call options. However, this requires the evaluation of a normal integral, which must be interpolated numerically. Soon after its introduction, it was commonplace for option traders to have the Black–Scholes–Merton formula literally at their fingertips on the trading floor, programmed into pocket calculators.[30]

The basic insight of Fermat in 1654 was that one could describe – and assign probabilities to – all possible future paths that might ensue in an unfinished game of chance. This is the essence of Monte Carlo analysis, which generates the data to describe a large number of possible future evolutions of the system. One goal is to reveal the odds of the various payoff outcomes for a complex security (see Glasserman, 2003). Because this approach frequently involves brute-force recursive and iterative algorithms, it has benefited enormously from advances in computing power. Monte Carlo is also able to calculate probabilities in many cases where more traditional closed-form solutions like the Black–Scholes–Merton model are inapplicable (e.g., the odds of a mortgage prepaying or defaulting). These probability estimates enable valuation approximations that in turn allow new markets to emerge in many complex products, such as structured mortgage-backed securities.

Program trading is another data- and computation-intensive technology with risk implications. In 1982, the Chicago Mercantile Exchange (CME) introduced futures contracts on the S&P500 stock index, a value-weighted average of the prices of the common stocks of 500 large firms. Upon maturity, the futures contract pays the buyer

[29] See Hoag (2006). The telephone followed in the 1870s, and was quickly commercialized, and the wireless radio arrived in the 1920s.

[30] Although the pricing formula is relatively recent, exchange traded options have a much longer history; see Gelderblom and Jonker (2005). The original paper by Black and Scholes (1973) also introduced the insight that limited liability equity is effectively a call option on the assets of the firm. Limited liability equity was first introduced for the Dutch East India Company in 1602, mentioned above. A slightly modified version of the Black–Scholes–Merton formula is the core of one popular approach to estimating corporate default risk (see Saunders, 1999). The compendium by Field (2003) provides a good survey of risk modeling and risk management innovations since the introduction of the Black–Scholes–Merton model (up to 2003 of course).

a simple multiple of the value of the index on that day. Index arbitrage is a program trading strategy that watches for pricing discrepancies between the index futures and the individual underlying stocks, buying stocks and selling futures when the former are relatively undervalued (or vice versa when stocks are overvalued). To implement this most effectively, the program must execute large volumes of transaction orders nearly simultaneously. The tactic would not be practical without network bandwidth and computing power to validate data and process messages. An important side benefit is that it penalizes deviations between the futures price and underlying stock prices, enforcing consistency in the price signals in the marketplace. Portfolio insurance is a related strategy that continuously rebalances a short position in index futures to emulate a put option for a portfolio of stocks. Again, frequent trading is required. Because it relies on continuous recalibration, abrupt jumps in prices can expose the portfolio to unbalanced moments. Both index arbitrage and portfolio insurance contributed prominently to the Black Monday crash of 1987, the largest daily percentage loss (−22%) in the history of the US equities markets.[31]

Program trading is only one example of a broader trend toward automation of all aspects of the trading process. In many cases, human brokers have given way to electronic brokerage, algorithmic trading has replaced human discretion in investment decisions, and electronic communication networks (ECNs) have attracted execution volume away from traditional trading floors (see Harris, 2002). Straight-through processing (STP) of transaction and settlement details has come to dominate back offices (see Weiss, 2006). There are obvious operational efficiencies to automating much of the repetitive work in the trading process. At one extreme, high-frequency trading designs streamlined decision algorithms to speed throughput, and co-locates them on machines at the exchange to minimize network latency. The “arms race” in this domain has generated extremely large trading volumes in recent years. On the other hand, while not the primary culprit, high-frequency trading was a factor in the Flash Crash of May 2010 (see CFTC/SEC, 2010). The same computing and trading power that monitors and enforces price consistency continuously across thousands of securities also has the potential to generate enormous operational events very quickly.

1.4 Complexity and professionalization

It has become commonplace to observe that financial markets have become more complex. “Complexity” is a multifaceted concept, with at least three common usages: systems complexity, descriptional complexity, and computational complexity. Although all three are relevant, we are interested especially in the last: the

[31] See Brady (1988), SEC (1988), and Carlson (2007) for analysis of the event.

computational burdens complexity places on participants in the financial system.[32] These burdens may be multiplied by the degree of interconnectedness among the participants. Financial complexity appears in several interrelated forms in the financial markets, visible at the level of the individual contract, at the level of the firm, and at the level of the system as a whole.

At the contract level, the development of option pricing theory and rapid expansion of derivatives markets starting in the 1970s marked a revolution in the practice of finance. Options are a form of contingent claim, meaning that the payoff to the option holder depends on whether a particular contingency occurs. Traditional insurance typically pays out for well-defined physical transitions like the death of the policyholder or sinking of a ship. In contrast, contingencies for financial derivatives are typically more abstract, and this abstraction enables considerable flexibility and creativity in designing these contracts.[33] An important benefit of this flexibility is the increased ability to trade risks, allowing exposures to flow (for a price) to those best able to bear them.

Of course, the seeming costless ease of this abstract creativity is partly a mirage. There are complexity costs. By definition, contingent claims create dependencies to be measured and managed among the values and payoffs of different securities. Goetzmann et al. (2007), for example, demonstrate that it is straightforward for investment managers to use options contracts to manipulate standard portfolio performance metrics. This is particularly significant for hedge funds and other managers who are compensated based on reported performance. Securitization contracts are an example of a useful innovation that manifests descriptional complexity as a side-effect. The offering documents for a typical structured securitization go on for many hundreds of pages, not counting the thousands of pages of paperwork for the underlying mortgage loans. Arora et al. (2011) demonstrate the possibility of hiding poorly performing loans within a complex structured securitization so that it is computationally infeasible to detect their presence.

The last four decades have also seen significant changes in market structure leading to increased complexity within and across firms. In the USA, the relaxation of interstate branching restrictions in the 1990s led to a massive consolidation of the banking industry. Mergers reduced the total number of commercial banks in the USA from over 14,000 in the mid-1980s to less than half that today (Mester, 2007),

[32] Simon (1962) provides an important early overview of complexity with an emphasis on hierarchical dependencies. Mitchell (2011) also highlights the systems aspect, emphasizing interconnectedness and decentralized decision-making and computation. Arora and Barak (2009) focus on computational intricacy. A third definition is Kolmogorov (a.k.a. "descriptional") complexity: the length of the shortest description that can capture the full detail of a given set of facts.

[33] A "derivative" is a contract for which the contingencies in question are *derived* from the properties (typically the price) of some underlying financial security. For example, a call option pays off if the price of an underlying security is above a threshold value on a particular future date. Derivatives exist for a vast range of underlyings and contingencies. See Hull (2008) for details.

and the industry as a whole has shed assets through securitizations into off-balance-sheet special purpose vehicles. At the same time, the very largest banks have grown into institutions that are now known to regulators as "large complex financial institutions" (LCFIs). Citigroup, for example, now has nearly 2500 subsidiaries.[34] Complexity has also increased at the systemic level, particularly through the growth of securitization vehicles and related markets. As wholesale-funded LCFIs have taken on a greater role, and as securitization and trading have increased in prominence, unregulated sectors such as the "shadow banking" markets for repurchase agreements have ballooned (Adrian and Shin, 2009). At the same time, the hedge fund sector grew roughly tenfold just in the decade from 1994–2004, not counting funds of funds (see Fung et al., 2008). The fact that securitization funding chains stretched around the world, from US mortgage borrowers, through mortgage-backed securities (MBS) pools and tranches, through repo collateral in the shadow banking sector, to pension fund investors in Europe and elsewhere, meant that the collapse triggered by the failure of Lehman Brothers in 2008 necessarily had significant systemic ramifications.

The increasing complexity of financial contracts and the corresponding growth in the use of computational technologies have created a need for professional risk managers. As with important legal and medical decisions, where crucial inputs must be entrusted to technical experts, a certification process has emerged to indicate subject-matter competence. For example, the most recent supervisory proposals from the Basel Committee would codify a requirement for a Chief Risk Officer in large banking firms, affirming what has been standard practice for many years (see BCBS, 2011a, p. 43). Separately, industry participants have evolved professional societies and formal processes for certification in risk management, as described below.

Within the field of financial risk, this trail was blazed by the actuarial profession in the late nineteenth century (see Hickman, 2004). The Institute of Actuaries (IA) was founded in London in 1848, following a successful experiment in pooling policy data across firms to better estimate mortality. By 1900 both the IA and the US-based Society of Actuaries were offering formal credentials of professional competence. Other technical financial specialties have also evolved to provide formal certification. In the USA, the Certified Public Accountant (CPA) credential similarly traces its roots back to the late nineteenth century. Among other roles, accountants and actuaries provided expertise to courts and creditors seeking an equitable distribution of assets in bankruptcy. The Chartered Financial Analyst (CFA) designation for experts in corporate securities research is more recent, with the first certification issued in 1963.

[34] See Herring and Carmassi (2010) for an analysis of subsidiary counts for LCFIs. See Group of Ten (2001) for a study of the recent history of consolidation.

The example of value at risk

More recently, the development of value-at-risk (VaR) models provides a useful case study of the expansion and evolution of a technical risk-management tool to a standardized industry practice and regulatory requirement. The Black–Scholes–Merton model for valuing derivatives originated in academia before spreading decisively throughout industry in the 1980s. By the early 1990s, derivatives markets had revolutionized financial services, creating a need for more transparent reporting of risk exposures. JPMorgan instituted the so-called "415 report" – a single page delivered at 4:15 pm daily to the CEO – to overcome the shortcomings of traditional accounting for risk measurement. The 415 report captured the risk exposure of each trading desk in a single value-at-risk (VaR) number, the estimated largest dollar loss that might occur within a 95% probability window. VaR demonstrated new possibilities in risk accounting and reporting, but implementing it on a firm-wide basis is a significant technical challenge.[35] While early VaR measures addressed market risk – the possibility of losses due to short-term fluctuations in market prices – the basic idea was soon extended to credit risk (the possibility of counterparty or borrower default). The data required to calibrate market-risk models primarily consist of various price histories. Credit risk, in contrast, deals with highly asymmetric distributions, where there is a small chance of a very large loss. Because interesting events are much rarer, calibrating these models typically relies on information on counterparty characteristics to augment actual loss histories.

Regulators quickly took note of the potential of the new VaR tools in the mid-1990s. In 1988, the BCBS proposed its first capital accord, Basel I, which required banks to hold a specific percentage of capital between 0 and 8% for every asset on the balance sheet, depending only on a broad categorization of the asset type as a proxy measure of credit risk (see BCBS, 1988). The goal was to harmonize regulatory capital requirements across countries, to prevent international competition in this dimension. In 1996, with the ascendancy of VaR, the BCBS amended Basel I to include VaR-based measures of market risk. By the end of the century, the shortcomings of the simplifications in Basel I were becoming clear. A major revision, Basel II, was proposed in 2004. It reached final regulatory approval in the USA just before the crisis struck in 2008. By the end of 2009, Basel III was under discussion, and by 2011 the BCBS had begun to propose the details (see Part III, Chapter 13 of this Handbook and BCBS, 2011b).

Under the most recent Basel III proposals (BCBS, 2011b), the practical reality of the CRO will become a regulatory requirement for many firms. In 1996, the Global Association of Risk Professionals (GARP) was formed and began to offer

[35] Finger et al. (1996) is the technical document JPMorgan provided to its clients. Laubsch (2000) offers one approach to the reporting of enterprise risks in a VaR context.

certification exams for risk managers. In 2002, a controversy over GARP's governance led to a schism, and the Professional Risk Managers International Association (PRMIA) split off. In addition, the older Risk Management Association (RMA), which traces its history to 1914, now offers certification for competence in credit risk. Today, GARP, PRMIA, and RMA all exist as non-profit organizations offering training, conferences, and certification examinations to risk professionals. The professional field of quantitative risk management is here to stay.

1.5 Conclusion

In tracing the evolution of financial risk and information over time, we have identified four fundamental trends. Institutionalization, technology, globalization, and complexity, with its attendant rise of risk management professionals, have all contributed to and been a part of the growth of financial markets. Not coincidentally, these developments have occurred in concert with the growth in human population and economic productivity. Through all these striking changes, the system has remained susceptible to financial crises, information-driven risk events of the first order. *Plus ça change, plus c'est la même chose.*

Acknowledgements

The author thanks Margarita Brose, Henry Gordon, Greg Feldberg, Jim Lothian, Larry Neal, Jim Overdahl, Charles Taylor, and Dave Wheelock for their helpful comments on drafts of this chapter. Any remaining errors are the responsibility of the author alone.

Bibliography

Abken, Peter, 1991, Globalization of stock, futures, and options markets, *Economic Review*, Federal Reserve Bank of Atlanta, July, 1–22.

Adrian, Tobias and Hyun Song Shin, 2009, Shadow Banking System: Implications for Financial Regulation, *Staff Report* No. 382, Federal Reserve Bank of New York, July.

Ahamed, Liaquat, 2009, *Lords of Finance: The Bankers Who Broke the World*, Penguin Books.

Arora, Sanjeev and Boaz Barak, 2009, *Computational Complexity: A Modern Approach*, Cambridge University Press.

Arora, Sanjeev, Boaz Barak, Markus Brunnermeier and Rong Ge, 2011, Computational complexity and information asymmetry in financial products, *Communications of the ACM*, **54**(5), May, 101–107.

Arlinghaus, Franz, 2003, Bookkeeping, double-entry, in: C. Kleinhenz (ed.), *Medieval Italy: An Encyclopedia*, Volume 1, Routledge.

Bagehot, Walter, 1873, *Lombard Street: A Description of the Money Market*, H.S. King.

Bank for International Settlements, 2010, Triennial Central Bank Survey: Report on Global Foreign Exchange Market Activity in 2010, Technical Report, December, `www.bis.org/publ/rpfxf10t.htm`.

Bank of England and Financial Services Authority (BoE/FSA), 2011, Instruments of Macroprudential Policy, Technical Report, Bank of England, December, `www.bankofengland.co.uk/publications/Pages/news/2011/160.aspx`

Basel Committee on Banking Supervision (BCBS), 1988, *International Convergence of Capital Measurement and Capital Standards*, Bank for International Settlements, July, `www.bis.org/publ/bcbs04a.pdf`

Basel Committee on Banking Supervision (BCBS), 1996, *Amendment to the Capital Accord to Incorporate Market Risks*, Bank for International Settlements, January, `www.bis.org/publ/bcbs24.htm`

Basel Committee on Banking Supervision (BCBS), 2009, History of the Basel Committee and its Membership, Technical Report, Bank for International Settlements, August, `www.bis.org/bcbs/history.htm`.

Basel Committee on Banking Supervision (BCBS), 2011a, *Core Principles for Effective Banking Supervision, Consultative Document*, Bank for International Settlements, December, `www.bis.org/publ/bcbs213.pdf`

Basel Committee on Banking Supervision (BCBS), 2011b, Basel III: *A Global Regulatory Framework for More Resilient Banks and Banking Systems*, Bank for International Settlements, June, `www.bis.org/publ/bcbs189.htm`

Baskin, Jonathan and Paul Miranti, Jr., 1997, *A History of Corporate Finance*, Cambridge University Press.

Becker, Joseph D., 1976, International insolvency; the case of Herstatt, *American Bar Association Journal*, **62**, 1290–1295.

Bernstein, Peter L., 1998, *Against the Gods: The Remarkable Story of Risk*, John Wiley and Sons.

Black, Fischer and Myron Scholes, 1973, The pricing of options and corporate liabilities, *Journal of Political Economy*, **81**(3), 637–654.

Blustein, Paul, 2003, *The Chastening: Inside the Crisis that Rocked the Global Financial System and Humbled the IMF*, Public Affairs.

Bordo, Michael, 1998, Commentary (on The Financial Crisis of 1825 and the Restructuring of the British Financial System), *Review*, Federal Reserve Bank of St. Louis, **80**(3), 77–82, `http://research.stlouisfed.org/publications/review/article/3109`

Bordo, Michael, Barry Eichengreen and Douglas Irwin, 1999, Is globalization today really different from globalization a hundred years ago? *Brookings Trade Forum 1999*, 1–72.

Bordo, Michael and Angela Redish, 1987, Why did the Bank of Canada emerge in 1935? *Journal of Economic History*, **47**(2), June, 405–417.

Brady, Nicholas, 1988, *Report of the Presidential Task Force on Market Mechanisms*, US Government Printing Office.

Braudel, Fernand and Frank Spooner, 1967, Prices in Europe from 1450 to 1750, in: E. Rich and C. Wilson (eds.), *Cambridge Economic History of Europe*, Volume 4, Cambridge University Press, 374–486.

Bruner, Robert and Sean Carr, 2007, *The Panic of 2007: Lessons Learned from the Market's Perfect Storm*, John Wiley and Sons.

Calomiris, Charles and Gary Gorton, 1991, The origins of banking panics: models, facts, and bank regulation, in: R. G. Hubbard (ed.), *Financial Markets and Financial Crises*, NBER and University of Chicago, ch. 4, 109–174, `www.nber.org/chapters/c11484.pdf`.

Carlson, Mark, 2007, A Brief History of the 1987 Stock Market Crash with a Discussion of the Federal Reserve Response, FEDS Working Paper 2007–13, Board of Governors of the Federal Reserve, April, http://papers.ssrn.com/sol3/papers.cfm?abstract_id=982615

Coase, Ronald, 1990, *The Firm, the Market, and the Law*, University of Chicago Press.

Commodity Futures Trading Commission and Securities and Exchange Commission (CFTC/SEC), 2010, *Findings Regarding the Market Events of May 6, 2010: Report of the Staffs of the CFTC and SEC to the Joint Advisory Committee on Emerging Regulatory Issues, September 30*, www.sec.gov/news/studies/2010/marketevents-report.pdf

Daston, Lorraine, 1988, *Classical Probability in the Enlightenment*, Princeton University Press.

Devlin, Keith, 2008, *The Unfinished Game: Pascal, Fermat, and the Seventeenth-Century Letter that Made the World Modern*, Basic Books.

Eichengreen, Barry, 1996, *Globalizing Capital: A History of the International Monetary System*, Second Edition, Princeton University Press.

Eichengreen, Barry, 1992, *Golden Fetters: The Gold Standard and the Great Depression, 1919–1939*, Oxford University Press.

El-Gamal, Mahmoud A., 2008, *Islamic Finance: Law, Economics, and Practice*, Cambridge University Press.

Federal Deposit Insurance Corporation, 1997, *History of the 1980s*, Volume I: *An Examination of the Banking Crises of the 1980s and Early 1990s*, www.fdic.gov/bank/historical/history/vol1.html

Feldman, Ron and Mark Lueck, 2007, Are banks really dying this time? An update of Boyd and Gertler, *The Region*, Federal Reserve Bank of Minneapolis, September, 6–14.

Ferguson, Niall, 2008, *The Ascent of Money: A Financial History of the World*, Penguin Press.

Field, Peter (ed.), 2003, *Modern Risk Management: A History*, Risk Books.

Financial Crisis Inquiry Commission (FCIC), 2011, The Financial Crisis Inquiry Report: Final Report of the National Commission on the Causes of the Financial and Economic Crisis in the United States, January, www.gpoaccess.gov/fcic/fcic.pdf

Finger, Chris, Scott Howard, Jacques Longerstaey and Peter Zangari, 1996, *RiskMetrics – Technical Document*, Morgan Guaranty Trust and Reuters, December, www.msci.com/resources/research_papers/technical_doc/1996_riskmetrics_technical_document.html

Flood, Mark, 1992, The great deposit insurance debate, *Review*, Federal Reserve Bank of St. Louis, July/August, 51–77.

Friedman, Milton and Anna Schwartz, 1971, *A Monetary History of the United States: 1867–1960*, Princeton University Press.

Fung, William, David Hsieh, Narayan Naik and Tarun Ramadorai, 2008, Hedge funds: performance, risk, and capital formation, *Journal of Finance*, **63**(4), August, 1777–1803.

G-30 Working Group on Global Derivatives, 1993, Derivatives: Practices and Principles, G-30 Special Report, July, www.group30.org/rpt_29.shtml

Galbraith, John K., 1955, *The Great Crash 1929*, Houghton Mifflin.

Garbade, K. and W. Silber, 1978, Technology, communication and the performance of financial markets: 1840–1975, *Journal of Finance*, **33**(3), 819–832.

Gelderblom, Oscar and Joost Jonker, 2005, Amsterdam as the cradle of modern futures trading and options trading, 1550–1650, in: W. Goetzmann and G. Rouwenhorst (eds.),

The Origins of Value: The Financial Innovations that Created Modern Capital Markets, Oxford University Press, ch. 11, 189–206.

Glasserman, Paul, 2003, *Monte Carlo Methods in Financial Engineering*, Springer.

Gleick, James, 2011, *The Information: A History, a Theory, a Flood*, Pantheon.

Global Association of Risk Professionals (GARP), 2011, *2011 FRM Examination Study Guide*, www.garp.org/frm/test-preparation/study-guide.aspx

Goetzmann, William, Jonathan Ingersoll, Matthew Spiegel and Ivo Welch, 2007, Portfolio performance manipulation and manipulation-proof performance measures, *Review of Financial Studies*, **20**(5), September, 1503–1546, http://rfs.oxfordjournals.org/content/20/5.toc.

Goetzmann, William and Geert Rouwenhorst (eds.), 2005, *The Origins of Value: The Financial Innovations that Created Modern Capital Markets*, Oxford University Press.

Goodhart, Charles, 2011, *The Basel Committee on Banking Supervision: A History of the Early Years 1974–1997*, Cambridge University Press.

Groenfeldt, Tom, 2011, A systemic risk story, *Risk Professional*, April, 20–26, www.riskprofessional-digital.com/riskprofessional/201104.

Grossman, Richard, 2010, *Unsettled Account: The Evolution of Banking in the Industrialized World since 1800*, Princeton University Press.

Group of Ten (G10), 2001, Report on Consolidation in the Financial Sector, Technical Report, January, www.bis.org/publ/gten05.htm.

Hacking, Ian, 2006, *The Emergence of Probability: A Philosophical Study of Early Ideas about Probability, Induction and Statistical Inference*, Second edition, Cambridge University Press.

Hald, Anders, 2003, *A History of Probability and Statistics and Their Applications before 1750*, Wiley-Interscience.

Hammond, Bray, 1991, *Banks and Politics in America from the Revolution to the Civil War*, Princeton University Press.

Harris, Larry, 2002, *Trading and Exchanges: Market Microstructure for Practitioners*, Oxford University Press.

Heeffer, Albrecht, 2010, On the curious historical coincidence of algebra and double-entry bookkeeping, in: K. François, B. Löwe, T. Müller and B. van Kerkhove (eds.), *Foundations of the Formal Sciences. Bringing together Philosophy and Sociology of Science*, College Publications, London.

Herring, Richard and Jacopo Carmassi, 2010, The corporate structure of international financial conglomerates: complexity and its implications for safety & soundness, in: A. Berger, P. Molyneux and J. Wilson (eds.), *The Oxford Handbook of Banking*, Oxford University Press, January, ch. 8, 195–231.

Hickman, James, 2004, History of actuarial profession, *Encyclopedia of Actuarial Science*, John Wiley and Sons..

Hoag, Christopher, 2006, The Atlantic telegraph cable and capital market information flows, *Journal of Economic History*, **66**(2), 342–353, http://journals.cambridge.org/action/displayAbstract?fromPage=online&aid=449528

Holmström, Bengt and John Roberts, 1998, The boundaries of the firm revisited, *Journal of Economic Perspectives*, **12**(4), 73–94, www.aeaweb.org/atypon.php?return_to=/doi/pdfplus/10.1257/jep.12.4.73

Homer, Sidney and Richard Sylla, 1986, *A History of Interest Rates*, Fourth edition, John Wiley and Sons.

Huffman, Wallace and John Lothian, 1984, The gold standard and the transmission of business cycles, 1833–1932, in: M. Bordo and A. Schwartz (eds.), *A Retrospective on the Classical Gold Standard, 1821–1931*, NBER and University of Chicago Press, ch. 10, 455–512.

Hull, John , 2008, *Options, Futures, and Other Derivatives*, Seventh edition, Prentice Hall.

International Monetary Fund (IMF), 2011, A Guide to Committees, Groups, and Clubs, Factsheet, August, www.imf.org/external/np/exr/facts/pdf/groups.pdf.

Jain, Pankaj, 2005, Financial market design and the equity premium: electronic versus floor trading, *Journal of Finance*, **66**(6), 2955–2985.

Johnson, Roger T., 2010, *Historical Beginnings ... The Federal Reserve*, Federal Reserve Bank of Boston, www.bos.frb.org/about/pubs/begin.pdf

Kane, Edward J., 1977, Good intentions and unintended evil: the case against selective credit allocation, *Journal of Money, Credit and Banking*, **9**(1), 55–69.

Kaufman, George and Larry Mote, 1994, Is banking a declining industry? A historical perspective, *Economic Perspectives*, Federal Reserve Bank of Chicago, May, 2–21.

Keynes, John Maynard, 1936, *The General Theory of Employment, Interest, and Money*, Macmillan, London.

Kindleberger, Charles, 1993, *A Financial History of Western Europe*, Second edition, Oxford University Press.

Kindleberger, Charles and R. Aliber, 2005, *Manias, Panics and Crashes: A History of Financial Crises*, John Wiley and Sons.

Knights, Mark, 2012, Fiscal-military state, in: *Oxford Bibliographies Online: Atlantic History*, Oxford University Press, www.oxfordbibliographies.com/view/document/obo-9780199730414/obo-9780199730414-0073.xml.

Kohn, Meir, 1999a, *Bills of Exchange and the Money Market to 1600*, manuscript, Dartmouth College, www.dartmouth.edu/~mkohn/orgins.html

Kohn, Meir, 1999b, *Risk Instruments in the Medieval and Early Modern Economy*, manuscript, Dartmouth College, www.dartmouth.edu/~mkohn/orgins.html

Kose, M. Ayhan Eswar Prasad, Kenneth Rogoff and Shang-Jin Wei, 2009, Financial globalization: a reappraisal, *IMF Staff Papers*, International Monetary Fund, **56**(1), http://prasad.dyson.cornell.edu/doc/research/imfsp200836a.pdf.

Lane, Philip and Milesi-Ferretti, Gian Maria, 2007, The external wealth of nations mark II: revised and extended estimates of foreign assets and liabilities, 1970–2004, *Journal of International Economics*, **73**(2), 223–250.

Laubsch, Alan, 2000, Risk transparency, in: M. Lore and L. Borodovsky (eds.), *The Professional's Handbook of Financial Risk Management*, Butterworth-Heinemann, ch. 25, 740–775.

Lee, Geoffrey, 1977, The coming of age of double entry: the Giovanni Farolfi ledger of 1299–1300, *Accounting Historians Journal*, **4**(2), 79–95.

Lothian, James, 2000, Capital market integration and exchange rate regimes in historical perspective, *Research in Banking and Finance*, **1**, 139–173.

Lowenstein, Roger, 2001, *When Genius Failed: The Rise and Fall of Long-Term Capital Management*, Random House.

MacKenzie, Donald, 2006, *An Engine, Not a Camera: How Financial Models Shape Markets*, MIT Press, http://mitpress.mit.edu/catalog/item/default.asp?ttype=2&tid=10841

Markham, Jerry, 2002a, *A Financial History of the United States, Volume I: From Christopher Columbus to the Robber Barrons (1492–1900)*, M. E. Sharpe.

Markham, Jerry, 2002b, *A Financial History of the United States, Volume II: From J.P. Morgan to the Institutional Investor (1900–1970)*, M. E. Sharpe.

Markowitz, Harry, 1959, *Portfolio Selection: Efficient Diversification of Investments*, John Wiley and Sons, http://cowles.econ.yale.edu/P/cm/m16/index.htm

Melamed, Leo, 2009, *For Crying Out Loud: From Open Outcry to the Electronic Screen*, John Wiley and Sons.

Merton, Robert, 1973, Theory of rational option pricing, *Bell Journal of Economics and Management Science*, **4**(1), 141–183.

Mester, Loretta, 2007, Some thoughts on the evolution of the banking system and the process of financial intermediation, *Economic Review*, Federal Reserve Bank of Atlanta, First and Second Quarters, 67–75.

Mitchell, Melanie, 2011, *Complexity: A Guided Tour*, Oxford University Press.

Morrison, Alan and William Wilhelm, 2007, Investment banking: past, present, and future, *Journal of Applied Corporate Finance*, **19**(1), 8–20, http://gates.comm.virginia.edu/wjw9a/Papers/JACF%20Morrison%20Wilhelm%20Final%20version.pdf

Mulherin, J. Harold, Jeffry Netter and James A. Overdahl, 1991, Prices are property: the organization of financial exchanges from a transaction cost perspective, *Journal of Law and Economics*, **34**(2), 591–644.

Munro, John H., 2003, The medieval origins of the 'Financial Revolution': usury, rentes, and negotiability, *International History Review*, **3**(25), 505–562.

Neal, Larry, 1990, *The Rise of Financial Capitalism: International Capital Markets in the Age of Reason*, Cambridge University Press.

Neal, Larry, 1998, The financial crisis of 1825 and the restructuring of the British financial system, *Review*, Federal Reserve Bank of St. Louis, **80**(3), 53–76, http://research.stlouisfed.org/publications/review/article/3108

Neal, Larry, 2005, Venture shares of the Dutch East India Company, in: W. Goetzmann and G. Rouwenhorst (eds.), *The Origins of Value: The Financial Innovations that Created Modern Capital Markets*, Oxford University Press, ch. 9, 165–176.

Neal, Larry and Stephen Quinn, 2001, Networks of information, markets, and institutions in the rise of London as a financial centre, 1660–1720, *Financial History Review*, **8**(1), April.

North, Douglass and Barry Weingast, 1989, Constitutions and Commitment: The evolution of institutions governing public choice in seventeenth-century Britain, *Journal of Economic History*, **49**(4), 803–832.

Obstfeld, Maurice, 1998, The global capital market: benefactor or menace? *Journal of Economic Perspectives*, **12**(4), Fall, 9–30, www.aeaweb.org/articles.php?doi=10.1257/jep.12.4.9

Obstfeld, Maurice and Alan Taylor, 2004, *Global Capital Markets: Integration, Crisis, and Growth*, Cambridge University Press.

O'Rourke, Kevin and Jeffrey Williamson, 2001, *Globalization and History: The Evolution of a Nineteenth-Century Atlantic Economy*, MIT Press.

Organisation for Economic Cooperation and Development (OECD), 2010, Measuring Globalization: OECD Economic Globalization Indicators 2010, Technical Report, `http://dx.doi.org/10.1787/9789264084360-13-en`.

Reinhart, Carmen and Ken Rogoff, 2009, *This Time is Different: Eight Centuries of Financial Folly*, Princeton University Press.

Risk Management Association (RMA), 2011, *RMA – Credit Risk Certified: Frequently Asked Questions*, The Risk Management Association, May, `www.rmahq.org/NR/rdonlyres/F4D56E87-9FE6-4BE1-9E13-FEF68C40D993/0/CRCFAQtoprint_051911.pdf`

de Roover, Florence, 1945, Early examples of marine insurance, *The Journal of Economic History*, **5**(2), 172–200.

Saunders, Anthony, 1999, *Credit Risk Measurement: New Approaches to Value at Risk and Other Paradigms*, John Wiley and Sons.

Scott, Hamish, 2009, The fiscal-military state and international rivalry during the long eighteenth century, in: C. Storrs (ed.), *The Fiscal-Military State in Eighteenth-Century Europe*, Ashgate, ch. 1, 23–53.

Securities and Exchange Commission, 1988, *The October 1987 Market Break*, SEC Staff Report, Division of Market Regulation, February.

Seligman, Joel, 2003, *The Transformation of Wall Street*, Third edition, Aspen Publishers.

Simon, Herbert, 1962, The architecture of complexity, *Proceedings of the American Philosophical Society*, **106**(6), 467–482.

Stephen, Leslie, 1885, Rothschild, Nathan Meyer, *Dictionary of National Biography*, Volume. 49, Macmillan, 306–309.

Sylla, Richard, 2005, The origins of the New York stock exchange, in: W. Goetzmann and G. Rouwenhorst (eds.), *The Origins of Value: The Financial Innovations that Created Modern Capital Markets*, Oxford University Press, ch. 17, 299–312.

Tracy, James D., 1985, *A Financial Revolution in the Habsburg Netherlands: Renten and Renteniers in the County of Holland, 1515–1565*, University of California Press.

Usher, Abbot P., 1934, The origins of banking – the primitive bank of deposit, 1200–1600, *Economic History Review*, **4**(4), 399–428.

Weiss, David, 2006, *After the Trade is Made: Processing Securities Transactions*, New York Institute of Finance.

Wennerlind, Carl, 2011, *Casualties of Credit: The English Financial Revolution, 1620–1720*, Harvard University Press.

Wheelock, David, 2008, The federal response to home mortgage distress: lessons from the Great Depression, *Review*, Federal Reserve Bank of St. Louis, **90**(3, Part 1), pp. 133–148, `https://research.stlouisfed.org/publications/review/08/05/Wheelock.pdf`.

Williamson, John, 2004, A Short History of the Washington Consensus, Working Paper, Institute for International Economics, September, `www.iie.com/publications/papers/williamson0904-2.pdf`.

Woo, Wing Thye, Jeffrey Sachs and Klaus Schwab (eds.), 2000, *The Asian Financial Crisis: Lessons for a Resilient Asia*, MIT Press.

2

Risk management

Robert Mark and Dilip Krishna

2.1 What is risk management?

Financial institutions (FIs) that deal in commercial banking or trading activities are exposed to a variety of risks. In fact, risk-taking goes to the core of their business. When a bank lends money to a borrower, it is taking the risk that the borrower may not be able to pay back the loan. The profits it earns on the loans can be considered to incorporate the price of accepting this risk. Likewise when a trading firm buys (or sells) securities in the marketplace, it is accepting the risk that the price of the security will go down (or up) resulting in a loss. Global FIs are often exposed to many risks on a truly massive scale. Complicating the picture for these large FIs is that some risks counteract others. In fact, it is commonplace for firms to accept new risks as a way to counter existing risks, a process called "hedging." It is therefore critical for banks to control this complex patchwork of risks across the firm in a measured and organized fashion. Risk management is the discipline that FIs use to do this.[1]

The first step in managing risk is to be able to measure it – perhaps the most significant achievement of modern risk management is the development of techniques to quantify risk in terms of a potential monetary loss. Risk can be measured in statistical terms as the loss that would be experienced with a particular level of confidence. All firms expect some small level of loss in the normal course of doing business, for example human errors in the settlement process that can never be fully eliminated. This is referred to as *expected* loss. The cost of expected losses can be made up by adjusting the price of the firm's products upward as an increasing function of the size of the expected loss. But on occasion businesses also experience

[1] This chapter focuses only on risk management in banks that deal in traditional credit products such as loans, mortgages and credit cards, as well as those institutions that deal with traded securities. In particular, the chapter does not address risks in insurance companies that offer property and casualty, life or health insurance products.

Handbook of Financial Data and Risk Information, ed. Margarita S. Brose, Mark D. Flood, Dilip Krishna and Bill Nichols. Published by Cambridge University Press. © Cambridge University Press 2014.

large, unanticipated losses. A business can endeavor to protect itself from taking on a significant amount of risk in its various activities through the accumulation of sufficient capital in an effort to protect against potential unexpected loss levels. Risk management encompasses managing both the expected loss as well as unexpected variability of financial outcomes for a business.

Risks can also be categorized into financial and non-financial risks. Financial risk is a major risk for financial institutions and encompasses credit, market and liquidity risk. *Credit risk* is the risk of loss following a change in the factors that drive the credit quality of an asset. These include adverse effects arising from credit grade migration, including default, and the dynamics of recovery rates. *Market risk* is the risk of losses arising from changes in market risk factors. Market risk can arise from adverse changes in interest rates, foreign exchange rates, equity prices and commodity price factors. For example, a long corporate bond position has market risk arising from both an increase in interest rates and credit risk arising from deterioration in credit quality. There are two kinds of *liquidity risk*. Trading Liquidity Risk is the risk of not being able to exit a financial position due to lack of market liquidity. Funding Liquidity Risk is the inability to raise cash, which could occur for several reasons (e.g., due to lack of availability of credit in the market).

There is a cycle of taking risk, managing risk and elimination of risk in the trading book and banking book. The key differences between the trading and banking books relate to holding intent and valuation practices (e.g., mark-to-market valuation). Banking book positions are typically held to maturity and subject to accrual accounting. Available for sale securities are carried at fair value. Trading book positions are typically marked-to-market and actively hedged by the firm. The pace of risk management techniques at each stage in the trading book and banking book is usually different. Also, the policies, methodologies and infrastructure (such as organization structures, data etc.) are different between both books.

Non-financial risk includes *operational risk, reputation risk, strategic risk* and *business risk. Operational risk* refers to financial loss due to a host of potential operational breakdowns related to people risks, process risks and technology risks. *Reputation risk* is another key risk that poses a special threat to financial institutions because the nature of their business requires the confidence of customers, creditors, regulators and the general marketplace. *Strategic risk* refers to the risk of loss when making a significant investment or a choice to enter or exit a business. *Business risk* refers to uncertainty about the demand for products, the price that can be charged for those products, or the cost of producing and delivering products. There remains the question of how business risk should be measured within a formal risk management framework. Strategic risk occurs due to the inability of a firm to create an effective strategy, while business risk is caused by the inability to execute such an otherwise sound strategy due to internal or external factors.

Individual business units (BUs) are subject to different kinds of risks depending on the types of business they are engaged in as well as the financial products that they sell. The types of risks BUs face, along with associated metrics that describe the potential impact of a particular risk, are called the BU *risk profile*. Individual risk profiles can be aggregated to develop a risk profile of the complete firm. Since financial organizations need to take risks to generate profits, they must determine the optimal balance of risk and reward given their particular situation. The level of risk that an organization is prepared to accept is called the *risk appetite* of the firm. Once the risk profile and appetite have been determined, firms must then manage their activities to ensure that the risks taken by their business units do not extend their risk profile beyond their risk appetite. To do this, firms delegate the task of managing risk to their various businesses through the use of *capital* – a financial measure of the risk that can be calculated at various levels of a business up to and including individual transactions. Capital is assigned to each BU to absorb losses that may occur from time to time. Capital is also assigned to each BU as a function of how confident management is that the risks will be actively managed in alignment with risk appetite. The amount of capital assigned to a BU also depends on how conservative senior management and the board wishes to be.

Classification is an important part of the effort to make an otherwise ill-defined risk measurable, manageable and transferable. Each category of risk calls for distinct sets of risk management skills which are often used to define, delegate and organize the risk management activities of an FI. At the enterprise level, Enterprise Risk Management (ERM) examines, integrates and manages these risks. Figure 2.1 shows an example of how ERM integrates risk management across a diversified bank (note that the classification of risk shown in Figure 2.1 is illustrative – other banks may use somewhat different risk classifications).

There are many potential approaches that can be used separately or together to manage risk throughout an FI. In the first place, an institution can limit its risk by

Overall Risk Position		
Economic Capital Usage in $ m.	12/31/2011	12/31/2012
Credit Risk	$ 7,488.10	$ 10,024.22
Market Risk	$ 12,937.28	$ 9,240.72
Trading Market Risk	$ 4,430.67	$ 6,212.44
Non-Trading Market Risk	$ 8,509.72	$ 3,530.28
Operational Risk	$ 3,536.90	$ 4,206.92
Diversification Benefit	$ (3,157.20)	$ (3,060.03)
Business Reputation & Strategic Risk	$ 967.28	$ 693.57
Total Economic Capital Usage	$ 21,775.47	$ 21,607.39

Figure 2.1 Map of enterprise risk.

defining its risk appetite in terms of limits set by the board and appropriately delegating those limits. These can be translated into a position limit, risk limits or loss limits in different activities. Then risks can be actively mitigated and managed within those limits using underwriting, syndication, diversification and hedging. A bank can also choose to avoid outright certain types of investments that exceed their risk appetite.

2.2 Structuring and managing risk in a banking environment

Risk managers have played an increasingly important and active role in ERM by developing analytical tools to measure the potential returns of financial positions and the potential losses associated with those positions. A key role of the risk manager is to make the risk at the BU and enterprise level transparent. The Chief Risk Officer (CRO) is at the center of integrating the management and governance of risk in a bank. The Board of Directors and the management committee of the bank increasingly rely on the CRO to integrate corporate risk governance responsibilities with the CRO's existing risk responsibilities. Several direct or indirect reports of the CRO look after specialized areas of risk.

There are several organizational structures for managing risk. For example, in some cases the head of market risk looks after all market risk related activities in both the trading book and the banking book, the head of credit risk looks after all credit risk related activities and the head of operational risk looks after all operational risk related activities. These functions may further divide into specialty areas. In some banks, for example, the head of credit risk is responsible for approving the credit risk for a derivative transaction whereas the head of market risk is responsible for calculating the potential future exposure for the derivative transaction. The reporting lines of the CRO vary across organizations. The CRO increasingly has a functional reporting relationship to the board and an administrative reporting relationship to the CEO.

The banking industry has not yet developed standards of practice (SOP) for risk management practices at the individual risk manager level. SOP at the individual level are typically a true sign of a profession and we expect that ultimately a set of standards will be developed at the individual level for bankers that will parallel what has already been developed for other professions such as actuaries.[2]

A primary responsibility of the board is to ensure it develops a clear understanding of the business strategy of the bank and the fundamental risks and rewards

[2] Actuarial standards are codified in the Actuary Standards of Practice (ASOP). The ASOP are evolving to provide guidance to actuaries when performing professional services with respect to risk management (see Mark, 2012).

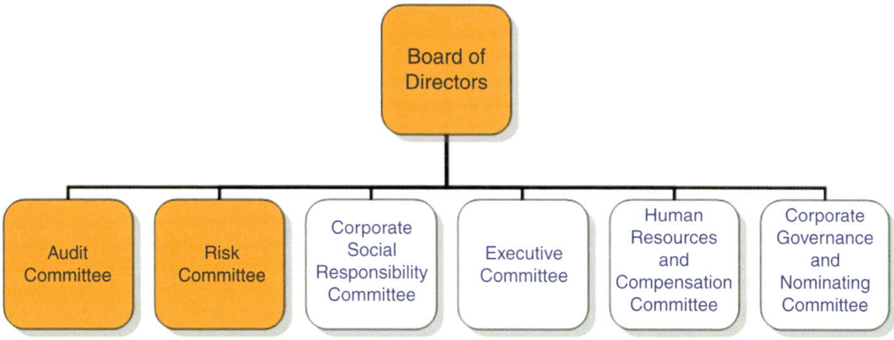

Figure 2.2 Risk management at the level of the Board of Directors (Board of Directors and Committees at Bank of New York Mellon).

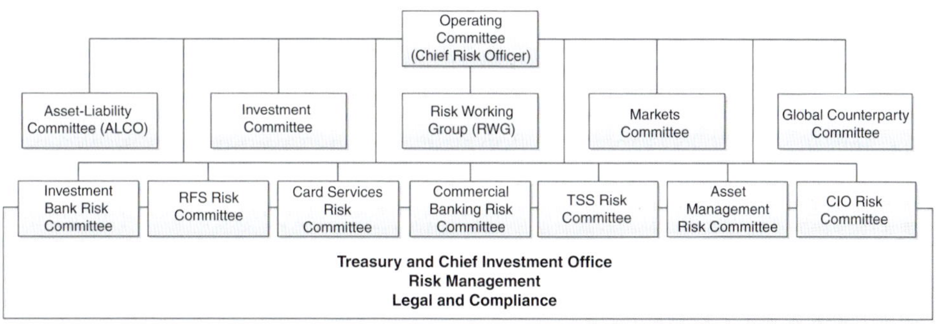

Figure 2.3 Sample risk governance structure (information drawn from JP Morgan Chase Annual Report, 2010).

associated with this strategy. The board also needs to make sure that risks are made transparent to managers and to stakeholders through adequate internal and external disclosure. The board examines whether the structure for managing risk is adequately designed and also reviews whether the risk and return relationships are appropriate. An illustrative structure of board subcommittees (including the Risk Committee and Audit Committee) is shown in Figure 2.2.

The CRO typically serves on the management committee and senior risk committee. The senior risk committee oversees the risk management practices and detailed reporting. Junior risk committees look after specific risk types such as the credit risk committee. The risk committees of the bank (see Figure 2.3) are, among other things, charged with ratifying the key policies and associated processes of the bank's risk management activities. The committees help to translate the overall risk appetite of the bank into a set of limits that flow down through the bank's executive officers and business divisions.[3]

The senior risk committee of the bank recommends to the risk committee of the board an amount at risk – the *risk appetite* for the whole institution – that is prudent for the risk committee of the board to approve. The risk committee of the board approves the bank's risk appetite each year based on a well-defined and broad set of risk measures. The risk committee of the board delegates authority to the senior risk committee of the bank. The senior risk committee of the bank delegates to the CRO the authority to make day-to-day decisions on its behalf, including the authority to approve excesses of limits provided to the bank's various businesses as long as these limits do not breach the overall risk limits approved by the board.

Market risk limits serve to control the risk that arises from changes in the absolute price (or rate) of a position. Credit risk limits serve to control and limit the number of defaults as well as helping to limit a downward migration in the quality of the credit portfolio. The bank will also want to set out limits for the Corporate Treasury function regarding exposure to both market risk and funding liquidity risk (a function also known as asset/liability management). The setting of the risk limit in terms of a particular metric should be consistent with the bank's risk appetite. The CRO reports on firm-wide risk and is responsible for independently monitoring and maintaining effective control of all risk limits. The CRO may well order that business units reduce their positions or close them out because of concerns about the risks. The CRO also delegates some responsibilities to the risk management heads of the various business units. The business managers are responsible for the risk management of their business and, in turn, may delegate limits to members of their business unit.

The bank's operations function prepares the profit and loss report (P&L), provides independent valuations, and supports the operational needs of the various businesses. The finance function develops valuation and finance policy, and strives to ensure the integrity of P&L. The audit function provides an independent assessment of the design and implementation of the bank's risk management. For example, the internal audit group would review the overall risk management process including the integrity of the risk management system, the organization of the risk control unit, the integration of risk measures into daily risk management, and so on. Internal auditors may also verify the accuracy of models through an examination of the back-testing process. Audit may also examine the documentation relating to compliance with the qualitative/quantitative criteria outlined in any regulatory guidelines.

[3] See SAS White Paper (2011).

2.3 Measuring and managing market risk

Market risk is the risk that changes in financial market prices and rates will reduce the value of a security or portfolio.[4] There are four major classes of market risk: interest rate risk, equity price risk, foreign exchange risk and commodity price risk.

Throughout financial history, firms have periodically experienced extremely large losses due to unexpected changes in market prices. Barings Bank was a public example of a major financial organization being forced into bankruptcy due to a combination of operational risk and market risk.[5] Market risk measurement has grown in sophistication over time, starting from simple measures of price sensitivity such as the duration and convexity of a bond.[6] As portfolios and securities became more complex, comparatively more sophisticated measures of risk were required, resulting in the development of a key measure of market risk called Value-at-Risk or VaR (though the passage of time has shown that VaR itself has potential for improvement – more on this later).

VaR offers an estimate of the potential change in the value of a portfolio resulting from a change in market factors over a specified period of time. The concept behind VaR can be illustrated via the distribution of daily returns shown in Figure 2.4. Observe that Figure 2.4 visually appears to be a "bell-curve" or normal distribution.[7] In the simple case of a portfolio consisting only of the Nasdaq Index, one measure of risk may be the predicted maximum loss at a certain level of probability – in the figure one can state with 95% probability that losses will not be greater than −3%. This probability is called the *confidence level.*

VaR can therefore be defined as the potential loss that might be expected from holding a security or portfolio of securities over a given period of time (say over a 1 day period) for a specified confidence level. For example, if we say that a portfolio "has a daily VaR of $1 million at the 95% confidence level" then the expectation is that daily losses will on average be no higher than $1 million, 95 out of every 100 trading days. Changes in VaR can be warnings of increased risk and are widely reported in financial institutions (see Figure 2.4 for an example[8]). It is important to note that calculation does not state by how *much* an actual loss is expected to exceed the VaR measure on those 5 in 100 days. Conditional VaR (CVaR) offers a way to estimate expected loss in the tail. If we assume that a

[4] See Crouhy et al. (2001) and Section 9.2.1 of this Handbook.

[5] A single trader in Singapore was able to establish positions on behalf of the bank that resulted in $1.3 billion of losses to the prestigious bank. While the fact that he was able to do so is actually an example of Operational Risk (see Section 2.6), Market Risk was the final cause of the demise of Barings.

[6] Duration measures the price sensitivity of a bond to interest rates but is not the same for the bond at all rates. Convexity measures the change in duration to interest rates. Duration is a poor measure for the price sensitivity of a bond for large changes in interest rates.

[7] As described later, looks can be deceiving. The actual distribution of prices in the markets is *not* normal, one of the principal criticisms of VaR.

[8] See Kelly (2007) for an example of how one bank used VaR measures during the financial crisis.

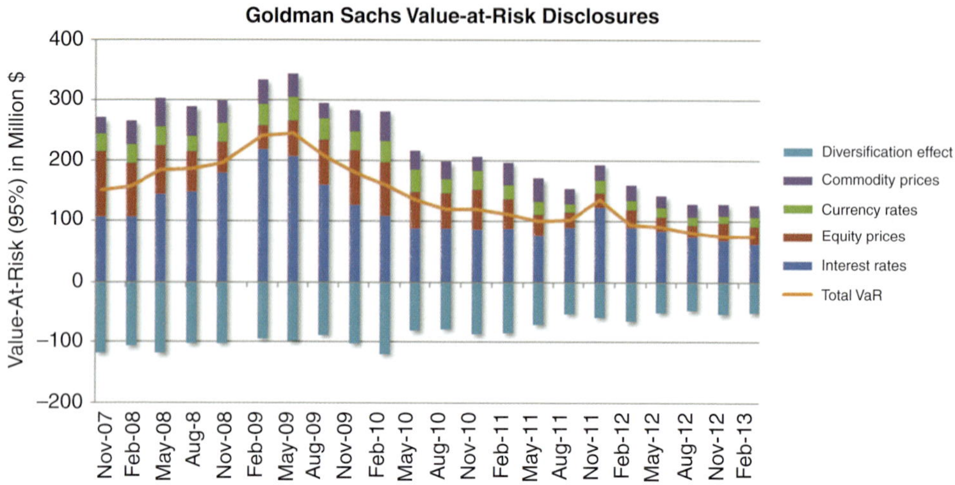

Figure 2.4 Example of daily VaR (data from Goldman Sachs, 2013).

distribution is normally distributed (bell shaped) then a VaR set at a 95% probability confidence interval is based on 1.64 sigma, which leads to a CVaR at 2.06 sigma. See Section 2.6 for a discussion on another approach to estimating these losses.

VaR calculations can be illustrated via the following steps. Figure 2.4 illustrates this calculation using the daily distribution of returns on the Nasdaq QQQ index (for 3422 days) ordered from worst to best. First, the forward distribution of returns on the portfolio needs to be calculated at the chosen *time horizon* (say 1 day). Second, the required percentile of this distribution needs to be identified so that a particular loss number can be read off. Since only 5% of the daily returns result in a loss greater than 3.29% of the portfolio, one can say that the 95% VaR is 3.29% of the QQQ portfolio (in dollar terms, 95% VaR on a $1 million portfolio would be $32,900). The CVaR associated with a 95% VaR is 4.65% of the QQQ portfolio (or $46,500). The 99% VaR would be higher at 5.46% of portfolio value (or $54,600 on a $1 million portfolio). Put another way, according to the VaR estimate, one can be 99% confident that this portfolio cannot lose over $54,600 on a given day from identified market risks.

There are three basic ways to calculate VaR. The first is the *historical method* illustrated in Figure 2.5. The second is the so-called *parametric* or variance-covariance technique. This technique assumes that the market obeys a simple statistical formula (for example, the so-called *normal* distribution) and uses the two factors of such a distribution – mean and standard deviation – to describe the returns. In this calculation, a confidence level of 99% corresponds to 2.33 standard deviations. The third method is the *Monte Carlo* simulation technique that randomly

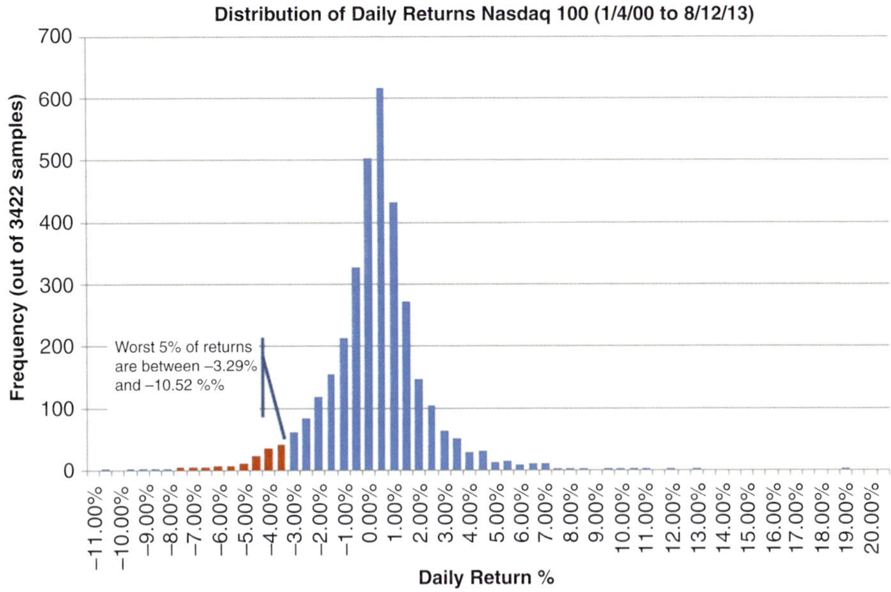

Figure 2.5 Calculating VaR (taken from David Harper).

generates market factors such as interest rates and calculates the expected return on the portfolio. This procedure is repeated many times (potentially hundreds of thousands of times). The resultant set of portfolio values form a distribution similar to Figure 2.5 from which VaR can be calculated in the way earlier described. Each method has benefits and drawbacks based on theoretical soundness, the intensity of computational requirements and the amount of data required.

In conformity with the Basel Accord (see Section 2.9), regulators initially set a time horizon of 10 days for the purpose of VaR calculations that are used to report regulatory capital requirements. Ideally, this this "10-day VaR" would be derived from a corresponding distribution of results over a 10-day horizon. Many banks derive an approximation of 10-day VaR from daily VaR data, by multiplying the daily VaR by the square root of time (here, 10 days). Over longer time periods (say 10 days) there is another computational issue to consider. Daily VaR ignores the fact that positions may generate returns during a shorter time horizon that may not be achievable in the longer term. *Absolute* VaR is the loss without considering return, whereas *Relative* VaR takes the return into account.[9]

[9] For instance, if a position returns 10% over the course of a year, and the Absolute VaR is calculated at 20% of the portfolio, then the Relative VaR would be mitigated by the expected return and be only 10% (i.e. Relative VaR = Absolute VaR − Expected Return).

The use and reliability of VaR have been criticized, by the Basel Committee among others, for many reasons. These include questioning its theoretical validity, the dangers of misinterpreting the metric and implementation challenges arising from the lack of data.[10] Such observations have included the following.

- A frequent comment on VaR as applied to the parametric method is that extreme events happen in markets with much greater frequency than one might expect from a statistically normal distribution. These "fat-tails" (i.e., tails on the distribution having higher probability and therefore being thicker) tend to generate higher losses in practice than predicted by VaR.[11]
- Another critique of traditional VaR is that it assumes that correlations between different positions in the portfolio are constant, which may only be true for relatively short time horizons.[12] This means that dynamic liquidity risks cannot be included in VaR analysis. The concept of a risk measure that is only right a certain percentage of the time can also be difficult to articulate and use.
- VaR implementations rely crucially on the availability of reliable and timely data. Lack of quality inputs such as correlations, security characteristics and positions can lead to unreliable VaR calculations.
- While the measure itself explicitly disavows such a claim, VaR can sometimes be confused with the worst-case loss.[13]

However, VaR continues to be a popular and useful measure in risk management when used with caution and combined with other risk measures. VaR offers a good way of representing the risk appetite of a firm in normal market conditions. Stress testing (see Section 2.6) is a supplementary methodology to help address some of the shortcomings of VaR methodologies. However, stress tests do not obviate the need for high-quality data. On the contrary, the need for data is heightened due to the need for new types of information required for this process.

VaR can be considered one measure of capital to be held by a firm to safeguard against market risk.[14] In fact, VaR represents the economic capital that the bank should hold to protect against unexpected, large losses. The levels of capital to be held are independently derived by internal management processes as well as by regulators (explained further in Sections 2.8 and 2.9) – these are referred to as Economic and Regulatory Capital respectively. Economic capital differs from

[10] These criticisms have been widely published in the literature (see Basel Committee for Banking Supervision, May 2012).

[11] For example, Basel (2012) states that "A number of weaknesses have been identified with using value-at-risk (VaR) for determining regulatory capital requirements, including its inability to capture 'tail risk'."

[12] See Linsmeier and Pearson (1996) which states "value at risk is not a panacea. . . risk managers . . . may look at portfolios' exposure to other factors such as changes in correlations."

[13] See Jorion (2003).

[14] Note that recent developments in risk management and regulation have put significantly more focus on stress testing (see Section 2.6) as a way to ensure that a firm is sufficiently capitalized.

regulatory capital because the confidence level and the time horizon chosen for VaR calculations are usually different. For example, when banks are determining their economic capital for market risk, they may choose a higher confidence level than the 99% set by the regulator. They may also vary the time horizon in economic capital calculations from one day for very liquid positions, such as a government bond, to several weeks for illiquid positions, such as long-dated over-the-counter equity derivatives. By contrast, as indicated above, the Basel Accord arbitrarily sets the time horizon to 10 days for any position in the trading book. If VaR is to be computed for a 10-day period then a simple approximation to compute a 10-day VaR is to multiply the daily VaR by the square root of 10.[15]

2.4 Measuring and managing credit risk

Credit risk arises from the potential for borrowers to default on loans made to them. Banks are exposed to credit risk across their entire spectrum of business. These range from loans made to corporations (also known as *wholesale lending*) to loans made to individuals (*retail lending*). In addition, credit risk can also arise from holding bonds of corporations, in securities financing deals with other banks, in counterparty-to-counterparty derivative transactions (so-called over-the-counter transactions) and in insurance contracts. In all cases, assessment of credit risk depends among other things on analyzing the creditor's ability and willingness to repay the loan. However, credit risk assessment for wholesale borrowers (or counterparties) differs significantly from the process for retail borrowers. This is due both to the amount of information that is usually available in each case (there is usually much more information available for individual wholesale borrowers than for the retail borrower) and to the necessity of dealing with scale in retail banking. A large bank may only have tens of thousands of wholesale counterparties whereas the largest banks have retail borrowers numbering in the tens of millions, so risk assessment techniques need to be cost effective in the latter case.

A loan, whether wholesale or retail, goes through a basic lifecycle shown in Figure 2.6. The first step is *Credit Origination*, where the bank acquires credit exposure through various means such as extending a loan to a corporate borrower, executing a trade with a counterparty or issuing a mortgage. Before the bank accepts the exposure, however, it must determine that the risk it is about to take on is acceptable and that the bank is being adequately compensated for it. *Risk Assessment* is therefore a key step in the process. Beyond examining the individual transaction itself (as described below), the bank should also determine whether it

15 Assume a portfolio of \$100 million is expected to lose no more than \$10 million in 99 out of 100 days. This means it has a *daily* 99% VaR of \$10 million. Over a 10-day period the VaR will be \$31.62 million, assuming a normal distribution.

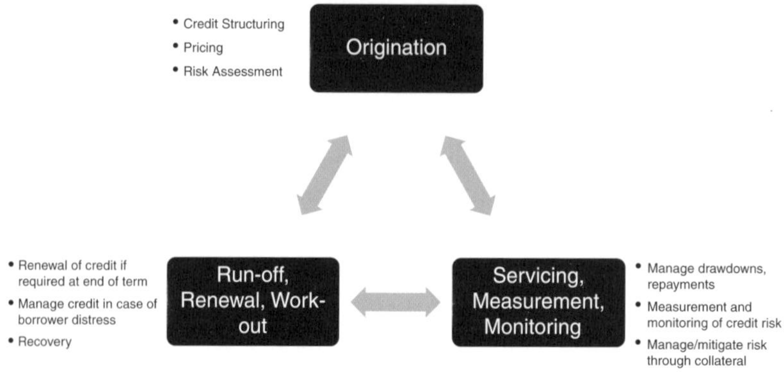

Figure 2.6 Steps in the credit lifecycle.

adds undue risk to the portfolio by creating concentrations. To do so banks create *Concentration Limits* of various kinds based on characteristics such as geography and industry. Another important step is to mitigate the risk during origination. One way is to consider the *collateral* for the loan, which is typical in retail and small-business loans as well as in capital market transactions. In larger, corporate transactions it is common for banks to spread the risk by *syndicating the loan* so that each bank takes on only part of the risk of any one transaction. Once the loan has been originated, it must be serviced as well as monitored on a continuous basis. Servicing processes differ widely depending on the product. For example, in contrast to retail markets where collateral (like residential property) is re-evaluated infrequently, in the capital markets margining and collateral transfer processes are commonly executed on at least a daily basis. Mortgage servicing involves ensuring timely payment of interest and principal from borrowers (typically biweekly or monthly), assessment of late fees and so forth. The final stage of the credit cycle is at the end of the loan. Most credits terminate gracefully, with the borrower repaying interest and principal in full. Often such loans are also renewed through a process that is similar to origination. In some cases however borrowers fall into distress, forcing the bank into a Work-Out process. The bank may restructure the loan by extending the period of repayment, or may try to recover the monies due to it by legal means by initiating collections or foreclosure proceedings.

Wholesale credit analysis depends on an intimate knowledge of the borrowing counterparty. Banks expect to make many large loans across a range of different products to such borrowers, making the cost of detailed credit analysis worthwhile. Credit analysts take into consideration many complex attributes of a firm in order to make an informed credit assessment. These involve financial, managerial as well as quantitative and qualitative assessments. The credit analyst should ascertain the financial health of the borrower, examine its liquidity position and take into account the nature of the industry to which it belongs as well as the potential effect of macroeconomic events.

More sophisticated measurement techniques review two separate aspects of credit risk. The first dimension calls for measuring the likelihood of a borrower defaulting and is called *Probability of Default (PD)*. The second dimension measures the loss if a default were to occur and is referred to as *Loss Given Default (LGD)*. The PD is dependent on the characteristics of the borrower, while the type of loan given to the borrower heavily influences the LGD. For example, a loan secured by good collateral is less likely to suffer loss than an unsecured loan, and is thus likely to have a lower LGD. The PD and LGD measures are used by many banks as key inputs into establishing allowances for losses and regulatory capital calculations. The expected *loss* of a particular transaction is based on the product of the amount of credit *exposure at default*[16] (say, \$100) multiplied by the *PD of default* (say, 4%) of an obligor (or borrower) and the *LGD* (say, 25%). In this example, the expected loss provision is $\$100 \times 0.04 \times 0.25 = \$1$.

Banks use credit rating systems in order to arrive at ratings in a systematic manner. While *external* credit ratings (e.g., from firms like Standard & Poor's, Fitch and Moody's) are useful, larger banks also often rely on their own internal credit risk rating systems to analyze borrowers. Internal ratings create a consistent framework and permit comparisons across the entire loan portfolio. A typical internal rating system of a bank combines quantitative information (e.g., financial ratios) with professional judgment to arrive at a credit rating. The goal of a rating system is that it must be consistent over time and be based on sound economic principles. Two kinds of ratings are assigned. Each borrower is assigned a PD. A borrower may have many lines of borrowing called "facilities" available with different terms and conditions. Therefore each facility has a different potential for loss and is independently assigned an LGD to identify the risk of loss in the event of default on the obligation.

The first step in an Internal Credit Risk Rating System typically starts with a financial assessment of the borrower (initial obligor rating). For example, the credit analyst examines the fundamentals of the business such as earnings and cash flow measures, asset values, liquidity and leverage measures, financial size, flexibility and debt capacity. Other non-financial measures are also examined such as the managerial capability of the borrower, the borrower's absolute and relative (tier) position within the industry, the quality of the financial information, country risk of

[16] During times of distress, borrowers will usually make all attempts to keep their business viable by drawing down on any available credit lines to the maximum extent. The level of borrowing near default is therefore typically much larger than the average borrowing and is referred to as *Exposure-At-Default* or EAD.

the borrower (in case of foreign companies) and reviewing the impact of the loan structure on PD.

Information on borrowers is a key input in to the wholesale loan risk management process. Complexity, quality and timeliness of data can be a major challenge. Each borrower can be expected to transact with the bank on a range of financial products. While the risk on any individual product may be tolerable, it is the aggregate risk across all products that is often the important measure. This can be further exacerbated by the fact that corporations often borrow through their subsidiaries, and have a range of agreements with the bank about how much, and via what product, each subsidiary can borrow. Therefore it becomes important to aggregate not only by the various products and transactions that a corporation has with the bank, but also across its various subsidiaries. Maintaining a current and accurate legal-entity hierarchy (i.e., the corporation, its subsidiaries, their subsidiaries and so on) is often important in these situations (a useful discussion on establishing a Legal Entity Identifier (LEI) can be found in Part IV, Chapter 15). In addition, banks lend money to a very wide spectrum of public and private companies. Keeping the financial data for private companies refreshed is an important logistical challenge since this is harder to do than for public companies. Finally, given the size and sophistication of wholesale transactions, timeliness of data is important. For example, it may be too late to wait till month-end to know that a major corporate customer is heading into default.

Wholesale portfolios contain relatively few, but large loans. A loss of any one corporate loan could have significant ramifications, while individual small value retail loans generally impact in the aggregate. Therefore retail risk management is often different from the wholesale process, depending largely on statistical techniques.[17] Credit scoring models can be used to assess the probability of default of an individual as well as to determine a cut-off score to extend credit. In contrast to corporate loans where internal ratings are the norm, banks rely heavily on external credit scores for retail borrowers. A popular score is the FICO score, developed by the Fair-Isaac Corporation.[18] A higher FICO score implies better credit-worthiness than a lower score.

A cut-off score can be used to distinguish "good" accounts that represent dependable borrowers from the "bad" accounts that are likely to default. A simple rule for determining creditworthiness can be to accept an applicant only if her score is higher than the cut-off score (say, a score of 720 against a cut-off score of 680).

[17] See Burns (2002).

[18] The FICO score is calculated based on various kinds of data in an individual's credit report. Five categories of information are used: Payment History, Amounts Owed, Length of Credit History, New Credit and Types of Credit Used. More details can be found at "What's in your FICO® score."

It must be noted that these scores are not definitive – there is some likelihood that borrowers below the cut-off are in fact good borrowers and vice versa. Therefore, while a firm using the scoring system would avoid lending money to customers judged not credit-worthy due to their lower than the 680 cut-off score, it risks foregoing the smaller set of "good" accounts below the cut off score. Moving the cut-off score higher (say from 680 to 720) will cut off a higher percentage of perceived "bad" accounts but simultaneously risk losing a higher percentage of perceived "good" accounts. The cut-off credit score thus represents a trade-off between profitability and the risk-taking. The bank can determine the cut-off score based on its actual experience, the loss rate and profitability for the retail product as well as its prediction of the state of the economy and the resultant impact on creditworthiness of retail borrowers. Banks can also adjust the cut-off score in order to optimize the profit margin by product. In practice, many variables are combined with the FICO score to separate good accounts from bad accounts such as loan-to-value ratios, documentation type, debt-to-income ratios, etc.

Data volume and quality are typically the major challenges in retail risk management due to the large number of customers and accounts that need to be analyzed and managed. This leads to very large datasets on which statistical analysis must be performed. Quality of data is an important issue – any data quality problem introduced at time of data entry can become intractable simply by the size of the resultant data. For example, dates wrongly entered by agents in a large branch-network could result in a major remediation effort across millions of loans.

Historically, wholesale and retail lending have been treated as separate activities by banks. In fact, even when bank-holding companies have owned both wholesale and retail banking franchises, they have been treated as different businesses with distinct management styles and cultures, often even based in different locations. In the 1980s and 1990s, numerous factors such as easing of regulations, increased demand for credit products and advancements in risk management itself served to push these businesses closer together through the mechanism of *Securitization* which creates a wholesale product by packaging retail portfolios. The benefit of securitization until recently was that banks could efficiently reduce the size of their balance sheets, enabling them to further extend credit to retail customers.[19] Banks thus had an incentive to securitize assets since this was an effective means to expand their book of business without increasing the size of their balance sheets. Generating an asset and then securitizing it eliminated the need for on-going funding by the bank for those assets. As with

[19] The issuance by the Financial Accounting Standards Board (FASB) of FAS 166 and 167 has significantly changed the way in which securitizations are accounted for (see Rosenblatt et al., 2010).

every financial innovation, this can be a double-edged sword and lead to potentially dangerous levels of over-leverage.

The securitization process is complex (see Figure 2.7) and is covered in detail elsewhere (see Chapter 6 and Part IV, Chapter 18), but a brief description will be useful here. The bank pools a set of loans (or assets on their balance sheets) into an entity that is "bankruptcy remote" called a Special Purpose Entity or SPE.[20] The SPE is legally incorporated to pool and buy the assets from the bank in such a way that the assets would be protected from creditors in the event of a bankruptcy of the bank. The entity finances the purchase of these assets by selling securities to investors that are backed by the cash-flows from the underlying assets. A simple example is illustrative. A bank wanting to reduce its credit-card portfolio would sell its credit-card assets to an SPE, which would issue Asset-Backed Securities (ABS) to investors (like pension and hedge funds) in exchange for cash. The cash is used by the SPE to pay the bank. As credit-card holders make their payments to the bank, these payments are channeled through the SPE to the holders of ABS. The risk of card-holders defaulting is likewise transferred to the holders of the ABS who would lose a portion of their investment determined by the amount of loss due to default on the retail credit-card portfolio.

The actual process is vastly more complicated than the above description. One reason is the introduction of "tranched structures" of securities. These tranches do not all share equally in the risk and return. Payments are divided into tranches, with the first tranche receiving the first set of payments and other tranches taking their turn. The structure of the tranches, their rating and the spread paid over LIBOR depend on the credit quality of the assets securitized. Other complexities are introduced by further innovations designed to fine-tune the risk-reward profiles of the various securities issued by the SIV. In addition, securitized products are themselves often incorporated into higher degree securitizations.

Securitized products incorporate the data challenges of both retail and wholesale lending.[21] The risk analysis needs to deal with complexity of securitization structures as shown in Figure 2.7 as well as the large volumes of loans that are part of the underlying retail portfolio. This has hitherto been a significant challenge as it could lead to the aggregation of the risk measures failing to capture the concentrations in securitized pools. One of the significant data challenges in risk management going forward therefore is to address the combined problems of complexity and data size.

[20] These entities are also described by various other names including Special Purpose Vehicle (SPV).

[21] See Krishna and Mark for further information on the data challenges of securitizations.

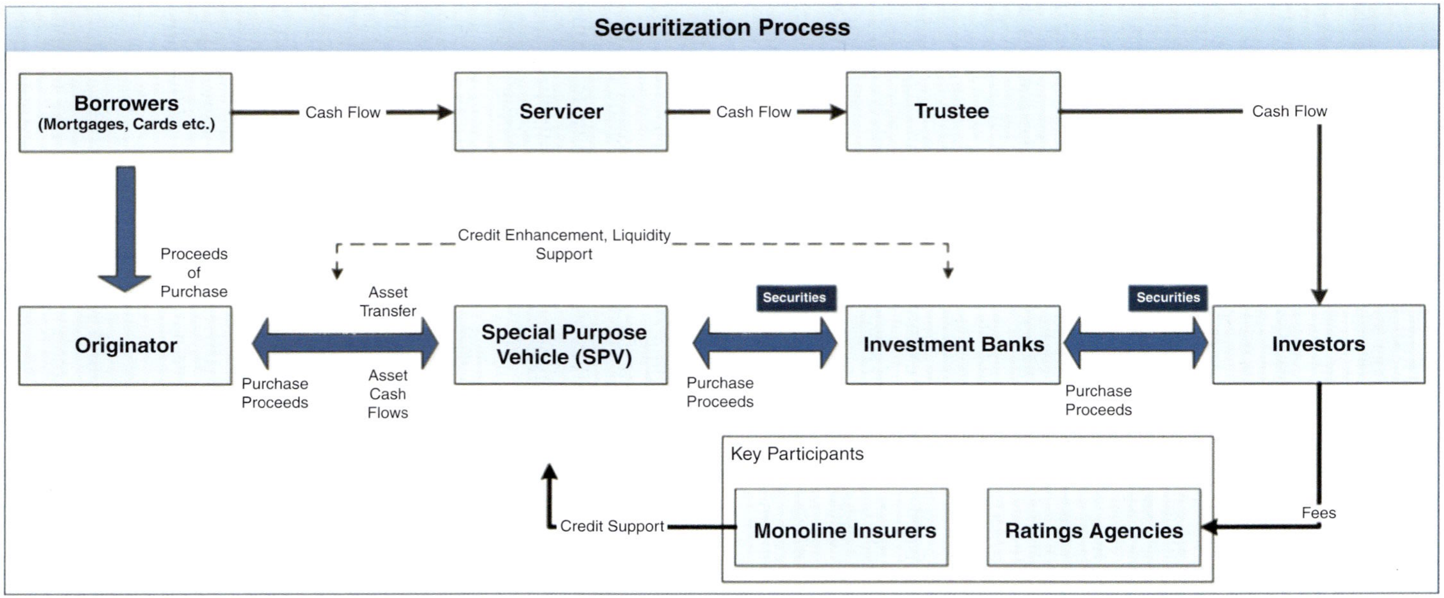

Figure 2.7 An example of the securitization process.

2.5 Measuring, managing and transfer pricing interest-rate and liquidity risk

Credit and Market risk measurement primarily focuses on the firm's ability to remain *solvent*, which roughly translates to the firm having a positive net worth. A positive net worth is a balance sheet concept and means that the value of assets is greater than the value of liabilities. A positive net worth does not, however, speak to the timing of when these assets can be converted to cash to pay off liabilities. Banks are expected to pay off liabilities in a timely manner, leading to the concept of funding liquidity risk. Funding liquidity risk relates to the ability of a bank to meet financial obligations as they come due. Funding liquidity risk has become a particularly important topic.

There is another related but distinct kind of risk called *Trading Liquidity* risk. Financial institutions hold portfolios of traded securities like equities, bonds, foreign exchange and commodities. The prices of these securities fluctuate for a variety of reasons (market sentiment and fundamentals for equities and interest rates and credit quality in the case of fixed-income for example). These prices only hold true in the case of a liquid market where the number of securities sold is small enough so that the act of selling itself does not affect its price. In an illiquid trading market, by contrast, even a small number of securities being sold into a market will depress its price. The result is that the actual value obtained for the securities could be far less than that projected from market prices before the sale. Trading liquidity risk is generally very hard to quantify but may drastically reduce an institution's ability to manage and hedge market risk as well as its capacity to satisfy any funding shortfall by liquidating its assets.

Funding liquidity risk can be more easily understood in the context of the way a bank does business. Banking revolves around the ability to acquire assets that generate returns. Depending on the type of banking organization, these assets may vary across equity and fixed-income instruments (where profits are made by a rise in their price) to retail and credit-card loans (where profits are made from periodic interest payments). The date an asset is acquired is referred to as its *acquisition date*. Conversely, the date it is liquidated for cash is its *maturity date*. The bank funds these assets by accepting short- and long-term liabilities in various forms ranging from retail demand deposits and CDs to inter-bank secured financing using repo and securities financing transactions. Liabilities must be *serviced* (i.e., interest payments and other obligations met on an on-going basis) and they must be repaid in full on their maturity date. Funding liquidity risk arises when there is an *asset-liability duration mismatch* or a timing difference between the maturity dates of assets and liabilities. Banks typically make long-term loans for assets like mortgages. On the other hand, short-term deposits may fund these loans which are required to be repaid on demand causing a duration mismatch. If there is a run on the bank, it may find that it has significantly less in liquid cash than it would need to pay out to depositors.[22] This example of retail banking has its analog in the wholesale banking world as well – banks have the ability to fund a variety of fixed-income instruments by short-term collateralized funding mechanisms in the repo market.[23] In either case, the bank depends on an orderly funding market to address its funding mismatch problem – the retail bank attempts to retain deposits from its customers, while the wholesale banks manage large operations where they "roll-over" their wholesale liabilities on a daily basis in order to fund their longer-term assets.

Funding liquidity risk can be managed through holding cash and cash equivalents, setting credit lines in place, and through monitoring buying power. A spectrum of funding sources is typically constructed so that a bank might assign a higher liquidity credit for "stable funds" than for "hot funds." "Hot funds" are funds supplied by depositors that could be quickly removed from the bank in the event of a crisis. Funds from dealers or brokered deposits are examples of such funds. A funding liquidity ranking process can be constructed to enable a bank to reward or penalize a business unit depending on whether it is a net supplier or net user of liquidity. A funding liquidity-ranking scheme can be quantified using a symmetrical scale. For example, a liquidity rank can be attributed to a product by multiplying the dollar amount of the product by its rank. If a business unit is both a supplier and user of liquidity then a net liquidity calculation is made. The approach is simply a heuristic tool that helps managers to control the liquidity profile of their institution. The next step is to charge each business unit for the liquidity risk that it generates via a funds transfer pricing process.

Funds transfer pricing (FTP) is a process used by banks to ensure that the profit contributions of both the funding (e.g., deposit-raising units in retail banks) and profit-generating units (e.g., lending units) are properly measured and recognized. For example, the transfer price is used to charge an interest rate for the money that lending business units borrow internally from Corporate Treasury and provide credit from Corporate Treasury to business units that attract money from depositors. This process is referred to as *Interest-Rate Transfer Pricing*. Bank depositors often deposit money for the short term, while the bank's loans may have a much greater maturity, leading to the need to also transfer price *funding liquidity*. In other words, a Transfer Pricing system recognizes that certain parts of the organization generate liquidity (e.g., branches generate liquidity by receiving deposits) that is in turn used by other parts of the organization to lend (e.g., a wholesale lending operation likely does not have a large deposit gathering mechanism). Each business unit is able to secure its profit margin through the transfer pricing system at the time of origination

[22] This is not just a hypothetical example. The history of banking is replete with examples of banking runs whose causes can be traced back to this phenomenon. Deposit insurance, where the government insures deposits via improved oversight of insured banks, is a way to alleviate this problem.

[23] See Gorton (2009) for a detailed discussion on the funding mechanisms in the wholesale market.

of its products and can focus on developing and managing the business side of its activity such as the credit quality of its portfolio.

The difference between interest rate and funding liquidity transfer pricing can be illustrated by the following example. If a business unit made a five-year fixed-rate loan then the business unit that made the loan is charged through the interest rate transfer pricing system the cost of raising five-year funds (typically higher than short-term funds). If on the other hand the five-year loan has a variable rate priced off three-month LIBOR then the business unit that made the loan is charged through the interest rate transfer pricing system the cost of raising three-month money with interest rates typically below longer-term rates. Similarly, if a business unit raised a five-year fixed rate deposit then the business unit that raised the deposit is credited through the interest rate transfer pricing system the credit for supplying five-year funds. If a variable-rate deposit is priced off three-month LIBOR then the business unit that raised the deposit is credited through the interest rate transfer pricing system the cost of three-month money.

Funding liquidity transfer pricing calls for analyzing the contractual maturities of liabilities and assets rather than the interest rates. For example, a five-year fixed-rate loan has a funding liquidity maturity of five years. A variable-rate five-year loan priced off three-month LIBOR has an interest rate sensitivity of three months but a liquidity maturity of five years. An asset-generating (lending) unit should be charged more for using longer-term liquidity while a liability-gathering unit (e.g., via deposit collections) should be credited for supplying such liquidity.

In response to the recent financial crisis the regulatory community has issued specific guidance on liquidity risk management under the Basel III framework. Aside from publishing principles on sound liquidity risk management, the Basel Committee has also developed two new minimum standards for funding liquidity.[24] The first, known as the *Liquidity Coverage Ratio (LCR)*, addresses the need for banks to hold highly liquid assets to survive a significant stress event lasting one month. The second ratio, called the *Net Stable Funds Ratio (NSFR)*, aims to promote a stable funding structure over a time horizon of one year. Banks are still implementing these recommendations. Liquidity risk is a complex risk that has yet to be understood and managed with the same rigor as market and credit risk and remains an area of significant focus in banks world-wide.[25] Liquidity risk management can be expected to be a fertile topic for development in the coming years.

[24] Basel Committee for Banking Supervision (2008, 2011).

[25] While liquidity risk has received significantly increased focus since the recent financial crisis, this is an issue that has been discussed for a long time (see Comptroller's Handbook, 2001).

2.6 Stress testing

Stress testing focuses on analyzing risks associated with extreme events. There are two basic approaches. *Sensitivity Analysis* tests the impact of changes in different factors on the portfolio. An example is to measure the sensitivity of the portfolio to swap spreads only, keeping all other factors the same. *Scenario Analysis* tests the behavior of the portfolio under severe but plausible scenarios. Such scenarios can be historical (e.g., the 1994 Mexican Peso crisis) or synthetic. Historical scenarios can highlight a point-of-failure that might arise due to a repeat of a historical event. While historical scenarios can be useful, it is also important to create scenarios of situations that have never been seen before.

Stress testing has become a critical supplementary risk measurement tool. Traditional risk management techniques, including using models like VaR, have been criticized for a number of limitations that could potentially be overcome by stress testing.[26]

- A common critique of traditional VaR models is that they make simplifying assumptions about market behavior (e.g., assume normal market conditions – see Section 2.3) that could understate the risks in a portfolio.[27] These models may understate volatility by using historical information. The actual market may be subject to large jumps in price encompassing the intersection of multiple risks (e.g., the effect of liquidity on market risk) and include markets that experience considerable regime switches.
- The assumptions made by traditional VaR methods have been criticized as especially unrepresentative with instruments having non-linear price movements (e.g., an option on an interest rate swap) and key risk factors (e.g., gamma of an option) that are not easily captured by the traditional VaR model.[28]
- VaR by definition makes no assessment of actual losses that a bank may experience in the tail of a market distribution (see Figure 2.4).[29] Stress testing can help remedy this situation.
- Stress tests can be good for identifying and quantifying the impact of a loss of either trading liquidity or funding liquidity.
- Stress testing can prevent the organization's collective experience being anchored in recent economic history by focusing on potential trouble-spots in the portfolio.

[26] See Linsmeier and Pearson (1996), Basel (2012), and Jorion (2003).

[27] As stated in Khindanova (1998), "parametric approaches are subject to 'model risk': distribution assumptions might be incorrect."

[28] As Pritsker (1996) states, "[The Delta method's] linear Taylor series approximation may be inappropriate for portfolios whose value is a non-linear function of the factors." See also Feuerverger and Wong (2000).

[29] Basel (2012) states, "A number of weaknesses have been identified with using value-at-risk (VaR) . . . including its inability to capture 'tail risk'." Also see Jorion (2003).

A long period of benign economic conditions can lead to underpricing of risk due to complacency.[30]

- Stress testing can also play a critical role in measuring risks in portfolios that contain new products for which little or no historical performance data are available.

Stress testing can help management determine whether the size of potential losses related to extreme events for the current portfolio is larger than the organization is willing to tolerate. If the decision is made to take the risk, it can also be used to develop contingency plans for dealing with stress eventualities. In addition to regular market and credit risk limits, banks place limits on the outcome of stress tests. These stress test limits should anticipate and compensate for an attempt to model a participant's (e.g., management or trader) behavior and potential actions to stay below a stress test limit in a stress environment. Stress tests can help in deciding in advance the actions to take in response to large market moves. Stress testing is sometimes used as a way to allocate capital (see Section 2.8).

2.6.1 Stress testing techniques

Sensitivity analysis is the more straightforward of the two types of stress testing. It involves moving one or more risk factors by a pre-set amount to measure the effect on the portfolio. A common type of sensitivity test is to move interest rates by a set amount (say 10 basis points in either direction) and calculate the worst-case loss. Owing to the relative ease of running such tests, firms usually run a large number of sensitivity analyses on a daily basis to get an approximation of the effect of one or more financial factors (interest rates, equity indices, etc.) on their portfolios. These tests can usually be run at the trading desk, line of business or the firm level. These tests are, however, simplistic since they do not consider correlations of risk factors in the market. Therefore, they can only offer limited information on the actual risks to the firm's position. While they can be improved by incorporating historical data to determine the worst-case impacts, such techniques often misstate actual risks.[31]

Scenario analysis, on the other hand, involves considerable judgment. Realistic but exceptional events need to be formulated. These events are used to stress the portfolios and yield a perspective of the portfolio's loss-potential. Scenarios can be created based on perceived risks to the portfolio (e.g., a scenario of sharply rising interest rates may be relevant to a yield-curve sensitive portfolio). Alternatively, historical or hypothetical events can be used to determine the effect on underlying risk factors (interest rates, equity indices, commodity prices, etc.), which can then be

[30] See Basel Committee for Banking Supervision (2009, p. 7).

[31] See Committee on the Global Financial System (2005, p. 14).

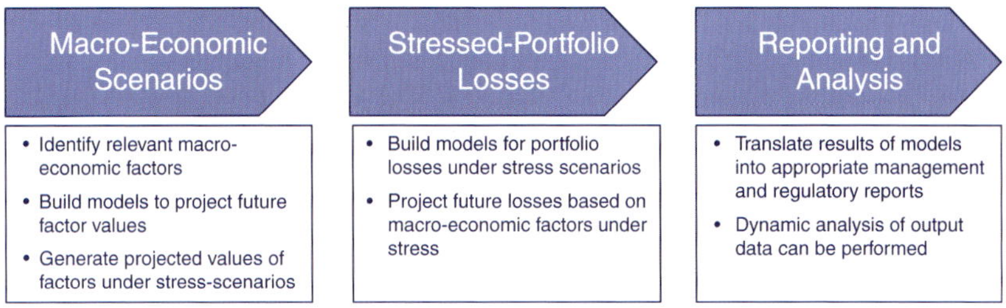

Figure 2.8 Stress testing process.

applied against the firm's portfolio. In practice, firms often use a hybrid approach by creating hypothetical scenarios informed by market behavior during real events.

Scenario analysis can be understood from the three-step process shown in Figure 2.8. First, a series of macroeconomic factors are identified to describe the economy under stress. These factors should be chosen to be relevant to the portfolio being analyzed (e.g., the USD/EUR exchange rate might not be relevant in analyzing a domestic-only portfolio, while this rate would be important in testing the portfolio of an international bank). Models of how these factors behave under stress are created and used to project stress values into the future (current practice suggests projecting out quarterly values for two or more years).

Another set of models is required to translate these estimates of macroeconomic factors into loss projections for the portfolios being tested. These models should consider the specifics of the portfolio under question. For example, stressing a residential mortgage portfolio in a certain region requires the use of home-price indexes for that particular region which may differ substantially from the national home-price index. Stress losses could be due to market, credit or funding liquidity risk, so models must pay attention to the factors that may induce any of these types of risk in a portfolio. Once projected losses have been calculated, they must be reported via internal management or external regulatory channels. The output of stress test models also provides a rich set of data for analysis.

Financial disruptions (such as the recent financial crisis) have sharpened interest in stress testing among regulators as well as inside large financial firms.[32] Within firms, there is increased recognition of the importance of its role in risk management and capital planning, as well as the realization of the need for senior management support in the process. Stress testing frameworks need to be improved in terms of the granularity and range of risks to be considered. One important area of interest is in improving the analysis of correlated risks across books and risk types (e.g., integrating market, credit and funding liquidity risk). Funding Liquidity stress

[32] See Basel Committee for Banking Supervision (2009, p. 6).

testing is a particularly important but complex type of stress test that has had renewed interest.

During the recent financial crisis, the US Federal Reserve asked major bank-holding companies to conduct comprehensive stress tests under the Supervisory Capital Assessment Program (SCAP).[33] The scenario in SCAP was specified by the supervisor and aimed at determining the resulting loss and revenue estimates on a comparative basis across each of the 19 firms asked to participate in the program.[34] These firms were asked to project a two-year, forward-looking view of revenues, losses, and loan loss reserve needs under two different scenarios – a "baseline" scenario which reflected consensus estimates of the economy at that point, and an "adverse" scenario which reflected a considerably more pessimistic view of the economy during that period. SCAP scenarios dealt with three variables – GDP growth, unemployment rate and depreciation in home-prices. The SCAP exercise resulted in a detailed report by the Federal Reserve that identified capital needs for some, but not all, the banks in the program.[35] In November 2010, the Federal Reserve followed up SCAP with a more comprehensive approach to supervising the capital management process within FIs – the Comprehensive Capital Analysis and Review (CCAR) program. One goal of CCAR – an on-going, annual program – includes analyzing an FI's capital planning processes and their stress testing capabilities. In contrast to SCAP, where supervisors provided baseline and adverse scenarios, FIs are required to create their own stress tests.

Both these programs are examples of bottom-up analysis where individual FIs produce results under stressed economic conditions, which are then aggregated by the supervisor. This approach allows for drill-down to understand the most vulnerable FIs as illustrated by SCAP. However, it does not explicitly consider problems such as a disruption to market liquidity due to network effects that are commonly observed under typical severely adverse market conditions. The absence of these effects often yields results from extreme scenarios that are simply scaled-up versions of moderate scenarios, unless FIs themselves take market disruption into account in their individual stress tests. Some regulators have now begun focusing on top-down stress testing where such effects are explicitly modeled into scenarios.[36]

The recent efforts by regulators to run stress tests have also highlighted opportunities for improvements in the stress-testing infrastructures across large financial

[33] As stated in Hirtle et al. (2009), "A fundamental conclusion drawn from the recent financial crisis is that the supervision and regulation of financial firms in isolation. . . are not sufficient to maintain financial stability . . . a macroprudential perspective . . . seems necessary. . . The recently concluded Supervisory Capital Assessment Program (SCAP). . . is one example of how the macro- and microprudential perspectives can be joined to create a stronger supervisory framework." Most of these were banks, but the group includes a life insurance company among other firms not commonly thought of as banks. Nevertheless, this discussion will broadly refer to this group as "banks."

[34] US firms with greater than $100 billion in assets on December 31, 2008 were eligible for SCAP.

[35] See Board of Governors of the Federal Reserve System (2009). [36] Jenkinson (2007).

institutions. As the pace of stress testing increases, driven both by regulatory and by internal needs, FIs may need to consider the benefits of integrating stress testing into their regular risk management processes (e.g., one of the goals of CCAR is ensure robust capital management processes). This may mean that data collection functions will need to be systematized to run frequent stress tests with many scenarios and risk factors. The data processing environment may need to be further developed to enhance the flexibility to allow FI analysts to perform a variety of analyses to arrive at the appropriate scenarios and add new economic factors when appropriate. Analysts may also need access to high quality data to construct realistic stress tests with properly estimated fat tail probabilities, correlations and other characteristics. An important consideration is the assembly and integration of data for purposes of combining risk measures in normal markets with risk measures in stress markets. For example, regulators now require a bank to calculate a "stressed value-at-risk" (sVAR) measure based on the 10-day, 99th percentile, one-tailed confidence interval value-at-risk measure of the current portfolio, with model inputs calibrated to historical data from a continuous 12-month period of significant financial stress relevant to the bank's portfolio.

Stress testing has become a topic of great interest among FIs and regulators across virtually all major markets. This is an area where much development is to be expected. One can also expect major improvements in infrastructure to meet the enhanced needs of stress testing in parallel with the development of stress testing practice.

2.7 Operational risk

Operational risk is defined as the risk of loss resulting from inadequate or failed internal processes, people and systems or from external events. Operational risk related failures include computer breakdowns, errors of judgment, deliberate internal or external fraud, and so on. Operational risks can be some of the most severe risks experienced by a financial institution. There have been some well-publicized examples in recent years.[37] The actual losses occurred as a result of changes in market risk factors but the underlying cause was fraud. In other words, these are considered examples of operational risk since the underlying cause was likely due to inadequate supervision rather than simply due to adverse market movements. While such high-profile losses are fortunately few and far between, all banks suffer from regular small losses. Securities trading operations daily suffer very small losses in the form of wrongly filled trades for example. Many credit-card operations have whole departments devoted to detecting and minimizing losses due to fraud.

[37] For a first-hand account see Leeson and Whitley (1996).

Operational risk is qualitatively a different kind of risk from financial risks (credit and market and liquidity risk). These financial risks can be thought of as the raison d'être for a financial firm – it is only by accepting, managing and optimizing one or more of these financial risks that a firm can make a profit. In contrast, a firm assumes operational risks simply by being in business but there is usually no upside for assuming operational risk.[38]

There are a variety of useful operational risk approaches such as key risk indicators (KRIs), likelihood versus severity approach, risk control self-assessment (RCSAs) approaches, etc. A KRI is a measure used to indicate how risky an activity is. It is typically used to provide an early warning signal to identify a potential operational risk event. KRIs are often based on a backward-looking view on risk events and therefore enable lessons to be learned from the past. KRIs can also be used to provide an indication that the risk appetite or tolerance for risk is reached.

Operational risk assessment based on a likelihood (or frequency) of a particular operational risk, as well as a review of its possible magnitude or severity, is a particularly popular approach. Operational risks can be classified into two kinds – those that are infrequent but high value and known as Low-Frequency-High-Severity such as the Barings bankruptcy, and those that are high frequency but low severity such as credit card fraud losses.[39] Operational risk measurement and management is often focused on addressing both these types of losses by measuring the potential severity of a risk set against its frequency and devising ways to manage the most damaging risks, which can be of either kind. Since banks cannot directly expect a higher expected return from assuming operational risk, it is fundamentally different from financial risk where risk-reward trade-offs can be made to balance expected rates of return and risk. Therefore operational risk management is focused on limiting it to the fullest extent possible.

Loss data are easier to collect than data associated with the cause of loss. This complicates the measurement of operational risk because each loss is likely to have several causes. The relationship between these causes, and the relative importance of each, can be difficult to assess in an objective fashion. A curve can be drawn to show the same expected loss based on multiplying likelihood by severity. Given an acceptable level of expected loss, management should take appropriate actions to mitigate the risk located above the EL curve. One can publish a graph displaying the potential severity of a risk set against the frequency of certain operational failure

[38] Note that there are however a number of firms that have made operational risk mitigation their business by successfully building a business model where certain essential functions may be out-sourced by financial firms at a lower cost than keep the function in-house and assuming the resulting operational risk. Examples include IT out-sourcing contracts, securities custody services and payroll processing.

[39] The other potential combinations are either transient or absorbed in the cost of doing business. High frequency high severity losses would quickly extinguish the financial institution in question and low frequency low severity losses are immaterial.

risks. This diagram allows managers to visualize the trade-off between severity and likelihood.

Operational risks have been an important risk for a long time.[40] Nevertheless, it is only recently that operational risk has been codified into an easy-to-understand framework in the Basel Capital Accord (see Section 2.9 and Chapter 11).[41] The Basel framework considers seven operational risk related loss event types.

1. *Internal fraud* is defined as losses due to acts of a type intended to defraud, misappropriate property or circumvent regulations, the law or company policy.[42]
2. *External fraud* is defined as losses due to acts of a type intended to defraud, misappropriate property or circumvent the law, by a third party. For example, credit card issuers lose billions each year in fraud.[43]
3. *Employment practices and workplace safety* is defined as losses arising from acts inconsistent with employment, health or safety laws or agreements, from payment of personal injury claims, or from diversity/discrimination events.
4. A key operational risk loss event type is also related to *clients, products and business practices* which include losses arising from an unintentional or negligent failure to meet a professional obligation to specific clients or from the nature or design of a product.
5. *Damage to physical assets* is defined as losses arising from loss or damage to physical assets from natural disaster or other events. Hurricane Katrina in 2005 dispersed the population in affected areas in such a chaotic manner that some banks were concerned about the whereabouts of their creditors and their ability to meet their obligations.
6. *Business disruption and system failures* are increasingly common in a technology-dominated industry. Several recent and well-known events, including Hurricane Katrina and the 2001 terrorist attacks in New York, resulted in widespread business disruption precipitating huge losses to firms.
7. The final category is *execution, delivery and process management* which are defined as losses from failed transaction processing or process management, from relations with trade counterparties and vendors. Model risk – where errors in quantitative trading models may lead to massive losses – is an example of such a risk. Data management risks also fall into this category (see Mark and Krishna, 2008).

[40] Losses that medieval bankers feared from shipwrecks can be considered an operational risk for example.
[41] Basel Committee for Banking Supervision (2004, Part 2, p. 137).
[42] A trader, Joe Jett, was accused by the SEC of having defrauded his firm resulting in a $210 million loss. For Joe Jett's perspective see BusinessWeek (1999).
[43] See Sullivan (2010).

Operational Risk (in $ m.)	31-Dec-2012	31-Dec-2010
Economic capital usage (for operational risk)		
Investment Bank	7235	8222
Asset Management	399	564
Private Bank	8	17
Total	7642	8803

Figure 2.9 Example of an operational risk report.

Some operational risks may not be easily classified within these seven categories, but nevertheless have significance and must be considered. An example is *Implementation Risk*. Most banks today are in a constant stage of change and run hundreds or thousands of projects to execute these changes. These run the gamut from very small projects with departmental impact, to extremely large projects of firm-wide, strategic importance. Careful management of these projects is essential not only due to budgetary considerations, but also due to the potential fall-out due to project delay or failure.[44] This topic is discussed further in Part VI, Chapter 25.

The basic objective of operational risk management is to make such risks as transparent as possible so that decisions can be made with full knowledge of the risks involved. A loss arising from an operational risk takes the form either of a direct external cost, or of a write-down associated with the resolution of an operational risk event, net of recoveries. For example, best practice calls for reporting on such items as the largest types of operational risk, the drivers of the risk and how the risk changes over time. Figure 2.9 shows an example of an operational risk report (where we assume for illustrative purposes a correlation of 1). Transparency can lead to better management of operational risk through the formulation of specific action plans and rigorous implementation schedules. Given their complex and interdependent nature, operational risks can be better managed as a partnership between business units, business infrastructure groups and corporate governance units such as internal audit and risk management.

Measurement of operational risks has proved difficult and controversial. The Basel Accord lists three measurement techniques.

1. *Basic Indicator Approach*: Calculates operational risk capital as 15% of average gross income of the firm over the past three years.
2. *Standardized Approach*: Calculates capital as a multiplier on income, with the difference that it allows for the bank to subdivide its business into eight categories with a different multiplier ranging from 12–18% for each line of business.

[44] The increasingly stringent regulatory environment makes managing projects even more important, with large fines and reputational impacts as potential adverse outcomes of a failed project.

3. *Advanced Measurement Approach (AMA)*: Allows banks to develop their own internal operational risk measurement systems to calculate operational risk capital and allocate these appropriately to business lines. Obviously, this means that supervisors of the bank (including supervisors in the host country in the case of international banks) need to be very active in reviewing and approving the operational risk measurement approach.

The Basic and Standardized approaches are extremely crude and unintentionally penalize success by requiring higher charges for banks with larger revenues. On the other hand, AMA approaches have not achieved standardization in risk measurement to anywhere near the level that has been achieved in market risk or credit risk. Operational risk remains a discipline where control and audit processes are as important as, if not more than, quantitative approaches. Nevertheless, one can expect continued focus by banks on measuring, managing and mitigating operational risk.

2.8 Capital allocation and RAROC

Economic capital (or risk capital) and regulatory capital (e.g., driven by Basel III) are two of the main types of capital. Economic capital provides protection against the various risks in a firm in order to maintain its financial integrity and remain a going concern even in the event of a "worst case" scenario (where the worst case is defined in terms of say a 99.97% confidence level[45]). The level of calculated economic capital should enable the institution to absorb unexpected losses up to a level of confidence in line with the requirements of the firm's various stakeholders. The exact choice of a confidence level is often associated with a desired target credit rating from a rating agency. For example, a bank wishing to maintain an AA rating from Standard and Poor's would want to target a confidence level of at least 99.95% (though such a bank would typically aim to go higher than this[46]). Regulatory capital is calculated according to a set of industry-wide rules and formulas and is intended as a *minimum required level* of capital adequacy by regulators charged with ensuring the safety and soundness of banks. Regulatory capital, discussed further in the next section, is a minimum standard required by the bank's regulators. Actual capital held by FIs is required to be greater than this regulatory amount.

Economic capital has a variety of uses beyond ensuring safety and soundness at a firm-wide level. It can also be used to manage the performance of the firm, its business units and its employees in granular detail through a process called Risk-Adjusted Performance Management (RAPM). In this way, economic capital can act

[45] See Section 2.3 for a discussion on the meaning of confidence intervals. [46] See Gordy and Howells (2004).

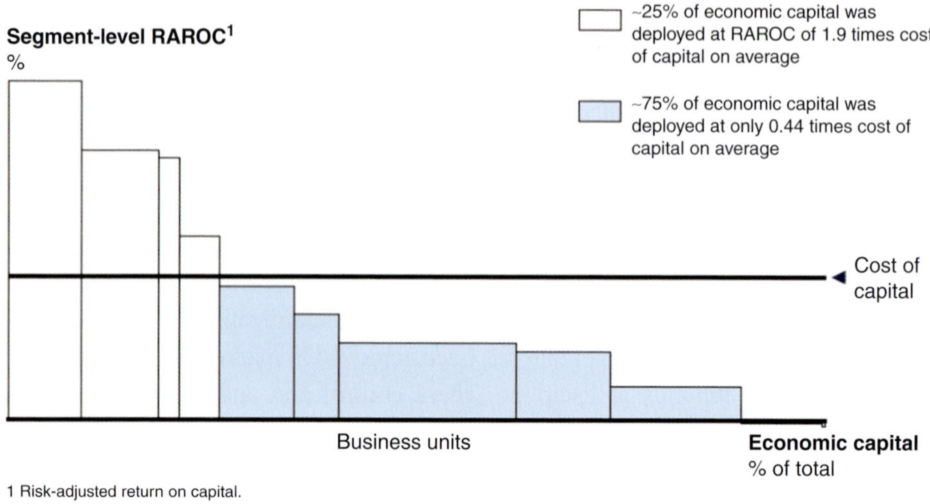

Figure 2.10 An example of capital allocation (taken from Baer et al., 2011).

as the medium of communication of the firm's goals throughout the organization. The first step in doing this is to allocate capital to and within individual business units (e.g., to trading business units and individual traders themselves). Allocated capital can then be used in performance measurement systems that determine employees' financial performance and can be tied to their compensation.

Certain business units (such as a trading business) are compensated heavily on the profits they generate. A sophisticated way of managing behavior is to normalize profits for the amount of risk taken to achieve them. A prominent RAPM method is called Risk-Adjusted Return-On-Capital (RAROC), which provides a uniform risk-adjusted measure of performance.[47] Not only can this generate a fair measure of profitability for a single trader, but it can also be aggregated upwards from the trader to compare the economic profitability of various trading desks and businesses. This measure can then be used to compensate not only traders but also business management for their contribution to shareholder value. Traders are usually prevented from taking too much risk through limits on the amount they can risk.

Figure 2.10 shows the RAROC values for each of nine business units. The four business units without shading generate the largest risk adjusted profits, while the other nine have lagged in this regard. This analysis becomes even more interesting when one overlays the cost of capital – a firm-wide measure by definition – on top of this chart. The firm as a whole can only succeed if it generates sustained profits above the cost of capital. The figure shows that only the first four units in fact meet

[47] For a comprehensive, if dated, case study of RAROC at a major US bank, see James (1996).

this criterion, while the other five units miss this target. Such an analysis can be used to make firm-wide decisions. For example, the firm may best position itself for the future by allocating new capital to areas that already generate high risk-adjusted profits by re-allocating capital from low profit generating areas. Alternatively, it could focus on bringing low-RAROC businesses back to profitability by applying RAROC measures at a deal level and ensuring that these businesses only accept deals that meet a minimum RAROC threshold.

Applying RAROC analysis to pricing deals is called Risk-Based Pricing (RBP), which applies the same sort of analysis as shown in Figure 2.10. The first step is to set a *Hurdle Rate* which is the minimum rate of profitability at which a deal may be concluded to take into account the firm's cost of capital. This hurdle rate then becomes a floor price to a customer so that negotiation on terms (interest rates, fees and other deal conditions) may occur only if returns to the firm are above the hurdle rate. RBP is used by some banks to set prices for retail products like mortgages and retail loans.

The generic RAROC equation is

$$\text{RAROC} = \text{Risk-Adjusted-Return}/\text{Economic Capital}$$

where

$$\begin{aligned}\text{Risk-Adjusted-Return} = {} & (\text{Expected Revenues} - \text{Costs} + \text{Expected Losses} \\ & + \text{Return on Risk Capital} +/- \text{Transfers}).\end{aligned}$$

The equation shows the components of profitability that are embedded in the equation. Expected revenues are the revenues the activity (the business as a whole or the individual deal) is expected to generate assuming no losses over say a 1 year period. Costs are the direct expenses associated with the activity. Expected losses are the average losses the business has come to expect from similar activities. These are usually losses from credit default but also include the expected loss from other risks such as market and operational risk. Taxes are the expected amount of taxes imputed to the activity. The Return-On-Risk-Capital measure adjusts for the opportunity cost of using the same capital in risk-free activities by calculating the return on securities such as government bonds. The Transfers term reflects transfer pricing mechanisms that are used to fund business units in banks (see Section 2.5). As described earlier, each business unit is assumed to borrow funds from the corporate treasury at a price called the Funds Transfer Price that takes into account inputs such as the cost of funds in the marketplace. The RAROC equation calculates the profitability of the deal in the numerator and adjusts it by Economic Capital in the denominator to calculate a percentage figure.

The utility of RAROC numbers often depends on the quality of the data about risk exposures and positions collected from the management systems that are used to forecast the input to RAROC. A rigorous process of data collection is necessary to help ensure reasonable estimates of risk and financial information for the RAROC equation. Banks typically set up a RAROC group to be accountable for the integrity of the data collection process, the calculations, and the reporting process. Nevertheless, business units and the finance group are the first line of defense in ensuring the integrity of the specific data they produce and feed into the RAROC system. RAROC results are used in a multitude of systems and therefore the RAROC system must also transmit granular results back to business units and the performance management systems in an efficient and timely manner.

Compensation plans should consider the value added contribution of employees to shareholder value, and there are often incentives for prudent risk-taking. However, exceeding the RAROC hurdle rate is only one of several important considerations towards rewarding performance through compensation, since there are other criteria for accepting a project. Several organizations reward employees as a function of their contribution to a strategically important area of the firm. Some firms construct a two-dimensional strategic grid for analyzing the strategic importance of a BU. The horizontal axis, say, corresponds to the RAROC return calculated on an ex ante basis. The vertical axis corresponds to (say) a qualitative assessment of the quality of the earnings produced by a BU. This measure takes into consideration the strategic importance of the activity for the firm, the growth potential of the BU, the sustainability/volatility of the earnings in the long run, and any synergies with other critical businesses in the firm. Priority in the allocation of balance-sheet resources should be given to the BU that ends up in the upper-right quadrant. At the other extreme, the firm may want to exit, scale down or fix the activities of a BU that fall into the lower-left quadrant. The BU in the category "managed growth," in the lower-right quadrant, is high-return activities that have low strategic importance for the firm. On the contrary, BUs in the "investment" category in the upper-left quadrant, are currently low-return activities but which have high growth potential and high strategic value for the firm.

RAROC can be used to justify compensation decisions connected with the Dodd–Frank Act (DFA). The DFA requires public companies to include a non-binding shareholder vote on the compensation of their named executive officers disclosed in their proxy statements at least once every three years. The DFA also mandates that any proxy solicitation that seeks shareholder approval of a merger must disclose any compensation arrangements that relate to the transaction and the aggregate total of all such compensation that may be paid. The DFA indicates that these shareholder votes are not binding on a company's board of directors. In other words, a negative vote cannot overrule any company or board decision. The "say on pay" rule provides public company shareholders with a means for opposing or supporting a company's pay practices. Public company boards will consider a shareholder vote on executive compensation packages. The DFA also requires the SEC to implement rules that prohibit national securities exchanges from listing the securities of any company that does not have a compensation committee formed solely of independent members of its board. The implementation of this provision of the DFA is expected to impose factors to be considered in determining the independence of a director to serve on the compensation committee.

2.9 Regulatory approaches to risk management

Banks are a critical component of any economy. Their well-being is therefore a matter of interest not only to their shareholders, but also to the communities that they serve. While no greater proof of this is required than the events during the recent financial crisis, earlier episodes going back at least two centuries have been making this point for quite some time. What has changed in the past couple of decades has been the increased globalization and interconnectedness of financial institutions, which has raised several issues. One of them is the need to level the playing field and ensure that the home domicile of a particular institution does not give it any particular competitive edge (for example the fact a bank is based in the USA should not give it any particular advantage or disadvantage over an EU or Japanese bank). Another issue of great importance is that many of these global banks have indeed become "too big to fail" on a global level, meaning that their safety and soundness is a matter of interest not only to their home supervisors but also to regulatory authorities in every country they have a significant presence. Chapters 5 and 11 discuss these issues in further detail, but this section will serve as a high-level overview.

The first concerted effort to rationalize international banking regulation was the Basel I Accord signed in 1988. The Basel Committee on Banking Supervision (BCBS) was established as the Committee on Banking Regulations and Supervisory Practices by the central bank Governors of the Group of Ten countries at the end of 1974 and now counts more than 25 countries among its members.[48] The Committee provides a forum for cooperation between its members on banking supervisory matters but does not have any formal supervisory authority. Rather, it formulates broad supervisory standards and expects its member regulatory authorities to implement them through detailed regulations in their individual jurisdictions.

Over a period of time, the Basel Committee began to focus on the topic of capital adequacy. In the early 1980s, the Committee became concerned with the need to achieve greater convergence in the measurement of capital adequacy across major banking corporations world-wide. Specifically, there was a need for a multinational

[48] Basel Committee for Banking Supervision (2009).

accord simultaneously to strengthen the stability of the international banking system and to remove a source of competitive inequality arising from differences in national capital requirements. The result of this effort was a weighted approach to the measurement of risk, both on and off the balance sheet. This capital measurement system was published in 1988 in the form of the Basel Capital Accord, now also known as the Basel I Accord. This system provided for the implementation of the framework with a minimum capital ratio of capital to risk-weighted assets of 8% by end-1992. Since 1988, this Accord has been progressively introduced not only in member countries but also in virtually all other countries with active international banks.

Basel I (also known as the 1988 Basel Accord) introduced a two-tiered approach to capital with Tier 1 consisting of stronger capital (such as stock issues).[49] The second tier consists of supplementary capital that was considered less capable of absorbing losses (such as gains on investment assets and deferred tax assets). Offsetting these capital cushions were calculations of credit risk. Only credit risk (see Section 2.4) of loans on banks' balance sheets was considered in calculating Risk-Weighted Asset (RWA) values for each of the loans according to a simple calculation. The RWA for a loan was determined simply by the size of the loan and the category of counterparty that it was made to – the latter determining a risk-weight multiplier that ranged from 0% to 100%. Counterparties deemed low risk such as sovereign governments were allotted 0% weights while high-risk unsecured loans to non-bank counterparties were assigned 100% weightings.[50] The risk-weight multiplier was intended to be a proxy for the credit risk inherent in the deal and reflect the deal's combined PD and LGD (see Section 2.4). The minimum capital requirement was simply 8% of the combined RWA values for all loans on the books.[51]

Basel I was an important step forward in harmonizing bank regulation across the globe, as well as in introducing risk as the key concept in the capital cushion. All banks in the original G10 countries rapidly adopted it.[52] Basel I had several limitations however. One was the arbitrary assignment of risk-weights to classes of loans, which was driven by politics rather than business reality. All corporations

[49] See Investopedia.

[50] To take a simple example, $1000 lent out to the Greek government would merit an RWA of $0 while the same $1000 lent to an AAA rated multinational company would result in an RWA of $1000. This (somewhat) contrived example illustrates the major drawback of the Basel I Accord. As of 2011, it is hard to argue that a major AAA-rated US corporation is a larger credit risk than the Greek government.

[51] A simple example will serve to illustrate this calculation. Suppose the bank has made a loan of $1 million to an OECD bank, which carries an RWA of 20%. The RWA will be 20% × $1 million = $200,000. The regulatory minimum capital requirement will be 8% × $200,000 = $16,000. If the same loan had been made to a non-bank corporate on the other hand the RWA would be $1 million with the minimum capital requirement being $80,000.

[52] The G10 (Group of 10) is actually a group of 11 countries: Belgium, Canada, France, Germany, Italy, Japan, the Netherlands, Sweden, Switzerland, the UK, and the USA.

(or all governments for that matter) are not equal credit risks. Another limitation was that the Basel I Accord was not designed to cover risk components like Market Risk. Finally, certain important subtleties in risk management – such as credit risk term structure[53] and portfolio diversification effects[54] – were not captured in its calculation. These limitations offered (unintended) incentives to banks to engage in *regulatory arbitrage* and make certain types of loans based solely on their regulatory benefits. For example, US banks raised their holding of government loans from 15% to 22% by 1992, driven in part by favorable treatment of such loans in the Accord.[55]

Market Risk was addressed via the so-called Market Risk Amendment (MRA) that went into effect in 1998. It was designed to incorporate capital requirements for the market risks arising from banks' open positions in foreign exchange, traded debt securities, equities and commodities. A key advance in this accord was the introduction of an alternative to the standardized measurements that were central to the original 1988 Accord. Banks were permitted, subject to standards imposed by their regulators, to use internal Value-at-Risk (VaR) models (see Section 2.3) to measure their market risk capital requirements. Some allowance was made to address the drawbacks of VaR (discussed in the earlier section) in measuring market risk by requiring banks to hold capital in excess of the minimum calculated VaR values.

Certain changes were made in the so-called Basel II Accord, which was finalized in 2004. Basel II built on the foundation established by Basel I, but also importantly drew from the concept of alternative internal risk measurements introduced by the 1998 MRA, resulting in significantly more complex calculations for risk-weighted assets. Essentially Basel II offered banks a choice of regulatory regime they wished to adhere to ranging from the so-called *Standardized Approach* to the *Advanced Internal-Ratings Based (AIRB)* approach. These employed the same RWA technique used in Basel I. The critical change was in the way RWA was calculated to better estimate the credit risk inherent in an asset. The Standardized Approach for credit risk was designed to be easy to calculate (reducing costs of regulatory compliance) but would presumably lead to higher capital requirements. Even this simple approach addresses some of the flaws of the Basel I calculations by using external credit ratings (from firms such as Standard and Poor's, Fitch, or Moody's) which act as proxies to PD and give a better gauge of credit risk. Banks could go much further with AIRB by using their own internal ratings systems for both PD and LGD as a

[53] The Term Structure of a credit refers to the duration of a loan. Self-evidently, there is a much higher chance that a loan made for a week will be paid back compared to a loan that will not be repaid for five years.

[54] The business climate usually favors one group of companies against another. For example, high oil prices will favor oil companies over airlines, while the converse holds true when oil prices are low. A bank that has made loans only to one group will therefore be more risky than one with a *diversified portfolio* that has made loans to both.

[55] See Jablecki (2009).

more reliable measure of how risky they consider their own assets.[56] AIRB can be used only upon supervisory permission. One of the key criteria that banks must satisfy to use AIRB is that their internal ratings must also be used in their internal decision-making processes when extending business. This is known as the "use-test" and is intended to ensure that Regulatory Capital is better aligned with the way that banks manage themselves internally using Economic Capital.

Basel II was a seminal piece of work and a huge advance over the existing 1988 Basel regime of bank supervision. It had important and salutary impacts on major banks world-wide. Banking supervisors across the globe, including those outside the original G10 countries, have adopted the Basel II accord as a way to ensure minimum acceptable levels of bank safety and soundness (though some jurisdictions, notably the US, have lagged in implementation compared to others). The Accord's focus on large, multinational banks has led to the refinement of approaches for risk measurement (also called "home-host" issues) in multiple jurisdictions.

Basel II introduced many other innovations. While the regulatory aspects of market risk were not advanced significantly in this accord, it did continue the practice of including more risks under regulatory supervision by explicitly bringing operational risk measurement and management into the regulatory domain (see Section 2.6). Basel 2.5 further refined the regulatory aspects of market risk (for example, introduced sVAR).

Basel also introduced a "three-pillar system" of regulatory supervision. The first pillar (pillar 1) called out explicit measurement techniques for Credit, Market and Operational risks via a series of calculations. Pillar 2 acknowledged risks for which commonly accepted measurement techniques did not exist, including concentration risks and liquidity risk, by relegating them to individual country supervisors to address in the best manner they deemed fit. Pillar 3 laid out the methods by which banks could disclose their capital and risk to the market without losing competitive advantage.

However, the pace of innovation in the banking industry meant that Basel II ignored several important topics in risk management that were associated with the recent global financial crisis (see Chapter 5 for a complete discussion of this topic).

[56] As an example, consider the AIRB calculation for risk-weighted assets for mortgages that are not in default. This is calculated by inserting the PD and LGD into a formula that first calculates a regulatory capital requirement K (expressed as a % of the exposure) and next sets RWA = $K \times 12.5 \times$ Exposure-At-Default (see Section 2.4). The $K \times 12.5$ portion of the RWA formula is called the risk weight. If PD = 1% and LGD = 25% for a \$200,000 mortgage (i.e., EAD = \$200,000) then the RWA = 31.33% × \$200,000 = \$62,600 (where K = 2.5064% and the risk weight is 31.33%). Regulatory capital for the mortgage is \$5012, which is obtained by multiplying the RWA (\$62,600) by 8% (\$62,600 × 8% = \$5012). While regulatory capital requirements for a single loan can also be obtained by multiplying K times EAD (2.5064% × \$200,000 = \$5012), the factor of 12.5 is required when aggregating the bank's entire loan-book. If the PD rises to 5% and the LGD stays the same (at 25%) then the regulatory capital rises to \$13,176. If the PD rises to 20% and the LGD rises to 45% then the regulatory capital rises to \$40,499.

The nature of the Basel Committee meant that the rules were sometimes influenced by political considerations.[57] Similarly, the fact that funding liquidity risk was addressed only in Pillar 2 meant a less specific focus in this critical area of risk. Basel II also did not address the issue of *procyclicality* or the tendency to relax rules via rosy predictions when things are going well, which results in behavior that is exactly the opposite of saving for a rainy day.

Notwithstanding all these problems, the Basel framework has now become an established framework for risk-based regulatory supervision world-wide. The deficiencies of the accord are now being addressed via *Basel III*.[58] The key tenets of Basel III focus on addressing the shortcomings of Basel II, including a stronger capital framework and addressing liquidity risk management. The items introduced by the committee to address the capital framework include the following.

1. Raising the quality, consistency and transparency of the capital base by requiring mainly common shares and retained earnings to raise the quality of Tier 1 capital, for example.
2. Increasing the coverage of risks to include losses for a variety of counterparty risks including taking into account *Credit Valuation Adjustment* (CVA) or the risk of loss due to the counterparty's decreased creditworthiness, addressing better risk measurement by stress testing (see Section 2.6) and better risk management by strengthening collateral management.[59]
3. Introducing a leverage ratio to supplement capital standards.
4. Reducing procyclicality by measures designed to increase capital and resiliency in favorable economic climates.
5. Finally, the committee has focused on systemic risk, which addresses the dangers that the financial system as a whole faces due to the high degree of interconnectedness in modern finance.

Basel III introduces a global funding liquidity risk (see Section 2.5) framework that focuses on two measures that disclose the firm's liquidity over time.[60] The Liquidity Coverage Ratio (LCR) focuses on surviving acute short-term stresses lasting less than a month, while the Net Stable Funding Ratio (NSFR) focuses on longer-term liquidity risks lasting one year.

The scale of the recent financial crisis was so great that individual governments have also been forced to take measures to increase regulation. The Dodd–Frank Act of 2010 passed by the USA has been a prominent example. It is a voluminous piece of legislation that covers a variety of topics thought to have been causes of the

[57] See Bjerke (2007). [58] See Basel III (2011). [59] See Basel Committee for Banking Supervision (2009).
[60] See Basel Committee for Banking Supervision (2011).

financial crises.[61] The 16 chapters (or *titles*) of DFA include provisions to address systemic risk and the "too-big-to-fail" problem via the creation of a Financial Stability Oversight Council (FSOC) and the Office of Financial Research (OFR). The goal calls for the creation of laws to allow orderly liquidation of failed banks, re-organizing the patchwork of regulatory agencies and expanding regulation to include hedge funds as well as a national insurance office. It introduces the so-called *Volcker Rule*, which prohibits banks from proprietary trading and pushes for electronic trading and clearing to reduce the lack of transparency in OTC derivatives. One of the most sweeping, and controversial, aspects of DFA is the focus on consumer protection via the establishment of the Consumer Finance Protection Bureau (CFPB) and anti-predatory laws aimed at the mortgage market. DFA is in its early stages of implementation and it is too soon to predict its long-term effectiveness (and indeed, given the political climate in Washington, its viability). DFA in some cases is at odds with Basel Accord rules, so the interpretation of the act remains a work in progress.

2.10 Conclusion

Risk Management as a formal discipline has made significant strides over the past three decades. New techniques for quantifying risks of different types have been developed and implemented in banks world-wide over this period. These developments have propelled the creation of new products and the growth of banks in many countries. The growth of banks has paralleled the globalization of the world economy. It is difficult to see how the creation of new products and the growth of the world banking system could have occurred without these improvements in risk management. International financial regulators have for their part used these same techniques to enhance the sophistication of the regulatory system.

On the other hand, shortcomings in risk management practices and technology have been blamed for allowing some firms to assume risks far in excess of their capacity to absorb the resulting losses (see Senior Supervisors Group, 2008). There has been a significant investment by banks to continually upgrade their methodologies and infrastructure (e.g., their risk management systems) over the past 20 years (see Bansal et al.).

It is easy, and perhaps appropriate in well-publicized cases, to fault risk management techniques for these unfortunate events. However, a little reflection will show

[61] The Dodd–Frank Act (officially the "Financial Stability Act of 2010") is a hefty 848 pages long (see Dodd–Frank Act, 2010). DFA also expands the scope of stress-testing described in Section 2.6 to many more financial institutions beyond those covered in SCAP (see Federal Register).

that overly criticizing risk management techniques ignores their very real benefits. Without risk management, much of the modern financial system would at least be inefficient and costly and it is even conceivable by extension that the global economy as we know it would not exist. What is required is a continued improvement in risk management policies, methodologies and infrastructure with a focus on those areas that are not yet sufficiently robust. The energy being devoted by banks and regulators to enterprise risk management, liquidity risk, stress testing, the integration of credit and market risk, and operational risk is indicative of this focus. The new Basel III framework, as well as national legislation (like the DFA in the United States), is illustrative of this trend.

Risk management theory and practice continues to be an area of fertile and significant development. In particular, much energy is being expended toward improving FIs' risk management infrastructure. There are many areas that need improvement, including implementing better risk management controls and risk measurement processes. One critical area of focus is better risk information management which encompasses the entire chain of data production, processing and distribution. All these areas can be expected to see heightened levels of activity in the next few years.

Acknowledgements

The authors would like to express thanks to the following people for their help in reviewing drafts of this chapter: Erin Carberry, Hugh Guyler, Bill Kirby, Christie Murphy, and Charles Taylor. Any remaining errors are the responsibility of the authors alone.

Bibliography

Baer, T., A. Mehta and H. Samandari, 2011, The use of economic capital in performance management for banks: A perspective, McKinsey Working Papers On Risk, Number 24.

Bank of New York Mellon, Board of Directors and Committees, at `www.bnymellon.com` (downloaded August 14, 2011).

Bansal, A., R. Kauffman, R. Mark and E. Peters, Financial Risk and Financial Risk Management Technology (RMT)-Issues and advances, at `http://papers.ssrn.com/sol3/papers.cfm?abstract_id=1289046`

Basel Committee for Banking Supervision, 2004, *Basel II: International Convergence of Capital Measurement and Capital Standards: a Revised Framework*, June, at `www.bis.org/publ/bcbs107.htm` (downloaded December 30, 2011).

Basel Committee for Banking Supervision, 2008, *Principles for Sound Liquidity Risk Management and Supervision*, at `www.bis.org/publ/bcbs144.htm` (downloaded October 20, 2011).

Basel Committee for Banking Supervision, 2009, *Principles for Sound Stress Testing Practices and Supervision*, May, at www.bis.org/publ/bcbs155.pdf (downloaded December 30, 2011).

Basel Committee for Banking Supervision, 2009, *History of the Basel Committee and its Membership*, August, at www.bis.org/bcbs/history.pdf (downloaded September 5, 2011).

Basel Committee for Banking Supervision, 2011, *Basel III: A Global Regulatory Framework for More Resilient Banks and Banking Systems*, June, at www.bis.org/publ/bcbs189.htm (downloaded October 20, 2011).

Basel Committee for Banking Supervision, 2012, *Fundamental Review of the Trading Book*, May.

Bjerke, M. B., 2007, *Experts, Banks and Politics: What explains the makings of Basel II?*, Norwegian Institute of International Affairs.

Board of Governors of the Federal Reserve System, 2009, *The Supervisory Capital Assessment Program: Overview of Results*, May, at www.federalreserve.gov/bankinforeg/bcreg20090507a1.pdf (downloaded December 20, 2011).

Burns, P., 2002, Retail Credit Risk Modeling and the Basel Capital Accord, January, Federal Reserve Bank of Philadelphia Payment Cards Center Discussion Paper No. 02–01, at www.philadelphiafed.org/payment-cards-center/publications/discussion-papers/2002/CreditRiskModeling_012002.pdf (downloaded December 30, 2011).

BusinessWeek, 1999, *Joseph Jett: Kidder Is Gone. I'm Still Standing*, at www.businessweek.com/bwdaily/dnflash/may1999/nf90512a.htm (downloaded September 3, 2011).

Committee on the Global Financial System, 2005, Stress Testing at Major Financial Institutions: Survey Results and Practice, January, Report by a working group established by the Committee on the Global Financial System, at www.bis.org/publ/cgfs24.htm (downloaded December 30, 2011).

Comptroller's Handbook, 2001, *Liquidity*, February, at www.occ.gov/publications/publications-by-type/comptrollers-handbook/_pdf/liquidity.pdf (downloaded May 20, 2011).

Crouhy, M., D. Galai and R. Mark, 2001, *Risk Management*, McGraw-Hill.

Crouhy, M., D. Galai and R. Mark, 2006, *Essentials of Risk Management*, McGraw-Hill.

Dodd–Frank Act, 2010, *Dodd–Frank Wall Street Reform and Consumer Protection Act*, at http://docs.house.gov/rules/finserv/111_hr4173_finsrvcr.pdf (downloaded December 30, 2011).

FDIC, Continental Illinois National Bank and Trust Company, in: *Managing the Crisis: The FDIC and RTC Experience*, Volume One, issued by the FDIC.

Federal Register, 2012, *Enhanced Prudential Standards and Early Remediation Requirements for Covered Companies*, Volume 77, no. 3.

Feuerverger, A. and A. C. M. Wong, 2000, Computation of value-at-risk for nonlinear portfolios, *Journal of Risk*, **3**(1), 37–55.

Goldman Sachs Media Relations, 2013, at www.goldmansachs.com/media-relations/press-releases-and-comments/current/index.html.

Gordy, M. B. and B. Howells, 2004, *Procyclicality in Basel II: Can We Treat the Disease Without Killing the Patient?*, at www.bis.org/bcbs/events/rtf04gordy_howells.pdf (downloaded September 3, 2011).

Gorton, G., 2009, *Slapped in the Face by the Invisible Hand: Banking and the Panic of 2007*, at www.frbatlanta.org/news/Conferen/09fmc/gorton.pdf (downloaded October 16, 2011).

Hirtle, B., T. Schuermann and K. Stiroh, 2009, *Macroprudential Supervision of Financial Institutions: Lessons from the SCAP*, November, at http://fic.wharton.upenn.edu/fic/papers/09/0937.pdf (downloaded December 30, 2011).

Investopedia, *How Basel 1 Affected Banks*, at www.investopedia.com/articles/07/BaselCapitalAccord.asp#axzz1X51b0dBG (downloaded September 5, 2011).

Jablecki, J., 2009, The impact of Basel I capital requirements on bank behavior and the efficacy of monetary policy, *International Journal of Economic Sciences and Applied Research*, **2**(1), 16–35, at www.ijesar.org/docs/volume2_issue1/a_impact_basel.pdf (downloaded September 10, 2011).

James, C., 1996, *RAROC Based Capital Budgeting and Performance Evaluation: A Case Study of Bank Capital Allocation*, at http://fic.wharton.upenn.edu/fic/papers/96/9640.pdf (downloaded September 4, 2011).

Jenkinson, N., 2007, *Developing A Framework For Stress Testing Of Financial Stability Risks, Comments to the ECB High Level Conference on "Simulating Financial Instability*," 12–13 July, at www.bis.org/review/r070716g.pdf (downloaded December 30, 2011).

Jorion, P., 2003, *Financial Risk Manager Handbook*, Second edition, GARP Library.

JP Morgan Chase Annual Report, 2010, available at http://investor.shareholder.com/jpmorganchase/annual.cfm (downloaded December 30, 2011).

Kelly, K., 2007, How Goldman Won big on mortgage meltdown, *Wall Street Journal*, December 14.

Khindanova, I., 1998, *Value at Risk*, University of California, Santa Barbara, CA.

Krishna, D. and R. Mark, *Risk Information Management for Complex Financial Products*, Teradata White Paper, available at www.teradata.com/article.aspx?id=4773 (downloaded December 30, 2011).

Leeson, N. and E. Whitley, 1996, *Rogue Trader: How I Brought Down Barings Bank and Shook the Financial World*, Little Brown and Company.

Linsmeier, T. J. and N. D. Pearson, 1996, *Risk Measurement: An Introduction to Value At Risk*, University of Illinois at Urbana-Champaign, IL.

Mark, R., 2012, *ERM Standards of Practice for Bankers, April 20, ERM Symposium*, at http://cas.confex.com/cas/erm12/webprogram/Handout/Session4927/bobM_ERM%20DC%20V3%20BRM%20ERM%20SOPs%20%20April%20%202012.pdf

Mark, R. and D. Krishna, 2008, How risky is your risk information?, *Journal of Risk Management in Financial Institutions*, 1(4).

Pritsker, M., 1996, *Evaluating Value at Risk Methodologies: Accuracy versus Computational Time*, The Wharton School, University of Pennsylvania.

Rosenblatt, M., J. Mountain and A. Kenyon, 2010, *Securitization Accounting: The Ins and Outs (and some Do's and Don'ts) of FAS 166, 167, and Counting...*", Eighth edition, Deloitte, available at www.deloitte.com/assets/Dcom-UnitedStates/Local%20Assets/Documents/AERS/us_deloittte_securitization_accounting_Jan2010.pdf (downloaded May 3, 2012).

SAS White Paper, *The Art of Balancing Risk and Reward: The Role of the Board in Setting, Implementing and monitoring Risk Appetite*, at www.sas.com/resources/whitepaper/wp_40392.pdf (downloaded December 24, 2011).

Senior Supervisors Group, 2008, *Observations on Risk Management Practices during the Recent Market Turbulence*, March 6.

Sullivan, R. J., *The Changing Nature of U.S. Card Payment Fraud: Industry and Public Policy Options*, Kansas City Fed, at www.kansascityfed.org/Publicat/Econrev/pdf/10q2Sullivan.pdf (downloaded September 3, 2011).

Viscusi, G. and A.-S. Chassany, 2008, Societe Generale Reports EU4.9 Billion Trading Loss, *Bloomberg*, January 24, at www.bloomberg.com/apps/news?sid=a8GBEB7UuuXc&pid=newsarchive (downloaded December 28, 2011).

What's in your FICO® score, available at www.myfico.com/crediteducation/whatsinyourscore.aspx (downloaded December 30, 2011).

3

Portfolio risk monitoring

Clifford V. Rossi

This chapter reviews the building blocks for an effective portfolio risk monitoring capability, emphasizing the reporting and governance elements needed for successful risk tracking within a financial firm.[1] To motivate the discussion and provide consistency throughout, the focus is on monitoring a mortgage portfolio, realizing that the monitoring function spans across assets, liabilities, business lines and risk types. However, the general lessons presented here apply to a broad range of portfolio types, including equities, fixed income, derivatives, etc.[2]

3.1 What is portfolio risk monitoring?

Performance monitoring is a risk manager's primary tool for understanding the quality of the risk profile of a portfolio and must be viewed as a dynamic process as economic and market conditions and obligor risk profiles change over time. Such a process must be tailored to the specific features of the portfolio and so requires a high degree of customization. For example, designing an asset-liability management (ALM) monitoring framework might entail development of interest rate shock scenarios against which the portfolio would be measured in terms of market value changes, and/or include a value-at-risk (VaR) analysis of individual asset types such as equities, foreign exchange exposures and fixed income positions. As a guidepost to developing such capabilities, the portfolio manager must determine the audience for the results and the timelines required for each reporting deliverable. From this, she must determine the metrics, performance window, cohorts of interest, key performance assumptions and indicators of performance. Portfolio monitoring

[1] The concepts described in this chapter are drawn from Rossi (2012).
[2] A more detailed introduction to these other portfolio types is found in Chapter 9 of this Handbook.

Handbook of Financial Data and Risk Information, ed. Margarita S. Brose, Mark D. Flood, Dilip Krishna and Bill Nichols. Published by Cambridge University Press. © Cambridge University Press 2014.

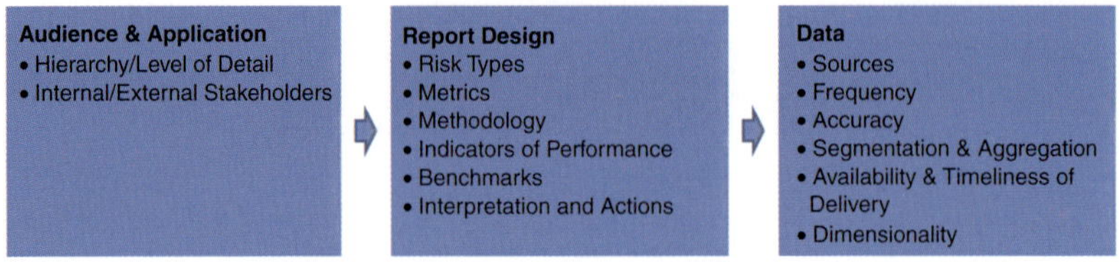

Figure 3.1 Building blocks for effective portfolio risk monitoring.

thus is not simply a passive exercise of reporting on the trends and condition of a risk portfolio. Rather it is intended to provide management and other stakeholders with ongoing insights regarding the level and direction of risk across key portfolio segmentations. In other words, portfolio reports are a structured way to provide key analytical insights rather than a disgorgement of data. A critical challenge for portfolio managers in this regard is determining how much and what type of information is appropriate for a particular audience as well as what steps are being taken to mitigate perceived risks. This process starts as shown in Figure 3.1 with understanding who the target audience is for a particular set of reports and analysis. This target audience ranges from the most senior levels of the organization, including the Board of Directors, to junior portfolio analysts charged with sifting through performance information and conducting analysis.

Board reporting, for example, requires a concise summarization of only the most critical aspects of portfolio risk exposures, while the portfolio analysis group benefits from highly disaggregated portfolio segmentations to perform their analysis. Too much detail provided at the Board level can be disastrous for a variety of reasons, as would be too little information for portfolio analysts whose job it is to dissect the portfolio and uncover important risk issues.

Closely linked to understanding stakeholder needs is report design and content. Portfolio monitoring takes place across different risk types including individual obligor credit, counterparty credit, market and interest rate risk, operational and liquidity risk, among others. The monitoring effort entails characterizing each risk at various levels of aggregation depending on the audience and requirements of the report.[3] Critical to this effort is deciding upon what to measure: default incidence, loss severity or loss when measuring credit risk exposures, or net interest income, duration gap and convexity profile for interest rate risk.

Important criteria for effective portfolio monitoring include the ability to provide early detection of risk issues, flexibility to report risks along various levels of data aggregation, integration of potential threats to the portfolio from

[3] For general coverage of the major risk types, see Crouhy et al. (2005), and Saunders and Cornett (2010). For a more focused overview of specific aspects of risk (market, liquidity, operational, etc.), see Chapter 9 of this Handbook.

macro- and other external drivers, understanding potential structural changes affecting portfolio quality, and providing effective benchmarks of actual to expected performance. Good portfolio monitoring also requires assessing potential ex ante risk and actual (observed) ex post gains or losses. But such efforts are of limited value without useful benchmarks and indicators of performance established to gauge portfolio risk exposures, align management incentives and employ risk mitigation strategies.

Finally, ensuring that the right data are captured and available on an ongoing basis establishes a lasting foundation for a sound risk surveillance process. Identifying valid sources and integration points with other related data, perhaps on other lending/asset platforms, is an important part of data management. Moreover, the reporting forum, whether board committee meeting, risk and business committees or other, determines the frequency of monitoring reports along with any inherent limitations in data availability. For example, if a board meets on the fifth business day of the beginning of each quarter and portfolio data are not available until the third business day of each month, it may be difficult to process and analyze the latest data in time for the meeting. During more stable market periods, not including these data may be a reasonable approach, however during more volatile periods, that new information could be critically important to detecting emerging risk trends. Closely aligned to understanding the audience for each type of report is the level of aggregation and segmentation strategy of the portfolio. Many a portfolio group has bogged down due to too fine a level of portfolio stratification. Striking a proper balance in data aggregation is as much an art as science. These topics are covered in more detail in the next sections along with key criteria and considerations in effective portfolio risk monitoring.

3.2 Understanding the intended audience and application

A challenge in developing a monitoring capability is keeping in mind the varied audience for reports and analysis. Monitoring infrastructure must accommodate junior analysts as well as board members and external analysts. Thus, the level of detail provided to each may be vastly different, and the system must be able to produce this seamlessly. In developing the monitoring system, considerable attention must be given to aggregation, management information systems (MIS) reporting and information accessibility. Starting from the finest level of aggregation, for example, the loan or transaction level, allows the portfolio analyst to aggregate along important risk dimensions. Product and/or business-line monitoring reports are logical reporting segmentations, and within such reports there can be a multitude of possible risk segmentations along key risk attribute cohorts, geography and product features, among others. Parsimony is valuable in developing the reporting segments. Without careful attention to what is most meaningful, the exercise can quickly devolve into

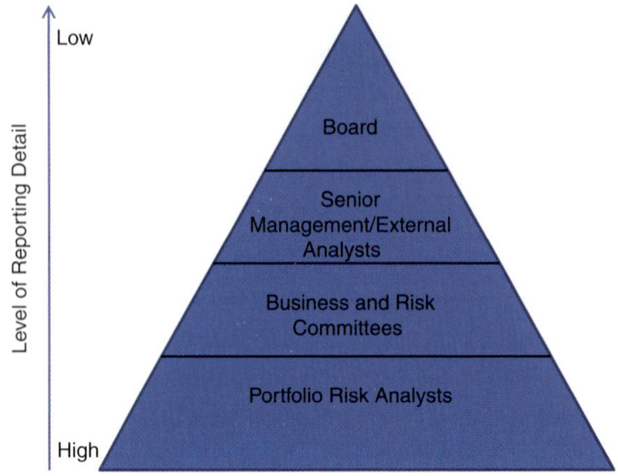

Figure 3.2 Portfolio risk monitoring reporting hierarchy.

hundreds of pages of reporting minutiae that obfuscate the key risks facing the portfolio. Too much disaggregation can also be distracting to analysts chasing down small risk anomalies with limited impact on the portfolio's risk profile.

Understanding the needs of each consumer of portfolio monitoring results helps shape the hierarchy and flow of information. This is evident from Figure 3.2, showing that the greatest level of detail in portfolio reporting occurs at the bottom of the pyramid. Here risk analysts sift through loan- and transaction-level information for important trends for review by risk and business managers. These in turn are distilled into key risks for various business and risk committees. As the seniority of the audience increases, the level of familiarity and expertise about risk tend to decline, requiring extreme care in developing reporting products that meet the needs of these audiences. By the time portfolio monitoring analysis has reached the board level, the level of data aggregation is at its highest and the risk summary must be concise and directive. It is at this point that portfolio risk managers need to temper their proclivity toward data and analytics with proper packaging and communication skills. Highly technical presentations to the board are likely to be of limited effect and could backfire if the board becomes disinterested. Messaging is key in these circumstances: communication of the top three to five risks along with associated risk mitigants is a useful device for leading a board through a summary of the portfolio's risk profile. In the end, striking the right balance between information content and clear communication is an art form that is critical to effective risk monitoring.

3.3 Report design

Report design and development first and foremost depends on the type of risks being monitored. Major risk classifications for reporting would include individual loan credit risk, counterparty credit risk, market and interest rate risk, liquidity risk and even operational risk; see, for example, Crouhy et al. (2005). Organizationally, separate groups may monitor different risk types. For example, credit and market risks may be monitored independently from each other along functional responsibility lines for credit risk and asset-liability management (ALM). If so, it is absolutely critical that both groups periodically share reporting strategy, methodology and data in order to ensure alignment to the greatest extent possible.

3.3.1 Guiding principles

Effective portfolio monitoring must adhere to several core principles. These include

- reporting transparency or clarity,
- operational tractability,
- prospective assessments,
- report integration and
- action-oriented reporting.

Transparency is critical to informed risk reporting. It must be clear exactly what risks are being monitored and why, as well as associated indicators of performance. Risk managers should develop these up front and in consultation with the business. If not, the exercise could devolve into a "check-the-box" reporting effort with limited impact on risk discussions and business strategies. As such, portfolio monitoring should facilitate action-oriented discussions across the organization. For example, reporting that credit losses for the portfolio in the current quarter reached 100 basis points of outstanding balances might be a useful piece of information in its own right, but what does this mean for the portfolio? Is this in line with expectations? If not, what actions should management take to adjust new originations or address potential risks on the back-end? Portfolio management is a dynamic exercise, and monitoring capabilities are the early warning system to motivate course corrections when appropriate. Consequently, effective benchmarks or milestones of performance must accompany monitoring of actual performance. Specific actions should address deviations outside established risk tolerances.

Monitoring efforts must also be **operationally tractable**. With monitoring results a key part of business, risk and board committee discussions, data processing, aggregation, analysis and report generation functions must be highly automated in order to meet tight reporting schedules.[4] Monitoring results that are stale due to reporting infrastructure limitations have little value in managing the portfolio.

[4] Straight-through processing (STP) is a core technique for report automation. See CPSS (2011) and CPSS-IOSCO (2001).

Highly manual processes are extremely inefficient, burning out talented staff and adding uncertainty to reporting deadlines. Consequently, portfolio managers must assess where their monitoring infrastructure gaps are relative to required reporting. Setting expectations with management early on as to what can be provided, as well as a plan for staging reporting enhancements, are critical for maintaining credibility within the organization.

Monitoring efforts must be **forward looking**. Relying only on historical views of performance is like "driving via the rearview mirror." To be sure, understanding performance trends over time is important, however, having some sense of where performance may be heading is essential to actively managing a portfolio. Market conditions and other factors affecting portfolio performance change over time so it is essential to monitor the portfolio dynamically and take requisite action to manage risk exposures within tolerance. Thus, portfolio monitoring must include projections of expected or likely performance over some designated period of interest to the risk manager as well as under stress scenarios. Underlying such estimates are key assumptions about economic and market conditions that drive asset performance. These assumptions must be clearly delineated and made consistent over time. A related criterion is that monitoring must be performed over the lifecycle of the asset. Effective reporting must build up a time profile of default for each asset type as the portfolio ages (i.e., "seasons") from origination. For portfolios such as mortgages having a longer time horizon on which to observe credit performance, seasoning is perhaps more important than for other asset classes such as credit cards, for example. For obvious reasons, this requires segmenting the portfolio by the year the loan was originated (i.e., vintage) and even within vintage it may be necessary to further disaggregate by quarter of origination. Some assets such as mortgages usually take longer to season, such that disaggregation by specific origination cohort is essential to provide a fair representation of performance. Other assets such as credit cards with much shorter seasoning profiles might not require the same level of disaggregation by time interval as mortgages.

Providing an **integrated view** of risks by asset class is recommended but is often difficult to accomplish. Nonetheless, such efforts can pay off handsomely by allowing the business to view risk trade-offs holistically that might otherwise result in suboptimal outcomes. An example of the value of an integrated view of risk is found in mortgage portfolios and is depicted in Figure 3.3. Consider a bank that originates mortgages and chooses either to hold some portion for investment long-term (held-for-investment, or HFI) or to make them available-for-sale (AFS) at some future time in a residential mortgage-backed securitization.[5] From the standpoint of the HFI disposition, the credit

[5] Fair value accounting treatment, including the rules for HFI and AFA, is an intricate subject. See Ryan (2007) for further details.

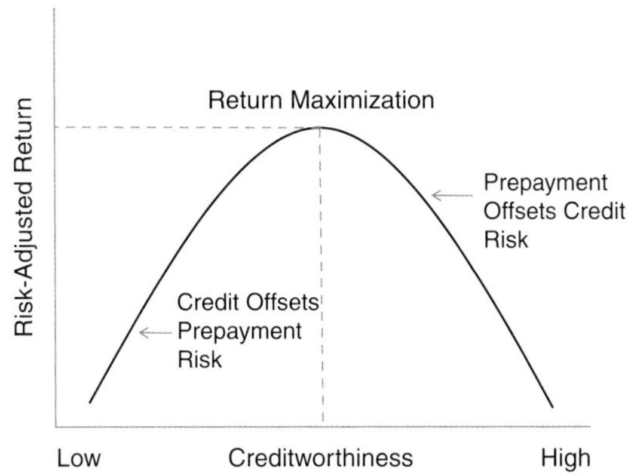

Figure 3.3 Competing risks trade-off on performance.

risk manager might want to establish portfolio standards limiting the creditworthiness of borrowers to a minimum credit score of 720. Alternatively, from the perspective of the asset-liability manager, who focuses on managing the interest rate risk exposure of the portfolio, the preference may be to limit the portfolio to borrowers with credit scores below 660 since borrowers with credit scores above this level exhibit much higher prepayment rates which tends to increase interest rate risk exposure. In this example, credit and interest rate risk are juxtaposed in terms of borrower creditworthiness. Strategies aimed at selling off highly creditworthy assets to maintain certain ALM targets could pose excessive credit risk to the portfolio. Consequently, a monitoring framework that portrays both the expected credit and interest rate risks together would optimize the portfolio profile across both risks. Finally, reporting must be **action oriented**. As mentioned earlier, the reporting effort is not simply an exercise in data summarization. Rather, in addition to providing thoughtful assessments of the direction and trends in the portfolio, the exercise must link performance to some set of management objectives or key performance indicators(KPIs). KPIs form the basis for how well the business and risk teams are doing in managing firm risk. Course corrections can then be made through changes in policies, product terms, risk mitigation strategies and other activities to bring performance in line with expectations. Reporting that provides no indication of how well the portfolio is doing against measured benchmarks significantly handicaps the management team and reduces opportunities to better align incentives of management with prudent risk taking.

3.3.2 Developing prospective risk views

Portfolio monitoring capabilities must be designed to allow the risk manager to **identify emerging trends** in the portfolio's risk profile and isolate causal factors for any deviations in performance. Depending on the asset type, a portfolio may require some time to season before meaningful performance can be established. During that time, waiting for actual performance to be established severely hampers the ability of the risk manager to understand important shifts in the mix of new additions to the portfolio which could result in demonstrably different risks over time. For example, consider a situation where it takes at least one year for delinquency performance to be established on a set of new loans. Without some way of measuring the ex ante risk during this interval, the risk manager would not be in a position to understand the extent to which market conditions and/or counterparty behavior could be changing the risk profile of loans originated by the firm. Assets purchased from counterparties according to specified credit terms could lead to adverse selection that increases the risk profile of the portfolio. Thus, having capabilities to gauge the level of expected risk early in the life of a set of assets is critical to the monitoring exercise. One way of implementing such a capability is to deploy a default risk model such as an automated underwriting scorecard that could assign an expected level of default risk to each asset coming in the door (see Mays, 2004).

To gauge what a "normal" level of risk looks like for new assets, a baseline level of risk could be established using the model over a number of years and a distribution of expected default or loss generated as shown in Figure 3.4. This graph depicts losses on the mortgage portfolio that have a skewed (long) right-hand tail, suggesting that there is a low probability of observing very high loss events. With that historical distribution, the expected loss could be estimated on a representative pool like that of the incoming assets. In addition, excessive risk layering and concentration in the tail of the distribution (denoted by losses to the right of the tail loss cutoff) could be assessed according to some specific target. For example, a monitoring

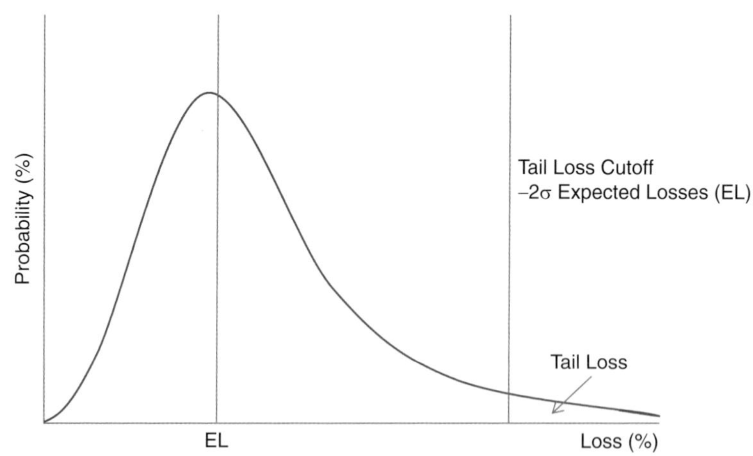

Figure 3.4 Monitoring risk buildups via ex ante measures.

objective could be twofold for this example portfolio. First, establish a range of expected losses around the historical average; for example, if historical lifetime expected losses have been 2%, then new loans of comparable risk should in aggregate have expected losses between 1.8–2.2%. Second, there could be a limit that no more than 5% of new loans can have expected losses more than 2 standard deviations over expected loss. This metric would help guard against adverse risk concentrations while the former metric would help maintain comparability of average losses on new pools consistent with previous expected performance.

Embedded in this example are a number of important concepts in risk monitoring including tail risk, risk layering, concentration risk and structural changes (see Rossi, 2010). The shape of the distribution, whether losses or default rates, is critical to understanding what is going on in the portfolio as a whole. During the recent financial crisis, many lenders suffered significant credit losses due to underestimation of credit risk on their portfolios, resulting in distributions that generated flatter overall distributions but with fatter tails. Monitoring cannot solve inherent weaknesses in the analytical tools used to measure risk; however, it can provide indications of where unacceptable risk concentrations might be building up.

Monitoring can also be used to judge the level of **risk layering** in the portfolio. Risk layering is the combination of higher risk attributes in a single loan. This is depicted in Figure 3.5. Product A could be thought of as a relatively plain vanilla loan that allows for one of three risk attributes as shown to be present. The incremental risk of each risk attribute is shown as the height of each box against the vertical axis denoting credit risk. In theory the aggregate credit risk of Product A would be represented by the level X based on a simple aggregation across all three

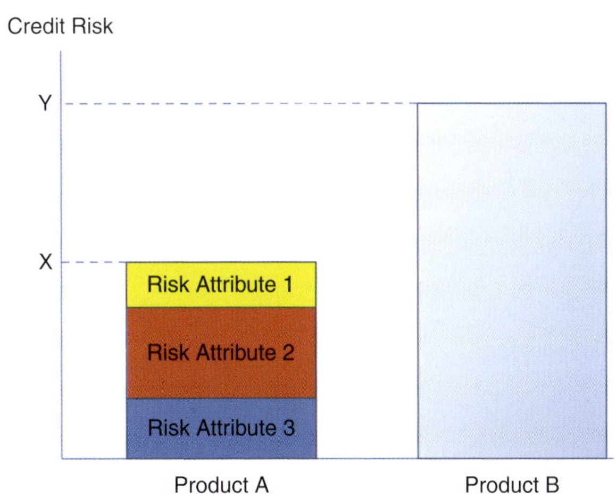

Figure 3.5 Illustration of risk layering.

risk attributes. However, no such product exists since only one of the three attributes may be present. Product B is a variation on Product A that now allows the inclusion of all three risk attributes. But the combination of all attributes significantly changes the type and risk profile of the borrower, leading to higher aggregate risk (point Y) than for Product A. During the mortgage boom leading up to the financial crisis, many lenders took otherwise standard mortgage products with relatively low risk features and began originating products with higher loan-to-value (LTV) ratios in conjunction with low credit scores and low income and documentation verification. While the individual risk attributes of the loan could perhaps be understood from the historical data, the combination of riskier attributes together reflected an altogether different type of borrower with different incentives ultimately affecting the risk of this product.

At a portfolio level, understanding the **concentration** of risk along certain dimensions is critical. Analogous to the benefits of diversification for investment portfolios, risk portfolios also benefit from diversification across key risk segmentations. This is illustrated in Figure 3.6. However, diversification only works as long as correlations do not converge in one direction. Concentrations come in many forms: geographical, loan quality, product, origination period, among others. Measuring such concentrations in consumer or commercial loan portfolios provides insight into where excessive credit exposure might lie over time, however, concentrations of risks in other asset types should be measured. Examples include industry sector concentrations in equity and corporate bond portfolios or convexity exposures in fixed-income portfolios, among others. Understanding where the footprint

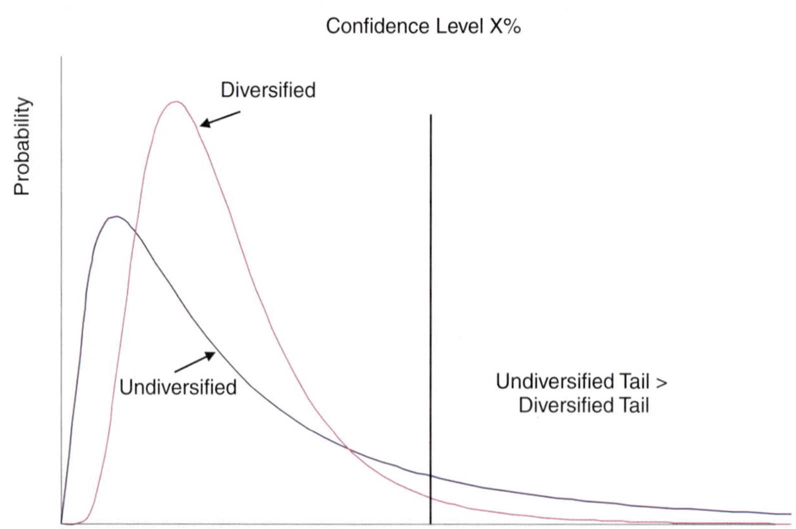

Figure 3.6 Concentration risk and portfolio diversification.

of the firm is relative to desired concentrations is critical to implementing a concentration monitoring framework reflecting the needs of both risk management and the business. For example, consider a bank with 65% of its retail branch offices in California. Even with other sourcing channels for loans, it is likely that the bank's portfolio will comprise a disproportionate share of assets from California. Establishing a portfolio risk limit on concentrations sets up important boundary conditions for shaping acceptable levels of risk in the portfolio. Concentration limits could thus be established on those segments of the portfolio where the variability of losses might be high. Implementing a limit that no more than 25% of the portfolio could be located in California, Florida or Nevada would have been one way for firms to have minimized mortgage loss exposure after the housing bust given the high loss rates experienced in these areas, however, securing senior management buy-in for those limits during the housing boom would have been extraordinarily difficult given historically low losses during that period.

Establishing limits is just one aspect of the risk manager's role, but limits, if effective, should be augmented by specific actions to be taken should a limit be breached. An example of this might be requiring the bank to sell a portion of its loans located in certain states that are in excess of the risk limit. Portfolio limits are not just relegated to credit portfolios but are also found in other risk positions, such as trading portfolios, in the form of position limits and even in asset-liability management (ALM) activities such as limits on funding sources. Risk management is responsible for establishing position limits and associated policies on the trading group, and for reporting of any breaches to management.

In monitoring the risk performance of the portfolio, establishing a set of **leading indicators** likely to affect performance is an important ingredient to understanding the direction of performance due to a variety of external factors. This would include macroeconomic factors such as GDP, interest rates, employment as well as industry or market-specific factors such as house prices and sales for mortgage portfolios, consumer sentiment for credit cards, and vacancy rates for commercial properties. Figure 3.7 provides an example leading indicators report for a stylized portfolio. Being able to quickly assess the overall impact of changes in leading indicators individually and in aggregate is essential to communicating expected effects to management.

Portfolio risk monitoring activities span a number of risk management, regulatory, accounting and financial areas. Within risk management, it is essential that the effectiveness of products and programs be assessed periodically along a number of dimensions. These could include pricing and volume trends, credit performance, characteristic analysis and mix issues. This is particularly important for new products where limited historical experience may be lacking. Widespread morphing of traditional mortgage products during the mortgage boom (see the description in

IndicatorType	Portfolio Indicator	Trend	Portfolio Impact
Macroeconomic	**Economic Growth:** Growth overthe period is expected to slow down in the 4th quarter from 3% to 2.5%. This is likely to put some upward pressure on default rates as business activity moderates	Deteriorating	Negative
	Interest Rates: Mortgage interest rates have been steadily declining over the last 4 quarters andthe fixed-rate 30 year mortgage rate is expected to fall 15bps more in the 4th quarter, which should improve credit quality	Slight Improvement	Positive
	Unemployment: Unemployment rates remain highat 8.2% which is a significant drag on housing demand and while delinquencies have stabilized over the last 4 quarters, delinquencies will remain elevated due to abnormally high unemployment rates	Stable	Negative
Housing Market	**Housing Supply:** Housing inventories continue to rise across most markets, and including the shadow inventory of foreclosed properties puts the average at just over 16 months.	Deteriorating	Negative
	Housing Demand: Newhome sales rose 5% from this time last year, however, it fell 2% between 3rd and 4th quarters owing to seasonal factors.	Sight Improvement	Neutral
Credit Standards	**Underwriting Standards:** The Federal Reserve's Senior Loan Officer Opinion Survey of Lending Practices indicates that credit remains tight generally in the mortgage market	Stable	Neutral
Borrower Capacity	**Affordability:** TheNational Association of Realtor Housing Affordability Index rose to 206 from 205.5 last quarter, an indication that housing affordability continues to improve since the crisis	Improvement	Positive
	Consumer Obligations: TheFederal Reserve Financial Obligations Ratio (FOR) for mortgage debt declined this quarter from 9.5 to 9.3 signaling continued improvement in consumer debt burdens	Improvement	Positive
Collateral	**Home Values:** The CSW 20-city composite index rose againthis quarter by 3 percent nationwide. Home prices remain well below 2006 levels but appear to be stabilizing	Stable	Slight Improvement

Figure 3.7 Leading indicators report.

Section 3.5) underscores the need for vigilant and regular monitoring coupled with new product limits. As lenders began expanding underwriting guidelines during the mortgage boom, they failed to realize that these changes would affect borrower behavior and ultimately product performance in a vastly different way than previously observed. For instance, in taking a negative amortizing (neg am) Adjustable Rate Mortgage (ARM) product designed for financially sophisticated borrowers and mass marketing it, lenders created a brand new product with credit performance unlike that of neg am ARMs originated before the boom. One of the pitfalls in monitoring these types of products is that projecting expected performance from standard products onto new products can understate the risk profile of the new product over time. New products should be closely monitored separately from other parts of the portfolio until adequately seasoned. Coupled with this should be position limits based on notional dollars or risk capital reflecting the underlying risk of the asset rather than book or regulatory capital, for example, that gradually relax as satisfactory performance history is built.

Other areas for portfolio risk monitoring include risk mitigation assessment. Risk mitigation involves supplementary arrangements to protect the portfolio, such as collateral, credit derivatives, and third-party guarantees (see BCBS, 2000). Dynamic or active portfolio risk management requires that risk managers implement a variety of strategies to adjust the performance of the portfolio. This could include collections and default treatments for borrowers in various stages of delinquency, rebalancing the portfolio for interest rate risk management, opportunistic asset disposition strategies, such as asset sales, and credit policy changes, such as active line management of credit cards and home equity lines of credit (HELOCs). Results from such activities should be assessed against management indicators of performance to ensure these actions are meeting their objectives and if necessary to guide adjustments in risk mitigation activities.

Efforts to credit enhance portions of the portfolio should also be closely monitored. Credit enhancements entail securing financial guarantees from other counterparties to absorb losses up to some contracted level should they appear in the portfolio. In return the portfolio lender pays a premium to the counterparty for this protection. This includes activities to secure reinsurance on portions of the portfolio outside desired risk tolerances, securitizations and related structured financing transactions (see, e.g., Banks, 2004). An important aspect of monitoring credit enhancements is to benchmark performance to some set of standards – either in the market or to peer institutions or securities/transactions. For example, asset-backed security performance should be benchmarked against similar classes of securities of comparable product type, tenor and risk profile to the extent that such information is available. Likewise, comparisons between held-for-investment (HFI) and available-for-sale (AFS) transactions of comparable assets provide a useful way of gauging whether the firm has achieved the desired balance in risk between these categories.

3.4 What to measure?

Among the most important questions a portfolio manager must first answer is what the specific metric of interest is. Ancillary questions include what unit of measurement to apply, and over what time period the measurement should take place. Measures will be unique to each risk type to be monitored. For credit risk measurement, decomposing loss into probability of default and loss given default (LGD) or loss severity is a useful way of understanding the loss profile of the portfolio (see, e.g., Glantz, 2002). For measuring interest rate risk exposure a variety of measures should be considered such as net interest income gap, leverage-adjusted duration gap, and market value changes over a range of interest rate shocks. For derivatives, portfolios metrics referred to as the "Greeks" measure changes in the value of options for changes in underlying asset prices, interest rates, asset price volatility, and time, among others.[6] Likewise, liquidity measures might include the number of months of liquidity available to the firm as well as stable funding and short-term

[6] The sensitivities are traditionally represented in equations using letters of the Greek alphabet. See Hull (2011) or Saunders and Cornett (2010) for a detailed treatment.

liquidity ratios as described in Basel III (see, e.g., Chapters 10 and 11 of this Handbook). Operational risk poses unique measurement challenges given the heterogeneity of operational risk events. While some advances in quantifying operational risks have been made in recent years, for most companies estimating operational losses is more art than science.[7] As a result, monitoring of operational risk might include summary information on subjective assessments of specific types of operational risks such as security breaches, fraud, non-compliance, systems breakdowns and other possible events.

From a measurement standpoint a preferred approach is to characterize each risk type in an appropriate way, for example, 90+ days past due (DPD) for credit risk, and to augment with some form of a risk-adjusted return measurement. Within each risk type, multiple measures may be warranted. For example, credit risk measurement may feature delinquency rates of various stages, for example, 30+, 60+, 90+, 180+ days past due or worse. This facilitates the development of a time profile of delinquency. Large 30DPD rates could signal higher rates on later stage delinquencies, thus providing management with an indication of the trajectory of default over time. Once a loan defaults, understanding the percentage or dollar amount lost (loss given default, or LGD) on an asset is informative. Actions to mitigate default frequency and loss severity can differ, so understanding how these loss components are changing in the portfolio is important. LGD could in turn be decomposed in reporting to outline important shifts between categories. For example, how have house prices affected sale prices for homes taken into real-estate owned (REO, i.e., properties returned to the lender after an unsuccessful sale at foreclosure auction) inventory as a result of borrower default? Such detail could enable companies to employ different sales strategies or even to modify negotiation tactics with distressed borrowers, for example, exploring the possibility of short sales.

Consideration must be given to the stability of the measure. In the case of a 30 day delinquency rate definition for a mortgage, such a metric tends to be a relatively unreliable measure of mortgage performance as other factors may be driving a 30 day late payment such as simply forgetting to mail the payment in on time. Including such metrics in a monitoring report provides a complete picture of delinquency so long as the limitations on the metric are understood.

A parallel concept for credit risk measurement is the application of roll or transition rate analysis to understand the migration of the portfolio from one stage of delinquency to another.[8] To illustrate the concept, consider Figure 3.8. For a

7 There are several widely used approaches to operational risk measurement, including the guidelines of the Basel Committee, the Committee of Sponsoring Organizations of the Treadway Commission (COSO), and the Risk Management Assocation (RMA). See Chapter 9 of this Handbook for a discussion.

8 Credit risk modeling is an intricate subject. See Saunders (1999), Bluhm et al. (2010), or Chapter 9 of this Handbook for an introduction.

	t+1						
	30DPD	60DPD	90DPD	120DPD	150DPD	180DPD	Total
30DPD	87	7	3	1.5	1	0.5	100
60DPD	6	76	8	5	3	2	100
90DPD	1	4	80	9	4	2	100
120DPD	1	2	3	73	12	9	100
150DPD	0	0	0	4	88	8	100
180DPD	0	0	0	0	5	95	100

Figure 3.8 Default rate transition matrix (%).

portfolio of loans as of time *t*, the objective is to understand how delinquencies migrate over time. For loans designated as 30DPD for instance in time *t*, 87% of these loans remain 30DPD in the next period, $t+1$. Likewise, 7% migrate to the next delinquency state of 60DPD. Note that the row sum for each delinquency state in period *t* must add up to 100% as shown. The shaded cells along the diagonal show the percentage of loans in each delinquency bucket that remain in the state in the next period. Such matrices are used for projecting delinquencies forward in loss reserving and forecasting exercises.

Once a performance measure is selected to meet some specific monitoring question, the analyst then needs to determine what unit of measurement is most appropriate. Again, applying the example from mortgages, multiple delinquency definitions may exist. Institutions operating under a thrift charter may follow an Office of Thrift Supervision (OTS) delinquency definition while other firms follow the Mortgage Bankers Association (MBA) definition.[9] Neither is incorrect but simply reflects differences in the timing of when a loan becomes past due. For some banks that originate mortgages in multiple legal vehicles, it is quite possible that over time both definitions are used in different subportfolios. The portfolio manager must determine which definition to apply against the entire portfolio to ensure consistency at the top of the house. For regulatory or business reasons it may be the case that both definitions are needed at times. However, when reporting the aggregate mortgage exposure of the firm, it is essential to use one definition.

Another consideration in establishing metrics is the time interval of interest. Most portfolio reporting exercises are performed monthly, however, for various reasons monthly data may not always be available for some subportfolios, in which case quarterly data would be reported. The challenge for the portfolio analyst is weighing the availability and ease of acquiring data on a shorter performance window and the information content it provides. Refreshing performance data on a quarterly basis when monthly information is available might lessen the value of the report if the data are too stale. For assets with short performance windows there is greater need to

[9] Thrift charters persist even though the OTS does not.

generate performance updates on a more frequent basis. In other cases, such as mortgages, under normal conditions, updating delinquency rates on a weekly basis may pose more operational costs to obtain than the value from providing delinquency data on a shorter window. During the crisis, moving to weekly reporting on delinquencies was important given the pace and magnitude of mortgage defaults. The analyst thus must consider a number of factors in making the time interval decision based on expected product performance and structure, current economic conditions, and operational capabilities.

3.5 Key performance indicators

Monitoring actual performance of a portfolio without comparing it to some benchmark for performance drastically limits the utility of such reporting. Such benchmarks, or KPIs serve as important guideposts helping management understand whether the portfolio is performing or underperforming with respect to their expectations. Further, KPIs can be tied to employee incentive compensation arrangements in the business so as to align their interests with company stakeholders. For this process to be most effective it is important that KPIs be established during strategic planning and that management sign on to such benchmarks. KPIs can take on a variety of forms such as simple point estimates of default or change in market value of an asset, to ranges and in some cases measurement of tail risk depending upon the complexity of the portfolio and technical affinity of management. Other possible measures include peer group analysis that benchmarks company performance against a group of comparable companies or asset types. Another common KPI entails comparing actual to expected performance.

For some specific objectives such as credit risk management, a simple KPI that limits the 90+DPD rate to some range based on historical or expected performance is relatively easy to implement. For instance, a model that predicts the one-year 90+DPD rate on a particular asset type might be 100bps with a standard deviation of 15bps. In that case, the portfolio manager might propose setting a KPI at some target range, perhaps +/− 1 standard deviation around the expected default rate of 100bps, or 85–115bps. Such a metric could be applied on an ongoing basis, or at the end of the year. If 90+DPD rates turn out to be less than 85bps, then this portfolio would be considered above plan and bonuses tied to this metric would be increased by some preestablished factor.[10] Alternatively, if 90+DPD rates turn out to exceed 115bps, the performance would be below plan and bonuses could be haircut. Introducing such a metric tied in this fashion to compensation will generate much interest among

[10] Kashyap et al. (2008) provide a further discussion of the incentive effects and agency problems created by risk measurement systems.

management and if reported on a routine basis can enable the business and risk teams to make course corrections such as tighten underwriting standards, turn off certain high risk sourcing channels or take other actions that would bring the default rate back in line with the KPI.

Another way a KPI could be established is by using a ratio. For example, a bank might wish to keep its loan loss reserve above a certain level relative to delinquent loans. Management could impose a KPI that the ratio of the allowance for loan and lease loss reserves (ALLL) to non-performing assets (defined as loans delinquent 90+DPD or in non-accrual status) must equal or exceed 1. Such a KPI would need to be balanced against GAAP accounting and regulatory policy to ensure it does not contravene those rules. However, the firm could use such a ratio to monitor its reserve levels relative to non-performing assets. If the ratio threatens to breach the KPI level, a root cause analysis could determine why the ALLL was not keeping pace with delinquencies and make appropriate adjustments consistent with accounting policy.

Some metrics might track tail risk or other adverse concentrations in the portfolio. Examples include geographical or product concentration limits. During the housing boom leading up to the financial crisis, home prices in many localities (Metropolitan Statistical Areas (MSAs)) rose to unprecedented levels. Identification of overheated markets based on trends in home prices along with other indicators of demand and supply for housing might have mitigated exposures to certain markets and product combinations. For example, capping the percentage of San Diego condominiums might have surgically protected a portfolio of mortgages in the 2004–2007 period. With management support, originations of loans in these hot markets could have been turned off. In many cases the perceived business need to continue to originate loans in these markets precluded the application of such dynamic portfolio countermeasures to the eventual detriment of the portfolios of many large mortgage lenders.

It is important to differentiate between KPIs that focus on how the portfolio is expected to perform on average and KPIs that focus on managing the tail of the distribution. Examples of tail risk management include the concept of limiting the "red tail" of a portfolio depicted in Figure 3.4, defined as some number of standard deviations above the expected performance of the portfolio. Such metrics could be applied to both new originations and the existing portfolio to guard against adverse changes in the mix of loans coming through the door. More complex metrics could include value-at-risk (VaR), requiring sophisticated modeling capabilities (see Jorion, 2006). VaR models became popular over the last several years for firms with trading books exposed to fluctuations in the value of their portfolios based on market trading activity. The basic concept of VaR is identifying a level of risk as defined as the worst loss observed over a specified period of time (e.g., 1 day) at a certain level of confidence (e.g., 99%). The red-tail concept mentioned above can be viewed analogously as a form of credit VaR. Despite the technical elegance of such constructs, VaR measurement is subject to a number of limitations. Such measurements are built upon historical information and thus may be unable to accurately assess the risk of an extreme event, as was evidenced by the financial crisis. In other words, VaR models using historical data before the crisis significantly underestimated tail risk because the likelihood of such an outcome was thought to be negligible based on the data. This fat-tailed risk problem requires frequent assessment of underlying data and assumptions driving the model results. Another limitation to such analysis is the common reliance on the assumption that losses are normally distributed. In fact, credit loss distributions tend to be asymmetric, with long right-hand tails signifying low frequency of high loss events. As a result, application of standard statistical estimates such as standard deviation may not accurately characterize the true underlying loss distribution. Despite these limitations, VaR models are useful if not perfect measures of the potential for extranormal loss events. Given the analytical complexity of such models, only firms with trading books or with specialized technical resources are likely to adopt such measures.

Other measurements of extreme outcomes include stress testing and scenario analysis, both of which should be considered in portfolio monitoring activities.[11] Stress testing entails subjecting the portfolio to shocks that simulate potential adverse outcomes well outside of expected or average performance. During the financial crisis, for instance, the largest banks were subjected to a regulatory-imposed stress test referred to as the Supervisory Capital Assessment Program (SCAP). In this exercise banks were required to apply certain stress tests to their portfolios. For instance, a stress test could subject mortgage portfolios to further losses by assuming that home prices would drop over the next few years by 20%. Imposing such a decline in home prices would generate higher defaults as borrowers respond to the value reductions relative to their existing mortgage balances. Stress tests provide a mechanism for management to evaluate their appetite for certain risk outcomes and make any necessary adjustments to guard against their occurrence. An important difference between stress tests and VaR analysis is that stress tests can be easily communicated to senior management because the loss outcome(s) being described can be attributed to concrete stress scenarios. For example, a board or executive committee can easily understand a scenario calling for a 20% decline in home prices but would have much greater difficulty understanding what a 99% event looks like. Stress tests can be constructed using actual scenarios such as the year-over-year decline in home prices during the early 1990s southern California real estate recession. By looking to severe events, either stylized or actual, stress tests can bring an element of realism to the analysis that seems more abstract and

[11] For further details on industry practices in stress testing and scenario analysis, see BIS (2005, 2009), ECB (2006), and Jones et al. (2004).

conceptual when applying a VaR framework. Stress testing suffers from the same limitation as VaR analyses, namely that even extreme historical events may not be sufficiently stressful. Hypothetical (i.e., non-historical) scenarios can be arbitrarily extreme but, whether hypothetical or historical, scenarios must be credible and achieve buy-in from management. It is hard to imagine management teams before the crisis placing much credence in stress test results based on a nationwide collapse in home prices as experienced in the years after the boom. Thus an important aspect of the portfolio manager's job is to socialize and gather support for various scenarios early in the monitoring process among key players on the management team.

Peer group analysis can be an effective part of a monitoring program. For instance, data available from 10-Ks, annual reports, investor presentations and similar materials can be used to compare performance along portfolio dimensions as well as other important areas such as loan loss reserves. In selecting peer institutions, it is important to consider the specific attributes that must be present. This might include asset size; if your company has assets over $10 billion, then firms of like size should be included in the comparison. Likewise, if your firm is a prime lender, then comparisons should be made to other prime lenders where possible. Even with these restrictions, it is virtually impossible to create a set of peer institutions that provide a perfect match of asset type and composition as your firm. Take for example a situation where an analyst compares the ratio of ALLL to non-performing assets of their firm to that of another firm of similar size. Data limitations may make it impossible to decompose the ALLL and non-performing assets to specific asset types, thus rendering direct comparisons of limited value. If the peer institution, for example, comprised 80% home equity loans and 20% first lien prime mortgages, that could greatly skew the ratio relative to the analyst's portfolio comprising 80% first lien mortgages and 20% home equity lines. But the level of reporting might not be below the product level, in this case mortgages. Nonetheless, augmenting firm portfolio performance with comparable peer portfolio performance over time can provide useful insights on the level and direction of risk-taking of your firm over time.

Monitoring may not be confined only to asset performance. With the use of analytical models increasing in popularity among portfolio managers, keeping an eye out for any deterioration of these tools is critical and also mandated by safety and soundness regulators.[12] A bewildering array of diagnostics is available for monitoring model performance. For statistically based models a simple approach is to compare actual performance with expected and compute a simple ratio. For example, if actual loss performance on a portfolio is 100bps on a year-over-year

[12] BCBS (2006) provides guidelines for risk "model validation" for commercial banks. FRB-OCC (2011) offers formal guidance for US banks.

basis, and the model's projection of losses is 80bps, then computing the ratio of actual/expected loss rate yields a result of 1.25, or that the model underestimates loss by 25%. Model tolerance could be established as KPIs to designate when certain actions for redevelopment might need to take place. Other simple metrics for measuring model risk include model R^2 or root mean square error (RMSE) of actual and model results, among more rigorous statistical goodness-of-fit tests available. Credit risk models are particularly vulnerable to shifts in underwriting processes or other structural changes going on in the market. Take for example mortgage loss models used before the crisis. Over time some lenders looked at providing mortgages with non-traditional features such as negatively amortizing option ARMs to customers. Such products had been originated for a number of years but under tightly controlled underwriting conditions. At the onset of the housing boom, historical performance of neg am products formed the basis of how the new neg am products would likely perform, controlling for the risk layering now introduced into these mortgages. What was not understood was that the risk layering that took place on these loans effectively morphed into a completely different profile of risk brought on in part by borrower selection issues. The models could not account for the changes in borrower behavior from the historical performance which reflected a very different set of underwriting circumstances. Relaxed documentation lending, for example, coupled with higher LTVs and lower FICO neg am mortgages severed historically strong psychic bonds between the borrower and the home. As a result, the models would be unable to extrapolate the higher loss rates on these products that ultimately materialized. Product morphing is a real threat to accurate portfolio performance monitoring given the difficulty in picking up behavioral changes from historical risk attribute combinations. This gives rise to the need to potentially classify major product changes into a subset of the portfolio for special monitoring that includes position limits and other portfolio management activities to modify the risk profile should it change.

Other areas where underwriting practices can affect portfolio monitoring activities is in the data used to measure portfolio performance. A good example of where this broke down was in computing debt-to-income (DTI) ratios during the housing boom when low documentation loans that did not require income validation became prevalent in the industry. Prior to this change in underwriting, the DTI-default effect was statistically significant though not as pronounced as for other risk factors such as credit score or LTV. As more low documentation loans were originated, the DTI-default relationship weakened, as shown in Figure 3.9.

This happened because borrowers tended to overstate their income, hence lowering their DTI captured in the origination data files. Over time, lower DTI loans showed up as having higher default rates, an artifact of the misrepresentation of borrower income. To augment the portfolio monitoring process, various safeguards

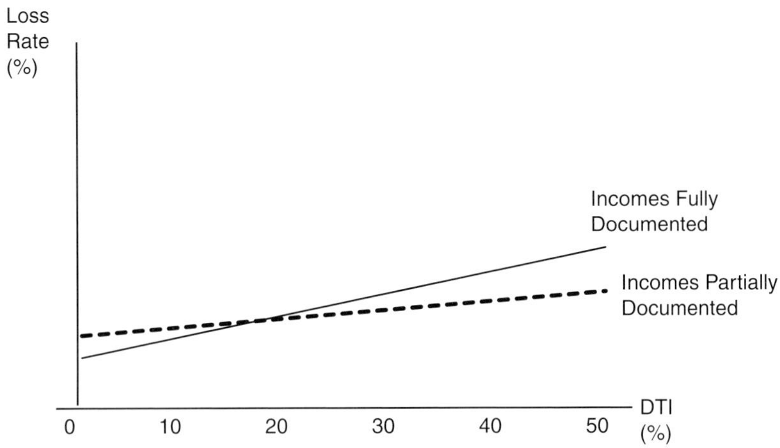

Figure 3.9 Relationship of loss to debt-to-income ratio (DTI) by documentation type.

around data accuracy should be an integral part of the process. Beyond standard data scrubbing and integrity checks, other ways to determine data accuracy are available. This includes leveraging quality control units that perform post-origination file reviews. In the DTI example, standard data integrity checks would have been insufficient to catch subtle underwriting changes on borrower behavior. However, reviews of loan samples would be able to uncover whether an unusually large percentage of borrowers overstated their income at origination, information which could be used to modify underwriting standards and/or address data deficiencies identified in the QC results.

Measuring and monitoring individual risk types by business line and subportfolio provides direct focus on risks of interest, allowing the company to take action to bring performance back in line before it is too late. However, integrating risks where possible and deriving risk-adjusted return metrics for each portfolio yields better decisions for the firm in allocating capital over the long run (see Bessis, 2002). Adjusting returns for the inherent risk associated with an asset allows the firm to efficiently allocate capital to its greatest value. Consider a simple example where a bank has \$1 of capital to allocate to either prime or subprime mortgage lending. Assume the prime loan generates 25bps of net income and the subprime loan 100bps. Under regulatory risk-based capital requirements each loan would require \$0.04 in capital. Using a simple return on equity (ROE) measure, the prime loan would generate a return of 6.25% while the subprime loan yields 25%. If the hurdle rate for the bank is 15%, clearly the decision would be to allocate the \$1 entirely to the subprime loan. However, this result would wind up being suboptimal as the ROE measure does not take into account the riskiness of each loan. Assuming that the bank now had a capability to estimate the risk of each loan using a credit-VaR analysis, and applying a 99% confidence interval to the problem, the prime loan has a credit-VaR of 1.5% while the subprime loan has a credit-VaR of 8%.[13] Using the income estimates from before, the risk-adjusted return on capital (RaRoC; see Crouhy et al., 2005), defined as net income divided by economic, or risk capital (in this case derived from the credit-VaR framework) for prime and subprime loans is now 16.7% and 12.5%, respectively. Although ROE is much easier to compute and explain to management and external stakeholders, the RaRoC measure more accurately reflects the relative risk of an asset. Introducing such metrics requires a significant amount of data and analytics most firms do not possess. Further, such metrics may be more appropriate for complex portfolios comprising a broad spectrum of assets along the risk continuum. A number of variations in risk-adjusted metrics exist, such as shareholder value-added (SVA) among others, that provide similar results. An advantage of these measures beyond individual asset allocation is that at a strategic level, management can more efficiently allocate capital across business lines. Business line RaRoCs can be computed by aggregating asset-level RaRoCs and performance comparisons made across the company. Lines of business not attaining corporate hurdle rates would be subject to scrutiny during the strategic planning period which may include paring back allocations to underperforming divisions and reallocating to higher performers.

3.6 Data considerations

3.6.1 Portfolio stratification

Portfolio stratification is one of the most important and challenging aspects of risk monitoring. At the individual business line level, the portfolio may consist of separate asset classes. For instance, consider the case of a large bank that has a consumer bank as one of its major divisions. Within the consumer bank are several business lines including credit cards, mortgages, auto and student lending. A natural segmentation to monitor performance would thus be by product line. However, depending on how the division is structured, some aggregation among products may be necessary in order to provide a profile of the business, if, for example, student and auto lending are operated as a single entity.

Determining the level of stratification is dependent upon available data, frequency of reporting, management appetite for detail, operational constraints and staffing. For instance, monitoring the credit risk profile of mortgages requires

[13] A similar mismeasurement problem also afflicts the regulatory treatment of risky assets. For example, the original Basel capital accord gave all mortgages the same risk weighting (50%) in calculating risk-weighted assets, creating opportunities for "regulatory capital arbitrage." See Kashyap et al. (2008) and Jones (2000).

understanding the distribution of credit scores in the portfolio. But what level of disaggregation is best? Should the portfolio be segmented in 10 point intervals of origination FICO score for delinquency reporting or are fewer intervals satisfactory? Again, there are no easy answers to this question. However, the delinquency profile of the asset can reveal useful segmentations and can be guided by two criteria: number of observations and performance differentials. Creating credit score segments in this example for scores between 610–620 where only 25 loans exist and where the performance of these loans is comparable to loans between 620–640 may warrant collapsing scores into larger cohorts. Segments of less than 50 loans, particularly less than a year from origination, may not provide meaningful information and hence take up extra time for analysis as well as become a distraction for both the portfolio analysis team and management whose eyes may ultimately be drawn to misleading results. We could find, for example, that a grouping of 25 loans in a score category has an unusually large default rate but that may simply be a function of one loan rather than a statistically important result.

Since each portfolio comprises a number of risk attributes, portfolio monitoring activities frequently entail a layered analysis, much like peeling the skin of an onion. This increases the complexity of the monitoring process since the portfolio manager must consider what combinations of attributes are important to examine in terms of their contributions to performance. To illustrate this problem, consider again the manager of a consumer loan portfolio. Having established that product level segmentation provides a reasonable way to break the portfolio down, the effort must then shift to parsing each product further into its core risk elements. For mortgages, this might require further stratification between product types, for example, fixed-rate versus adjustable, first and second liens, prime and subprime. But within each mortgage product, performance differentials by credit score, loan-to-value (LTV) ratio, documentation type, occupancy type and other attributes are important to report.

There is a "curse of dimensionality" at work: the size of the reporting effort becomes very large based on the combinations from just a handful of attributes. The trade-off between parsimony and risk detail becomes critical and dependent again on operational capacity and ability to synthesize important trends to a representative level for reporting. Referring to our example, it may be essential to understand the performance of prime, fixed-rate first lien mortgage along credit score stratifications as well as other risk attributes. And because of underwriting and seasoning differences, monitoring of this product should be further segmented by origination vintage. Couple this with variations in geographical performance due to local market and economic drivers, and the monitoring effort quickly blossoms into a monumental effort where hundreds of reports could be easily generated. This unfortunately can lead to analysis overload where important trends are overlooked by focusing on report minutiae and/or incorrect inferences drawn regarding the portfolio from reaching too far down into portfolio subsets. The benefits of computing power and speed in this instance can become a drawback to the process.

At this point, some way to filter the data must be developed and an effective way to approach this problem is through the use of standard data mining techniques available in most analytical software packages. Data mining can provide a structured way of identifying appropriate risk attributes as well as the level of segmentation (e.g., Chen and Chen, 2010). A useful technique is cluster analysis, which identifies important differences in performance by evaluating various combinations of risk attributes or clusters that provide statistically different results. While these types of tools can help isolate important risk combinations from hundreds of possible alternatives, they must be applied cautiously and not used as replacements for sound risk judgments. It is critical that the portfolio manager augment the tool with intuition and insights drawn from experience and not rely solely on analytical tools.

A good example of this is drawn from monitoring of mortgage portfolios during the housing boom. During the boom, traditional characteristics of mortgage performance woefully understated the true risk on these portfolios due to important structural changes going on in the market that would be undetectable using standard risk attributes and metrics. The traditional psychic attachment of housing for borrowers as a home rather than as an investment fundamentally broke down in the years leading up the housing crash as leverage rose to excessive levels and standard underwriting practices loosened, thereby allowing considerable layering of risk attributes. Distributions of individual risk attributes such as credit score and LTV shifted toward riskier concentrations that were picked up in portfolio monitoring activities, but what was missing was any sense of whether these changes would actually change borrower behavior. The idea that with sufficient negative equity resulting from a decline in home prices leaving the value of the home less than the mortgage, a borrower might strategically default on the home, rather than continue to pay, was not understood, nor would it be easily discerned from monitoring reports. Consequently, any portfolio monitoring activity must consider the potential for structural shifts to compromise historical relationships. This leads back to the issue of leveraging actual performance history to profile the performance of the portfolio. However, it may be misleading as new combinations of risk factors are originated with no history to guide expected performance. Stress testing and scenario analysis can provide some insight, but even these activities cannot guarantee an accurate representation of behavioral shifts by borrowers. In these instances, the portfolio manager should ring fence these new and untried products, working with the business to establish new product limits and performance targets that could be used to gradually adjust the flow of this product as satisfactory performance history is built up. A problem, however, with such a strategy is that it may take years, as in the case of the mortgage boom, before adverse changes are identified and by that time management may have little patience to wait for further evidence before deciding to expand a new product offering.

3.6.2 Managing risk attribute intertemporal and structural shifts

Another challenge for portfolio managers is that important risk attributes also change over time, thus affecting portfolio performance. Drawing on the example above, it is a well-established fact that credit scores of mortgage borrowers will change over time from origination. Thus, while origination credit score is an important factor in determining credit risk, how and to what level the borrower's score migrates is also important. Likewise, origination LTV will affect default performance, but updated LTVs based on current home prices in the local market provide valuable insight into the borrower's incentive to default on the mortgage. Obtaining refreshed data on key risk attributes such as credit score and LTV should be considered as part of the monitoring effort but it comes with additional cost and effort. In the case of credit score, this would require procuring updated scores on the portfolio from various third-party credit repositories and being able to demonstrate that permissible purpose exists among meeting other compliance requirements. This can be a costly exercise for a large portfolio, thus it may be useful to take a periodic random sample of the portfolio and obtain updated credit information. Updating house prices requires access to third-party vendor sources of home price data and/or property valuation models. While these capabilities pose costs, they can yield critical insights on prospective portfolio performance.

In the case of credit score, it may be important to use detailed credit attributes to profile the risk of the portfolio. Credit scores such as FICO are developed on the performance of a broad array of consumer products, and not just mortgages. Consequently, while predictive of mortgage default, credit scores may not be as precise a risk factor as a custom mortgage score that uses a subset of detailed credit attributes reflected in a borrower's credit profile. This might include such elements as number of tradelines and inquiries for credit, and late payments, among others. This information can be obtained from credit repositories for a cost and increases the analysis for monitoring the portfolio. Given the importance of leveraging data to discern subtle changes in borrower behavior that may not be picked up relying on standard attributes such as credit score, providing for some capability to periodically profile detailed credit attributes can be a useful monitoring activity.

During the years preceding the financial crisis, risk management made greater use of quantitative analytics for assessing risk performance. Unfortunately, in many instances important qualitative risk assessments such as quality control (QC) units for some asset types were not put on a comparable footing with new model-based risk analysis, to the detriment of effective risk management. For some assets, such as residential mortgages and commercial real estate, post-origination loan reviews are essential to understand the quality of the loan manufacturing process and to identify underwriting defects. Because sampling all loans in the portfolio would not be practical, random and targeted samples are usually drawn. From detailed loan level file reviews, QC departments can quickly get a handle on where underwriting issues might be occurring and alert risk managers to potential exposures for remediation. QC findings can also be used to make qualitative adjustments to portfolio and loss reserving models. For example, in the absence of hard empirical evidence on their performance, QC determinations of fraud or other forms of misrepresentation of loan files could lead to materially higher losses than expected. During the housing boom, the advent of such low documentation products as stated-income stated-asset (SISA) were fraught with misrepresentation where in some cases QC units were finding significant percentages of their samples overstating their incomes by 50% or more. This information would be vital to risk analysts as well as to senior management in deciding whether such a program's credit and pricing terms should be modified or the product shut down altogether.

Portfolio monitoring must be set up to address the needs of the firm first rather than be reactive to what regulators might require. However, regulatory and compliance monitoring requirements will feature prominently in any portfolio risk monitoring capability (see Comptroller of the Currency (OCC), 1990). A challenge to portfolio managers is weaving in regulatory reports with standard risk management monitoring in a way that does not impair the clarity and focus on underlying risk issues. Moreover, with the advent of the Dodd–Frank Act, regulatory requirements on data and reporting will increase, making it more important that monitoring capabilities are in place to modify and augment their processes flexibly for any new reporting requirements. For the largest institutions, as agencies such as the Office of Financial Research establish industry standards on financial data, and interagency working groups determine the type and frequency of portfolio stress testing, data and systems capabilities will need to be greatly enhanced. This may become challenging in situations where assets of the same type are held in different legal vehicles on multiple data platforms due to acquisitions over the years. The demands for consistent and timely risk performance reporting will increase over the next few years, making it essential that portfolio managers take the steps within their companies to develop integrated data repositories that have the ability to routinize standard risk reporting as well as provide for ad hoc and customized reporting and analysis as required.

Beyond safety and soundness reporting, renewed emphasis on compliance and associated consumer issues feature prominently in the Dodd–Frank Act, most visibly in the form of the Consumer Financial Protection Bureau. The recent foreclosure crisis also underscores the need for companies to invest in data, systems and staff to provide accurate and timely monitoring of compliance with federal and state consumer regulations. In this regard, portfolio managers need to be able to integrate origination and servicing platforms in order to understand specific borrower circumstances in real time. Chronic underinvestment in such capabilities over many years hampered mortgage servicers' abilities to streamline loan modification and foreclosure activities. Greater use of tools to triage borrowers based on delinquency status such as tailoring specific treatments to statistically based behavioral scores augmented with expert judgment by default and collections specialists enhances risk management through the application of surgical mitigation strategies as well as efficiency. Ensuring such capabilities are built on strong processes that provide for consistent treatment of borrowers based on regulatory requirements is an essential ingredient in this process.

Related to regulatory required portfolio monitoring are a host of accounting and financial required monitoring activities. These include fair value accounting, loan loss reserving, product pricing and financial performance assessment. For example, one of the most visible and important risk management functions surrounds the allowance for loan and lease loss reserves (ALLL) to be reported under generally accepted accounting principles (GAAP). Establishing the ALLL requires a combination of quantitative and qualitative assessments of the portfolio. Given the importance of the ALLL in bank financial reporting processes it warrants special attention and resources (see Comptroller of the Currency (OCC), 1998). As a result, it is critical that any deviations in the estimates of reserves and portfolio performance be reconciled. Although GAAP accounting rules do not permit forward-looking views of losses to be used in generating the ALLL, understanding the components of credit loss and their intertemporal dynamics is critical to establishing an accurate representation of risk trends. At no time was this more urgent for portfolio managers than during the recent crisis when due to inherent procyclicality of loan loss reserving rules, as portfolio delinquencies escalated, these firms accelerated their loan loss reserves which sapped net income at the worst possible moment for the industry (see Kashyap and Stein, 2004). Not surprisingly, heightened attention to the ALLL and emerging loss trends meant portfolio managers not only had to refine their ability to project future losses in relation to reserve estimates but also had to justify their core assumptions driving those losses.

3.7 Summary

Portfolio monitoring, if performed correctly, provides risk management with an effective means to understand portfolio performance and to make course corrections in the profile of risk when warranted. While such processes are data and systems intensive, portfolio monitoring requires considerable judgment regarding reporting and analysis. Over the years, the financial services industry has tended to underinvest in data management and processing capabilities as evidenced first by the financial crisis and later underscored by the foreclosure crisis. Data underlying financial transactions are dynamic, reflecting changes in borrower or counterparty financial condition and behavior and the economic environment. Thus it is critical that management understand and make the required commitment to architecting and building a risk data management infrastructure that provides flexibility in monitoring performance across risks, business lines and products, not only leverages insights from historical performance, but has the tools to provide forward-looking assessments of portfolio trends, leverages and integrates analytical tools with risk expert judgment, and offers accurate and timely risk analysis to various stakeholders.

Report design and development requires balancing risk data availability with metrics that can be used to manage portfolio risk. Part of this effort requires thoughtful and comprehensive segmentation strategies that provide sufficient detail without overwhelming the analytical process. Consideration must be given to devising stratifications that align as much as possible across risk types, thus facilitating an integrated view of risks for the enterprise. Portfolio monitoring needs to be viewed from the perspective as structured analysis rather than as a simple reporting process. With that in mind, portfolio monitoring must be actionable, able to be repeatable on a regular basis and be transparent to consumers of such information. Such processes therefore benefit greatly from development of a set of KPIs that relate to each risk and can be used to inform management of important shifts in the portfolio, align management incentives and facilitate risk mitigation activities. Finally, portfolio monitoring requires understanding the needs of the various internal and external consumers of the process. The wide variation in knowledge and perspective of audiences for monitoring creates the greatest challenge for designers of such capabilities, as the scope and content must be adapted for the most strategic levels of the company as well as for junior analysts. Efficient summarization of essential information and analysis tailored to the specific needs of each stakeholder is an essential ingredient for successful portfolio risk monitoring.

Bibliography

Alexander, Carol, 2009, *Market Risk Analysis* (four-volume set), John Wiley and Sons.

Bank for International Settlements (BIS), 2005, *Stress Testing at Major Financial Institutions: Survey Results and Practice*, Committee on the Global Financial System (CGFS) Publication, No. 24. January.

Bank for International Settlements (BIS), 2009, Principles for Sound Stress Testing Practices and Supervision, Technical Report 155, May.

Banks, Erik, 2004, *Alternative Risk Transfer, Integrated Risk Management through Insurance, Reinsurance and the Capital Markets*, John Wiley and Sons.

Basel Committee on Banking Supervision (BCBS), 2000, Industry Views on Credit Risk Mitigation, Technical Report, Capital Group, January.

Basel Committee on Banking Supervision (BCBS), 2006, International Convergence of Capital Measurement and Capital Standards: A Revised Framework, Comprehensive Version, Technical Report, June, `www.bis.org/publ/bcbs128.htm`

Bessis, Joël, 2002, *Risk Management in Banking*, Second edition, John Wiley and Sons.

Bluhm, Christian, Ludger Overbeck and Christoph Wagner, 2010, *Introduction to Credit Risk Modeling*, Second edition, Chapman and Hall.

Board of Governors of the Federal Reserve System, and Office of the Comptroller of the Currency (FRB-OCC), 2011, Supervisory Guidance on Model Risk Management, Technical Report, OCC 2011-12, April, `www.occ.gov/news-issuances/bulletins/2011/bulletin-2011-12a.pdf`

Chen, Tsung-Hao and Cheng-Wu Chen, 2010, Application of data mining to the spatial heterogeneity of foreclosed mortgages, *Expert Systems with Applications*, **37**(2), 993–997.

Committee on Payment and Settlement Systems (CPSS), 2011, Principles for Financial Market Infrastructures, Consultative Report, Bank for International Settlements, March, `www.bis.org/publ/cpss94.pdf`

Committee on Payment and Settlement Systems (CPSS), and Technical Committee of the International Organization of Securities Commissions (IOSCO), 2001, Recommendations for Securities Settlement Systems: Report of the CPSS-IOSCO Joint Task Force on Securities Settlement Systems, Consultative Report, January, `www.bis.org/publ/cpss42.pdf`

Comptroller of the Currency (OCC), 1990, *Comptroller's Handbook: Review of Regulatory Reports, Narrative and Procedures – March 1990*, `www.occ.treas.gov/publications/publications-by-type/comptrollers-handbook/regreport1.pdf`

Comptroller of the Currency (OCC), 1998, *Comptroller's Handbook: Allowance for Loan and Lease Losses, Narrative – June 1996, Procedures – May 1998*, `www.occ.treas.gov/publications/publications-by-type/comptrollers-handbook/alll.pdf`

Crouhy, Michel, Dan Galai and Robert Mark, 2005, *The Essentials of Risk Management*, McGraw Hill.

European Central Bank (ECB), 2006, Country-level macro stress-testing practices, *Financial Stability Review*, June, 147–154.

Glantz, Morton, 2002, *Managing Bank Risk, An Introduction to Broad-Base Credit Engineering*, Academic Press.

Hull, John C., 2011, *Options, Futures, and Other Derivatives*, Eighth edition, Prentice Hall.

Jones, David, 2000, Emerging problems with the Basel Capital Accord: Regulatory capital arbitrage and related issues, *Journal of Banking and Finance*, **24**, 35–58.

Jones, M. T., P. Hilbers and G. L. Slack, 2004, Stress Testing Financial Systems: What to Do When the Governor Calls, Working Paper No. 04/127, International Monetary Fund, `www.imf.org/external/pubs/cat/longres.aspx?sk=17517.0`

Jorion, Philippe, 2006, *Value-at-Risk*, Third edition, McGraw Hill.

Kashyap, Anil and Jeremy Stein, 2004, Cyclical implications of the Basel II capital standards, *Economic Perspectives*, Federal Reserve Bank of Chicago, 1Q/2004, 18–31, `www.economics.harvard.edu/faculty/stein/files/basel-chicago-fed-04.pdf`

Kashyap, Anil, Raghuram Rajan and Jeremy Stein, 2008, Rethinking capital regulation, in: *Maintaining Stability in a Changing Financial System*, Federal Reserve Bank of Kansas City, 431–471, www.kc.frb.org/publicat/sympos/2008/KashyapRajanStein.09.15.08.pdf

Mays, Elizabeth (ed.), 2004, *Credit Scoring for Risk Managers: The Handbook for Lenders*, Thomson South-Western.

Rossi, Clifford V., 2010, Anatomy of Risk Management Practices in the Mortgage Industry: Lessons for the Future, Technical Report, Research Institute for Housing America, May.

Rossi, Clifford V., 2012, *Fundamentals of Risk Management*, John Wiley and Sons.

Ryan, S., 2007, *Financial Instruments and Institutions: Accounting and Disclosure Rules*, John Wiley and Sons.

Saunders, Anthony, 1999, *Credit Risk Measurement: New Approaches to Value at Risk and Other Paradigms*, John Wiley and Sons.

Saunders, Anthony and Marcia Millon Cornett, 2010, *Financial Institutions Management: A Risk Management Approach*, Seventh edition, McGraw-Hill Irwin.

4

Frameworks for systemic risk monitoring[1]

Alan King, John C. Liechty, Clifford V. Rossi and Charles Taylor

4.1 Introduction

This chapter reviews where the debate on systemic risk frameworks now stands. The unprecedented costs and the complexity of the 2008 financial crisis have spurred academics and policy experts to look afresh at systemic risk and a number of different approaches have emerged with different practical implications for policy-makers. After some discussion of the background to this rapid rise in intellectual interest on systemic risks, this chapter groups different analytical approaches into three broad categories and then examines the data and policy implications of each category.

4.1.1 The 2008 crisis as a spur to systemic analysis

The cost of the 2008 crisis was enormous. For example, in the USA, the incremental loss above what one might think of as a normal downturn was about 3% of GDP, according to Phil Swagel of Georgetown University (Swagel, 2010). Andy Haldane, Bank of England Executive Director for Financial Stability, put the total first year loss of world output in 2009 at 6.5% of global GDP. If only 25% of that loss proved to be permanent, the net present value of the total loss would be close to one entire year's worth of global GDP (Haldane, 2010).

Economists from across the theoretical and political spectra have offered up various explanations of the crisis and generated conflicting proposals for how to prevent another one in the future. No dominant consensus has yet emerged on what was the fundamental cause.

[1] None of the opinions or positions described in this chapter are necessarily those of the Comptroller of the Currency.

Handbook of Financial Data and Risk Information, ed. Margarita S. Brose, Mark D. Flood, Dilip Krishna and Bill Nichols. Published by Cambridge University Press. © Cambridge University Press 2014.

At the time of writing, the global crisis is not over. As well as a glacially slow recovery in North America, a new threat of instability has emerged in the Eurozone. But this threat originated in a more conventional set of problems than the problems that characterized the 2008 meltdown. Weak fiscal performance and poor banking practices are at its heart, rather than a surprising network or contagion effects or unintended consequences of financial innovations. Restoring stability requires some new thinking – mostly about the European institutional superstructure – but not perhaps to the extent that is needed to address potential systemic challenges like the ones posed in 2008.

4.1.2 Systemic risk defined

If no dominant framework for systemic risk monitoring has appeared yet, a consensus of sorts has emerged recently on the nature of systemic risk and how to define it. To paraphrase the IMF, it is the risk of a financial system breakdown with (1) serious negative consequences for the real economy (IMF, 2011). Such a breakdown will often be associated with (2) the failure of one or more so-called "systemically important financial institutions" (SIFIs), with (3) contagion among institutions and markets, (4) a collapse in asset values and (5) positive feedback loops exacerbating problems across the system. To be systemic in scope, an event has to satisfy (1). To be systemic in character, it has to satisfy the majority of (2)–(5). So, for example, a single SIFI failing might not be a truly systemic event, but if it is combined with a severe collapse in prices of a major class of assets (4) and significant consequences for the real economy (1), then it would be.

4.1.3 The challenge for policy

Therefore, the public policy challenge is threefold. The first challenge is to toughen and strengthen systemically important institutions, activities and markets and to evolve a network structure that is relatively stable, robust and resilient. The second challenge is to shepherd the financial system and its participants away from circumstances and practices where systemic risks rise to unacceptable levels. The third challenge is, in the event that a crisis starts, to be ready to manage it as effectively as possible.

Finally, costs should be taken into account. While it is true that some of the changes that are required to meet these challenges may have positive externalities – for example, spurring automation that leads to services that are more reliable and more economic in the long run, or improving information for institutional risk managers as well as systemic risk managers – other changes will raise the costs of financial and risk intermediation in normal times. To the extent possible, these costs should be weighed against the benefits of increased financial system stability.

4.1.4 Policy responses to date

In addition to intellectual and economic developments, the past few years have been marked by new legislation and regulation across the globe. In the USA, the Dodd–Frank Act created the interagency Financial Stability Oversight Council (FSOC) and tasked it with monitoring and managing systemic risks. It also created a new and independent Office of Financial Research (OFR) within the US Treasury to support FSOC and to monitor systemic developments. The Federal Reserve has been given new authority to regulate SIFIs and the Federal Deposit Insurance Corporation has new powers to require living wills from these institutions and then to liquidate them in an orderly way if they fail. In the UK, prudential supervision – including responsibility for systemic stability – has been taken away from the old Financial Services Authority and handed over to a new Financial Policy Committee and Prudential Regulatory Authority in the Bank of England. The European Community has created the European Systemic Risk Board (ESRB) and European Supervisory Authorities (ESAs). And globally, the Financial Services Board (FSB) has emerged as an influential force for regulatory overhaul, reporting to the G20 and providing new direction to the International Monetary Fund (IMF) and the established standard-setting bodies – the Basel Committee on Banking Supervision (the Basel Committee), the International Accounting Standards Board (IASB) and the International Organization of Securities Commissions (IOSCO). All these agencies and offices – and others in other countries and jurisdictions – are supporting the development of new frameworks for understanding and monitoring systemic risks.

One early overview of the issues facing these institutions is given in Acharya and Richardson (2009). Acharya et al. (2010) analyze the ensuing Dodd–Frank legislation, highlighting as central issues: financial and regulatory architecture; measurement and regulation of systemic risk; the impact of shadow banking markets; the importance of securitization and credit markets; and the accounting and compensation aspects of corporate control. French et al. (2010) engage in a similar overview. In addition to addressing the issues of shadow banking, micro-prudential regulation, and corporate compensation, they dedicate chapters specifically to systemic regulation and the information infrastructure of financial markets.

4.1.5 The interplay between theory and data

The problem of financial system stability is so complex that we cannot expect any single dominant explanation to emerge any time soon. If systemic risk theory and practice follows the pattern of previous scientific and policy discovery, there will be a gradual convergence on a dominant framework as advances in data and in theory leapfrog one another repeatedly. New data will spur new frameworks and theories, and then new theoretical insights will create new demands for data and that in turn will produce new theoretical insights, and so on.

However, data collection and policy cannot wait indefinitely, and both policy and data efforts have to start somewhere. All other things being equal, early policy decisions and early data collection efforts should be as robust as possible in the sense that they should be likely to add value regardless of which frameworks eventually come to dominate the systemic risk landscape. Basic efforts to improve the quality of financial data, such as the international Legal Entity Identifier (LEI) initiative, meet this criterion.

4.1.6 The chapter structure

Our approach is to divide explanations and analyses into three broad families – high level, network and institutional, and granular. The next three sections review recent examples of each.

- *High Level Analyses* These focus on a few causes and effects. Analysis relies on a few key macroeconomic phenomena such as macroeconomic imbalances, or a few key features of the financial system such as overall levels of leverage and capital adequacy. They include analyses that attribute most systemic problems to systemically important financial institutions (SIFIs) and their regulation.
- *Network and Institutional Analyses* These concentrate on specific features of markets and institutions, processes and practices, how the relationships between institutions and markets form a network, the resultant network properties and how details of the structure of the financial system can exacerbate stresses and strains when they emerge.
- *Granular Analyses* These look for the seeds and amplifiers of instability in the characteristics and behavior of individual agents and in the patterns of individual transactions and the incentives facing both financial system agents and regulators.

This taxonomy is not completely "clean" for a number of reasons. For example, SIFIs could be under "institutional analyses" as easily as "high level analyses." Some themes of analysis appear in all three families – such as the importance of incomplete information and information asymmetries. Each family of analysis can accommodate positive feedback loops, contagion, herd behavior and liquidity collapse. Each can benefit from simulations and stress testing.

Bisias et al. (2012) have completed a survey of broad approaches to systemic risk measurement in the scholarly literature, including in-depth examination of some specific models. This pioneering work provides code for alternative models to help researchers in the future check their performance over time.

Still, the appeal of this three-way classification system is that analyses within each family are likely to impose similar data requirements on any framework for monitoring systemic risk, whereas those in different families tend to have quite different implications for data collection and systemic monitoring.

The last section of this chapter draws together the implications of each analysis for data collection and for policy.

4.1.7 A summary table

We end this introduction with a summary table of policy issues and references. We hope this is useful for the reader who wants to probe deeper into current thinking about systemic frameworks, data and policy issues.

Policy issues

Name	Description	References
High level analyses		
Macroeconomic shocks	Financial firms are affected by macroeconomic conditions, and the crisis demonstrated again that the financial markets can feed back into the macroeconomy. There is a consensus on the need for macroprudential policy.	Borio (2011), Huang et al. (2009), Greenlaw et al. (2012), Borio et al. (2012)
Moral hazard	Deposit insurance gives banks funding that is not risk sensitive. This encourages over-investment in risky loans.	Allen et al. (2011), Demirgüç-Kunt and Detragiache (2002)
Bubbles	Asset bubbles are self-perpetuating price expansions. When bubbles pop, they can impose large sudden losses on investors.	Brunnermeier (2001), Evanoff et al. (2012)
Systemically important institutions	By definition, failure of a systemically important financial institution (SIFI) can threaten broader distress. SIFIs warrant special attention.	IMF, BIS and FSB (2009), Elliott and Litan (2011), Tarullo (2011)
Bankruptcy	Freezing the assets of a large financial institution in traditional bankruptcy proceedings traps creditors as well. DFA provides for an orderly resolution regime, yet to be tested in practice.	FDIC (2011a), FDIC (2011b)
Fire sales	Forced asset sales can artificially reduce prices, creating a self-reinforcing need for additional sales. Regulatory	Shleifer and Vishny (2011), Duffie (2010), Kashyap et al. (2011)

Policy issues		
Name	Description	References
	intervention may help alleviate such pressures.	
Credit crunches	Uncertainty about asset values dries up inter-bank lending and threatens global liquidity. Aggregate constrictions in credit can have broad impact on financial institutions, and can be self-reinforcing. Monetary, fiscal, and regulatory policy can help limit extremes.	Brunnermeier (2009), Tong and Wei (2011), Chor and Manova (2011), Mizen (2008)
Illiquidity	Regardless of how the underlying imbalances and fragilities arise, panics tend to play out as a rush for liquidity and safe assets. Regulation can require internal liquidity and assure access to emergency facilities, such as a lender of last resort.	CGFS (2011), BCBS (2012), Acharya et al. (2010), Tirole (2011)
Leverage cycles	Leverage rewards risk-takers in a boom, encouraging further risk-taking until a reversal of the process occurs. Regulators should monitor leverage and intervene against extremes.	Adrian and Shin (2010), Minsky (1982), Geanakoplos (2010a, 2010b)
Procyclical capital	Simplistic risk-based capital rules can force banks to raise capital in a bust, when they are most vulnerable, and when new capital is most costly. Capital buffers should be raised ex ante, when the economy is booming.	Hanson et al. (2011), Heid (2007), Kashyap and Stein (2004), Demirguc-Kunt et al. (2010)
Institutional and network analyses		
Network complexity	The financial network architecture is intricate and evolving. Supervisors need tools to assess this complexity and understand its implications.	Haldane (2009), IMF (2009, ch. 2), Cont et al. (2010)
Information asymmetries	Information asymmetries can create hidden exposures and allow regulatory arbitrage. These help increase imbalances and risk concentrations.	Allen et al. (2009), Brunnermeier (2001)
Concentrated default risk	Concentrated exposures can result in multiple simultaneous defaults. Regulation should discourage concentrations by industry, geography, market and asset type.	Gai et al. (2011), Rossi (2010)

Policy issues		
Name	Description	References
Contagion	Large defaults can propagate like dominoes through the financial system, and 2008 proved that the lender of last resort may not be enough. Other techniques of monitoring and intervention may help.	Eisenberg and Noe (2001), Billio et al. (2010), Moussa (2011)
Securitization markets	The originate-to-distribute model can diminish incentives by separating credit analysis from its consequences, while structured securitization can introduce complex chains of exposure that are difficult to unwind. A better understanding of the risk mechanisms and implications is needed.	Jaffee et al. (2009), Purnanandam (2011), FCIC (2011), Mizen (2008)
Opaqueness and ratings	Complexity (opaqueness) hampers investors' and regulators' ability to assess risk and valuations accurately, creating a role for rating agencies. The incentives for third-party agencies may not align well with those of investors and regulators.	Flannery et al. (2010), Cukierman (2011), Morgan (2002), Packer and Tarashev (2011)
Volatility	Volatility surprises can provoke failures and panicked reactions if firms are exposed and unprepared. Ironically, episodes of low volatility can encourage the accumulation of exposures and unpreparedness.	Brownlees and Engle (2011), Brunnermeier et al. (2011a), Bean (2011)
Financial innovation	Innovation in products and institutions can frustrate supervision and complicate investor due diligence. This can result in undetected microprudential and macroprudential risks.	Hu (2011), Flood et al. (2012), Adrian and Shin (2009), Pozsar et al. (2012)
Evolution and diversity	The populations that make up the financial system are continuously evolving. For those populations to remain healthy, policy has to encourage diversity and allow failure. But it also has to shepherd evolution away from heightened systemic risks.	Taylor (2011)

Policy issues		
Name	Description	References
Granular analyses		
Incentives	Institutions and individuals respond to incentives. If poorly designed, they can lead to suboptimal levels of risk-taking.	Richardson and Walter (2009), Acharya (2009)
Due diligence	Diligence is a costly and highly skilled activity. Low levels of diligence encourage negligence and fraud.	Clauss et al. (2009)
Terrorism	The September 11 attacks demonstrated the ability of terrorism to disrupt financial markets. Response will require coordination between financial regulators and national security experts.	DHS (2007), IARPA (2011)

4.2 High level analyses

Several macroeconomic analyses have focused on the business cycle.

4.2.1 Monetary policy missteps

It may be tempting to dismiss a purely macroeconomic explanation for the crisis, since the long build-up in home prices took place against a background of sustained growth and low inflation. Nevertheless, some respected analysts lay blame at the feet of monetary policy. John Taylor of Stanford has argued that, if monetary policy had been steadier and tighter in the middle part of this decade, the housing bubble would not have occurred and there would have been no crisis (Taylor, 2009). Perhaps there would have been a smaller recession at an earlier date, but that would have been a reasonable price to pay.

4.2.2 Macroeconomic imbalances

Another type of analysis with explanatory power focuses on macroeconomic imbalances. In particular, low savings rates in the USA and high savings rates in Europe and Asia brought funds into the US capital markets looking for yield but with cursory attention to risk. This inflow began in the 1990s and continued right up to the beginning of the crisis in 2006–2007. According to this line of analysis, this capital inflow drove securitization and was behind the deterioration in underwriting and other standards that eventually led to the collapse.

4.2.3 A credit cycle

Some analysts argue that booms and busts, whether they are minor or major, are nearly always credit driven (Rogoff and Reinhart, 2009). Low volatility in markets and rising prices causes investors to underestimate market and credit risks as a boom takes hold. Underwriting standards decline and lenders become increasingly tolerant of income projections that are dependent on future asset appreciation. In 2008, the explosion of securitization compounded this effect because many loan originators became indifferent to repayment prospects, whether they were based on forecasts of asset appreciation or not. But in any boom, this argument goes, at some point asset supply outstrips demand, price rises pause, sentiment becomes uncertain and then a reverse occurs and a bust ensues.

4.2.4 A leverage cycle

Other analysts put more emphasis on leverage than credit, particularly in systemically destabilizing cycles. When a boom is taking hold, leverage gives optimists more buying power and, at the same time, encourages a degree of recklessness because the downside risk is limited. The marginal buyer shifts from being a relatively less optimistic person to a relatively more optimistic person. Most importantly, leverage increases investors' purchasing power, accommodating increases in asset values during the upswing. When asset prices do hesitate or start to decline, the most leveraged and optimistic investors get wiped out first, leaving behind a pool that is on average less prone to believe in price rises anyway. At the same time, any asset price falls associated with non-recourse transactions that exceed borrower's equity will impose losses on lenders too, limiting available financing (and of course, making lenders a good deal more cautious). Asset price declines and declining availability of financing reinforce one another through changes in both lender and investor behavior – creating a powerful and damaging positive feedback loop.

The leverage cycle traces its ancestry back through Minsky's (1982) financial instability hypothesis (FIH) to Fisher's (1933) debt-deflation theory of the Great Depression. In particular, the FIH explains that early in a boom, the success of leveraged "bets" encourages additional leverage. More recently, this sort of leverage dynamic is laid out formally in the models of Bhattacharya et al. (2011) and Geanakoplos (2010a, 2010b).

4.2.5 Financial innovation and leverage

Much of financial innovation is designed to increase leverage one way or another. The positive characterization is that it puts capital to work. The negative characterization is that it makes it easier to gamble with other people's money.

Derivatives, securitization (and the accompanying advances in financial modeling) stretch scarce capital and collateral. This can be socially beneficial when they facilitate trade and investment in the real economy, or when specific risks are offset, diversified or shifted to those who are best able to hold them. Provided there are no implied guarantees as there have been in the recent past, shifting risk from insured depository balance sheets to private equity and institutional investors can be a good way of getting risk away from the core of the financial system. However, if the ultimate investor is highly levered, it may nevertheless be systemically destabilizing. Moreover, pyramiding can be destabilizing, where collateral is used to back a loan which is then used as collateral for another loan and so on.

Leverage was carried to an extreme in the recent crisis. Securitization was supported by an implicit government guarantee in the US housing market. CDS markets emerged in time to allow tranching and pyramiding to take hold on a grand scale. The ABS subprime market collapsed a bit before the housing market. Delinquencies had gone up and the CDS market was the only available vehicle for shorting the mortgage market – creating more supply and, incidentally, depressing prices and further spurring the housing market collapse.

4.2.6 Leverage and inequality

The leverage cycle also contributes to inequality. Rich investors can lever up earlier in a boom and ride it longer than poorer investors who have to wait for underwriting standards to decline before they can borrow in the upswing and are more likely to default and bear the associated costs during a downswing. Rich investors who survive the bust can then go bottom fishing. Since 2008, equity markets have roughly doubled in value which is perhaps an indicator of the fortunes of the well off, while many poorer households have lost their savings, are mired in debt and struggling under the threat of, or the actual effects of, foreclosure (Geanakoplos 2010a, 2010b).

4.2.7 A complex cycle

Other high-level explanations bring liquidity effects into their account of the cycle, on top of credit and leverage. For example, the authors of the Geneva Report "The Fundamental Principles of Financial Regulation," argue that as booms develop financial institutions expand their balance sheets, lower their funding costs by shortening borrowing maturities, causing liquidity ratios to worsen as well as increasing their leverage (Brunnermeier et al., 2009). Institutions that do not follow the trend are punished for under-performance. When the boom ends, assets prices fall, solvency becomes a concern, lowered profit prospects rule out raising additional capital and assets generally have to be sold. This creates a positive feedback loop that intensifies the asset decline and leads to wide-scale liquidity hoarding and withdrawals. Over the last couple of decades, the role of current market prices has increased, exacerbating both the boom and the bust, and risk management systems have become more sensitive to recent volatility, providing false comfort during the upswing and then precipitating widespread destabilizing activity during the subsequent downswing. That in turn meant that individual institutions have tended to demand more and better quality margin and collateral as volatility measures increased during the downturn, adding fuel to the collapse of liquidity and prices.

4.2.8 High-level indicators of systemic risk

In addition to the rise in asset prices and leverage, others have noted the concentration of corporate profits in the financial sector is an indicator of under-pricing of risk. Since the 1970s the share of US corporate profits originating in the financial sector rose to the point where it accounted for over 40% by the early 2000s. And over the same period, compensation in the financial system rose from about the average for the economy to about 1.7 times the average, a level not seen since the period preceding the Great Depression at the beginning of the last century. This suggests that any deviation from trend in financial sector compensation relative to compensation in the economy as a whole might also be a macroeconomic indicator of systemic vulnerability.

4.2.9 The interbank repo market

A somewhat different high-level approach is taken by Gary Gorton at Yale (Gorton, 2010). Gorton believes the cause of the recent crisis was the spread of "banking" outside of the banking sector. He argues that repos are the wholesale inter-institutional equivalent of cash deposits in the retail system. Before the crisis, the asset as well as the cash interchanged in a repo transaction had come to be viewed as informationally insensitive in the sense that MBSs and CDOs seemed to be protected by diversification from any shortcomings in the performance of the underlying and often unrealistically high credit ratings. The massive repo market provided liquidity to financial institutions in much the same way that deposits provide it for the real economy. However, in the absence of an equivalent of deposit insurance, there was the equivalent of a nineteenth century bank run in the repo market when diversification failed, and the strength of counterparties and the quality of assets fell into question simultaneously.

Gorton's analysis focuses on one aspect of the 2008 crisis that it shared with the Great Depression and no subsequent financial cycle. In his view, it was not a matter of degree but rather a matter of kind. 2008 and the Great Depression were close siblings and neither was related to anything that happened in between. This goes some way to explaining the severity of the 2008 crisis – and the effectiveness of the monetary and fiscal response.

4.2.10 "Too big to fail"

Another high-level approach for analysis and policy action is to concentrate on institutions that are "too-big-to-fail" – those systemically important financial institutions (SIFIs) whose failure would threaten to trigger a cascade of failures across the financial system. Stern and Feldman (2004) analyzed the economics of scale and Sorkin (2011) also examined the political aspects of SIFIs. Johnson and Kwak (2010) provided an historical analysis going back to the founding fathers that looks at the pernicious effects of the concentration of both economic and political power. In their view, President Jefferson, and later President Jackson, were right to distrust large financial institutions (Johnson and Kwak, 2010). During the last three decades of deregulation, large financial institutions have diversified and expanded their opportunities for risk taking, which in their view has just made the problem worse.

Their account treats the failure and rescue of Continental Illinois in 1984 as a watershed event. Since then, big financial institutions have been consistently "bailed out." Sometimes some shareholder wealth and all SIFI creditor claims have been protected by forced sales to bigger banks. Failures managed this way have exacerbated the problem of concentration. On other occasions, SIFI shareholders, creditors and managers have been protected from outright failure, as in the case of the repeated rescues of Citibank, most recently in 2008.[2] Either way, investors have been willing to lend to SIFIs at lower rates – conferring a potentially significant competitive advantage. Such strategies may have been understandable disaster management practices, but they foster extraordinary moral hazard in the long run.

According to this line of analysis, SIFIs influence the policymaking process directly through lobbying, and indirectly and perhaps more powerfully through their substantial impact on the way that policymakers think about risk and finance. This thinking has tended to emphasize the benefits of large financial institutions – economies of scale and scope compared with existing smaller competitors – while characterizing the problems – such as management information systems challenges and potential intra-institutional contagion – as things that can be managed. The indirect but very potent result is that their management and shareholders have

[2] Citibank shareholders did end up paying a price for the crisis. For the four years after the crisis, its stock traded down 90% from its average value for the preceding seven years. Shareholders in other large banks suffered similarly. Bank of America shares dropped roughly 80% in value from 2004–2008 to 2009–2012.

tended to enjoy the rewards of success in good times, while insurance schemes and taxpayers bail them out and bear the costs of failure when times turn bad.

In the debate over SIFIs, the term "bailout" is a loaded one. Provisions under the Dodd–Frank Act, if rigorously implemented, would not bail out shareholders, management, and most general creditors of a failed SIFI – just the opposite in fact. But many conservative commentators believe that designating some firms as SIFIs guarantees their special status and, when one of them is on the brink of failure, political pressures plus the uncertainties of even putting an institution through orderly FDIC liquidation would lead policymakers to guarantee general creditors. That would be enough, of course, to give SIFIs cheaper access to credit in the meantime.

Finally, in some of his work Markus Brunnermeier has also focused on systemically significant institutions. He and Tobias Adrian developed the concept of CoVaR to internalize systemic risks (Adrian, 2011). The idea is to estimate the incremental capital required of financial institutions related to certain risk factors such as size, interconnectedness, maturity mismatch, and leverage that predict the scale of the effect that the failure of the institution might have on the system as a whole.[3] These capital charges would create an incentive for institutions to adopt business strategies that reduced the risks they posed to others – as though fire insurance premiums reflected not only the costs of your house burning down but also the risk and the cost of your fire causing your neighbor's house to burn down too.

4.3 Network and institutional analyses

Other analyses drill down a little to look at the interconnections between institutions and how they function as a network. Some look a little deeper still at topics as specific as the price discovery process in a given market, the clearing process for a specific class of instruments, or the structure of a particular process that threads its way across the financial network.

4.3.1 Network analyses

Modeling the financial sector as a network helps us understand contagion and positive feedback effects. Typically, nodes are used to represent financial institutions and other market participants, and the links are used to represent bilateral credit or counterparty exposures between them. Other indirect relationships, due for example to a common exposure to a particular asset class or the danger of

[3] The difficulty that other firms would have in substituting for the services of a failing firm – the contestability of its services – has been suggested as an additional criterion of "SIFIness."

reputational contagion, could be represented through network analysis but, to date, these sorts of linkages have received less attention than credit links.

Measuring credit exposures is not straightforward: default correlations may rise and collateral values decline as systemic risks increase, resulting in net OTC derivatives exposures that are quite unstable. Insufficient capital or liquidity (or an operational breakdown) at a major node may send losses or margin payment demands rippling through the system so that other nodes face simultaneous stresses from links that are usually independent.[4]

The default of a single institution can start a cascade across the network, generating a total default impact equal to the cumulative losses of all of the affected institutions. Combining this default impact with an estimate of the probability of default of the first institution creates a measure of that institution's likely contagion impact. If the default impact is measured on just a set of core institutions that are particularly important systemically, then this measure becomes a good index of systemic significance – a systemic risk index (SRI). Like the degree of interconnectedness and in a way closely related to CoVAR, SRIs are distributed across institutions with a fat tail. This suggests that regulators can focus their attention on a few institutions and capture most of the potential for contagion.

This SRI analysis is not a dynamic model in all respects. Exposures, for example, do not change as successive institutions default in the model. Moreover, in the real world during a crisis, feedback effects can become very important quite quickly. Firms take actions in the short term as prices change – such as buying or selling credit protection – that can change the network configuration significantly. The contribution of contagion effects to the SRI varies but can be very high – more than 90% for some large institutions. In other words, interconnectedness is a major determinant of systemic significance.

Credit default swaps (CDS) modify the network topology, replacing one exposure with another by introducing contingent long-range links. Bought for insurance, they shift exposure; bought speculatively, they add a new exposure. Either way, they will likely greatly increase the small world character of the network (whereby there are numerous shortcuts across the financial system as a whole).

Central counterparties (CPPs) reduce credit linkages and the risk of contagion between banks and other financial institutions. However, they do so at a price. They become more systemically significant themselves, increasing the vulnerability of the system to any credit or operational problems that might arise in a CCP. They also

[4] If an institution is demonstrably solvent, it should be able to borrow using its net assets as collateral and so not be illiquid. But unfortunately, institutions are sometimes unable to convince lenders that they are solvent either because they cannot produce strong enough evidence, because the lenders do not want to hear, or because there is real uncertainty about the solvency of the institution.

reduce the scope for cross product netting, although this can be partly offset by collateral sharing arrangements among CCPs.

In "Rethinking the Financial Network," Andrew Haldane (2009) argues that the financial system is an adaptive network that, over the last decade, has become increasingly complex and homogeneous. The network structure became denser and more complex, as securitization proliferated and lengthened linkages and as OTC derivatives spawned a huge variety of conditional obligations. As large institutions individually sought to diversify their risks and their business lines, they came to resemble one another more and more. The market became both more robust when faced with normal shocks and peculiarly fragile in the face of greater stress. What had appeared to be a self-regulating and self-repairing system in the 1990s and early 2000s fell apart in the 2008 crash.

The final issues of network structure worth highlighting are redundancy and substitutability. In networks that exhibit redundancy, any two points are connected by many pathways that do not intersect. In a financial network with a high degree of redundancy there would, for example, be many quite distinct ways for a trade or investment transaction to be financed between two countries. Substitutability or ready contestability means that if a market or an institution fails, its services to other financial system participants and to non-financial customers are readily picked up by others. So, for example, when small banks in the USA fail, they are often bought by other neighboring institutions in purchases arranged by the FDIC over a weekend. Both redundancy and substitutability (or contestability) greatly add to the stability of any economic network – and the stability of the services that it provides to the rest of the economy. Examples of network models of the financial system include Billio et al. (2010), Nier et al. (2007), Cont et al. (2010), Haldane (2009), Haldane and May (2011) and IMF (2009, ch. 2).

4.3.2 Populations and evolution

The issue of population diversity has not been studied much in economics, let alone finance, but has been a major area of study in the natural sciences for a long time. It is potentially an important dimension of the financial network. Shocks, whether endogenous or exogenous, which strike a diverse population are less likely to lead to a general collapse than a shock striking a homogeneous population. If a large institution is replaced by a number of nearly identical small institutions, the benefits in terms of systemic stability are likely to be small compared to the costs in terms of lost economies of scale and scope – so policy strategies aimed at reducing the prevalence or importance of SIFIs need to be supplemented by policies aimed at encouraging institutional diversity. In principle, the importance of diversity for systemic stability applies to processes and industry practices as well as in business strategies and institutional structures. In practice, there is much work to be done to characterize, assess, measure and monitor different types of financial system diversity and to understand their implications for stability.

Amongst economic and financial historians, it is a common observation that financial systems are inherently unstable. In nature an analogous phenomenon arises when populations that have thrived in equilibrium with their environment for millennia sometimes suddenly disappear. Evolutionary theorists have developed insights into the origins of this sort of systemic fragility, the roots of which can be traced back to the basic processes of mutation, reproduction and selection – processes that are at work in the financial system. Some insights are readily available from thinking of the financial system as an evolutionary system – for example, that the continuing "death" of institutions and diversity amongst institutions can be desirable phenomena so far as the continuing "health" of the financial system as a whole is concerned. In time, evolutionary theory may help explain the emergence of financial system interconnectedness, complexity and periodic fragility. For the moment, however, there is a dearth of empirical work supporting these ideas, and it is unclear what the payoff would be from such a substantial shift in thinking about financial stability (Taylor, 2011).

4.3.3 Repo market processes

The recent report of the Task Force on Tri-party Repo Infrastructure provides an example of institutional analysis. This, however, is far more detailed and confined to a single market. "First and foremost, the Task Force has focused on specific actions needed to fundamentally strengthen the operational arrangements at the heart of the tri-party repo market. These actions are necessary to reduce the market's reliance on intraday credit provided by the Clearing Banks and clarify the credit and liquidity risks borne by market participants" (Hendricks et al., 2010).

In good times, the extension of intraday credit was merely an operational convenience. But in a sufficiently severe period of stress, these very large exposures became extraordinarily consequential and at the same time quite uncertain in an almost Knightian way, threatening to seize up liquidity in ABS, MBS and Treasuries. When one thinks of the dependence of the Federal Reserve on the tri-party repo market as a place in which to execute monetary policy, this operational convenience had serious systemic implications.

4.3.4 Securitization processes

One more example of an "institutional" analysis of systemic significance is a paper entitled "Understanding the Securitization of Subprime Mortgage Market Credit" (Ashcraft and Schuermann, 2008). In this analysis Ashcraft and Schuermann provide a quite detailed description of the pre-crisis subprime mortgage securitization process from origination through to distribution. They identify and document seven key points at which informational asymmetries and moral hazards arise – places where the separation of duty from interest eventually led to the breakdown of the system and to the subprime mortgage crisis. They discuss both predatory subprime borrowing and lending and also the process whereby the credit rating agencies assigned and monitored ratings for mortgage-backed securities.

This sort of analysis of the integrity of processes that stretches across several nodes of a network is familiar enough in engineering systems and network analysis. It is unusual in economics. It may have an important role to play in understanding the stability of financial networks in the future.

4.4 Granular analyses

Granular frameworks look at individual behavior at the portfolio, trading strategy, instrument and trade level. In principle, they can be more detailed than any other types of analyses. In practice, institutional and even high-level analyses have to date captured specific financial processes and market mechanisms more subtly.

4.4.1 Simulation-based analysis

Many of today's granular analyses would not be possible without agent-based simulation, because simulation provides a method for dealing with path dependency and stochastic behavior at the level of the individual. Also, detailed and comprehensive data on many aspects of the financial system are not available, at least for the moment. As more data are developed, either simulation will diminish in importance or, alternatively, simulation may be used to explore more specific policy issues and become more nuanced and realistic.

In their paper "Leverage Causes Fat Tails and Clustered Volatility" Stephan Thurner, Doyne Farmer and John Geanakoplos build a simple model of leveraged asset purchases with margin calls that depends on just one style of investment – value traders. Banks allow investors to increase leverage when times are good, but require them to reduce it when times turn bad. In the absence of leverage, analysis of simulations shows price movements are normally distributed. With leverage exacerbating selling pressure in a downturn and accommodating buying pressure in an upturn, the authors develop the fat tails observed in practice. There is no appeal to strategy-switching or irrational behavior to explain fat tails. For policy the implication is that something equivalent to a counter-cyclical loan loss reserve requirement for banks, applied to margin requirements, is worth investigating (Farmer et al., 2010).

Provided simulations are detailed and realistic enough to calibrate and simulate intelligently, they can provide useful policy insights. Bullock (2011) considers some of the issues. Farmer and Geanakoplos (2009) argue that the additional detail afforded by simulation techniques gives them an advantage over more traditional equilibrium modeling in answering many questions. In Farmer et al. (2012) we will see this proposition put to the test. Similarly, in Khandani et al. (2012) we see the results of a more focused systemic simulation emphasizing the housing sector.

4.4.2 Simulation as a tool

One simulation-based perspective on systemic risk is to develop a system dynamics model comprising several subsystems for different parts of the economy. Such a model would link these systems in a way that would allow an analyst to simulate the workings of different theories and the effect of different policy options. Unfortunately, the data requirements of this kind of analysis are sometimes large, if the results are not to be hard to interpret and sensitive to assumptions. Still, the advantage of such a simulation strategy is that it can be used to explore the implications of almost any theory or policy.

Using this approach, one place to start to understand the recent crisis would be to simplify the economy into just three linked systems – a banking system, a housing market system and a macroeconomic system.[5] The banking system would be characterized by stocks such as cash balances, mortgages and capital and flows such as mortgage lending and repayments; the housing system would be characterized by the stocks of available houses, foreclosed houses, housing sector workers and available buyers and flows between the different states of housing and the new creation of homes, employment and income; and the macroeconomic system would be characterized by macroeconomic variables such as unemployment, interest rates and gross output. The three systems would then be linked by flows, stocks, prices and elasticities. It should then be possible to test system sensitivity to structural and parametric changes as well as to exogenous shocks. An et al. (2009) have made a start at exploring such a systems dynamics model with a simple model comprising housing, banking and government sectors.

One of the strengths of this approach is its extensibility. While this particular three-subsystem model does not account for many important aspects of the economy, the general approach is amenable to almost unlimited elaboration. Heterogeneity of investors, households, or businesses; the impact of leverage; the formation of price

[5] If this seems like a gross over-simplification, it is worth remembering that, extraordinary as it may seem, most macroeconomic models used for policymaking over the past 50 years did not include the financial sector. Interest rates, for example, were determined in markets where savers and investors in plant and equipment met without the benefit of banks, mortgage companies, institutional investors or broker-dealers.

expectations; the effect on the behavior of households of going underwater; the effect of adding other non-financial sectors; and the dynamics of a bank run: these are all examples of aspects of the economy that might be built into a system dynamics approach.

Even in its most basic form, this sort of system dynamics model can be used to illustrate how lags at different points within and between each of the three systems – banking, housing and macroeconomic – can generate cycles that are potentially unstable. For example, it takes some time for builders to meet new demand for houses. Mortgages are financed and homes are bought. Demand is met. House building drops. Unemployment picks up. Some home owners cannot meet their payments. Banks foreclose. Insolvency becomes a possibility for banks too. The characteristics of the positive feedback loop between employment, demand and the financial sector can be explored because each of these steps is explicitly represented in the model. Even this simple a model can create multiple equilibria showing how, in response to a shock, the economy can shift from a locally stable high level of activity to a locally stable – and much less desirable – low level.

4.4.3 Transaction level data and the LEI

To observe and understand the mechanics of the financial sector in detail and to see at the most granular level the way in which changes in behavior and relationships play out across the financial system, it would be useful to have much more entity and transaction level data. For example, position and market data on most instruments and entities in the capital markets could be valuable to a systemic risk monitor. It could also be useful to identify every entity uniquely – such as structured investment vehicles, special purpose vehicles and subsidiaries – and to understand the conditional obligations, rights and triggers associated with individual instruments and contracts. This could inform how liquidity might sometimes drain out of parts of the system, for example, and how relationships between entities might change in periods of financial distress. It could also be useful to have granular data from outside the capital markets, for example, to detect early signs of potential stress such as new processes emerging akin to the originate-to-distribute process or changes in the composition of populations like the changes in home-owner populations and in the types of mortgages.

A first step in this regard has been taken by the OFR and the FSB in promoting the development and widespread use of unique legal entity identifiers (LEIs) for different entities engaged in transactions in established financial markets. Today, different firms, markets and regulators have different conventions and systems, making the use of data from multiple sources for firm level risk management or for systemic risk management difficult, time-consuming and unreliable. During the resolution of Lehman Brothers, it took the receivers many months to establish how many entities comprised Lehman and since then much of their efforts – perhaps most – have been devoted to sorting out what the claims between these entities are, as opposed to what the claims of third parties are on Lehman.

4.4.4 Behavioral instability

Another type of granular approach to systemic issues is to analyze the behavior of individual actors in the financial system. Herd behavior during a panic is clearly something of which policymakers have been acutely aware since the Teapot Dome Scandal and the Tulip Bubble. And much more recently, there has been a good deal of work done in behavioral economics and the intersection of neuroscience and economics to understand ways in which mental functions and real human behavior are at variance from the naive predictions of Homo Economicus. Given the importance of confidence and expectations in financial system stability, it is quite surprising that more has not been done to understand behavior during times of stress generally, and financial stress in particular.

This is a potentially fruitful line of enquiry not just for market participants. We know that in periods of financial excess, financial regulators have often found it difficult to take timely corrective action. This is the punchbowl problem – how to encourage regulators to act together to take away the punchbowl from the financial party (Martin, 1955). At one level, it is not difficult to sketch the main dimensions of the dynamics at work here: the prevailing attitudes of optimism during these periods; the difficulty of identifying the precise moment when action should be taken; uncertainty about policy calibration; the danger of legal and political challenge; and the complexities of policymaking and coordination mechanisms that have developed historically and now delay policy responses. Still, there are deep legal and cultural barriers within and between regulatory agencies in different parts of the world that inhibit resolute and timely action. These are fundamentally behavioral in nature and it is possible that behavioral economic insights might help senior policymakers and politicians address these cultural issues.

4.5 Policy and data implications

4.5.1 Why policy is hard

Despite the legislative and institutional changes affecting the regulatory framework, it may well prove difficult to get regulators and policymakers to focus on systemic stability. Many regulators believed, and continue to believe, that such a focus is not really necessary. They think that the best defense against systemic risks is to focus regulation on the safety and soundness of individual institutions. And, indeed, toughening regulations on institutions is an important part of the policy challenge. But strong institutions and robust markets do not necessarily mean the system as a whole is stable and resilient (Danielsson and Shin, 2003). Given that positive feedback loops may drive rising margin calls and asset fire sales and rip through capital buffers of any size, it is not an answer to say that banks must simply have more and more capital. Likewise, institutions must depend on one another for liquidity. No liquidity buffer will ever protect all institutions from a sufficiently protracted period of general liquidity hoarding. Still, there is a tendency to revert to the viewpoint that more capital and more liquidity is such a large part of the answer that we would do well just to get that far and anything more is of second order importance.

Moreover, even where there is deep regulatory commitment today, any policy innovation that addresses infrequent risks is bound to be hard to sustain. The costs will be ongoing and memories will fade about the benefits. There may be more appetite for monitoring risks than for preventive measures. But even something as innocuous as data collection could well provoke sustained opposition. Vendors and the industry itself make money from selling information. Privacy advocates are suspicious of any government department collecting more data. And the costs to the industry, especially in the near term, may not be trivial.

Beyond that, to meet the second challenge of policy at all – to shepherd the system away from heightened systemic risks – may be very difficult politically. In the past, it has been difficult for regulators to "take away the punchbowl" when almost everyone involved is enjoying the party during a credit expansion and an asset bubble. Bank regulators have typically faced widespread resistance from the financial industry, politicians, and economists who want the expansion to continue. Looking forward, some respected policy experts have expressed skepticism about the ability of the FSOC in the USA to ever invoke the counter-cyclical capital buffer built into the new Basel Committee standards and the Dodd–Frank Act, simply because they cannot imagine the US Treasury Secretary and all the regulators that make up the FSOC being able to reach agreement to act forcefully at a single point in time.

4.5.2 The evolving status of data

Different requirements at different times

The requirements for monitoring financial system stability change, depending on the condition of the system. In normal times, data monitoring has to help define and calibrate policy options, provide early warning of changes in condition and support research on system dynamics. When times become abnormal and the financial system fragility increases, uncertainty is rising and the time available for policy impact is falling so that data needs grow more intense. Then, in a crisis, the primary need is to monitor more frequently the state of the system – to know what exactly is going on – the shifting exposures, vulnerabilities, interrelationships and behaviors across the system. This includes running "what if" scenarios so that policymakers can understand the likely consequences of doing more or less, more quickly or more slowly, to the various actors in the financial markets

An historical perspective

For decades, regulatory agencies in the USA and elsewhere have required markets to report detailed price, volume and other condition data and have required depository institutions to report balance sheet and income statement data. The recent crisis showed, however, that these data were insufficient to reveal systemic risk buildups. For example, Call Report data in the USA were not detailed enough to reveal the growth of exposures to subprime and exotic mortgages. They were not comprehensive enough to reveal shifts in risk created (or reduced) by credit default swaps (CDS) and mortgage-backed securities (MBS). Moreover, they were not consistent enough, with different enterprises – or even different businesses or desks within an individual financial institution – reporting counterparty exposures and other risk concentrations differently, frustrating risk aggregation for regulators and risk managers alike. While brokerages and insurance companies also report some information on their condition to regulators, reporting from other parts of the financial system has been far less comprehensive – hence the expression "the shadow banking system" to describe money market mutual funds, private equity, venture capital and hedge funds. Whereas shifts in risk used to be associated with an initial cash flow in banks, nowadays risks are routinely taken outside the banking system – or inside but without any cash changing hands.

After the Great Depression new statistics were created on GDP, inflation and unemployment that became very hard for policymakers to ignore. Perhaps comparably powerful financial sector data frameworks can be developed going forward. To the extent that such data can be accessed widely, there is a chance that over time they will become the foundation of a strong research ecosystem, attracting new talent and helping to drive the interplay between data and theory toward convergence on a general and reliable framework for systemic risk management.

Data and their effect on behavior

As in particle physics, it can be difficult to observe behavioral systems without eliciting a response and it can be even more difficult to avoid behavioral feedback if what is observed is then publicly reported. For example, FICO scores began to be manipulated by mortgage brokers as soon as lenders started to pay attention to them. Also, information on the deteriorating state of an institution may accelerate that deterioration if, for example, it leads others to withdraw credit. That is a challenge for systemic risk data collection and representation. Not only must it be dynamic in the sense that what is collected and analyzed has to evolve in parallel with the evolution of the financial system, but also policymakers need to be alert to the possibility that collecting certain sorts of data will affect the behavior of individuals and institutions. Behavioral feedback can be gradual and adaptive – the adaptation to FICO scores – or it can be sudden and non-linear – the response to news about an institution in distress.

The role of the Office of Financial Research (OFR)

The OFR has crucial collaborative roles as risk monitor, research sponsor and data utility. While the Dodd–Frank Act puts the OFR right in the middle of systemic risk data collection and analysis, it will have to coordinate and collaborate effectively with the other federal financial regulatory agencies if it is to achieve its potential. The Act requires the OFR to monitor system stability, standardize and collect basic source data on financial transactions and entities, conduct short- and long-term research on systemic risk and develop tools for risk measurement and monitoring. The other federal financial regulatory agencies – particularly the Federal Reserve – already have responsibilities for parts of these activities. The OFR will have to fill in the gaps and lay the foundation for the development of strong macroprudential policies. This will involve:

- establishing reference databases and standards of reporting for legal entities and financial instruments;
- defining and starting to monitor a core set of risk indicators – exposure maps, capital and liquidity trends; leverage and credit cycles; developments in concentration and trends in asset prices – with a view to monitoring the progress of and threats to financial recovery; and
- developing tools for data collection, management, visualization, analysis and reporting.

In time, establishing reference data and standards for granular data reporting may well reduce the operational costs and risks for individual markets and institutions.

While the Dodd–Frank Act requires the OFR to set standards for entity and transaction data collection by the federal financial regulators and empowers it to collect these data, it does not actually require the OFR to collect all of the data all the time. Still, over time, it may be cheaper and easier for institutions and markets to create systems to report reliable source data comprehensively than to prepare specific data summaries and aggregations to meet changing regulatory requirements.

A catch-all strategy

A catch-all approach to collecting new data is one option. In principle, granular data can then be aggregated in a variety of ways and, if it is well organized and feeding sophisticated analytical engines, this aggregation could be carried out in response to a demand to analyze a new emerging risk, or a sudden potentially disruptive market development, or to monitor a crisis as it unfolds. There is a danger, however, that so much data becomes available that regulators are overwhelmed by it, even when data standardization has become well advanced. Understanding terabytes of new data every hour generated by a rapidly evolving system would present some daunting challenges. What broad direction can policymakers and economists give to the IT community to help them begin to prioritize and to organize large amounts of data in useful ways?

New data: a two-step approach

These questions begin to be addressed in Brunnermeier et al. (2011b). They suggest a two-step approach to risk monitoring which would simplify not only data collection, but also data distribution and use. They suggest using institutional risk managers to estimate the first round effects of shocks and then employing a combination of regulators, academics and industry experts to look at the second round effects – the interactions between banks and other market participants that can amplify institutional risks into systemic crises.

For the first step, regulators need to specify stress tests and ask risk managers to estimate what happens to the value of their firms if, say, house prices decline by 10, 15 and 20%. Similar questions should be posed for interest rates, exchange rates, growth rates and so on. Using multiple values for the scenario will allow regulators to get some idea of the deltas, and some cross-scenarios (involving say interest rates and exchange rates) should be run to get at co-variances. Scenarios should cover liquidity risks as well as credit and market risks – such as the impact of a doubling of margin requirements – and scenarios should generate estimates of the impact on both solvency and liquidity.

After risk summaries have been estimated for a number of firms, the second step would be to explore how each of these firms would then react to the loss in value and liquidity of its counterparties – and what the consequences of those reactions would be. So for example, would a loss of 10% in house prices lead to so many foreclosures that prices would be further depressed – a positive feedback loop with negative consequences.

Tests should be designed not only to reflect changes in system circumstances but also to track the impact to certain standard stresses – a set of core scenarios – over time to build up panel data, to provide good data for general equilibrium research and to give some idea of trends in systemic stability. This set of core scenarios would be one difference between such a program and the SCAP – the February 2009 stress testing exercise conducted by the Federal Reserve Board and other federal banking agencies. The other difference would be coverage: it should cover liquidity as well as solvency and do so for most of the financial system and not just the top 19.

4.5.3 High level analysis implications

Macroeconomic imbalances

For some macroeconomists, managing systemic risk requires nothing more than judicious use of well-established policy instruments. John Taylor of Stanford has argued that consistently following a rule for monetary policy, avoiding intemperate government intervention in the housing market and implementing higher standards of fiscal discipline would have limited imbalances and constrained the bubble that led to the 2008 crisis.

Others, however, believe that additional policies are needed to address systemic risks for two reasons. First, structural aspects of the financial system can change independently of such broad policy tools and, unattended, can contribute to financial fragility over time. Second, monetary policy cannot target two independent price levels at once, whether it is via short-term interest rates or the money supply. Monetary policy can either stabilize the general price level or it can stabilize asset prices, but it cannot do both.

Imbalances caused by large and underpriced international capital inflows call for a different kind of policy response. International policy coordination is required so that countries with savings surpluses stimulate domestic demand and countries with savings deficits curtail public and private consumption. The IMF and other international agencies and observers advocated such coordination in the years leading up to the crisis, but neither surplus nor deficit countries had the political will to respond.

Measuring macroeconomic developments

As a general matter, macroeconomic policy frameworks for macroprudential stability call for very little if any additional data collection and analysis. Traditional tools tend to work based on readily available data.

Leverage and credit cycles

Leverage and credit cycles create challenges for policy of both the first and second kind – how to toughen the system; and how to shepherd it away from heightened systemic risks.

An example of the first kind of policy response is to take steps to toughen up financial supervision, making it more difficult for underwriting standards to decline in future booms. Directly or indirectly, financial regulators can influence underwriting practices across most financial institutions. In the USA, for example, regulators not only supervise bank underwriting practices directly but also can influence non-bank lender practices by influencing the standards for securitization and other forms of financing that come through the markets or from the banking industry. The problem is that, during a boom, it has been very hard for supervisors to insist on maintaining high standards when it appears as though risks are falling and there are strong competitive pressures for institutions to expand their lending. It is difficult to sanction a bank for poor underwriting when a high proportion of its loans are performing.

To address this, bank regulators are beginning to look at ways to monitor and measure the quality of underwriting more objectively. This should make it much harder for politicians and the industry to argue for or pursue lenient practices with impunity. Some of the data needed will be available within the regulatory agencies and the challenge will be to aggregate them into meaningful indicators. Other data will likely require new collection efforts.

Since the crisis, policymakers have been developing and in some instances implementing other policies aimed at meeting the second kind of policy challenge for containing credit and leverage cycles – leaning against the wind by reducing leverage and making credit more expensive when a bubble is beginning to build up. For example, counter-cyclical capital requirements for banks are now a part of the Basel III international capital regime. Under Basel, additional counter-cyclical buffers can rise to an additional 2.5% of risk-weighted assets. Countries are expected to raise capital requirements during an upswing to inhibit credit expansion and reduce them during the downswing, to slow down the contraction in credit. Spain has experimented with counter-cyclical capital requirements with some success. Many other countries, including the USA, have introduced legislation and regulations to allow their regulators to use them.

It is unclear how widely counter-cyclical capital requirements will be used. Many policymakers are uncertain about when to invoke them and how to calibrate them. And they also worry about the fact that they are such a blunt instrument. So, for example, not all US banks expanded subprime lending and related businesses in the run-up to the crisis. It would have been unreasonable and inefficient at that time to have imposed an additional capital requirement on all banks, regardless of their involvement in housing, if the aim of the policy was to inhibit the housing market bubble.

Some countries have experimented with a more targeted alternative. Maximum mortgage loan-to-value ratios have been lowered in China and Israel to fight real-estate bubbles with some success. To have an impact it appears that the changes have to be quite large – of the order of 20–30%. In principle, the same approach could be applied to margin requirements in capital markets so that, potentially, there is a range of macroprudential tools available to policymakers to manage the credit and leverage cycles. The Bank of England issued a discussion draft early in 2012 outlining different possible macroprudential tools (Bank of England, 2011). In the

USA, SEC and CFTC policymakers have sway over margin requirements in capital markets. Since the Dodd–Frank Act become law, their efforts at creating new policy have been focused on migrating over-the-counter derivatives to exchanges, central counterparties, and registries – measures intended to reduce interconnectedness and uncertainty in times of stress regarding the pattern of counterparty exposures across the financial system.

Regulating bank balance sheet leverage, bank loan-to-value ratios, mortgage loan-to-values and capital market margins counter-cyclically will not completely solve the problem. The more aggressive these policies, the more that leverage will tend to migrate to the shadow banking system when times are good – that is when the risks in the shadow banking system seem contained – to an extent that, for the moment, is unclear.

Measuring the leverage and credit cycles

Data are needed for early warning and tracking of cycles and for calibrating policy responses as they unfold.

Some early warning indicators are available today. In the most recent crisis, amongst the median leveraged home buyers in the USA, leverage peaked at about 37-to-1. Average margin levels and requirements are available from many capital markets.

Banks look at borrower leverage on a loan by loan basis when setting initial collateral requirements. They also track leverage and other condition indicators through to loan maturity. It would be difficult to aggregate and interpret these data, but most of them are created in some form or other, at least within the banking sector. And then of course there is a good deal of information about the leverage of banks themselves – the ratio of their capital to their assets – and this is available for many different definitions or capital and adjustments to assets to reflect their riskiness and incorporate proxy measures for exposures from derivatives and other non-balance-sheet risks.

More generally, leverage can be attached to transactions or to the balance sheets of individuals and institutions. Some economists have suggested that aggregation across transactions will be easier for the purposes of tracking trends. However, to relate these trends to systemic vulnerability would seem to require instrument-type leverage to be related to financial institution vulnerability, which might be just as difficult as tracking institutional leverage in the first place.

Trading CDS on exchanges wherever there was sufficient liquidity might also help in interpreting leverage to the extent that it could help estimate the share of leveraged investment conducted by pessimists and optimists.

More work has been done to track credit than leverage cycles. While there are gaps resulting from private lending and there are no doubt quality issues on individual series, there are data on the extension of credit from most of the financial system. However, it is surprising how often high level trends in credit expansion are not analyzed in conjunction with any analysis of asset prices and prospects for asset earnings. Mortgages, rents and house prices were tracked and analyzed together by some economists in the recent crisis, notably Robert Schiller (Akerlof and Schiller, 2009), but most economists paid very little attention to the relationship between rents, incomes and house prices.

Managing liquidity cycles

Thinking about liquidity cycles has not progressed as much as for credit and leverage cycles. International capital standards for banks have been around for over 30 years whereas international liquidity standards – and national ones too – are less than five years old. So modifying as-yet-to-be-developed liquidity standards counter-cyclically is a brand new topic of current conversation amongst regulators.

An institution is liquid if it has enough liquid assets to hand to be able to meet future contractual requests for funds on demand. A holding of assets is liquid if it can be sold for cash at short notice without changing prices in the market. The Basel Committee has proposed standards for two bank liquidity ratios: a short-term one – the Liquidity Coverage Ratio (LCR) – and a long-term one – the Net Stable Funding Ratio (NSFR). Almost all regulatory efforts thus far have been directed at establishing the details for the LCR, the availability of liquid assets divided by the expected requirements for liquidity over the next 30 days. The NSFR is the same ratio – of available liquidity divided by requirements – measured over the next year rather than the next month. While the Basel Committee has set a minimum value for both ratios at 100%, the exact measurement of available and required funds has yet to be worked out in detail (Basel, 2010).

However, a standard of this kind can be highly procyclical. Penalizing a stressed institution that might be subject to a liquidity run could precipitate the failure that the policy in more normal times is intended to avoid. Equally, if markets are becoming stressed, there is likely to be a tendency for institutions to hoard liquidity anyway. If banks know that their regulators require them to hold some minimum amount of liquidity, this tendency may become more pronounced, aggravating hoarding.

Liquid assets are not just used to meet funding requirements by institutions. They are also used extensively as collateral. So, declines in the availability of acceptable liquid assets may have a doubly negative effect by reducing leverage. For this reason, one more policy that may be worth considering is that central banks should reduce their collateral requirements when lending to banks during periods of general financial stress, thereby offsetting somewhat the contradictiory effects of declines in liquidity/collateral availability among private financial institutions. Equally, central counterparties might reduce their rates of calls for member capital during contractions. Both policies, of course, depend upon setting standards that are sufficiently conservative during normal times, in much the same way as counter-cyclical capital buffers can offset credit contractions prudently only if the balance of capital remaining once the buffers are used up is still sufficient for safety and soundness purposes.

Measuring liquidity

Which assets should be considered liquid is a current policy issue. If that can be settled, the LCR and the NSFR will provide two measures of funding liquidity that could be aggregated or averaged across a financial system. However, each of these measures is only about liquidity at a single future point in time. An alternative measure which captures funding risk through time is maturity mismatch – the sum of positive differences at different time horizons between liabilities that will have come due and assets that will have matured. Maturity mismatch measures can also be averaged or aggregated to track changes in system liquidity.

However, the liquidity of everything except cash is always in doubt. It depends on the effective demand for an asset, which in turn depends on a variety of factors including the concentration of market participants, the financial health of potential buyers, market size, the size of recent trades and market volatility. (Even government debt may not be liquid if the creditworthiness of the government is in doubt or it has issued so few securities that they do not trade much.) All of these are elements of asset liquidity that should be monitored and factored into overall assessments of liquidity trends.

With the advent of liquidity standards, it is quite possible that aggregate demand for liquid assets for funding liquidity, combined with the demand for assets for use as collateral, will run ahead of the available supply. Although in some circumstances this may have a desirable restraining effect on leverage, it could also cause or exacerbate a shock to the financial system. These possibilities suggest that systemic regulators should also track the aggregate supply and demand for liquid assets across the real economy and the financial system.

Liquidity measurement presents another challenge that credit and even leverage will exhibit to a lesser extent: that is, volatility is very unpredictable. When liquidity disappears, it can do so very rapidly because of the positive feedback effects caused by collective changes in perception. Liquidity seems to present an extreme example of the past being a poor guide to the future. Therefore, it is arguable that much higher frequencies of data collection might be warranted to monitor and manage liquidity, to reduce the chances of small adverse developments appearing for the first time and then maturing to full blown crisis between observation times.

Runs in wholesale markets

Gorton's analysis suggested that the unique aspect of the 2008 crisis that explained its severity was that there was a run in wholesale financial markets – the equivalent of runs on deposits before the advent of deposit insurance (Gorton, 2010). His policy prescription is that senior tranches of qualifying securitizations should be government insured. The quid pro quo would be that the choices of asset class, portfolio and tranching should be subject to regulatory examination, just as depository institutions are examined today. Attractively, a rather general characterization of the problem leads to a rather specific cure.

Measuring vulnerability to wholesale runs

The data requirements for measuring and monitoring securitization markets to support government guarantees might look very like the data requirements that are emerging for the wholesale mortgage market today in the USA. These are aimed mostly at containing the operational risks created by low standards in mortgage servicing. Still, standardization and the requirements for far better data management at individual institutions would be likely to meet a good part of the needs for wholesale market government guarantees.

"Too big to fail"

Johnson's analysis leads to the conclusion that today's SIFIs should be broken up; anything else would be unconvincing and ineffective (Johnson and Kwak, 2010). Others take more nuanced views: that possible economies of scale should be weighed against the costs of raising systemic risks; and that incentives, such as capital surcharges on systemically important institutions, should be used to discourage increases in scale beyond a certain point – but where exactly that point is is uncertain, while how institutions should reduce their scale below it should be something left to the markets to decide.

Capital surcharges have in fact been introduced but not yet implemented in most countries. Basel III calls for global systemically important banks (G-SIBs) to be subject to a capital surcharge of between 1 and 3.5% not so much to encourage management to break up their institutions to realize shareholder value as to make the institutions stronger in the face of any shocks to the financial system. Basel is also setting guidelines for how countries should apply heightened standards of regulation and supervision to institutions that are domestically systemic – so-called D-SIBs – with the same aim of strengthening key financial institutions and the system as a whole.

A potentially important form of additional capital is contingent capital. It is an attractive idea to have a built-in capability to reduce debt and increase capital when it is most needed. Contingent capital convertible bonds should be cheaper than equity but more expensive than debt. Equity should be diluted on conversion, to give a share-compensated management team an incentive to keep the institution strong. To minimize the chance of market manipulation by short-sellers, the trigger that causes conversion must be designed correctly. Potential options include triggers that are automatic and related to an equity price of the individual institution or of, say, the top ten banks, or that are at the discretion of prudential regulators. For the moment, US and international policymakers have drawn back from requiring banks to issue contingent capital, preferring to see what sorts of options are developed by the private sector. Expectations are low because, without a regulatory capital offset, the additional costs of contingent capital may outweigh any overall advantage in funding strategies.

The second element of the US and the international strategy to address "too big to fail" is orderly resolution – whereby regulators have been given the power to wind down or break up failing SIFIs without disrupting markets. For US SIFI banks, a key element of this is that the FDIC requires US SIFIs to prepare for their own failure – to develop a liquidation plan. If this plan is inadequate, it can require several sorts of changes up to and including requiring the institution to change its business strategy, reorganize its legal structure or even downsize and withdraw from specific business lines. The importance of these powers is that certain knotty problems, like international cooperation over resolution and aligning the legal structure of large complex financial institutions to their business structure, become shared problems – ones which the institutions have to solve as well as their regulators.

The FDIC strategy is, roughly speaking, to recapitalize a failing institution by wiping out shareholders and paying long-term creditors cents on the dollar so that assets are then worth more than remaining liabilities. They have also been clear that they intend to fire the Board and all senior management in the event of a SIFI failure. After that, they can use a variety of policy tools over time to break an institution up. Until they have actually executed this strategy on a large institution in trouble, skeptics will question whether the political and economic environment will allow it.

One policy response to the crisis has exacerbated the "too big to fail" problem. That is, Dodd–Frank and legislation in other jurisdictions have encouraged, and in some instances mandated, derivatives dealers to use central counterparties for clearing instead of clearing them bilaterally "over the counter." The effect of this is to make central counterparties more systemically important. Regulation to increase their robustness and resilience is under development, but the fact remains that financial system stability will be much more vulnerable to disruptions at central counterparties in the future than it has been.

Measuring "too big to fail"

A first step in all of these policies to deal with "too big to fail" is to establish criteria for institutions to be designated systemically important, establish indicators for each criterion and then to set thresholds above which an institution should be designated. Combinations of size, connectedness to the rest of the financial system, and low contestability of key businesses and activities have all been widely accepted as important criteria. Some have also argued for a criterion of complexity: more complex institutions should be considered more systemic, because complexity may raise the chances of failure and the difficulty of containing its effects.

4.5.4 Network and institutional analysis implications

Networks

The main policy implication of CoVaR and Systemic Risk Index analysis of the network of SIFIs is that minimum capital standards for banks should be higher than they have been to date. Moreover, the two methodologies do give a rationale for which institutions should be considered SIFIs and then which ones should hold large SIFI surcharges and which should hold smaller surcharges (Adrian and Brunnermeier, 2011). They are less clear in their implications for the absolute size of surcharges but, assuming they were not insignificant, they could have two implications for the network structure: either key nodes would be strengthened – "superspreaders" be inoculated – making the network more robust; or they would induce SIFIs to change their business strategies and perhaps divest businesses, altering the configuration of the network. If this led to more contestability and to more options for "crossing the network" then it would also lead to greater network stability.

Increasing capital at "superspreaders" was the first of Haldane's three main policy prescriptions (Haldane, 2009). The second was to propose more data and better communications about network dynamics, so regulators could "map the global financial network and communicate to the public about its dynamics." This should have two effects: in normal times, to heighten private market sensitivity to systemic dependencies; and in abnormal times to quell the spread of panic based on wild speculation. His third proposal was to restructure the network through more widespread use of central counterparties and intrasystem netting arrangements, to reduce the financial system's "dimensionality," that is, to change the shape of the power law governing the size of financial collapses (see also IMF, 2009 and ECB, 2010).

The strength of CCPs depends on the strength of their members, the quality of their risk management, their conservatism and their own capital structure. Typically, a CCP will minimize possible losses by setting standards for membership and then monitoring the behavior and the strength of its members. It will require members to post collateral against any open positions. And then, in the event of loss, it has its own capital to run down, following which it can impose a draw on its members to cover any loss. This waterfall of options will make CCPs strong in the face of credit or counterparty risks if the details of the various processes and policies are worked out and executed proficiently. Ensuring they are well run is a newly important policy priority for IOSCO, and the national central banks and regulatory agencies that directly oversee them.

Measuring network structure and dynamics

Network mathematicians have developed several summary measures that can be used to characterize and track the complexity, density, and other structural features of networks over time. Some of these are likely to be useful for monitoring systemic stability in the financial network, and work is just beginning on identifying which. Other measures, like the aggregate exposure of large institutions to one another and the relationship of these exposures to their capital, are quite simple by comparison and very useful in understanding network vulnerabilities, but even they pose serious data collection and management challenges, particularly when the institutions and the network are global. National laws govern who may collect, manage and see different sorts of data and some data will be commercially sensitive, so that data security and access policies become matters for protracted international negotiation. A network perspective leads to the insight that an institution will be systemically significant if its counterparties are large, leveraged and susceptible to its default and are themselves connected to many other institutions. Analysis of chains of exposures-to-capital ratios will shed light on the importance of these sorts of possible knock-on effects.

Evolution

Thinking about the financial system as a set of populations that co-evolve has several immediate policy implications. First of all, diversity should be encouraged and homogeneity should be discouraged as much as possible. Not only does homogeneity of bank business strategies, for example, make them more vulnerable to a common shock, but they are also more likely to engage in herd behavior if their businesses are organized similarly and their risk management practices are similar (Taylor, 2011).

This general observation is a reason to be cautious about excessive regulation. Too many rules will tend to force institutions to become similar. By contrast, supervision, which can accommodate local differences in business circumstance and strategy, is capable of accommodating, if not positively encouraging, diversity.

The second implication is the importance of fostering evolution. When the environment for a population is changing – the technological, business and consumer environment in the case of the financial system – and the population does not evolve with the changing environment, then the odds of catastrophic failure will increase over time. This points to an additional cost associated with CCPs (other than the concentration of risks in a simple node of the network): they tend to freeze in place a particular set of network arrangements that may be very difficult to change in the future, and thereby may have the side-effect of inhibiting evolution.

Fostering evolution is a little different from fostering innovation, although that is part of it. Evolution includes the idea of death as a positive phenomenon, making way for change. This leads to the third implication: successful regulation is not about preventing financial failures but rather ensuring as far as possible that when institutions do fail they do so without unduly disrupting the smooth working of the system as a whole. If any extra support was needed, evolution provides it for policies that allow failing "too big to fail" institutions to be resolved, broken up and sold off rather than married to other large institutions and thereby laying down further difficulties for the future.

Measuring evolution

Techniques and measures of population diversity were first developed for tracking biological populations several decades ago. Economists and financial analysts have shown far less interest in diversity – partly because their neoclassical training has encouraged them to think of all populations as approximately homogeneous. But when policymakers are concerned with population stability, then diversity becomes an important aspect of the financial system to characterize and track. Portfolio, business strategy and risk management diversity are all candidates for measurement. So too is risk concentration. It would be worthwhile tracking aspects of institutional policy that are closely regulated, to be sure as far as possible that tightly implemented regulations do not inadvertently create undue uniformity in business policies, practices and strategies.

Risk concentration can be hard to characterize in advance. Put-back risk – the risk that banks that were involved in securitizations would be required to take them back onto their books if they failed to honor the terms of the securitization – is an example of a risk that materialized on a system-wide basis after the crisis, the nature of which would have been hard to anticipate even with the benefit of an evolutionary perspective. In other economic sectors where there is a low tolerance for risk, common practice is to investigate anything unexpected and to pay particular attention to areas of activity where several small unusual things may be happening at once – such as slipping underwriting standards, opacity in documentation, new kinds of customer and so on. Signs like this may provide some indication of emergent risk concentration.

4.5.5 Granular analysis implications

Regardless of the final analysis schemes, a relatively aggressive strategy for collecting, accessing and analyzing good quality data is justified. The more granular the data, the more flexibility there is for ways in which it can be rolled up. Consistent and reliable data should help improve the choice and calibration of policies. And perhaps more important, daily or even intraday data on changes in such things as valuations, exposures and business strategies, should support far better crisis management.

Granular analyses focus on incentives among other things. There are new provisions in the Dodd–Frank Act of 2010 (DFA) to realign incentive structures and hold those accountable for taking excessive risks. Bondholders, directors and managers of TBTF firms are more at risk than they were, as a result of the orderly liquidation authority of the FDIC. Moreover, the culture and the incentives regulators face is an enduring issue. So motivation is not just a subject of concern for policymakers looking at the private sector, but for policymakers considering how the public sector works as well.

A large scale economic simulator, which is calibrated to existing market conditions, offers a potential tool for policy investigation, preparation for crisis and real-time crisis management. This type of capacity allows for the development of an interactive system, a financial "war gaming" system, where regulators, academics and market participants (or former market participants) can explore and refine the system's capabilities and then use it for policy and crisis analysis.

4.5.6 Conclusions

Regulators will never be able to anticipate every risk to the financial system. From time to time, dynamic innovation may take forms that seem benign until it is too late. But that is no reason not to try to understand systemic risk. Our understanding of financial systemic stability is advancing. Policymakers should be held to the minimum standard of avoiding old mistakes. And, thanks to theoretical and data improvements, some new mistakes should be avoidable too. That is not a bad objective for a systemic policy framework.

Tommaso Padoa-Schioppa observed at the inaugural INET Conference at Kings College Cambridge in 2010 that he had the hardest time persuading academic economists to take the plans for the Euro seriously in its earliest stages of development because they were in the habit of taking national sovereignty as a given. Padoa-Schioppa was operating from the premise that sovereignty could, and would, be given up. Here too there may be some resistance to breaking the mold of reductionist analysis. The tendency has been to make do with what is available for analytical purposes. The challenge for systemic analysis will be to ask what could be done were more data available and shaping thereby the data development strategy for the future. (Padoa-Schioppa, 2010).

At the time of writing, it has been four years since the height of the crisis. Much remains to be done to implement effectively Dodd–Frank in the USA and similar legislation elsewhere in the world. The danger is that momentum will be lost and efforts to foster greater financial stability will peter out. A great deal will depend on the determination of regulators and the cooperative efforts of researchers in academia and elsewhere. Hopefully these efforts will bear fruit.

Bibliography

Acharya, Viral, 2009, A theory of systemic risk and design of prudential bank regulation, *Journal of Financial Stability*, **5**, 224–255.

Acharya, V. and M. Richardson (eds.), 2009, *Restoring Financial Stability: How to Repair a Failed System*, Wiley, www.wiley.com/WileyCDA/WileyTitle/productCd-0470499346.html

Acharya, Viral, Thomas F. Cooley, Matthew P. Richardson and Ingo Walter, 2010, *Regulating Wall Street: The Dodd–Frank Act and the New Architecture of Global Finance*, Wiley.

Adrian, Tobias and Liaquat Ahamed, 2009, *Lords of Finance: The Bankers Who Broke the World*, Penguin Books.

Adrian, Tobias and Markus Brunnermeier, 2011, CoVaR, Working Paper, Princeton University, September, www.princeton.edu/~markus/research/papers/CoVaR.pdf

Adrian, Tobias and Hyun Song Shin, 2009, The shadow banking system: implications for financial regulation, *Banque de France Financial Stability Review*, **13**, September, 1–10, www.banque-france.fr/fileadmin/user_upload/banque_de_france/publications/Revue_de_la_stabilite_financiere/etude01_rsf_0909.pdf

Adrian, Tobias and Hyun Song Shin, 2010, The changing nature of financial intermediation and the financial crisis of 2007–2009, *Annual Review of Economics*, **2**, 603–618, www.annualreviews.org/doi/abs/10.1146/annurev.economics.102308.124420

Akerlof, George and Robert Schiller, 2009, *Animal Spirits: How Human Psychology Drives the Economy And Why It Matters for Global Capitalism*, Princeton University Press.

Allen, Franklin, Ana Babus and Elena Carletti, 2009, Financial crises: theory and evidence, *Annual Review of Financial Economics*, **1**, 97–116, www.annualreviews.org/doi/pdf/10.1146/annurev.financial.050808.114321

Allen, Franklin, Elena Carletti and Agnese Leonello, 2011, Deposit insurance and risk taking, *Oxford Review of Economic Policy*, **27**(3), 464–478.

An, L., Alan King and D. Subrhamanian, 2009, On modeling some essential dynamics of the subprime mortgage crisis, *Proceedings of the 27th International Conference of the System Dynamics Society* (July 2009).

Ashcraft, Adam B. and Til Schuermann, 2008, Understanding the securitization of subprime mortgage credit, *Foundations and Trends in Finance*, **2**(3), 191–309.

Bank of England, 2011, Instruments of Macroprudential Policy, Discussion Paper, December, www.bankofengland.co.uk/publications/news/2011/160.htm

Basel Committee on Banking Supervision (BCBS), 2010, Basel III: International Framework for Liquidity Risk Measurement, Standards and Monitoring, December 2010, www.bis.org/publ/bcbs188.pdf

Basel Committee on Banking Supervision (BCBS), 2011a, Basel Committee on Banking Supervision reforms – Basel III, Internet resource, downloaded February 21, 2012, www.bis.org/bcbs/basel3/b3summarytable.pdf

Basel Committee on Banking Supervision (BCBS), 2011b, Progress Report on Basel III Implementation, Technical Report, October, www.bis.org/publ/bcbs203.pdf

Basel Committee on Banking Supervision (BCBS), 2012, Compilation of documents that form the global regulatory framework for capital and liquidity, Internet resource, downloaded February 28, 2012, www.bis.org/bcbs/basel3/compilation.htm

Bean, Charles, 2011, Joseph Schumpeter Lecture: The Great Moderation, the Great Panic, and the Great Contraction, *Journal of the European Economic Association*, **8**(2–3), 289-325, http://onlinelibrary.wiley.com/doi/10.1111/j.1542-4774.2010.tb00505.x/abstract

Bernanke, Ben, 2011, Implementing a macroprudential approach to supervision and regulation, speech given at the 47th Annual Conference on Bank Structure and Competition, Chicago, IL, May.

Bhattacharya, Sudipto, Dimitrios Tsomocos, Charles Goodhart and Alexandros Vardoulakis, 2011, Minsky's Financial Instability Hypothesis and the Leverage Cycle, Working Paper, London School of Economics, September, http://papers.ssrn.com/sol3/papers.cfm?abstract_id=1773946

Billio, M., M. Getmansky, A. W. Lo and L. Pelizzon, 2010, Econometric Measures of Systemic Risk in the Finance and Insurance Sectors, NBER Working Paper 16223, NBER.

Bisias, D., M. Flood, A. Lo and S. Valavanis, 2012, A survey of systemic risk analytics, *Annual Review of Financial Economics*, **4**, 255–296.

Borio, Claudio, 2011, Rediscovering the Macroeconomic Roots of Financial Stability Policy: Journey, Challenges and a Way Forward, BIS Working Paper 354, September.

Borio, Claudio and Piti Disyatat, 2011, Global Imbalances and the Financial Crisis: Link or No Link? BIS Working Paper 346, http://papers.ssrn.com/sol3/papers.cfm?abstract_id=1859410

Borio, Claudio, Mathias Drehmann and Kostas Tsatsaronis, 2012, Stress-Testing Macro Stress Testing: Does it Live Up to Expectations? BIS Working Paper 369, January, www.bis.org/publ/work369.htm

Brownlees, Christian T. and Robert Engle, 2011, Volatility, Correlation and Tails for Systemic Risk Measurement, Working Paper, New York University, June, http://papers.ssrn.com/sol3/papers.cfm?abstract_id=1611229

Brunnermeier, Markus K., 2001, *Asset Pricing under Asymmetric Information: Bubbles, Crashes, Technical Analysis and Herding*, Oxford University Press.

Brunnermeier, Markus K., 2009, Deciphering the liquidity and credit crunch 2007–08, *Journal of Economic Perspectives*, **23**(1), 77–100, www.princeton.edu/~markus/research/papers/liquidity_credit_crunch.pdf

Brunnermeier, Markus K., Andrew Crockett, Charles A. Goodhart, Avinash Persaud and Hyun Song Shin, 2009, *The Fundamental Principles of Financial Regulation*, Centre for Economic Policy Research, July.

Brunnermeier, Markus K., Thomas M. Eisenbach and Yuliy Sannikov, 2011a, Macroeconomics with Financial Frictions: A Survey, Working Paper, Princeton University, July, http://scholar.princeton.edu/markus/publications/macroeconomics-financial-frictions-survey

Brunnermeier, Markus K., Gary Gorton and Arvind Krishnamurthy, 2011b, Risk Topography, Working Paper, Princeton University, March, www.princeton.edu/~markus/research/papers/risk_topography.pdf

Bullock, Seth, 2011, Prospects for Large-Scale Financial Systems Simulation, Technical Report, Foresight Project: the Future of Computer Trading in Financial Markets, London, August, www.bis.gov.uk/assets/bispartners/foresight/docs/computer-trading/11-1233-dr14-prospects-for-large-scale-financial-systems-simulation.pdf

Chor, Davin and Kalina Manova, 2011, Off the cliff and back? Credit conditions and international trade during the global financial crisis, *Journal of International Economics*, **87**(1), 117–133, www.sciencedirect.com/science/article/pii/S0022199611000493#FCANote

Clauss, Pierre, Thierry Roncalli and Guillaume Weisang, 2009, Risk management lessons from the Madoff fraud, *International Finance Review*, **10**, 505–543.

Committee on the Global Financial System (CGFS), 2011, Global Liquidity – Concept, Measurement and Policy Implications, CGFS Papers Number 45, November, www.bis.org/publ/cgfs45.htm?qt=201112

Committee to Establish the National Institute of Finance (CE-NIF), 2009a, A Call to Create a National Institute of Finance, Internet Resource, downloaded February 22, 2012, www.ce-nif.org/component/content/article/29-the-nif/147-nif-general-1-page-summary

Committee to Establish the National Institute of Finance (CE-NIF), 2009b, Argument for Data Management in Support of the NIF Mission for Improved Oversight and Safeguarding of the Modern Financial System, Technical Report, www.ce-nif.org/images/docs/ce-nif-generated/nif_data_management_executive_summary_final.pdf

Committee to Establish the National Institute of Finance (CE-NIF), 2009c, Argument for Research and Analytics in Support of the NIF Mission for Improved Oversight and Safeguarding of the Modern Financial System, Technical Report, www.ce-nif.org/images/docs/nif_research_and_analytics_executive_summary_final.pdf

Cont, Rama, Amal Moussa and Edson Bastos e Santos, 2010, Network Structure and Systemic Risk in Banking Systems, Working Paper, Columbia University, December, http://papers.ssrn.com/sol3/papers.cfm?abstract_id=1733528

Cukierman, Alex, 2011, Reflections on the crisis and on its lessons for regulatory reform and for central bank policies, *Journal of Financial Stability*, **7**(1), 26–37, www.sciencedirect.com/science/article/B7CRR-4YS9RPM-1/2/6db4dca55145d6a5731cd0d5ea09674c

Danielsson, Jon and Hyun Song Shin, 2003, Endogenous risk, in: *Modern Risk Management: A History*, ch. 20, Risk Books, 297–313.

Demirgüç-Kunt, Asli and Enrica Detragiache, 2002, Does deposit insurance increase banking system stability? An empirical investigation, *Journal of Monetary Economics*, **49**(7), 1373–1406.

Demirguc-Kunt, Asli, Enrica Detragiache and Ouarda Merrouche, 2010, Bank Capital: Lessons from the Financial Crisis, Policy Research Working Paper 5473, World Bank, November, www-wds.worldbank.org/external/default/WDSContentServer/WDSP/IB/2010/11/10/000158349_20101110113157/Rendered/PDF/WPS5473.pdf

Department of Homeland Security (DHS), 2007, Banking and Finance: Critical Infrastructure and Key Resources: Sector-Specific Plan as input to the National Infrastructure Protection Plan, Technical Report, May, www.dhs.gov/xlibrary/assets/nipp-ssp-banking.pdf

Duffie, Darrell, 2010, Presidential address: asset price dynamics with slow-moving capital, *Journal of Finance*, **65**(4), 1237–1267, www.darrellduffie.com/uploads/pubs/DuffieAFAPresidentialAddress2010.pdf

Eichengreen, Barry, 1992, *Golden Fetters: The Gold Standard and the Great Depression, 1919–1939*, Oxford University Press.

Eisenberg L. and T. Noe, 2001, Systemic risk in financial systems, *Management Science*, **47**, 236–249.

Elliott, Douglas J. and Robert E. Litan, 2011, Identifying and Regulating Systemically Important Financial Institutions: The Risks of Under and Over Identification and Regulation, Brookings Working Paper. January 2011, www.brookings.edu/~/media/Files/rc/papers/2011/0116_regulating_sifis_elliott_litan/0116_regulating_sifis_elliott_litan.pdf

European Central Bank (ECB), 2010, Financial networks and financial stability, *Financial Stability Review*, June, 155–160, www.ecb.int/pub/fsr/shared/pdf/ivdfinancialstabilityreview201006en.pdf

Evanoff, Douglas, George Kaufman and A. G. Malliaris, 2012, *New Perspectives on Asset Price Bubbles*, Oxford University Press.

Farmer, J. Doyne and John Geanakoplos, 2009, The virtues and vices of equilibrium and the future of financial economics, *Complexity*, **14**(3), 11–38, http://onlinelibrary.wiley.com/doi/10.1002/cplx.20261/abstract

Farmer, J. Doyne, John Geanakoplos and Stefan Thurner, 2010, Leverage Causes Fat Tails and Clustered Volatility, Cowles Foundation Discussion Paper, No 1745, January.

Farmer, J. Doyne, John Geanakoplos, Peter Howitt and Robert Axtell, 2012, *Grant: An Agent Based Model of the Current Economic Crisis*, Internet Resource, Institute for New Economic Thinking, downloaded February 22, 2012, http://ineteconomics.org/grants/agent-based-model-current-economic-crisis

Federal Deposit Insurance Corporation (FDIC), 2011a, Resolution plans required, *Federal Register*, **76**(211), 67323–67340, www.gpo.gov/fdsys/pkg/FR-2011-11-01/pdf/2011-27377.pdf

Federal Deposit Insurance Corporation (FDIC), 2011b, The orderly liquidation of Lehman Brothers Holdings Inc. under the Dodd–Frank Act, *FDIC Quarterly*, **5**(2), www.fdic.gov/bank/analytical/quarterly/2011_vol52.html

Financial Crisis Inquiry Commission (FCIC), 2011, *Financial Crisis Inquiry Report: Final Report of the National Commission on the Causes of the Financial and Economic Crisis in the United States*, January, www.gpoaccess.gov/fcic/fcic.pdf

Financial Stability Board and International Monetary Fund (FSB-IMF), 2011, The Financial Crisis and Information Gaps: Implementation Progress Report, Technical Report, June.

Fisher, Irving, 1933, The debt-deflation theory of great depressions, *Econometrica*, **1**, 337–357.

Flannery, Mark J., Simon H. Kwan and Mahendrarajah Nimalendran, 2010, The 2007–09 Financial Crisis and Bank Opaqueness, Working Paper 2010–27, Federal Reserve Bank of San Francisco, September, www.frbsf.org/publications/economics/papers/2010/wp10-27bk.pdf

Flood, M., A. Mendelowitz and W. Nichols, 2012, Monitoring financial stability in a complex world, in: V. Lemieux (ed.), *Financial Analysis and Risk Management: Data Governance, Analytics and Life Cycle Management*, Springer Verlag, ch. 2, 15–46.

French, Kenneth R., Martin N. Baily, John Y. Campbell, John H. Cochrane, Douglas W. Diamond, Darrell Duffie, Anil K. Kashyap, Frederic S. Mishkin, Raghuram G. Rajan, David S. Scharfstein, Robert J. Shiller, Hyun Song Shin, Matthew J. Slaughter, Jeremy C. Stein and René M. Stulz, 2010, *The Squam Lake Report: Fixing the Financial System*, Princeton University Press.

Gai, Prasanna, Andrew Haldane and Sujit Kapadia, 2011, Complexity, concentration and contagion, *Journal of Monetary Economics*, **58**(5), 453–470.

Galati, Gabriele and Richhild Moessner, 2011, Macroprudential Policy – A Literature Review, BIS Working Paper 337, February, www.bis.org/publ/work337.pdf

Geanakoplos, John, 2010a, The leverage cycle, in: D. Acemoglu, K. Rogoff and M. Woodford (eds.), *NBER Macroeconomics Annual 2009*, University of Chicago Press, Volume **24**, 1–65.

2010b, Solving the present crisis and managing the leverage cycle, *Federal Reserve Bank of New York Economic Policy Review*, August, 101–135.

Greenlaw, David, Anil K. Kashyap, Kermit Schoenholtz and Hyun Song Shin, 2012, Stressed Out: Macroprudential Principles for Stress Testing, Working Paper 12–08, Chicago Booth Initiative on Global Markets, February, http://papers.ssrn.com/sol3/papers.cfm?abstract_id=2004380

Gorton, Gary, 2010, *Slapped in the Face by the Invisible Hand: Banking and Panic of 2007*, Oxford University Press, 32–38.

Haldane, Andrew, 2009, *Rethinking the Financial Network*, Speech delivered at the Financial Student Association, Amsterdam, April, www.bankofengland.co.uk/publications/speeches/2009/speech386.pdf

Haldane, Andrew, 2010, *The $100 Billion Question*, Bank of England, London, March, page 16, Table 1, www.bankofengland.co.uk/publications/Pages/speeches/2010/433.aspx.

Haldane, Andrew and Robert M. May, 2011, Systemic risk in banking ecosystems, *Nature*, **469**, 351–355, www.nature.com/nature/journal/v469/n7330/abs/nature09659.html

Hanson, Samuel G., Anil K. Kashyap and Jeremy C. Stein, 2011, A macroprudential approach to financial regulation, *Journal of Economic Perspectives*, **25**(1), Winter, 3–28, www.aeaweb.org/atypon.php?return_to=/doi/pdfplus/10.1257/jep.25.1.3

Heid, Frank, 2007, The cyclical effects of the Basel II capital requirements, *Journal of Banking and Finance*, **31**, 3885–3900.

Hendricks, Darryll et al., 2010, Task Force on Tri-Party Repo Infrastructure Payments Risk Committee Report, under the auspices of the New York Federal Reserve Bank, New York, May 17.

HM Treasury, 2010, A New Approach to Financial Regulation: Judgement, Focus and Stability, Technical Report, July, www.hm-treasury.gov.uk/consult_financial_regulation.htm

Hu, Henry T. C., 2011, Keynote address: systemic risk and financial innovation: towards a 'unified' approach, in: J. Haubrich and A. Lo (eds.), *Quantifying Systemic Risk* (proceedings of a conference, November 6, 2009), NBER and University of Chicago Press, ch. 1, forthcoming, www.nber.org/books/haub10-1

Huang, Xin, Hao Zhou and Haibin Zhu, 2009, A framework for assessing the systemic risk of major financial institutions, *Journal of Banking and Finance*, **33**(11), 2036–2049, www.sciencedirect.com/science/article/pii/S0378426609001216

Intelligence Advanced Research Projects Activity (IARPA), 2011, MATCHES Program Proposers' Day, conference announcement for the Markets Analysis and Testing of Contextual Hypotheses Enhancement System (MATCHES) Program, Internet Resource, downloaded February 28, 2012, www.iarpa.gov/solicitations_matches.html

International Monetary Fund (IMF), 2009, *Global Financial Stability Report, World Economic and Financial Surveys*, www.imf.org/External/Pubs/FT/GFSR/2009/01/pdf/text.pdf

International Monetary Fund (IMF), 2010, *Global Financial Stability Report, World Economic and Financial Surveys*, April, www.imf.org/external/pubs/ft/gfsr/2010/01/index.htm

International Monetary Fund (IMF), 2011, Macroprudential Policy: An Organizing Framework, IMF Policy Paper, October, www.imf.org/external/np/pp/eng/2011/031411.pdf

International Monetary Fund (IMF), Bank for International Settlements (BIS), and Financial Stability Board (FSB), 2009, Guidance to Assess the Systemic Importance of Financial Institutions, Markets and Instruments: Initial Considerations, Report to the G20 Finance Ministers and Central Bank Governors, October, www.financialstabilityboard.org/publications/r_091107c.pdf

Jaffee, Dwight, Anthony Lynch, Stijn Van Nieuwerburgh and Matthew Richardson, 2009, Mortgage origination and securitization in the financial crisis, in: V. Acharya and M. Richardson (eds.), *Restoring Financial Stability: How to Repair a Failed System*, Wiley, ch. 1.

Johnson, Simon and James, Kwak, 2010, *13 Bankers*, Random House, p. 25.

Kashyap, Anil and Jeremy C. Stein, 2004, Cyclical implications of the Basel II capital standards, *Federal Reserve Bank of Chicago Economic Perspectives*, **1Q**, 18–31.

Kashyap, A., R. Berner and C. Goodhart, 2011, The macroprudential toolkit, *IMF Economic Review*, **59**(2), 145–161, www.palgrave-journals.com/imfer/journal/v59/n2/full/imfer20114a.html

Khandani, Amir, Andrew Lo and Robert Merton, 2012, Systemic Risk and the Refinancing Ratchet Effect, Working Paper, MIT, February, `www.argentumlux.org/documents/refi18_3.pdf`

Liechty, John, 2012, Scientists and bankers – a new model army, *Nature*, `www.nature.com/news/scientists-and-bankers-a-new-model-army-1.10399`

Lopez-Espinosa, German, Antonio Moreno, Antonio Rubia and Laura Valderrama, 2012, Short-term Wholesale Funding and Systemic Risk: A Global CoVaR Approach, IMF Working Paper WP/12/46, IMF Institute, February, `www.imf.org/external/pubs/cat/longres.aspx?sk=25720.0`

Martin, William McChesney, 1955, Address to the Investment Bankers Association, October 19, Federal Reserve Archive, `fraser.stlouisfed.org`

Minsky, Hyman, 1982, *Can "It" Happen Again? Essays on Instability and Finance*, M. E. Sharpe.

Mizen, P., 2008, The credit crunch of 2007–2008: A discussion of the background, market reactions and policy responses, *Federal Reserve Bank of St. Louis Review*, **90**(5), 531–568, `http://research.stlouisfed.org/publications/review/article/6663`

Morgan, Donald, 2002, Rating banks: risk and uncertainty in an opaque industry, *American Economic Review*, **94**(2), 874–888.

Moussa, Amal, 2011, *Contagion and Systemic Risk in Financial Networks*, Ph.D. Thesis, Columbia University, `http://academiccommons.columbia.edu/download/fedora_content/download/ac:131475/CONTENT/Moussa_columbia_0054D_10092.pdf`

Newman, Mark, 2010, *Networks: An Introduction*, Oxford University Press.

Nier, Erlend, Jing Yang, Tanju Yorulmazer and Amadeo Alentorn, 2007, Network models and financial stability, *Journal of Economic Dynamics and Control*, **31**(6), 2033–2060.

Packer, Frank and Nikola Tarashev, 2011, Rating methodologies for banks, *BIS Quarterly Review*, June, 39–52. `www.bis.org/publ/qtrpdf/r_qt1106f.htm`

Padoa-Schioppa, Tommaso, 2010, Closing remarks at the inaugural conference of the Institute for New Economic Thinking, Institute for New Economic Thinking, April, `www.youtube.com/watch?v=W3lggNnTjbM`

Pew Trusts, 2012, *The Pew Financial Reform Project*, Internet Resource, accessed February 11, 2012, `www.pewtrusts.org/our_work_detail.aspx?id=327442`

Pozsar, Zoltan, Tobias Adrian, Adam Ashcraft and Hayley Boesky, 2012, Shadow Banking, *Staff Report no. 458*, Federal Reserve Bank of New York, July, `www.newyorkfed.org/research/staff_reports/sr458.pdf`

Purnanandam, Amiyatosh, 2011, Originate-to-distribute model and the subprime mortgage crisis, *Review of Financial Studies*, **24**(6), 1881–1915, `http://rfs.oxfordjournals.org/content/24/6/1881.abstract`

Reinhart, Carmen M. and Kenneth S. Rogoff, 2009, *This Time is Different: Eight Centuries of Financial Folly*, Princeton University Press.

Richardson, Matthew and Ingo Walter, 2009, Rethinking compensation in financial firms, in: V. Acharya and M. Richardson (eds.), *Restoring Financial Stability: How to Repair a Failed System*, Wiley, ch. 8.

Rossi, Clifford V., 2010, Anatomy of Risk Management Practices in the Mortgage Industry: Lessons for the Future, Technical Report, Research Institute for Housing America, May, http://riskmanagementcommunity.com/servlet/JiveServlet/previewBody/1101-102-1-1066/9946_Research_RIHA_Rossi_Report.pdf

Roxburgh, Charles, Susan Lund, Tony Wimmer, Eric Amar, Charles Atkins, Ju-Hon Kwek, Richard Dobbs and James Manyika, 2010, Debt and Deleveraging: The Global Credit Bubble and its Economic Consequences, Technical Report, McKinsey Global Institute, January, www.mckinsey.com/Insights/MGI/Research/Financial_Markets/Debt_and_deleveraging_The_global_credit_bubble_Update

Shleifer, Andrei and Robert Vishny, 2011, Fire sales in finance and macroeconomics, *Journal of Economic Perspectives*, **25**(1), 29–48, www.aeaweb.org/articles.php?doi=10.1257/jep.25.1.29

Sorkin, Andrew Ross, 2009, *Too Big to Fail: The Inside Story of How Wall Street and Washington Fought to Save the Financial System – and Themselves*, Viking Press.

Stern, Gary and Ron Feldman, 2004, *Too Big to Fail: The Hazards of Bank Bailouts*, Brookings Institute Press.

Swagel, Phillip, 2010, *The Cost of the Financial Crisis: The Impact of the September 2008 Economic Collapse*, Pew Charitable Trusts, Washington, DC.

Tarullo, Daniel K., 2011, Regulating Systemically Important Financial Firms, Speech at the Peter G. Peterson Institute for International Economics, Washington, DC, June 3.

Taylor, Charles R., 2011, Evolution and macro-prudential regulation, *Center for the Study of Financial Innovation*, London, October, http://csfi.org/index.php?option=com_content&view=category&id=35&Itemid=60

Taylor, John B., 2009, *Getting Off Track: How Government Actions and Interventions Caused, Prolonged, and Worsened the Financial Crisis*, Hoover Institution Press, xi–xii, www.hooverpress.org/productdetails.cfm?PC=1342

Tirole, Jean, 2011, Illiquidity and all its friends, *Journal of Economic Literature*, **49**(2), 287–325, www.aeaweb.org/articles.php?doi=10.1257/jel.49.2.287

Tong, Hui and Shang-Jin Wei, 2011, The composition matters: capital inflows and liquidity crunch during a global economic crisis, *Review of Financial Studies*, **24**(6), 2023–2052, http://rfs.oxfordjournals.org/content/24/6/2023.abstract

US Congress, 2010, Dodd–Frank Wall Street Reform and Consumer Protection Act, Enrolled Bill, H.R. 4173, July, www.gpo.gov/fdsys/pkg/BILLS-111hr4173enr/pdf/BILLS-111hr4173enr.pdf

5

Data-driven regulation and financial reform: one perspective from industry on the financial crisis

John C. Pattison

5.1 Introduction

The information basis for regulation and supervision has evolved to one driven more by structured data, often formulated in concept in a multinational context for comparability and equity. *Data-driven regulation* means that the information set guiding official supervision is guided and dominated by formal collections of well-defined, primarily numeric data reported by financial institutions.[1] While data-driven regulation introduces many operational efficiencies into supervisory processes, it also introduces layers of abstraction and quantification that have the potential to distract attention from the underlying financial reality. Supervisors may be misled if the collected data do not fully and accurately convey the underlying situation in a timely fashion.

Information and data issues are intimately connected to recent events. This was a modern, global financial crisis.[2] While it had its origins in housing markets, housing policy and securitizations, complex network effects operated through traded instruments and counterparty risks. It highlighted risks that were known in the literature but were neither ranked highly nor pursued by the official sector. The crisis occurred during the implementation of Basel II methods of regulation based upon quantified risks using proprietary data and internal models of institutions. These methods were in the process of being restructured and recalibrated. In any event they did not prevent failures.

The regulation of financial institutions and markets is not a science, although it borrows statistical techniques and theory from portfolio models and microeconomics.

[1] For example, Federal Reserve (2011, p. 75) appears to adopt this usage in connection with stress testing.
[2] See Financial Crisis Inquiry Commission (2011) or Federal Reserve Bank of St. Louis (2011) for a chronology of the crisis. This article will not repeat those factual details.

Handbook of Financial Data and Risk Information, ed. Margarita S. Brose, Mark D. Flood, Dilip Krishna and Bill Nichols. Published by Cambridge University Press. © Cambridge University Press 2014.

It is not normative in that there is disagreement on the standards, trade-offs and weights among goals. It is difficult to posit regulation as a positive study in that there is little in the way of standardized data and there are no robust models of general applicability.

Models such as those used in the Basel II framework are simplified, only asymptotically correct, and their risk curves were altered by administrative *fiat* to reduce procyclicality. Regulators traded off risk sensitivity for cyclical macroeconomic benefits (Pattison, 2008b). Tarullo (2008) provides a cogent analysis of the shortcomings of the advanced internal ratings based approach as well as placing it in the difficult context of international regulatory cooperation. In this environment it is necessary to look to public policy to understand how regulation evolves.

5.1.1 A very modern crisis

The crisis is often viewed as a US housing crisis in origin. However, as it unfolded it demonstrated weaknesses in the management of financial institutions, in risk management and regulation in many countries. The crisis had macroeconomic and international roots. It began with excess global liquidity and low interest rates.[3] Individuals, governments and financial institutions in the USA and other countries borrowed more, financed by inflows of capital from countries with trade surpluses. These inflows kept interest rates lower than they would have been. The creation of complex financial instruments that promised increased yield without increasing risk as indicated by their credit ratings attracted many investors, while risk ratings obscured underlying complexities and potential future risks. These instruments included collateralized debt obligations (CDOs) and other securitizations.

Weaknesses in liquidity management and funding of financial institutions exacerbated the crisis. The failure of Northern Rock in the UK raised many neglected supervisory issues, as Shin (2009) demonstrates. When Northern Rock failed, regulators had only required one week's liquidity buffer for a firm that relied upon periodic but infrequent large securitizations for a significant proportion of its funding. Firms frequently increased risks by raising leverage. The requisite debt was often short term, from interbank or commercial paper markets, leaving borrowers vulnerable when funding sources dried up. For example, funding to many European banks, particularly in dollars, disappeared or shrank to maturities of days rather than months.

Funding issues compounded the crisis through network effects. Funding involves not only access to interbank markets but also the use of foreign exchange, swaps and derivatives markets, all of which involve counterparty risks. Counterparty risks also

[3] See for example Bank of England (2007, p. 8) or European Central Bank (2007, p. 176). The latter also examines links between movements in risk aversion and market liquidity.

loom large through the trading and settlement of customer and proprietary transactions. Counterparty risk is the chance that another institution will not fulfill its obligations. Transacting institutions ordinarily do not know their counterparties' transactions with third parties. Network effects thus preclude firms from assessing their own risks accurately. When Lehman Brothers failed, interbank markets dried up because institutions could not judge which firms might be caught by first round or second round counterparty losses. Liquidity-constrained firms then sought to sell their most marketable assets, further driving down prices, and provoking new margin calls. A self-reinforcing spiral of fire sales spread network effects to other, healthy firms.[4]

The conclusions are clear: while all financial crises share many common features, this was a modern crisis involving greater network effects of global scope. To comprehend them is difficult, as chains of counterparty effects are not directly assessable. Also instruments that facilitate such risks are complex, often leveraged and created in unique tranches for resale and are actively traded. Risks are thus dispersed, opaque, and substantially difficult to identify and quantify accurately. The inherited regulatory paradigm focused on *individual* institutions, but network effects, counterparty risks, fire sales and market failures proved decisive. The whole is not simply the sum of the parts. Current regulation, data collection, supervision and resolution did not take adequate account of these interdependencies.

The crisis demonstrated weaknesses in regulation, risk management, and methods of management, supervision and oversight by both the official sector and financial institutions. It illustrated inadequacies in regulatory tools, in models to measure and manage risk, and in resolution methods to deal with troubled institutions. Central banks resorted to creative expedients to deal with illiquidity and insolvencies, in some cases requiring changes in the law. Management and boards of directors either did not have the information to manage their strategies and the attendant risks, including liquidity risks, or failed to exercise oversight wisely.

5.1.2 Data and regulation

Crises are possible but less likely to occur in a world of full and transparent information. Not only do markets and regulators lack data, but insufficient work has been carried out to identify the data that might be required. Regulation cannot be calibrated or tested, nor can costs and benefits be discerned without data. Prior to the Basel II Accord and the use of internal models, the information assembled by central banks, financial regulators and statistical agencies was largely accounting data rather than economic. Data were often poorly related to regulatory or supervisory needs.

4 The above description is a summary. Readers are referred to Brunnermeier (2009), Masson and Pattison (2009a) or the Financial Crisis Inquiry Commission (2011) for detail.

Financial institutions capture data at the customer or transaction level. Some, but not all, of this reconciles to the general ledger. Thus firms would generally use different types of data for different management purposes. These differences would give rise to data integrity issues, uncertain quality of information and discrepancies in information available to management and board members. There are supplemental needs for regulatory and risk management data. These may differ from economic or accounting data. Regulators and supervisors need to define their short- and medium-term goals, link these to their methodologies and map these to data requirements in a network environment of interdependent institutions. Data and model complexity can overwhelm the ability of management and supervisors to digest, interpret and act expeditiously.

Challenges do not stop here. Cross-border subsidiaries and branches necessitate international review and coordination of data requirements. When regulatory requirements, definitions and calibration differ across borders, the internal data management issues multiply as firms consolidate data from several jurisdictions. This consolidation may occur several times: for example, for subsidiaries within European groups, at a European holding company and then at a foreign head office. All three tiers may follow different capital and other regulatory rules.

The International Monetary Fund and the Financial Stability Board have begun to address data needs and gaps and the IMF mandate has been upgraded to deal with financial networks, data, data gaps and its surveillance function (see IMF, 2010a). However, this is a large and complex topic that will take years to refine.

Data are an input to decision making by institutions, regulators and supervisors. They are a *variable* input, not a fixed one. Data are not free. They are costly to capture, and reconciling risk, economic and accounting data is necessary. They must be used in conjunction with judgment, models, algorithms, risk aggregation and reporting to achieve their full value for corporate and supervisory decisions. Not all data can or will be communicated: judgment, aggregation, accounting systems and different management information frameworks will determine rightly and wrongly what management and boards of directors see. There are limits to the capacity of management, boards and supervisors to absorb and understand information. Some data will preempt others with an impact on decisions and performance. This is equally true for supervisors and is a constraint on creating an understanding of risk aggregations across even one financial system. A turning point was reached in the past decade. Basel II fundamentally changed the regulatory game from a top-down imposition of rules and simple ratios, to one where internal data were used for a few key regulatory functions, mainly for the determination of regulatory capital. As Reinhart and Rogoff (2009) note, there was a general lack of data prior to the crisis, for large categories of risk at the macroeconomic level.[5]

[5] This may be changing with the G20 "Data Gaps" initiative. See FSB and (2010), as well as the IMF (2010a).

5.2 The evolution of regulation

There are three types of regulation: safety and soundness (or prudential) regulation, conduct of business regulation, and regulation with social or political goals. Some conduct of business regulation is for prudential reasons. Other conduct of business rules are for consumer protection, market conduct, disclosure and similar reasons. The history and international context of the regulatory framework are covered elsewhere in this Handbook.[6] Here the focus will be on prudential regulation and supervision, which covers a wide front. International agreements generally focus on negotiating a common approach to one item at a time. The 2007 crisis necessitated a wider approach. Until recently, the largest resources under the Basel framework were devoted to capital regulation to the detriment of other, arguably more important, issues such as liquidity. Leverage, where regulated at all, was managed at a national level. Leverage varied widely between financial institutions and among different national regulatory environments. The disconnect to capital regulation arises from a number of factors including inadequate definitions of regulatory capital, where reported capital included types of instruments which bore more attributes of debt instruments than equity, and the calculation of risk weighted assets that understated risks.

The goals of Basel II were laudable. These included eliminating or reducing opportunities for institutions to arbitrage regulations, a process that tended to increase risks rather than reduce them. Secondly, regulators wanted to reduce the gap between economic and regulatory capital. Economic capital is the capital that financial institutions require of themselves for the risks that they take. Regulatory capital limits are imposed from without, and typically are based on a standard formula. There are many reasons why these differ, including different probabilities of default, and diversification benefits from correlations among risks in diverse business lines. Moreover, regulation is designed for a wider array of public policy goals, and may be calibrated to different probabilities of individual and joint failure. For example, individual institutions do not typically hold capital for systemic risks. History also demonstrates that institutions often misjudge risks and thus how much capital they require.

Basel II necessitated standardized data. The process was rushed relative to its challenges, and flawed on several fronts. It began with a simplified model and data requirements, while calibration was adjusted on an ad hoc basis. Basel II enforced short-term data requirements that were expensive to implement and often contentious between regulators and institutions. The effect was for institutions collectively to spend billions of dollars on data capture, reconciliation, warehousing and

[6] Chapter 10 provides a history of regulatory financial risk monitoring in the USA, and Chapter 1 is a history of financial risk and innovation.

calculators for different types of risk categories. While there were benefits to this process, the data were insufficient and its calibration did not perform well, particularly for market risk and related areas. Stress testing, Pillar Two assessments and other risk mitigants were applied unevenly and were difficult for supervisors to assess or compare across institutions. There is also a view in financial institutions that some national authorities allow more permissive methods of calculating risk weighted assets and thus lower the effective capital ratios for institutions under their authority.

The data-driven environment of Basel II no longer rests exclusively on capital adequacy for credit, market and operational risks. New liquidity rules require data to measure, calibrate and manage minimum liquid asset holdings. However, this is merely the tip of the data iceberg. There are far more extensive data requirements for business units to assess and manage risks, make decisions and inform customers. Demands on data are growing to deal with the needs of the official sector. First, monetary policy is not sufficient for financial stability. Regulation also has an important role to play. Second, regulation and supervision of *individual* institutions is a necessary but not a sufficient condition for stability. Thus regulators are seeking "macroprudential" policies to supplement monetary policy in stabilizing economies. We will return to these systemic issues below.

The data-driven approach that Basel II introduced may be criticized but it was revolutionary. The future will continue the same path. The obvious alternatives are either to refine the regulation and management of multiple complex risk types or retreat to simple ratios as in the past. Yet regulators cannot add capital or liquidity requirements without an understanding of what they are measuring, how to calibrate requirements, and a framework to assess performance. This does not mean that regulation becomes a slave to data. It was a mistake to assume that models and more data meant better decisions. Recall the emphasis on data as a variable input. The crisis illustrated that data need to be considered in conjunction with other indicators, models, experience and judgment to reach decisions.

5.2.1 Accounting issues

In the past two decades a new accounting emphasis on mark-to-market values and then fair value has meant that different types of data are relevant and more than one type must be retained in many circumstances. Fair value is related to mark-to-market measures in that it is an amount that would be agreed in an arms' length transaction between knowledgeable counterparties to exit a position. Fair value has been contentious, as amply demonstrated by Benston (2008) and Plantin et al. (2008). Whether or not it played a role in the financial crisis is complex, since fair value is intended to indicate a price at which an asset can be liquidated. Thus fair value measures may have forced painful reductions in valuations on many firms but this alone does not create problems in markets. Whether to use fair value data is critical for data needs in some aspects of the implementation of Basel II and III. One debate is whether accounting issues caused or followed from the market volatility. As a measure of the concern, in the USA the *Emergency Economic Stabilization Act of 2008* gave the Securities and Exchange Commission (SEC) the authority to suspend mark-to-market accounting if this was in the public interest.

Historically banks held most assets until maturity, although increasingly over time more were held for trading. Assets held to maturity would be valued at amortized cost while traded assets would be marked to market. However, this was an uneasy truce with credit assets traded and often held in the trading book for regulatory arbitrage, since traded assets had a lower capital requirement. If all markets were continuous, with market prices freely reported from a large number of transactions, there would be few difficulties. However, in the crisis many markets not only became illiquid but suffered fire sales as institutions raised cash from whatever assets could be sold. Even assets that remained in the portfolio required an "exit price" which would invariably be punitively low.

The movement to International Accounting Standards and International Financial Reporting Standards (IFRS) created challenges.[7] Transparency would suffer, because the information required to understand financial statements grows with the expertise required. Secondly, the mismatches resulting when some assets and liabilities are measured at fair value while others are at amortized cost can hinder understanding of financial institutions' balance sheets. A third issue is whether risk management processes at financial institutions are capable of managing changes to accounting standards involving credit customers and counterparties. A fourth issue is that Basel II and III are based upon credit data measured at historical cost and yet could move (or have moved) to fair value for new transactions in the same categories.

The issues are clear. First, regulators and supervisors have greater operational and data challenges due to changed accounting standards; see BCBS (2005) and Landsman (2006). Second, management has not only greater data requirements but also larger interpretive demands. This may lead to errors of judgment or errors of risk management. Third, analytical techniques are data dependent. A number of operational issues flow from this: (a) models and methodologies may be corrupted by changes in reporting conventions; (b) financial institutions may be forced to retain more historical data under different definitions at considerable expense; and (c) asset classes that should be subject to fair value measurement may be denied this treatment if it conflicts with methodologies that rely on historical data such as

[7] See IFRS Foundation (2011).

Basel II. Most of these are transitional issues that will be resolved with time in the transition to IFRS.

5.3 The context of re-regulation

The 2007–2009 crisis was the perfect storm for the contemplation of risk and regulation by the official sector and for hastily crafted regulations. It was a critical period for institutions to identify control weaknesses, including risk management, risk reporting, data aggregation and model issues.

This crisis led to a more fundamental reassessment of regulation than earlier crises. The factors involved are important to the constituencies that influence policy. In addition to the systemic risk issues referred to above, some of the policy issues include the following.

1. A strong sense in social and political terms that trust had been violated by the financial sector, and thus some forms of redress were necessary.
2. Deficits ballooned from bailouts of financial institutions as well as from the stimulus to address unemployment created by the crisis.
3. Governments and regulators now assess higher probabilities of crises. They have adjusted their methodological frameworks to increase the probabilities of adverse outcomes and lower the social costs of crises.
4. Existing risk models, including those used for Basel II capital, were inadequately structured, for example for market risks and securitizations, while also being inappropriately calibrated.
5. A trend began in the 1970s to deregulate liquid asset holdings, believing that such matters were best left to management. This led to "liability management" which reduced positions in cash and other liquid assets. This contributed to the crisis in important ways.
6. Time frames for resolving troubled institutions have changed. Some well-capitalized institutions went directly, and quickly, from health to state aid or insolvency. Many rescues were put together over weekends prior to global markets opening on Monday, or Sunday night in the USA. Historically governments had more time to put rescue operations in place.
7. Theories of efficient markets and rational expectations guided policy makers for many years. These are now discredited for purposes of regulation.

In response, authorities have reworked regulatory requirements, including a leverage ratio as an international requirement, the recalibration of capital rules, and a reintroduction of modest liquidity rules. Of the factors that were critical to the health of financial institutions, regulators focused on capital and liquidity issues in addition to structural reforms to address future risks. The latter included new rules for over-the-counter derivatives, shifting clearing and settlement to central counterparties, and segregating proprietary trading.

5.3.1 Conflicts: the crisis, management and the official sector

Many institutions did survive the crisis. This gave rise to a disconnect between these institutions and the impact of emerging policy goals. In general, survivors do not see the utility of new regulations, more capital and liquidity.

This put both regulators and the financial industry in a difficult place. Worse, changes in regulation have been rushed. Regarding implementation, coordination between financial centers has been lauded but in reality has been lacking. Implementation details have generally not been written, and the resulting uncertainty is expected to last for a period of time. The greater than ten years' experience from Basel II suggests that it would take time to finalize changes to rules affecting data capture, data warehouses, processing systems and accounting. Moreover, these rules will certainly change in coming years. Implementation processes are not only expensive but risky. One element of risk is the need to make firms' data and risk reporting to regulators common across all firms and congruent with loading and aggregation into standardized regulatory data frameworks. Firms face regulatory penalties for failure to provide data in a strict format and on the time frame demanded by regulators in different countries. These national time frames can be incompatible, given the need to consolidate financial statements for subsidiaries in different countries.

Regulation moves risks among institutions and will alter behavior. With so many new rules and regulations, the danger of unintended consequences is large. Cost-benefit calculations to assess regulations are difficult and often simplistic, neglecting some costs or impacts. The difficulty in modeling complex interactive processes is a common challenge. These include the responses of financial institutions or their customers to changing credit spreads contingent upon changed capital or liquidity rules. New regulations are typically evaluated piecemeal, and the cumulative effects (including changes to international accounting standards) are difficult to discern, particularly when detailed rules remain to be written. This is not to say that supervisors and the Basel Committee on Banking Supervision (BCBS) did not attempt to do an analysis and to publish the results (see, for example, BCBS, 2010d, 2010e, and Bank of Canada, 2010). Rather it is a comment on the difficulties and the fact that the quantitative impact studies do not mean that any individual bank will have similar results given their diverse balance sheets and funding structures. Quantitative impact studies are not cost-benefit studies, nor do they model behavioral changes. While official studies have established net macroeconomic benefits, risks remain as incentives are modified by regulatory changes such that categories of risks are shuffled within the regulated and unregulated components of the financial system.

As a result the industry remains cautious. While acknowledging the need for changes, large sections of the industry have resisted sharply higher capital requirements and leverage rules. The modest liquidity proposals were unwelcome in many quarters. Crook (2010) notes that "Wall Street has conducted a formidable lobbying effort to neuter costly aspects of the bill." The *Financial Times* (2009) also reported about the UK that "many in the City would like to return to the *status quo ante*." The debate has been around acquiescing to modest changes while questioning greater, more uncertain ones. Leverage is a central factor.

Both the official sector and the financial industry have trepidation as to the outcome of ill-defined, unquantifiable, often political compromises. For corporations, translating regulatory changes to product lines, risks and earnings is not easy. Regulation changes competitive factors. The official sector focuses on impacts that go beyond the financial sector and include negative externalities of crises such as unemployment and deficits. The financial sector focuses on the difficulties of the regulations, quantification, implementation issues, strategy and uncertainties and the competitive impact. There is inevitably a sizeable gap between these two positions, with network and systemic risks difficult to evaluate in the process.

5.4 The framework for financial regulation

Financial institutions exist by virtue of their authorization, the granting of a charter, letters patent and other binding legal devices whereby governments confer the right to carry on business. Questions naturally arise about the economics of how this relationship – including regulation – is managed. Theories range from the regulator being "captured" by institutions as Dal Bó (2006) reviews, to the improvement of the public good by correcting market imperfections as Stigler (1975) discusses. Common arguments for regulation include correcting market failures, externalities in general, and irrational decisions. While subject to debate, the issues of externalities and irrational decisions are relevant to this crisis. Benston and Kaufman (1996, p. 688) challenge one argument for regulation, namely that banks are "inherently unstable," as they keep low cash reserves and are very highly leveraged. The view that vulnerability from leverage does not automatically imply instability is common in the financial industry, as exhibited in resistance to the imposition of a leverage ratio. Prior to this crisis there was also a view that management and markets would act in self interest to manage and mitigate risks from leverage.

Originally some forms of regulation facilitated the implementation of monetary policy. When financial innovation created substitutes for bank deposits such as money market funds, commercial paper and other products, the formulation of monetary policy became more difficult. One of the themes in the literature is that regulation was needed to protect the open-ended government commitment to deposit insurance. Benston and Kaufman (1996, p. 694) note that "in the absence of government-provided deposit insurance, banks should not be regulated for purposes of achieving objectives such as efficiency, safety and monetary control." However, in the current crisis, governments have pushed beyond deposit insurance to provide a far wider array of support for failing or even failed institutions. Although Lehman Brothers was allowed to fail, its insolvency created great concern for counterparty risk and resulted in reduced liquidity and further drying up of interbank markets. Thus Lehman led to considerable rethinking of the tools available to financial authorities.[8] Benston and Kaufman point out that regulation happens for other reasons, one being to limit competition, and to provide revenue and power to government officials, allowing them to carry out a social agenda. One of Stigler's primary hypotheses (1975, p. 114) is that, "as a rule, regulation is acquired by the industry and is designed and operated primarily for its benefit." For example, Wagster (1996, p. 1321) points out that "the ostensible goals of the Basel Accord were to reduce the risk of the international banking system and to minimize competitive inequality." But he emphasizes that its ulterior goal was "to eliminate the funding-cost advantage of Japanese banks that allowed them to capture more than one-third of international lending." He notes (1996, footnote, p. 1322) that by 1988 seven of the top ten banks in the world were Japanese. He measured the wealth effects of the Basel Accord on shareholders of international banks, finding that Japanese bank shareholders gained from the Basel Accord and that the Accord failed to eliminate the pricing advantages of Japanese banks. In sum, "the regulation was shaped in such a way as to benefit shareholders of Japanese banks" (Wagster, 1996, p. 1343).

There is other support for Stigler's hypothesis. In recent years, commercial and investment banking regulation has been voluminous and costly but not restrictive on operations. The crisis illustrated that regulatory capital was often less than required by the risks. Liquidity, which was at the heart of the crisis, was substantially unregulated in most countries. Institutions were allowed to use internal models to determine regulatory capital notwithstanding the shortages of data, the fact that their data were derived from good times, and that models often performed poorly. At the creation of Basel II, critics pointed out the dangers of banks being allowed to use internal models, see for example Peterson (2001). Recently regulators have become concerned that the risk weights calculated by individual banks need scrutiny, are not comparable and may inflate measured capital ratios (see Jenkins, 2011).

[8] This is one reason why regulators have focused on a number of tools, including "living wills," and an array of "resolution tools" for winding down, transferring assets and liabilities to a bridge bank or a buyer or closing insolvent institutions.

Regulation is at an inflection point. The official sector, from politicians through central banks and regulators, has a changed calculus, including probabilities of default, for future policy. Their short-run difficulty is to implement their preferred strategy while not diminishing growth during the transitional period. Those looking for support of Stigler's proposition that regulation is acquired by an industry would note the following weaknesses of the new proposals.

1. The time frame for the new Basel III rules stretches to at least 2019.
2. The demands particularly for funding liquidity are not onerous and still focus on a survival period of only 30 days.
3. The proposed leverage ratio is weak and not final but merely for future assessment by supervisors.

The literature on government intervention has been dusted off as a result of the crisis. While Benston and Kaufman (1996) cogently presented their arguments for non-intervention in the absence of safety nets, there are other arguments for the role of governments. Many of the arguments since Benston and Kaufman relate to financial networks and the incompleteness of markets to deal with future states of liquidity. Allen and Carletti (2008) argue the case for regulation to deal with market failure, and specifically an inefficient provision of liquidity due to incomplete markets. Markets are incomplete in the sense that financial firms are unable to hedge all risks, for example liquidity needs in future states of the world. As a result, asset sales became the primary method of correcting shortages of liquidity. The interbank funding market suffers from excess demand during a crisis as well as a free rider problem. They note how there is a subsidy on bank debt through underpriced deposit insurance and government guarantees. Haldane (2009) argues that financial institutions underestimate the probabilities of severely adverse outcomes. He points out that network externalities, the complexity of networks and connections through the risks associated with counterparties' risk to their own counterparties cannot be assessed by individual institutions. Finally he notes how incentives are misaligned, contributing to greater risks to the financial system than those assessed at the level of individual institutions. He notes a problem of time inconsistency: that a pre-commitment by authorities to avoid moral hazard by letting institutions fail lacks credibility. Thus the official sector has recalculated higher probabilities of adverse outcomes and has a renewed agenda for greater regulation.

5.4.1 Providing benefits to governments

The public choice approach (e.g., Buchanan and Tullock, 1962) treats political actors as maximizing some function in a self-interested manner. They do not maximize public welfare but it is less clear what they do maximize. Alternatives include maintaining political tenure and maximizing the value of post-political employment. Short-term manifestations may include maximizing total votes, votes in marginal districts, and political contributions where allowed by law.

Pattison (2008a, pp. 38–43) illustrates that there are many actors in the creation of regulation, ranging from legislatures, individual politicians, cabinet ministers, officials in finance ministries, regulators, deposit insurers, central banks and others. They are often differently affected by regulations and may have duties and costs imposed on them which they are ill equipped to manage, let alone master. Collectively they either form or deliver government policy which is often incomplete or contradictory, especially when politicians react to short run crises, and in the absence of a coherent set of policy prescriptions.

The financial industry must understand the politics of regulation. The industry, arguably, must provide net benefits to governments, but how? Individual voters have scarce incentives to inform themselves of complex policy issues. Meanwhile, politicians respond to key interest groups or identifiable voter classes to allocate the political benefits of regulatory changes. Such groups, including consumer associations, small business, trade unions, large corporations, lawyers and accountants, should have a significant collective incentive to invest in new laws, rules and regulations.

This creates a coordination challenge for the beneficiaries of regulation. Different beneficiaries may have opposing interests. For example, shifting over-the-counter (OTC) transactions to centralized clearing would have winners in terms of safety and soundness. Yet many corporate treasurers prefer the status quo as they would need less funding to support OTC transactions than managing collateral with a clearing house. From the financial institutions' point of view, better safety and soundness regulation for banks confers competitive advantages to insurance companies and less regulated firms such as hedge funds.

Finally, the solution to this bargaining problem will depend partly on the initial distribution of wealth as well as its redistribution with re-regulation. One distributional issue is between shareholders and employees over the treatment of bonuses and remuneration generally. Another is between consumers of financial services and the shareholders in financial companies.

5.4.2 The social contract

The concept of a "social contract" has been used to describe an implicit, unwritten agreement between governments and financial institutions. Typically this has centered on the banking industry, as it controlled the payments system and was the major contact between voters and the industry. The complete social contract would include laws written by legislatures, rules and regulations promulgated by regulators and other agencies as well as an unwritten "code of conduct" for industry and government. It is difficult or impossible to codify fully the relationship between the financial industry and government, including the exchange of benefits, risks and costs. Historically, control of the payments system was essential to the social contract as governments' limited participation in the payments system. As Bryan states (1990, p. 115), "in every major nation in the world, banks were more than willing to be used as instruments of national policy, given the importance of their respective governments to their success." He also points out, consistent with Benston and Kaufman's views on competition issues in regulation, that such deals work only if government actions are effective at managing competition.

In the UK, moral suasion was used to achieve political, social or industrial goals outside of legal requirements. This could be used to rally bank support for an institution in difficulty or to give instructions as to the preferences of the official sector, for example on a potential corporate insolvency. Paul Tucker (2009, p. 1), Deputy Governor of the Bank of England spoke of "the environment – the 'rules of the game' or Social Contract – within which banks will need to operate in future." He described this as being about economic substance not legal form. There were three elements of the social contract in the UK: that banks profit from government-provided liquidity in exchange for submitting to regulation; they have access to liquidity insurance at the Bank of England; and they must finance depositor protection. Tucker noted that all three elements of this contract need to be redrawn. He said social contracts are about what society demands of banks to make their failures more tolerable. The more general point seems to be that the social contract reflects an implicit, unwritten cost-benefit relationship.

Harkening back to Benston and Kaufman, Tucker notes that deposit insurance changes the nature of regulation. Regulation must minimize risks to the deposit insurer. There is a significant data requirement for this. For example, banks must maintain current, detailed information on each retail deposit to facilitate rapid payouts under the EU's Financial Services Compensation Scheme.[9]

The social contract remains a useful umbrella for both written and unwritten rules. The concept retains a close relationship of regulation to deposit insurance, and the economics of the financial system and safety nets. It works best where there is an oligopolistic structure that facilitates discipline on the contract by government and coordination by the industry. In these circumstances financial institutions have large values in their brand, which facilitates support for, and compliance with,

[9] The European Union directive specifies a target for a payout within 20 days. The reason for the difficulty with data is that depositor protection is: (a) only for specific eligible deposits; (b) many customers may have multiple accounts with minor differences in the way the name appears on the account; (c) there are many trust accounts with a large number of beneficiaries, many of whom may have a small economic interest; and (d) there is a cap on the total protection and thus excessive payouts could easily occur.

government initiatives. This breaks down for smaller financial institutions which benefit less, and the unregulated or less regulated institutions such as hedge funds or private equity. Where it breaks down there is a risk that regulation will follow.

5.4.3 Social and private costs and benefits

The debate over regulatory change takes place at a level of abstraction as there is no agreed model of regulation. Costs, benefits and calibration will vary from system to system and from institution to institution. Costs and benefits, and differences among them, figure prominently in conflicts over public policy in financial regulation. This debate rests on complex costs and benefits to the government in power, especially one that has suffered a financial crisis, the regulator who must define and enforce the new laws and regulations, the regulatee, and to competing institutions who may be better or worse off because of less or more regulation on competitors. There are anticipated costs and benefits and unanticipated ones. It is impossible to model all the behavioral changes that will occur through altered incentives and changed competitive conditions. Many of these are opaque and difficult to quantify.

Three types of costs should be distinguished:

1. social costs and benefits, reflecting society as a whole;
2. individual private costs and benefits reflecting these for individual financial institutions; and
3. industry (private) costs and benefits reflecting these for the industry as a whole.

Industry costs and benefits do not distinguish the differential impact of common regulatory changes across firms as a result of their particular business strategies and balance sheets. One example in the current regulatory debate is the plan to increase regulation, and specifically capital requirements, for systemically important financial institutions (SIFIs). Customers might reasonably assume that the extra scrutiny makes them safer than other institutions and might then place more of their business with them. Alternatively, additional regulation might make SIFIs less competitive, which may be consistent with a goal of reducing their relative sizes. In this circumstance one might expect smaller financial institutions to support greater regulation for SIFIs. Quantitative impact studies (e.g., BCBS, 2010e), although they show distributions of outcomes, represent the outcome for the industry, not firms.

Individual private costs and benefits in the financial industry reflect costs and benefits at the firm level. They represent the unique loss distribution of each firm, reflecting its business mix, categories of risks and their correlations. They imply an amount of capital required to protect against failure at a specific probability level such as 99.9%. These will differ from industry costs and benefits. In a world of data-driven regulation, any new suite of regulations will have a considerable differential impact as noted in the quantitative impact study (BCBS, 2010e). Each firm will have hundreds or thousands of "risk buckets" (specific categories of risk weighted assets arising from different probabilities of default, loss given default, exposure at default and other parameters) that differ from those of competitors. It is analytically difficult for institutions to assess regulatory impacts, especially of complex Basel III rules, as firms with different risk buckets, models and data will be uniquely affected.

Social costs and benefits and how they differ from individual and industry costs and benefits illustrate the political and social dimensions of the regulatory challenge. Aggregate social costs of financial failures are greater than individual or industry private costs. Of course, industry does not factor these in to their costs. Social costs include welfare and wealth losses from bank failures, unemployment costs and the large risks from destabilizing macroeconomic policy and additional public sector borrowing to deal with deficit financing. These are particularly onerous for countries such as Iceland or Ireland, but also for the UK, potentially Switzerland and others (see Reinhart and Rogoff, 2009). In addition there is a misallocation of resources in financial crises. For example, maintaining low interest rates for long periods may conceal the true rates of return to capital investments. Thus private costs and benefits are divorced from public or social costs and benefits. Public policy, including regulation and sometimes taxation, is used to bridge the gap and to alter behavior.

Miles (2010) points out that the decline in real output in the UK relative to its path would be approximately 10% from 2007. However, in the case of Ireland, with a much worse financial crisis, GDP was 13% lower, but with the loss of a 3% annual growth rate foregone the total output loss would be more than 20%. Miles et al. (2011) assess the marginal benefit of additional equity capital relative to the cost of financial crises. This is discussed further below.

5.4.4 Collective action problems

Rational behavior by individual competitors may produce outcomes that are collectively negative for firms, industry and society. Many situations could be solved by collective action. That is, the strategies of individual competitors must be analyzed *together* to determine their interaction and outcomes. However, competition laws prohibit competitors from discussing and agreeing coordinated strategies. This type of analysis differs from the microeconomic view where atomistic competition determines the outcome from uncoordinated decisions. In competitive markets, no individual competitor can affect the equilibrium solution or price. If players cannot enforce agreement, they will choose rational strategies that yield an inferior outcome to choices which they could have selected with communication. What is best for each financial firm acting alone may not be best for all of them together.[10]

There are potential solutions to collective action failures. Competitors may learn from mutual behavior through signaling or repeatedly playing the same game. Competing firms could consider the public good and determine individually to sacrifice short-run earnings and thus deviate from industry practices, for example risk or liquidity behavior. Without cooperation or regulation, game theory illustrates that public benefits will be under-produced by competitive behavior.

Collective action issues occur where there are disincentives for private actors in taking decisions that are optimal to them but which competition might unravel. These disincentives often arise from earnings or market share reductions relative to industry practice and are exacerbated by analysts' focus on quarterly earnings. These situations are most prevalent in highly competitive situations. Examples include the proposition that "as long as the music is playing," banks must "get up and dance" (see New York Times, 2007). It has also been a long-standing observation that there is "lemming-like" behavior on the part of financial players. No bank can afford to lend to higher credit standards when these are diminishing in the marketplace. Equally no bank can afford to hold excess liquidity which lowers earnings when the industry holds few liquid assets. The marketplace cannot solve them and thus regulation is used to replace private decisions. This may well be welcomed by industry, as suggested by Stigler's analysis.

The Government of Canada has imposed regulations on the mortgage industry for this reason: to tighten lending standards that no individual institution would do on its own (Masson and Pattison, 2010, p. 9). Much international coordination that occurs through the BCBS or the FSB can be seen as agreements to capture collective action solutions across borders. Individual countries are under pressure to lower capital or liquidity requirements to benefit local competitors, to the detriment of other jurisdictions (Masson and Pattison, 2009b). Competition could lead to a race to the bottom in terms of regulatory requirements.

5.5 Network effects and systemic risks

An overriding issue is the fallacy of composition: the overall risk of the system exceeds the sum of the risks of the individual financial institutions and markets. This crisis demonstrated that risks in the financial sector are larger than the sum of individual risks. It is not sufficient to assure the safety and soundness of individual

[10] In game theory, this suboptimal outcome is typical in non-cooperative games, such as the Prisoner's Dilemma.

banks without looking at network and systemic effects. This is not captured in the economic capital and risk assessments of individual institutions.

Regulatory concern has moved from a focus on the safety and soundness of individual institutions to embrace additionally the collective and interactive effects of entire financial systems and networks. Systemic risk is related to contagion, changing correlations among markets and risks, and reductions in market and funding liquidity.[11]

Interconnectedness is a key driver of systemic risks, either from the impact of individual institutions to systemic events or where each bank not only adds to systemic risk through its unique exposure to exogenous shocks but affects the probability of transmitting losses through networks, largely but not only the interbank market. Both approaches must work together. The data requirements to assess such risks are unclear.

Data needs are not restricted to the official sector and their international counterparts. The private sector needs to manage counterparty risks with systemic components. Pillar Two of Basel II and III considers the congruence of bank risks with capital and risk mitigants. This cannot be based only on a single institution's data. Pillar Three, market transparency through enhanced risk and capital disclosure, is controversial but potentially at least as important. Pillar Three today cannot shed much light on the important exposure to network risks. Over time, as network effects become more important in the lexicon of financial risk and regulation, these concepts will need to move from the official sector to corporate disclosure. While many network effects are not visible to isolated participants, there are indicators and proxy measures including different types of settlement risks. These issues and data are critical on several levels: methodologies of regulation and supervision in general; the availability and robustness of diagnostic tools, such as stress testing; implementation of capital and liquidity rules for different types and sizes of banks with different businesses and balance sheets; international standards given the overflow of systemic risks from one country to another as illustrated in this crisis. Individual financial institutions focus on their balance sheets and risk profiles. Interactive effects do not figure prominently in discussions at the industry level, aside from current concerns over concentrating risks in central counterparties. To hold additional capital or liquidity for the larger policy reasons for uncertain systemic risks may not be commercially sound to individual managements and boards, hence the need for regulation.

While this suggests problems in achieving "buy-in" for systemic risk regulation by financial institutions, the IMF (2010b) raises an issue with the incentives of

[11] The Federal Reserve Bank of New York (2007, p. 7) notes that systemic risk is difficult to define precisely. Participants in the associated conference did not share a consensus on the definition. Chapter 4 of the Handbook covers systemic risk in greater detail.

regulators. Regulators have a history of forbearance as opposed to early intervention and prompt corrective action. While additional monitoring responsibilities could be imposed on regulators, and additional systemic risk tools employed, regulators have various incentives, including political ones, to forgive systemically important institutions while taking possibly timely intervention for smaller institutions. Thus it is necessary to consider how the official sector as a whole addresses systemic risks, including decision criteria, emergency liquidity, resolution, and deposit insurance.

There is a wide range of idiosyncratic risks in institutions and markets that can grow via contagion into systemic risks. Many issues have been raised about hedge funds, similar questions apply to mutual funds, exchange traded funds, clearing houses and others. There are questions about how to treat industrial companies that use swap contracts that may or may not be supported with collateral. There are issues about collateral management itself and its capital treatment. Thus this section reinforces two basic points: (a) re-regulation will continue to pursue these complex issues that are not yet resolved at the analytical or policy level; and (b) this work will call for new data from institutions and markets to evaluate risks that are not well understood or quantified.

5.5.1 Re-regulation and the 2007–2009 crisis

As noted above, Stigler (1975) explains why regulation is often desired by financial firms. The more competitive an industry, the more likely it is to seek regulation to achieve what it cannot do competitively. The original Basel Accord is an example (see Tarullo, 2008). It was designed to restrain competition through capital requirements that could not be achieved by competitive action or national regulation. The recent crisis revealed other anomalies. Financial regulation differs in important respects from the regulation of other industries such as pharmaceuticals, airlines or trucking. Systemic and macroeconomic impacts are two differences. Another variable that makes the financial industry unique is not simply the data-driven nature of regulation but the extent, volume and lengthy historical record of data that must be retained. In addition, only the financial industry offers safety nets that can bankrupt governments.

This crisis demonstrates the difficulties in managing market forces in addition to credit risks. Bryan (1991, p. 9) had already argued this point about a previous crisis. Now trading and derivatives markets create counterparty risks that can bring markets to a halt: destroy liquidity, create fire sales, affect funding markets, and create risks for otherwise healthy firms through network externalities, market failures and inabilities to settle transactions. These risks are not visible or predictable to individual firms in a quantifiable form. When triggered, the effect is rapid: interbank markets and some types of trading tend to shut down quickly as fire sales render markets less liquid. Many markets are differently regulated, with varying methods and capabilities of supervision, insolvency and bankruptcy laws. In a crisis uncertainty grows rapidly. Uncertainty works against stability.

What made the regulatory response different this time? While there was extensive international monetary and exchange rate cooperation, there was little international regulatory or supervisory cooperation prior to the Herstatt crisis in 1974. That failure ultimately resulted in the creation of the Basel Committee on Banking Supervision. For this crisis, governments rapidly invigorated international coordination through the Group of 20. In many ways, this made a resolution much more difficult. Regulatory and supervisory models may have shared common goals and techniques across countries but differed in such things as financial instruments, coverage of laws, legal interpretations, insolvency laws and practices, calibration and supervision. Determining the desired international and domestic outcomes and negotiating the bargain were constrained by these differences and lobbying. For example, the UK introduced new liquidity requirements prior to an international agreement that followed shortly thereafter. Switzerland introduced higher requirements than the BCBS because, as a smaller country, the size of its financial institutions was too great relative to GDP for the country to afford its safety net obligations or future bail outs. Canada continued more stringent leverage controls than the new proposals. Many countries, including Canada and Australia, advanced the implementation timetables compared with the international agreement.

Second, re-regulation swept up institutions and activities that did not directly contribute to the crisis or meet the underlying requirements for new rules. Non-credit derivatives did not contribute immediately to the crisis and yet figured prominently in plans to shift transactions from over-the-counter (OTC) to central clearing, although this was reasonable to address. Another example is the renewal of demands to recreate the Glass–Steagall Act creating a separation of commercial from investment banking. Yet while investment banks failed, such as Lehman Brothers, or required support, such as Bear Stearns, these were not *integrated* with commercial banks. It is true that mortgage backed securities created enormous difficulties but not the integration itself. European regulators put in place regulation and reporting around hedge funds even though their contribution to the crisis was unclear. Some hedge funds ceased trading or failed in a non-threatening manner without knock-on effects. Europe also planned to introduce new restrictions on commodity markets even though commodity markets per se did not contribute to the crisis.

The above points are not to suggest that because certain risks did not figure prominently in the crisis they should be ignored. Rather the point is that the crisis was used as the rationale to regulate certain activities or institutions whether or not they were involved in a material respect.

5.5.2 The importance of data-driven regulation for managing change

The fact that regulation is increasingly driven by formulas and data takes power away from the political system and moves it to a more technical plain. Data-driven regulation pushes regulators and supervisors as well as financial institutions to become data- and model-centric which, given limited budgets, limits their ability to allocate incremental resources to focus well on other issues. This poses risks to all parties. Politicians control regulatory bodies but cannot directly influence the models or data requirements. The fact that the framework for many of these is now negotiated internationally means that technical aspects are less amenable to political discussion or persuasion. This presents an intense challenge for governments.

Moreover, there will be regular updates to model and data requirements and periodic recalibrations. Events will show weaknesses in regulatory decisions around models and data as well as weaknesses in institutions' models. Regulatory additions to requirements such as cyclical factors will continue. This raises fixed costs for financial institutions and shapes requirements for their risk data, accounting and related architectures. Lobbying of the political system puts indirect pressure on international bodies as well as domestic regulators. A data-driven regulatory system changes the political nature fundamentally. However, the net contribution, effectiveness, consistency and enforcement of international financial regulatory cooperation have not themselves been assessed.

5.6.2 Regulatory catharsis

Re-regulation is a cathartic process, in addition to one decided by cost-benefit analysis, pricing social costs and external effects, or the capture theory (Dal Bó, 2006). Catharsis is defined as the process of releasing tensions or emotions. It is rare for legislatures to re-regulate when things are working well, even if there are identifiable needs for new rules. The many demands on available time and resources constrain legislative timetables. There are strong forces, both formal (e.g., legal standard for review) and informal (e.g., lobbying), that encourage deference to precedents and discourage regulatory change. Few countries have a regular review of financial legislation. Canada is an exception where bank charters cease on a specific date, requiring the government of the day to introduce new banking legislation.

International coordination takes years of negotiations. The intermediate result is "soft law," requiring implementation in each national context. The BCBS spent much of the 2000s dealing with detail on modeling capital for credit and trading risks, relatively neglecting more critical issues such as liquidity. The BCBS agenda changed priorities for national regulators.

Thus under a cathartic model of regulation, governments, politicians and regulators accumulate potential reforms, awaiting opportune moments, such as a crisis, to introduce debate. The fixed allotment of legislative time can dilute the focus on critical issues. The nature of crises, combined with fixed legislative time, shapes the political process for regulatory reform, and biases it against optimum outcomes.

Numerous examples exist of reforms introduced opportunistically in the wake of the crisis, supporting the notion of catharsis.[12] Scholtes (2009) notes that both consumers and legislators linked reform of overdraft charges to the bail-outs of banks. Similarly, politicians in some countries included long favored hedge fund regulation and restrictions on short-selling in reform proposals. The introduction of rules on unrelated commodity markets is one of the clearest examples of opportunistic reform predicted by catharsis. This is not to argue for or against these changes but merely that the financial crisis allowed legislative horse-trading to ensue on a diversity of issues.

An analysis of the regulatory proposals that ensued from the recent crisis could fill many volumes. We approach them on three levels: international, "soft-law" proposals, mainly from the BCBS; domestic proposals such as those in the Dodd–Frank Act; and things left undone, leaving risk gaps and arbitrage opportunities.

International coordination began strongly but lost momentum on several fronts. Individual countries regulated independently and with higher standards in many cases, such as Switzerland for capital and the UK for liquidity; Canada maintained a higher leverage ratio that had served it well during the crisis. Domestic lobbying watered down the final international agreement and stretched out implementation timelines. For example, short-term liquidity requirements and the framework for calculating them only provide for 30 days – barely sufficient for supervisors to put in place a resolution plan. Another coordination gap is the absence of an international resolution framework. Situations such as Lehman Brothers remain a threat and would perpetuate too-big-too-fail and challenge domestic safety nets.

In terms of major reforms negotiated in legislatures, as opposed to decided by regulators, the Dodd–Frank Act cannot be assessed as the final rules are not yet written. The ultimate outcome could range from an acceptable compromise to ineffective. Many countries did housekeeping in their domestic laws on such matters as resolution authorities and new forms of central bank intervention on liquidity and asset purchases.

What neither international coordination nor national laws did was establish a framework for effective regulation that was broadly accepted. Nor were there common views on calibration. Neither the international policy nor data frameworks support widely agreed risk and regulatory methodologies, or improved market

[12] See Table 1.1 of this Handbook, for example.

discipline in the context of network effects and systemic threats to stability. What was done was ad hoc and spread over an extended time frame.

The crisis did provide proposals for more radical reform. Countries such as the UK looked more widely at options than did the USA. Dodd–Frank was not radical, with the possible exception of the separation imposed on proprietary trading. However, there were limits to this separation that undermine its effectiveness and leave opportunities to farm out such risks. The verdict on Dodd–Frank remains to be written.

The key issue for much regulation is calibration: how much capital and liquidity are needed to balance private risk and social costs. The arguments of Miles (2010) and Miles et al. (2011) are not radical (although some may see their conclusions in that light). They begin with the proposition that rare, negative impacts on GDP from historic financial crises demonstrate large social benefits from financial institutions holding much more capital. Miles et al. (2011) argue that optimal capital ratios are much higher than current or planned amounts and are likely within a 16–20% range. Such capital ratios would be regarded as radical by the industry. Similarly Admati et al. (2010) make a compelling case for higher capital ratios.

5.7 Conclusions on regulation and the crisis

We entered a new era of "data-driven regulation" after 1999. The crisis interrupted this process and demonstrated how limited early benefits were. This is largely because the Basel II model was based on a risk sensitivity trade-off, short-run data, arbitrage opportunities, simplified correlations among risks, and reliance on historical cost, accounting-driven capital and risk data. Basel II was not appropriately calibrated for trading and securitization risks. Thus this crisis may have the silver lining of reformulating over a longer time frame a regulatory plan that was haphazard, and reform packages that had limited ambitions. However, this is a job for regulators, not politicians. The defects were those of structure, quantification and calibration. These issues are not well suited to political negotiations.

However, once regulation is founded on officially reported data from individual institutions for determination of capital, leverage, liquidity and other measures, there is no turning back to simpler forms of unquantifiable regulatory decisions. While much acclaim was given to the "market discipline" benefits of Basel II and III, little of this has been in evidence. Systemic risks and network effects raise the challenge for market discipline and both Pillars Two and Three of Basel III.

Data and judgment must work harmoniously. The crisis was founded on poor data, and poor judgments in institutions and official sectors, including central banks. It is reasonable to expect that supervisors will continue to demand much more data, specifically formatted for regulatory purposes. Institutions should organize their risk, accounting, information technology and other departments for a constantly changing demand for new, altered and historical data and their retention in data warehouses for many years. Data-driven regulation is incompatible with the status quo. The related demands of refining model requirements, specifying capital formulas and calibration have been underway for a few years, but there is much more to do, as set out in the work program for the Basel Committee. There is a shift in power in important areas from legislatures to regulators but also a battleground for future confrontations between the two, and for conflicts between international coordination and national regulatory independence.

Bibliography

Admati, A., P. De Marzo, M. Hellwig and P. Pfleiderer, 2010, Fallacies, Irrelevant Facts and Myths in the Discussion of Capital Regulation: Why Bank Equity is Not Expensive, Working Paper, Stanford Graduate School of Business, 2065.

Allen, F. and E. Carletti, 2008, Financial system: shock absorber or amplifier, *BIS Working Papers*, **257**, Bank for International Settlements.

Bank of Canada, 2007, *Financial Stability Review*, **22**, October.

Bank of Canada, 2010, *Strengthening International and Liquidity Standards: A Macroeconomic Impact Assessment for Canada*, Bank of Canada.

Basel Committee on Banking Supervision, 2005, Supervisory Guidance on the Use of the Fair Value Option by Banks under International Financial Reporting Standards, *Consultative Document*, Bank for International Settlements.

Basel Committee on Banking Supervision, 2010a, *Basel III: A Global Regulatory Framework for More Resilient Banks and Banking System*, Bank for International Settlements.

Basel Committee on Banking Supervision, 2010b, *Basel III: International Framework for Liquidity Risk Measurement, Standards and Monitoring*, Bank for International Settlements.

Basel Committee on Banking Supervision, 2010c, *Guidance for National Authorities Operating the Countercyclical Capital Buffer*, Bank for International Settlements.

Basel Committee on Banking Supervision, 2010d, *An Assessment of the Long-Term Economic Impact of Stronger Capital and Liquidity Requirements*, Bank for International Settlements.

Basel Committee on Banking Supervision, 2010e, *Results of the Comprehensive Quantitative Impact Study*, Bank for International Settlements.

Benston, G., 2008, The shortcomings of fair-value accounting described in SFAS 157, *Journal of Accounting and Public Policy*, **27**, 101–114.

Benston, G. and G. Kaufman, 1996, The appropriate role of bank regulation, *The Economic Journal*, **106**, 688–697.

Brunnermeier, K., 2009, Deciphering the liquidity and credit crunch 2007–2008, *Journal of Economic Perspectives*, **23**, 77–100.

Bryan, L., 1990, The role of banking in society, *The McKinsey Quarterly*, **3**, 113–126.

Bryan, L., 1991, *Bankrupt: Restoring the Health and Profitability of Our Banking System*, Harper Business.

Buchanan, J. and G. Tullock, 1962, *The Calculus of Consent: Logical Foundations of Constitutional Democracy*, University of Michigan Press.

Crook, C., 2010, Clive Crook's blog, *Financial Times*, June 21, 11.

Dal Bó, E., 2006, Regulatory capture: a review, *Oxford Review of Economic Policy*, **22**, 203–225.

European Central Bank, 2007, *Financial Stability Review*, December.

Federal Reserve Bank of New York, 2007, *New Directions for Understanding Systemic Risk*, Federal Reserve Bank of New York.

Federal Reserve Bank of St. Louis, 2011, *The Financial Crisis: A Timeline of Events and Policy Actions*, Internet site, downloaded August 28, 2011, http://timeline.stlouisfed.org/index.cfm?p=home

Federal Reserve Board, 2011, *97th Annual Report 2010*, Federal Reserve Board.

Financial Crisis Inquiry Commission, 2011, *The Financial Crisis Inquiry Report*, Public Affairs.

Financial Stability Board and International Monetary Fund, 2010, *The Financial Crisis and Information Gaps*, May.

Financial Times, 2009, Too much of a very good thing, *Financial Times*, September 1, 6.

Haldane, A., 2009, Why Banks Failed the Stress Tests, Speech given at the Marcus Evans Conference on Stress Testing, Bank of England, February, 9–10.

Haldane, A., 2011, Capital Discipline, Based on a speech given at the American Economic Association January 9, 2011, Bank of England.

IFRS Foundation, 2011, IFRS Technical Summaries, Internet resource, www.ifrs.org/IFRS+technical+summaries/Summaries+Eng+11.htm.

International Monetary Fund, 2010a, *Financial Sector Surveillance and the Mandate of the Fund*, March 19.

International Monetary Fund, 2010b, *Global Financial Stability Report*, International Monetary Fund, April.

Jenkins, P., 2011, Drive for global standard on banks' lending risks, *Financial Times*, May 3, 17.

Landsman, W., 2006, Fair value accounting for financial instruments: some implications for bank regulation, *BIS Working Papers*, **209**, Bank for International Settlements.

Masson, P. and J. Pattison, 2009a, *International Financial Policy Reform and Options for Canada*, Conference Board of Canada.

Masson, P. and J. Pattison, 2009b, Financial regulatory reform: using models of cooperation to evaluate current prospects for international agreement, *Journal of Economic Asymmetries*, **6**, 119–136.

Masson, P. and J. Pattison, 2010, *The Financial Crisis, Regulatory Reform, and International Coordination – What Remains to be Done*, Conference Board of Canada.

Miles, D., 2010, Leverage and Monetary Policy, Speech delivered at ESRI/FFS Conference, Dublin, October 12, Bank of England.

Miles, D., Y. Jing and G. Marcheggiano, 2011, Optimal Bank Capital, External MPC Unit, *Discussion Paper* 131, Bank of England.

New York Times Dealbook., 2007, Citi Chief on Buyouts: 'We're still Dancing', July 10.

Pattison, J., 2008a, *Introduction to Financial Services Regulation A Global Approach*, G7 Books.

Pattison, J., 2008b, Macroprudential conflicts in Basel II capital calculation, *Journal of Risk and Regulation North Asia*, **I**, 33–40.

Peterson, M., 2001, Basel gives banks the whip hand, *Euromoney*, March, 48–53.

Plantin, G., H. Sapra and H. Shin, 2008, Fair value accounting and financial stability, *Financial Stability Review*, Banque de France, October, 85–94.

Reinhart, C. and K. Rogoff, 2009, *This Time is Different*, Princeton University Press.

Scholtes, S., 2009, US banks set to be called to account for overdraft charges, *Financial Times*, August 10, 3.

Shin, Hyun S., 2009, Reflections on Northern Rock: The bank that heralded the global financial crisis, *Journal of Economic Perspectives*, **23**, 101–119.

Stigler, G., 1975, The economists' traditional theory of the economic functions of the state, in: G. Stigler (ed.), *The Citizen and the State, Essays on Regulation*, University of Chicago Press, ch. 7, 103–113.

Tarullo, D., 2008, *Banking on Basel: The Future of International Financial Regulation*, Peterson Institute for International Economics.

Tucker, P., 2009, Regimes for Handling Bank Failures: Redrawing the Banking Social Contract, Remarks to the British Banking Association Annual International Banking Conference, Bank of England, June 30.

Tucker, P., 2011, Macroprudential policy: building financial stability institutions, 20th Annual Hyman P. Minsky Conference, Bank of England.

Turner, A., 2011, Reforming Finance: Are We Being Radical Enough? 2011 Clare Distinguished lecture in Economics and Public Policy, Financial Services Authority, February 18.

Wagster, J., 1996, Impact of the 1988 Basel Accord on international banks, *Journal of Finance*, **LI**, 1321–1346.

PART II

REQUIREMENTS AND SOURCES FOR FINANCIAL RISK MANAGEMENT

EDITORS' INTRODUCTION TO PART II

Bill Nichols

The data environment in the financial industry is expansive, complex, rapidly changing, and riddled with inconsistencies and localized variations. The principal product of the industry being information, data are ubiquitous. Given that financial products are composed of data, the only barriers to manufacturing additional products reside in the ability of the relevant market participants to process that data. Data are not just actively created by every player in the markets, but are also energetically modified across the fabric of the marketplace.

Modifications happen for at least two reasons. Data about transactions are used in calculations to generate new, derived data. Risk and financial management metrics are an example. The derived data are also used to create new, derived transactions which give rise to their own set of data. This second process can be repeated several times to yield a chain of transactions that are linked only by data shared across multiple intermediaries.

The financial industry juxtaposes the real financial economy with the capital markets. The real financial economy includes retail banking (with products such as credit cards and mortgages), insurance and retail investment products. It also includes the financial activities of large corporations including such activities as hedging financial risks to their operations. Capital markets attempt to combine and make financially fungible these disparate activities. Understanding the intersection of these worlds is critical to analyze the stresses in the marketplace.

This section provides a high level overview of supply-chain data both in the capital markets and in the real economy, and describes the intersection and transmission of data across their common boundaries. This section thus acts as a foundation for the rest of the Handbook, which describes issues and problems that firms have historically had in collecting and managing these data, and possible solutions to some of these problems.

For the sake of simplicity, the descriptions of data in the real economy here focus on banking only. Not only was the recent financial crisis precipitated by actions in this sector, this is also the sector that is most closely intertwined with the capital markets, giving us the ability to discuss a broad swath of data problems. In Chapter 6, Jefferson Braswell and Robert Mark provide a thorough structural view of the landscape and the intersection between the banking system, the real economy, and the capital markets sector. In Chapter 7, Martijn Groot follows with a more in-depth look at the actual data created in capital markets activities, and provides guidance on how to keep your bearings in these dark woods. Continuing in this vein, in Chapter 8 Marc Alvarez outlines the world of capital markets reference data – and takes a stand to provide a clear working definition, something many have struggled with. Following on from the previous chapters, his discussion of the importance of synchronization across datasets should be taken as fair warning to all who wish to understand the information flowing in these systems. Finally in Chapter 9, Margarita Brose, Mark Flood and David Rowe provide a high-level review of the current state of practice in risk management, and provide a cogent discussion of the trends and approaches in emerging operational risk practices coming out of the recent financial market crisis. Their look at operational risk practices evolving in the hedge fund industry is especially timely.

6

Banking and financial activities in the real economy

Jefferson Braswell and Robert Mark

This chapter begins with an Overview and objectives (Section 6.1) that summarizes the challenges facing the financial industry in keeping up with the acceleration of the pace of change, the degree of interdependence and volatility in the global economy. More than just IT horsepower, the importance of timely, accurate and appropriate data is key to being able to meet these challenges.

Perspectives on the banking industry (Section 6.2) describes the banking industry from the point of view of a widening set of perspectives, beginning with the types of products, services and customers which form the core of the business of banking, and expanding in scope to examine banking from an institutional and industry sector perspective.

A discussion of the domains and properties of the underlying and inter-related categories of data that comprise the foundation of banking is provided in Foundations of banking data infrastructure (Section 6.3).

We look at what to do from a business strategy perspective to address these issues in Financial risk information management planning (Section 6.4). Central to that assessment is the fact that regulators (and in particular agencies[1] that are tasked with systemic risk management and oversight) ultimately need, and will expect to receive, consistent, accurate, timely and digestible data from tens of thousands of financial institutions.

Institutions comprising the source of all of those industry data must therefore previously have managed to implement their own firm-wide financial information infrastructure with internally consistent, accurate, timely and digestible data.

[1] For example, the Financial Stability Board (G20), Financial Stability Oversight Council (FSOC), and the Office of Financial Research.

Handbook of Financial Data and Risk Information, ed. Margarita S. Brose, Mark D. Flood, Dilip Krishna and Bill Nichols. Published by Cambridge University Press. © Cambridge University Press 2014.

Along with underscoring the importance of data standardization and accuracy, some recommendations pertaining to the types of measures and policies needed to respond effectively to the current challenges are presented.

For the sake of simplicity, the descriptions of data and their domains in the real economy will focus principally on banking. Not only was the recent financial crisis precipitated by actions in this sector, but this is also the sector that is most closely intertwined with the capital markets at the current time, giving us the ability to discuss a broad swath of data issues.

Banking is a broad and extensive component of the real economy, one which encompasses a huge and diverse set of actors, and one which involves a wide variety of financial products and facilities addressing economic horizons from the very short term to the very long term.

For the purposes of this discussion, the banking customer covers not just the consumers, be they students or heads-of-households, but also includes small businesses, larger corporate finance activities and the financing of large-scale commercial development.

In this view, the distinction between retail banking and wholesale banking is made based largely on whether the counterparty of the bank is an external customer of the bank or is instead another banking institution or the central banking system.

Indeed, most banks have organizational lines of business in which retail banking is focused on the individual consumer and commercial banking is focused on business customers. These organizational distinctions make sense operationally from the perspective of the types of staff and activities required to support differences in the type, size and behavior of the customer and the associated risk exposure.

Both consumer and business customers, however, impinge on and interact with the same fundamental mechanisms of banking in the real economy: liquidity-on-demand, management of short-term cash flow timing, asset protection in the form of savings or investments, and financing of longer-term development and growth via management of debt service in relation to revenue and profitability.

One the one hand, organic banking comprises the realms of shorter-duration, interest-bearing customer deposits and savings (liabilities of the bank). On the other hand, organic banking also comprises longer-duration, interest-earning mortgages and commercial loans (assets of the bank). The consumer spending makes up nearly 70% of the US Gross Domestic Product (GDP),[2] and the residential housing market and commercial financing have, despite the recent financial crisis, historically constituted a significant enabler of the creation of wealth through growth (and the management of risk). This alone puts banking squarely in the mix of financial sectors and infrastructure about which accurate and timely data knowledge and

[2] Bloomberg (2012).

subsequent analysis are crucial, from the consumer on up through the institution and on to government oversight and regulation.

When the intersection of retail banking with investment banking and capital markets is considered, the role of banking – specifically in the manner in which residential home mortgages are syndicated, leveraged and sold into the mortgage-backed securities capital markets – creates a nexus which draws together the realm of traditional banking with that of global financial markets.

6.1 Overview and objectives

Rapid change in banking, in the broader financial system and the global economy, demands more immediate and actionable risk management capabilities derived from comprehensive and timely access to accurate and detailed data

6.1.1 Acceleration of change in the financial markets and the real economy

Rapid rates of change demand a series of responsive changes by firms in order to adapt and survive. The rapid evolution and growth of banking products, activities and financial services in the last quarter century accelerated in the run-up to the recent financial crisis.

The resulting boom-and-bust "G forces," coupled with the follow-on regulatory and economic reactions to these events, have had a noticeable impact on the level and volatility of economic capital and equity – not only that of consumers, businesses and financial institutions – but also that of sovereign governments and the global financial system.

The challenges to financial institutions to adjust successfully to the rapidly changing financial landscape and to navigate the dynamic conditions and behavior of all of the economic players involved have become more daunting. In turn, this requires a more focused, diligent and appropriate set of responses on the part of consumers, corporate management and industry regulators alike.

Business and technology

These necessary responses are not confined to any one specific area, department or part of an institution or agency, but are spread across a wide variety of roles and responsibilities that span the full breadth and depth of an organization. The measurement and management of risk across an organization has, increasingly and of necessity, become a strategic and operational priority of corporate governance.

In order to be efficient and effective, risk management practice must incorporate and be predicated on an informational and analytical infrastructure that provides comprehensive and timely access to accurate and detailed data. The data are sourced and generated from a wide variety of input and reference sources, systems of record, intermediate result sets of analytical scenarios and decision support systems.

The integration of better risk management practices and informational awareness should first start with attention to the quality of the detailed data at the contract-level of customer accounts and positions. Next, it proceeds through the intermediate processing and operational stages of the organization up to the level of executive management and the board of directors. Subsequent transparent disclosure and reporting stages of customer and institutional data will, in turn, inform the public and support regulatory efforts to monitor and maintain an accurate systemic overview of the financial sector.

Financial system interdependence

Traditional distinctions between investment banking and retail banking have become blurred and entangled, due in part to the systemic interaction and integration of retail banking, home ownership and capital markets pipelines which are in turn partially driven by the investment demand for mortgage-backed securities.

Banking assets and liabilities are also becoming increasingly interconnected and co-dependent with securities markets. This has occurred not only because of the mortgage-backed asset pipeline on the asset side of the bank, but also because of the need to invest liabilities (customer deposits) in money market instruments which must avail themselves in one degree or another of financial markets. These linkages have increased the need for correspondingly more complex activities and instruments in corporate treasury and trading departments in order to manage banking products, customer behavior and economic impacts effectively.

Accounting issues

The evolution of General Accepted Accounting Principles (GAAP), and in particular the Financial Accounting Standards Boards (FASB) standards 115[3] and 157,[4] specify the accounting treatment for assets based upon whether a financial institution intends to hold an asset to maturity or is holding the assets principally for the purpose of selling them. If assets are *held to maturity*, the assets are booked on an amortized cost basis. If assets are *held for sale*, the assets are reported based on the *fair value*, or current market price, of the assets.

Mark-to-market accounting has been used in accounting practices for market-based securities trading operations for decades. The market prices of frequently traded securities in liquid markets change constantly, such as the equity shares and bonds of public companies listed on major exchanges. The difference in the market value of a security position, assessed at the beginning and at the end of an accounting interval

[3] FASB, Statement 115. [4] FASB, Statement 157.

(be that a day, week, month, etc.) generates a profit or loss in the income statement at the same time that the current market value is updated on the balance sheet. For example, if the market value of a position at the start of the month is $100 and the value of the position at the end of the month is $90, then a $10 *unrealized* loss would be booked to the income statement as the value of the position on the balance sheet was revised downward. If the value at the end of the month were, instead, $115, then $15 of income would be recorded to the income statement as *unrealized* gain.

Financial Accounting Standard 157 (FAS 157), which was issued in 2006, clarifies and qualifies the basic fair value accounting treatment prescribed for held-for-sale assets by FAS 115, which was issued in 1993. FAS 157 stipulates that the market price used to set the fair value of a held-for-sale asset would be the **bid price**, not the **ask price**. In other words, the fair value would be determined based upon the price at which a buyer would be willing to purchase the asset, not the value that the holder of the security, perhaps optimistically, was *asking*.

Given the increased amount of mortgage-backed securities which were held-for-sale on the balance sheets of financial institutions – particularly institutions which were underwriting, syndicating and retaining mortgage-backed securities in inventory – changes in the market value of a sizable mortgage-backed portfolio have had a dramatic effect on the equity of many financial institutions, and therefore on regulatory capital requirements.

At the height of the liquidity crisis these market value changes were both a cause and a symptom of the financial collapse. Amid the growing concerns surrounding the increasing incidence of default in the loans underlying mortgage-backed securitized products, the fair value of these securities became highly unstable, due largely to the dramatic fall in the number of willing buyers, and price discovery was seriously disrupted.

As a result of all of these factors, portfolios of these securities became "toxic": they could not readily be sold, except at "fire sale" prices, whereas holding on to them created significant hits to both economic and regulatory capital of institutions owning the portfolios – and thus the solvency of the institutions was threatened. Ultimately, as part of the emergency measures taken during the crisis, central banks such as the Federal Reserve acquired many of these so-called "toxic assets"[5] in order to bolster the stability of the financial system by removing them from the institutions' balance sheets and thereby arresting the downward spiral of valuation and risk capital.

This volatile episode has naturally focused much attention on the application of the Fair Value accounting treatment of mortgage-backed securities and mortgage

[5] Federal Reserve Bank Troubled Asset Relief Program (TARP) Information `www.federalreserve.gov/bankinforeg/tarpinfo.htm`. See also the Wall Street Journal interactive graphic, showing the evolution of the assets on the FRB balance sheet (by asset category) over the period 2007–2011.

loan assets in general. Historically, community and regional banks would originate mortgages with the intent of holding them to maturity (or, until they were prepaid by the customer). Such assets earn interest and traditionally have been the primary source of income for banking institutions. As such, the assets are recorded on the balance sheets based upon their (original) book value, or cost, and not marked to market as prices fluctuate. The equity of the institution may certainly grow as the value of loan portfolios grows, based on the net present value of the expected future cash flows from the portfolios, but the balance sheet itself is not subject to the volatility and uncertainty that results from constantly revaluing the assets as is done for those in the held-for-sale or trading classification.

Domestic and international regulatory pressures

Heightened regulatory responses spurred by the recent financial system collapse, such as the *Basel Committee on Banking Supervision*'s recently revised and more demanding capital adequacy requirements and expanded stress testing guidelines (Basel III) will add to the challenges of financial risk management within institutions. Furthermore, the provisions of the Dodd–Frank Wall Street Reform and Consumer Protection Act (Dodd–Frank Act), which empower regulators to require institutions to report significantly greater amounts of financial information at greater levels of detail, will place additional compliance requirements on financial institutions.

In addition to regulatory developments affecting specific parts of the operations of the financial sector (e.g., the requirement to trade OTC derivatives electronically through an exchange and register them in a contract depository) other standards setting efforts address such matters as accounting treatment. As in the case of risk management oversight rulemaking by US and other international regulatory bodies, guidelines of national and international standards boards regarding accounting treatment are not necessarily consistent. There are some differences, for example, between the Financial Accounting Standards Board (FASB) and the International Accounting Standards Board (IASB) stipulations regarding the reporting of one or both of amortized cost (book value) and fair value (mark-to-market) accounting valuation of financial instruments.[6]

Management challenges

Senior management of banks and other financial institutions face a variety of challenges in identifying and implementing policies and strategies that support several key goals and objectives of the bank. These challenges include:

[6] Accounting for Financial Instruments Constituent Outreach Summary May 2010, Financial Accounting Standards Board, www.fasb.org/cs/BlobServer?blobcol=urldata&blobtable=MungoBlobs&blobkey=id&blobwhere=1175820818351&blobheader=application%2Fpdf

1. Maintain and grow the business of the institution, responding to competition while managing the multiple risks inherent in operating a financial institution in the real economy;
2. Manage cash flow and liquidity demands while observing accounting rules which run the gamut from long-term accrual and amortization time frames to short-term mark-to-market treatment;
3. Accurately measure and manage both economic capital needs and regulatory capital requirements;
4. Establish effective procedures and policies that will ensure the integrity and efficiency of internal governance processes and risk management mechanisms;
5. Integrate the capabilities and skills of the information technology infrastructure with the opportunities and needs of the business lines by promoting communication, planning, and a shared vision among both vertical and horizontal segments of the organization;
6. Minimize the non-productive administrative overhead costs of compliance and regulatory reporting by taking the initiative to implement an internal data management, analysis and reporting framework that becomes a strategic benefit and provides competitive advantage of the institution – and which is then poised to deliver more detailed regulatory reporting data in a manner that flows readily from operational systems of record.

Fortunately, these are key goals and objectives that can be met and challenges that can be overcome using approaches that integrate the needs and expertise of the business with the capabilities and benefits of information technology in concert with an enterprise-wide data management infrastructure.

It must be stressed, however, that attention to such things as data accuracy, efficient and reliable infrastructure, adherence to regulatory rulings and even superior risk management methodology and analytics, while necessary, are not sufficient to achieve effective and desirable best practices when it comes to risk management in the financial industry. For that truly to occur, the *corporate culture from the top down* must set the tone and lead the way by treating risk management with the seriousness it deserves. In other words, it needs to be more than just a check box for regulatory compliance or testimonials in quarterly reports to shareholders.

6.1.2 The pervasive importance of data

Appropriate levels of granularity and aggregation

The scope, pace, and detail of requirements for financial measurement and management have changed dramatically over the past 25 years. Quarterly, even monthly, financial statements are often too aggregated and infrequent to make the risks of a firm transparent. They generally are derived from General Ledger and accounting systems, which, despite their notable evolution, are based primarily on balance amounts. As such, they do not contain sufficiently disaggregated information about the range of contractual parameters and structural characteristics of individual transactions that lie behind the figures on the balance sheet. Hence, traditional financial statements do not reflect the detail, diversity and true extent of a firm's risks. The loss of information and the introduction of data "noise" caused by inappropriate – and often misleading – aggregation will be discussed in more detail in a subsequent section.

In terms of frequency and timeliness, the volume and dynamic aspects of the composition and activity of customer accounts (and corresponding corporate treasury positions) are driving the movement toward enterprise management. Enterprise management calls for increasingly up-to-date measurement and flexible response performance requirements. Management policies and operational procedures – including the supporting infrastructure and underlying data management procedures – must change accordingly to keep pace.

Data risk

An increasingly important component of risk, especially given the scope and objectives of this discussion, relates to uncertainty regarding the comprehensive and timely access to accurate and detailed *data* – in all of its sources, forms, and uses – both throughout an institution as well as in the financial system as a whole.

Inaccurate or insufficient data as well as poor data management procedures can result in incomplete or incorrect information for decisions at all levels – from mortgage origination and through credit approval up to executive management, across financial markets and government regulators to central banks. *Management's view of the dynamic state of the institution is the result of an ongoing cycle of accumulation and distillation of data from the roots and branches of the organization.*

The processes and procedures by which detailed data move up through the organization to the top, and likewise the means by which management strategy and risk criteria are transmitted downward to all staff for governance and risk management purposes, are key to the success and optimal performance of the firm. This is true in any business or industry, but it is particularly true in banking and finance.

If data in different parts of the organization are not standardized then they cannot be viewed or integrated consistently when merged. If procedures for transforming and aggregating data follow different rules and methods in different business lines or divisions, then important connections among different dimensions of an organization's business will be missed. An obvious example of this type of data risk is the case where customers with credit products in different business lines (e.g., a mortgage and a credit card) are not recognized as the same customer because of the absence of standardized customer identification codes.

Furthermore, if independent departmental applications and methodologies are not predicated on consistent assumptions and input conditions, then the data and information which flow up the organization structure will not provide a reliable, accurate and manageable view of either the customer's profile or the institution's total risk profile.

The timeliness and accuracy of this process is important at the level of senior management. It is also important at the lowest levels of the organization. An operational process or workflow that allows incomplete or erroneous data to enter the system will eventually cause an accumulation of flawed detail that could, at some point, cause failures in the aggregate.

An example of this type of data risk is the inaccurate information that was submitted with a large number of mortgage applications during the height of the inflation of the housing bubble. The inaccurate information, whether entered by prospective borrowers or by mortgage originators, one application at a time, produced a higher-than-usual frequency of inaccurate data with respect to key criteria needed to determine loan suitability. The cumulative effect of a growing number of loans with inaccurate data led to a higher incidence of inclusion of lower quality loans in securitized instruments whose default risk was subsequently not correctly appreciated.

Given the high velocity and volume of financial transactions in the modern financial system, it is increasingly important for senior management of financial institutions to have a very accurate view of the current state of the institution – including its customers and counterparties – in order to chart a stable course through increasingly dynamic markets and economies.

Data leverage

"Leverage" is a term which, when mentioned in a financial context, pertains to the practice in which capital assets are used to secure the financing for investments of higher value than the capital pledged as collateral. An example is the leveraging of $22 billion of Lehman Brothers stockholders equity to secure $691 billion of total assets on the Lehman Brothers balance sheet in 2007.[7] The *accounting leverage* of Lehman Brothers at that time was therefore nearly 31 (691/22 = 30.7).

Data leverage is more of a concept than a metric. However, it is one that could inform the management of financial systems and institutions, in particular as it pertains to

- the reliance on (and degree of confidence in) the accuracy and timeliness of data that are rolled up from detailed operations in order to be . . .
- strategically analyzed, summarized, and ultimately acted on by the board of directors as well as presented to regulators.

[7] Lehman Brothers Holdings Inc., Annual Report for year ended November 30, 2007.

Big sky, small radar Given that financial contracts, inventory and transactions in the banking industry are, with the exception of cash, expressed and manipulated strictly in terms of their informational elements and structures, the total "informational capital" of a bank can be viewed as a roughly equivalent representation for the material assets and liabilities of the bank.

Data leverage, as it applies to information used to manage an institution, could be thought of in terms of the amount of information upon which management depends in order to make decisions, compared to the total "informational capital" of the bank in the form of the total data comprising all contracts, cash flows, credit exposures, sensitivity to changes in the macroeconomy and markets, off-balance-sheet transactions, counterparties, shareholder equity, corporate bonds and any other financial rights and obligations.

From an information theory perspective, data leverage could be defined as the ratio N/D that examines the amount of information in

D = the total data of an institution, compared to the amount of information in
N = the data produced after intervening aggregation, transformation, compression, and summarization steps, and which is then provided to senior management.

To the extent that a relatively small set of management information is relied upon to supervise and manage an institution (or a collection of institutions, in the case of a regulator) with a large and diverse total informational portfolio, then that constitutes a higher level of informational leverage than the case where a similar amount of management information is used to control a smaller and less extensive underlying data portfolio.

One could rightly point out that an equivalence exists between a single data element representing the total indebtedness of an entity vis-a-vis all of the numerous individual debts that make up the total. While this is true relative to a snapshot in time of the firm's total obligations to those other than its equity holders, the figure says almost nothing about the magnitude or sources of volatility in this total. It is insights into how sensitive the firm's financial position is to a wide variety of external circumstances that lie at the heart of effective management of risk. For this purpose, the single aggregate of total indebtedness is almost worthless. This remains true even assuming there are no inaccuracies or errors in the data or process of aggregation. For risk management purposes, the informational content of that sum is far less than the sum of the informational content of all the contracts or positions whose balances have simply been summed.

Completely lost in arriving at a single aggregate balance are all of the disparate and unique terms, conditions, and parameters – including the external dependencies on customers, counterparties, markets, economies, and even geopolitics – which are part and parcel of each and every account-level contract or position. Hence, it is more from the perspective of information theory that assessing the amount of *information* contained and summarized in management reports compared to the total information in the organization one arrives at a measure of data leverage. Seen this way, the degree of information or data leverage is potentially very large.[8]

Large supply, small funnel The global interconnectedness of financial markets further contributes to increasing informational leverage via information technology, communication services and Internet-enabled media. The confluence of these change drivers produces an accelerated flow and volume of information.

The rate of growth of financial, economic, and market data generated by theincreasingly dynamic and global financial markets and economies is exceeding the rate of growth of the extrapolated capacity of most existing systems to analyze, extract, and summarize those data effectively. The result is increases in an already high level of data leverage. The time frames for actionable information to be presented to management on a timely basis have been compressed and the rate of accumulation of data in the input hopper has grown noticeably with the increase in the velocity of information in the financial system today.

The so-called "Flash Crash" of May 6, 2010 is an example of a process, fueled by automated high frequency trading systems, that exceeded the ability of the financial system to react appropriately in a timely fashion to correct the unstable behavior of the apparatus. Fortunately, the system recovered nearly as rapidly as it crashed, but it was not without some considerable losses incurred as trades that were executed near the bottom of the cycle were canceled.

Consider also that, whereas the event iself took place over a matter of hours on one trading afternoon, it required nearly five months for experts at the Securities and Exchange Commission (SEC) and the Commodity Futures Trading Commission (CFTC), with full access to the data streams, to try to piece together what occurred, and why, in a post mortem fashion. Even then the picture was not definitively or causally clear.[9]

[8] See the discussion of the *data leverage* concept that we introduced here in the April 2012 column by David Rowe in the journal *Risk*, available at the following link: `www.dmrra.com/publications/Risk%20Magazine/201204%20Beware%20of%20Data%20Leverage.pdf`

[9] The most significant issue faced by those analyzing the trading activity was the inability to determine precisely the chronological sequence of transactions because of the lack of a synchronized time-stamp which was consistent system-wide (across all exchanges) and accurate enough in terms of the smallest unit of time resolution required to discern the sequence of trades between high-frequency trading systems whose speed of execution overloaded the resolution of the timestamp!

The importance of informational and analytical bandwidth It is important to assess accurately the full contractual obligations, financial risks, informational structures and transactional stream of events in order to understand the banking industry effectively. It is also critical to have an understanding of the analytical and informational technology and bandwidth required to do so.

With regards to the options markets alone, one industry analyst has recently estimated the rate, volume and size of data activity thusly:

> Within the options markets, the migration toward higher speed automated trading is causing a massive increase in market data. Peak market data message rates are increasing steadily, going from ~700,000 messages per second (mps) in 2008 to over 3.4 million mps in 2011. Similarly, storage requirements have also been going through the roof. Participants looking to store all of this information have had to increase their storage capacity from 18 to 52 terabits just to store one years' worth of data.[10]

6.2 Perspectives on the banking industry

6.2.1 Banking in the financial and economic arena

Mix of products and services

Retail banking serves both businesses and consumers. It includes the business of accepting consumer deposits, providing consumer credit, funding residential mortgages and commercial loans and providing lines of credit and liquidity for both consumers and businesses. The credit can either be unsecured (whose ability to be repaid is based on net income) or is otherwise secured by some form of collateral.

Banks also provide a range of other financial services to both consumers and businesses alike. These services include electronic funds transfer, foreign exchange currency transactions, online banking and bill payments. These services also include a variety of shorter-term investment vehicles in the form of money market shares, certificates of deposits and savings accounts.

Retail lending encompasses a variety of mortgage products. These include fixed-rate mortgages and adjustable rate mortgages (ARMs). The mortgages are secured by the residential properties financed by the loan. The risk associated with these retail products calls for analysis of certain factors such as a credit score, documentation type, loan-to-value (LTV) information and the ratio of mortgage payment divided by income. The LTV ratio represents the proportion of the property value financed by the loan and is a key risk variable. A home equity loan is a hybrid between a consumer loan and a mortgage loan and is secured by residential properties.

[10] See Tabb (2011).

Installment loans are revolving loans such as personal lines of credit. Installment loans may be used up to a specified limit. These include such products as credit card loans and other consumer loans for personal property and are usually secured by such things as automobiles, residential property, personal property and financial assets.

Small Business loans and lines of credit are secured by the assets of the business or by the personal guarantees of the owners. Business loans of up to a fixed amount such as $300,000 are usually considered as part of the retail portfolio.

Banks also provide loans and financing to corporations that need both ongoing liquidity and the commitment of capital spanning longer time frames, such as construction and big-ticket manufacturing contracts. In these cases, the loans and lines of credit are funded based on the eventual cash flows and sales that are *expected* to result from the successful performance of the corporations under contract.

In order to maximize market share, grow capital and sources of funding and diversify risks across product lines, banks have sought to provide a wide range of customer types with a full gamut of financial services across all product lines in order to maximize their ability to target and cross-sell to customers and businesses.

Operational nexus

Tight coupling with real estate and the economy There is a tight coupling between the banking and real estate sectors in the economic nexus, depicted in Figure 6.1, due to the need for debt financing of both residential mortgages and commercial development. In the case of residential mortgages, loans made by banks to finance home sales are secured by the value of the home and reflected in the important LTV ratio.

Commercial lending also usually requires collateral, but is further approved based on contracts or other indications of future cash flows (sales or leases) from completed developments. Decisions regarding the approval of commercial loans require significantly more analysis and evaluation than is typically needed for residential mortgages due to their size and relative complexity

The return on the investment of the loan asset (in addition to the underlying value of the real property used to secure the funding of loans) depends heavily on the future financial performance of the customer who is the obligor in contract with the bank. Banks are very dependent on the ongoing financial state (and actions) on the part of customers, whether consumers or businesses.

Banking customers, whether retail or commercial, are themselves greatly affected by the overall state of the economy. Periods of economic growth and expansion naturally make for fewer incidents of default and delays or failures to make

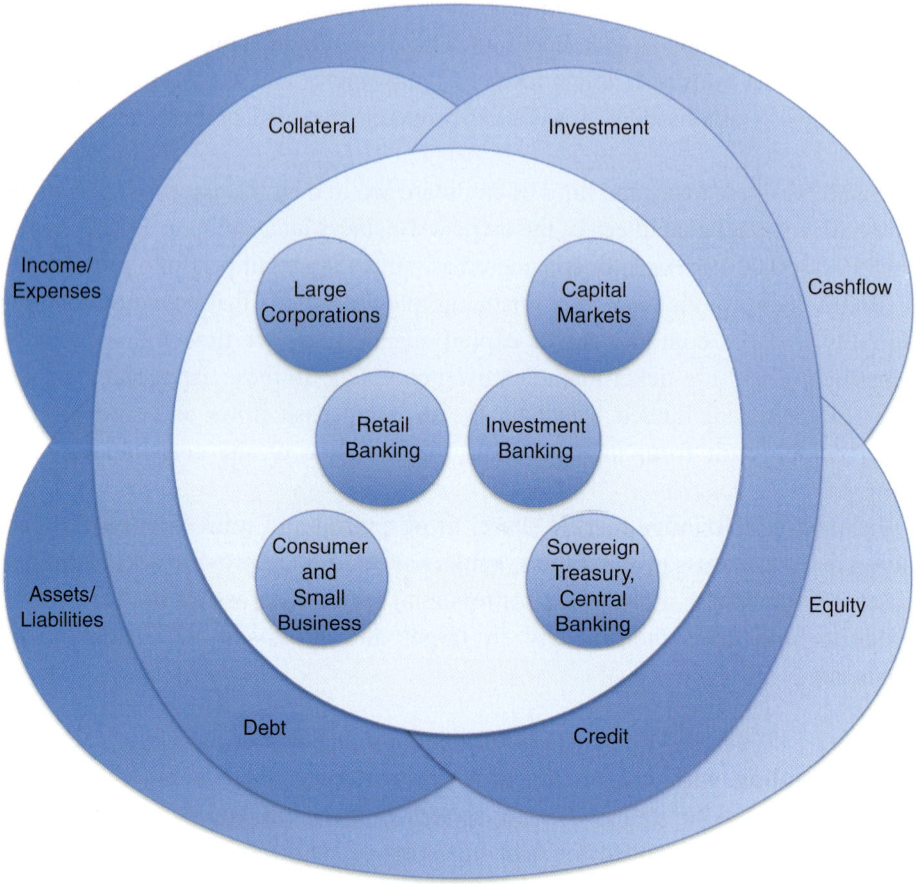

Figure 6.1 The operational nexus of banking in the real economy.

scheduled payments ("non-performance"), whereas periods of economic recession or contraction will naturally increase losses due to customer default. Furthermore, the demographics of both customers and the factors of their regional market areas will give rise to significant variations in the behavior of mortgage contracts over the life of the relationship with the customer.

Intersection with capital markets Retail products intersect with wholesale products in such areas as wealth management. Wealth management services include working with retail clients to invest in money market instruments, bonds, stocks, foreign exchange, commodities and derivatives.

Money market instruments include term deposits and loans, non-domestic loans or deposits, repurchase agreements (repos), and structured loans (one year). These also include treasury bills (government securities), commercial paper (short-term corporate paper similar to T-bills), banker's acceptances (BA) and certificates of deposits (CD).

Bonds are debt instruments to cover the financing needs of governments (federal, state, and city), government agencies and corporations. The market consists of borrowers (e.g., corporations), lenders (e.g., banks) and intermediaries. Stocks (equities) represent ownership in an organization. Stockholders are last in queue for recovery of an investment, hence the high volatility in stock prices. The market consists of the primary market (the initial launch of a share) and the secondary markets (the subsequent trading in those shares). Equity financing is a way of raising capital and spreading ownership (and risk).

The interbank foreign exchange (FX) market is an Over-the-Counter (OTC) decentralized, continuous and open bid market. FX rates vary according to the economic state of the country (fundamentals), the balance of payments, short-term market activity, international events and central bank intervention. Currency can be bought spot (for immediate delivery and settled one or two days after trade date) or forward (delivery for up to one year in the future).

Commodities include a diverse set of products. There are basically four broad commodity groups: (1) *grains, seeds and fibres* (wheat, soya bean, cloths, etc.), (2) *metals* (e.g., gold, silver, copper), (3) *livestock and other*, and (4) *"softs"* (e.g., sugar, rubber, etc.). Energy is a specialized commodity market. Energy trading (e.g., crude oil, petroleum, gas, electricity, coal, naphtha, carbon credits and renewable energy credits) began with oil futures and is now established in both regulated and OTC markets.

A derivative is a financial instrument that is derived from an asset. It can also be derived from a measurable future index, event or value. Derivatives include swaps (e.g., interest rate swaps), forwards, future contracts (e.g., interest rate future contracts), options (e.g., an option on a stock), and combinations of all the above (e.g., derivatives on derivatives such as an option on a swap (swaption)). Futures are exchange traded, whereas forward-based derivatives are contracts between two parties to a future action (i.e., the delivery and payment of an asset). There is a buyer and seller, a maturity/expiration date, and a future defined payment.

6.2.2 The product view

There are many of the types of financial products that are common to banking (as shown in Table 6.1). The off-balance-sheet financial instruments of the bank are used primarily to manage the various risks and exposures in the assets and liabilities of the balance sheet, as will be described in the discussion of the *Institutional view* in Section 6.2.4 that follows the *Customer view* in Section 6.2.3.

Table 6.1 *Banking products (illustrative)*

Assets	Liabilites	Off-Balance-Sheet
I *Customer-facing*	I *Customer-facing*	
A Residential mortgage	A Deposits	I Options
1. Fixed rate mortgage	1. Checking	A Puts
2. Adjustable rate mortgage	2. Saving	B Calls
3. Pay-option mortgage	3. Money market	II Swaps
4. Interest only loan	B Certificate of deposits	A Interest rate swaps
5. Neg-Am mortgage	1. Fixed rate CD	B Currency swaps
B Home equity loan	2. Variable rate CD	III Foreign exchange contracts
C Credit card	3. Callable CD	IV Exchange-traded futures
D Consumer loan	4. Zero-coupon CD	V Credit default swaps
E Commercial mortgage	C Investments (not FDIC insured)	
F Business credit lines	1. Annuities	
G Commercial loan	2. Mutual funds, securities	
II *Bank-owned*	II *Bank-owned*	
A Investment securities	A Bank-issued debt	
B Treasury stock	B Central bank borrowing	
C Trading inventory	C Inter-bank borrowing	

6.2.3 The customer view

Consumer/retail customers

Individual accounts and households Identifying the set of particular accounts and financial products which fall under the purview of a single household has long been a primary opportunity, goal, and challenge for banking institutions. Historically, this challenge was faced by the operations and information technology functions. This effort primarily involved the maintenance and grooming of the Customer Master File (CMF).

Grouping numerous accounts together under a single umbrella which corresponds to a household has been a difficult challenge for many institutions. One problem is that broad use of unique identifiers which could facilitate such grouping (such as social security numbers or tax IDs) is largely stymied by privacy restrictions and regulations designed to prevent the disclosure of identity information.

Historically it was very difficult for banking information systems to have the capability and technology to hide (or protect) access to such information while trying to use customer identity data to associate accounts for the purpose of CMF consolidation. This has changed somewhat, due in part to the data regimens put in place to protect identity data, such as social security numbers, so that these data can now be used to group accounts. Nevertheless, this remains a large Information Technology (IT) and information security issue. It is also fraught with considerable operational and reputational risks should the data be accessed or obtained improperly by external third parties.

Second, if unique and consistent customer identity information was not available or not obtained when accounts were originally opened, then names and addresses entered multiple times are often spelled or represented differently, even when they refer to the same individual or household.

Customer relationship management A considerable amount of effort and expense has been dedicated to identifying households and has subsequently become a large part of the mission of Customer Relationship Management (CRM) systems. A primary objective of a CRM system is to give the bank a more complete picture of a customer's relationship with the bank for the purposes both of targeting cross-selling marketing efforts and of having a more complete assessment of the creditworthiness of each customer.

CRM systems have become tightly integrated with customer service functions of a bank. These include capturing "branch banking" interactions with platform tellers and the identification of customers who would qualify for premium private banking and wealth management services which banks must offer in order to retain the business of their most desirable customers. CRM systems, by their very nature, must be refreshed with information from multiple business lines in the bank. It should not be surprising that the implementation of CRM systems became one of the early drivers that prompted banks to try to integrate data from disparate parts of the organization. Another early driver was to make the data available to multiple customer-facing business channels, including robust Internet-enabled online payment services, credit card activity and basic account maintenance and transaction processing. (See the related discussion of Derivation of Data Pools for Decision Support and Business Intelligence in Section 6.3.)

Consumer credit risk scoring A retail "credit risk scoring" model typically converts information about a credit applicant (or an existing account holder) into numbers that are then combined (usually added) to form a score. The scoring method is structured so that the higher the credit score, the lower the credit risk. This credit score is then regarded as a measure of the credit risk of the individual concerned, (e.g., the probability of default, PD).

Retail credit scoring allows banks to avoid the most risky customers. It helps banks assess whether certain kinds of retail businesses are likely to be profitable by comparing the risk adjusted profit margin that remains once operating expense, expected default expense and the costs of allocated risk capital are subtracted from gross revenues. Retail credit scoring is also important for reasons of cost and consistency. Banks can automate the adjudication process for small credits and credit cards as much as possible by using a credit-scoring model. Some applications may call for as many as 150 items.

A retail credit scoring system consistently weighs and treats the information items (**characteristics**) they extract from applications and/or credit bureau reports. The answers given to the questions in an application, or the entries of a credit bureau report, are known as **attributes**. For example, "rents" is an attribute of the characteristic "residential status."

The weighting of the values associated with each answer (or attribute) is derived using statistical techniques that look at the **odds** of repayment based on past performance. For example, an applicant with 24:1 (24 good and 1 bad) odds has a probability of 1 in 25 of being a "bad" customer (by which we mean delinquent or the subject of a charge-off).

There are three types of models for scoring consumer credit applications.

Pooled Models are built by outside vendors using data collected from a wide range of lenders with similar credit portfolios. For example, a revolving credit pooled model might be developed from credit card data that are collected from several banks, and therefore are not company specific.

Credit Bureau Scores in the USA and Canada are maintained and supplied by companies such as Equifax and TransUnion. This kind of *generic credit score* offers a broad overview of an applicant's creditworthiness. For example, credit bureau risk scores can be **tailored** to the preferences of a financial institution.

Custom Models are usually developed in-house using data collected from the lender's own unique population of credit applications. They are tailored to fit a specific profile of applicant for a specific lender's product.

In the early stages of credit scoring models, applicants were placed in *ranked order* in relation to their *relative* risk. This was because lenders used the models not to generate an absolute measure of default probability, but to choose an appropriate cut-off score.

Given the cut-off score, the bank can determine, based on its actual experience, the loss rate and profitability for the retail product. Over time, the bank can adjust the cut-off score in order to optimize the profit margin product by product. The Basel Capital Accord requires that banks segment their retail portfolios in subportfolios with similar loss characteristics (e.g., similar prepayment risk). Banks estimate both the PD and the loss-given-default (LGD) for these portfolios. This is achieved by segmenting each retail portfolio by score band. Each score band corresponds to a risk level for each score band. The bank can estimate the loss rate using historical data. Then, given an estimate of the LGD, the bank can infer the implied PD. Banks must use stress tests to gauge how devastating each plausible retail worst-case scenario might be.

There are significant challenges to developing a retail credit-scoring model. First, not all retail credit products can be associated with enough historical loss data to make their risk assessments reliable. Second, even well-understood retail credit products might begin to behave in an unexpected fashion under the influence of a sharp change in the economic environment. For example, risk factors tend to all get worse at the same time in stressed environments (correlations move toward unity). Third, the tendency of consumers to default is a product of complex interactions (e.g., a strategic default on a mortgage) that continually change. Fourth, any operational issue (e.g., fraudulent data) that affects the credit assessment of customers can have a systematic effect on the whole consumer portfolio.

Credit bureau data consist of numerous "credit files" for each individual who has a credit history. Each credit file contains five major types of information. **Public records** (legal items) come from civil court records and include bankruptcies, judgments and tax liens. **Trade line/account information** is compiled from the monthly "receivables" tapes that credit grantors send to the credit bureaus. **Collection information** is reported by debt collection agencies or by entities that grant credit. **Inquiries** are placed on file every time a credit file is accessed. **Identifying information** is personal information and therefore is not considered credit information as such, and is not used in scoring models.

Business/commercial Customers

Business products and services ***Lines of Credit***: In the normal ebb and flow of operations, businesses need to advance cash payments to procure raw materials, inventory, goods and/or services. These are all then integrated, manufactured or marketed as value-added products subsequently offered to customers. There is an inherent latency in the cycle between the expenses incurred to produce goods and services and the revenues to be received upon their sale or delivery and after billing or invoicing. This can add 30–45 days after delivery to receive payment if the business is part of the supply chain and is not able to collect immediately upon delivery (as in retail merchandizing).

It is imperative that businesses have the funding liquidity to span the cash flow gap between expenses paid and revenue received. If a business does not have the necessary cash on hand to do so, credit lines provided by banks are an essential part of the real economy. Business credit lines are qualified by the bank based upon an evaluation of the nature of the business. For example, these include its past performance history and the quality of commercial contracts which the business has if the work is of a commissioned or contracted nature.

Commercial Lending: If the ongoing "organic" cash flow of a business is not sufficient to support development initiatives (say, in order to grow, develop new capital infrastructure, invest in new facilities and staff or pursue competitive opportunities that involve degrees of risk), then businesses need debt financing of the sort that provides up-front capital which will be repaid (e.g., repaid using future cash flows receipts).

Payroll: Businesses that employ more than just a small number of employees and who are not large enough to afford dedicated in-house administrative and accounting staff typically outsource the payroll function. Banks typically offer payroll services that work in concert with the commercial banking account maintained by a business with the bank, since businesses have commercial relationships with banks for the types of facilities described above. The benefit to the business customer is the simplicity of not having to move money constantly to other accounts or services in order to make recurring payroll payments. The benefit to the bank is that the bank is able to charge additional fees for performing the service for the business. The bank also gains valuable insight into a large part of the ongoing finances of the business, as revenues are deposited with the bank and expenses drafted from the same account of the business customer.

Commercial credit risk A commercial credit risk system is designed to provide a structured approach to arrive at a commercial credit risk rating (CCRR). The steps in the CCRR consists of an *obligor rating* (OR) which is used to provide insight into the PD. The steps also include a *loss given default rating* (LGDR)

The first step toward arriving at an OR usually begins with a financial assessment of the borrower (initial obligor rating). A series of six further steps are used to arrive at a final OR. There are many financial measures that need to be analyzed as part of Step 1, such as the ratio of Earnings Before Interest and Taxes (EBIT) to Interest Expense.

The second step includes analyzing the managerial capability of the borrower. A third step includes examining the borrower's absolute and relative position within the industry. Step 3 calls for an analysis of the industry in terms of factors such as competitiveness, trade environment, regulatory framework, restructuring, technological change, financial performance, long-term trends affecting demand and vulnerability to the macroeconomic environment.

The analysis also calls for reviewing the quality of the financial information in Step 4 and analyzing country risk in Step 5. Country risk is the risk that a counterparty, or obligor, will not be able to pay its obligations because of cross-border restrictions on the convertibility or availability of a given currency. It includes an assessment of the political and economic risk of a country. Country risk exists when there is more than a prescribed percentage (say 25%) of the obligor's gross cash flow (or assets) located outside of the local market.[11]

[11] Country risk may be mitigated by hard currency cash flow received by the counterparty (where the hard currency cash flow refers to revenue in a major international currency).

One should work to compare internal ratings to any available external ratings in Step 6 and the impact of the loan structure in Step 7. The components of the loan structure that may affect risk are the financial covenants, the term of the debt, its amortization scheme and change of control restrictions. The LGDR is derived in a final Step 8. The LGDR is independent of the OR. Collateral can have a major effect on the final LGDR, as the value of securities used as collateral is usually a function of movements in market rates.

Additionally, the revised Basel II Framework published by the Basel Committee on Banking Supervision requires that the Loss Given Default parameters used (in the Pillar 1 capital calculation) must reflect economic downturn conditions to capture the relevant risks. The Downturn LGD parameters used for Basel Pillar 1 calculations are defined.[12]

A bank must have a rigorous and well-documented process for assessing the effects of downturn conditions on LGD. The process should include identification of downturn conditions for each asset class within each jurisdiction. The process should include identification of adverse dependencies between default rates and LGD. The LGD parameters should reflect the adverse dependencies between default rates and recovery rates.

The need to have access to accurate and consistent historical data for business lines and asset defaults is a factor that may not be fully appreciated when compiling a robust assessment of issues surrounding data in the banking sector.

The measures of recovery rates should reflect the costs embedded in the LGDs of holding the defaulted assets over the workout time period, including the appropriate risk premium. In other words, the NPV calculation must reflect the time value of money and a risk premium appropriate for risk (specifically the non-diversifiable risk). If there is no uncertainty in the recovery stream then a risk-free interest rate can be used to make the NPV calculation.

There are many data challenges in terms of estimating the LGD. First the historical peak loss rates may differ from product to product and from bank to bank. Second the peak of LGDs may not correspond to the peak of default rates. The LGD estimates should be no lower than the long run LGD weighted average loss rate and therefore the long run LGD weighted average loss rate should be compared to the downturn LGD. Also, for some products there is little empirical evidence that LGD is cyclical. Further there is a natural diversification effect associated with downturn LGDs.

[12] See *International Convergence of Capital Measurement and Capital Standards* Revised Framework Comprehensive Version, June 2006, paragraphs 468–473, (vii) Requirements specific to own-LGD estimates, p. 103.

6.2.4 The institutional view

Basic business model

Maintaining and improving the ongoing stability and financial health of a bank in the real economy requires a structured decision-making process. Matching (and deliberately mismatching) the mix of assets and liabilities is part of an integrated process across the entire firm.

In the banking industry, the underlying business model is based on key objectives that are to:

1. Ensure that the bank does not assume too much risk from the mismatching of interest rate sensitivities and maturities between assets and liabilities;
2. Maximize shareholder value, or net worth (NW);
3. Stabilize net interest income (NII), that is, the difference between the amount the bank pays out in interest for funding and the amount it receives from holding assets (as measured by accounting earnings).

Net worth is the net present value of assets minus the net present value of liabilities, plus or minus the net present value of off-balance-sheet items. The impact of interest rates on NW is considered over a relatively long time horizon.[13]

NII is traditionally a key indicator of bank profitability. NII is affected by the interest rate sensitivity mismatches of assets and liabilities (on- and off-balance-sheet). The impact of interest rate volatility on NII is usually analyzed over a short time horizon such as a quarter or a year, and is referred to as Earnings-at-Risk (EaR). Banks use analytical techniques to determine the impact on NW and NII of numerous interest rate scenarios, balance sheet trends and strategies over various time horizons. Foreign exchange risk can also be an important component of balance sheet risk management.

Funding liquidity risks in the retail bank can stem from both external market conditions and from structural problems within the bank's balance sheet. There are many examples where investors began to worry about the condition of the overall bank's credit portfolio and cut off short-term funding, which in turn created a funding liquidity crisis for the bank.

Asset liability management (ALM)

ALM decisions are driven by trade-offs (e.g., a trade-off between liquidity and interest rate risk in the retail bank). Short-term assets and liabilities carry less interest rate risk and offer more liquidity risk, while long-term assets and liabilities incur more interest rate risk but less liquidity risk.

[13] Interest rate risk also has an effect on a firm's capital-to-asset ratio, business volume and product mix as well as on the pricing of assets and liabilities, liquidity risk and credit risk.

There is also a trade-off between cost and risk. For example, in order to mitigate liquidity risk, institutions in a positively sloped yield curve environment can increase the maturity of their funding liabilities but this will cost them more than will cheaper shorter-term funding. The components of ALM policy (such as interest rate risk management, liquidity risk management, profit planning, product pricing, and capital management) need to be integrated and must be part of an overall firm-wide approach to balance sheet management.

There are many types of risk involved in ALM management. If the sensitivity of an asset (say a long-term asset) to interest rate changes is not matched to the sensitivity of a liability (say a short-term deposit), then a firm is said to be mismatched from an interest rate sensitivity point of view. Risks also include basis risk, timing of rate changes, embedded options risk and so on.

Basis risk is where the interest income on the on the asset side is driven by a different interest rate (e.g., the US Treasury bill rate) than the one that drives interest cost on the liability side (e.g., the bank CD rate). When these two rates do not move in lockstep, this will affect the net interest margin. The timing of rate change issue refers to the idea that the impact of an increase in rates on the liability side may be immediate, whereas the impact of the rate change on the asset side may come later. Embedded options risk refers to the case where retail products offer customers different types of options, such as a prepayment option on mortgages.

Funds transfer pricing (FTP) centralizes the management of interest rate risk. The objective of FTP is to remove the non-controllable interest rate risk from business results, and to charge each business for the cost of funding their activity and for hedging their interest rate risk. Each business unit needs to secure its profit margin at the time of origination of its products (say, mortgages) and can focus on developing and managing the business side of its activity as well as the credit quality of its portfolio.

Banks often benefit from an upward-sloping shape of the yield curve. The upward-sloping curve means that the interest rate on a long-term asset (such as a residential mortgage) exceeds the rate on a short maturity liability (such as a time deposit). During certain periods, however, an increase in short-term rates can push up the cost of funds, generating *negative* net interest margins relative to existing long-term assets (such as the aforementioned residential mortgage). Banks can manage interest rate risk using derivatives. Off-balance-sheet items such as interest rate swaps, futures, forwards, options, caps and floors can represent both assets and liabilities to the bank.

An interest rate swap is based on a "notional principal." The notional principal is an amount stated purely for calculation purposes that does not entail an actual credit or debit balance position. An interest rate swap that receives a sequence of cash flows based on a short-term floating rate and which in turn pays a sequence of cash flows based on a long-term fixed rate can be used to hedge a long-term fixed-rate asset and a short-term liability based on matching the off-balance-sheet interest rate sensitivities with the corresponding on-balance-sheet asset and liability management positions. A swap can be viewed as a combination of a long-term fixed-rate liability and a short-term floating-rate asset from the perspective of the effect of the resulting interest-rate-sensitive cash flows, without actually involving positions on the balance sheet.[14]

Banking risks

The risks generated by retail and commercial banking are quite diverse. For example, a defining feature of retail exposures is that the risk of a single customer in isolation is never large enough to threaten a bank. Another key feature is that retail customers tend to be financially independent of one another in normal markets. Corporate and commercial credit portfolios often contain concentrations of exposures to corporations that are economically intertwined in particular geographical areas or industry sectors.

A variety of risks will be present in varying degrees and sizes depending on the size and overall business mix of a bank.[15]

Market risk is the risk of losses arising from changes in market risk factors. For example, market risk for mortgages can arise from adverse changes in interest rates. The valuation and the hedging of retail assets subject to prepayment risk is complex because it relies on assumptions about customer behavior that are hard to validate and may not be stable across different economic environments. **Interest rate risks** are generated on both the asset and liability sides whenever the bank offers specific rates to both borrowers and depositors. This retail risk is generally transferred from the retail business line to the Corporate Treasury where they are managed as part of the bank's asset/liability and liquidity risk management.

Credit risk is the risk of loss following a change in the factors that drive the credit quality of the retail asset.

Operational risk refers to financial loss due to a host of potential operational breakdowns related to people risks, process risks and technology risks. For example, operational risk for a mortgage can arise from reasons that range from fraudulent documentation to political risk and can impact other categories of risk. For example, political discussions around the US debt ceiling can have an impact on interest rate risk.

[14] The term "off-balance-sheet" can be confusing. It does NOT mean that the instrument has no effect on the balance sheet whatsoever. Rather, where netting is legally enforceable, it effectively means that only the net values of such contracts with each counterparty appear on the balance sheet rather than, for example, both the pay and receive sides of an interest rate swap appearing as two separate gross amounts.

[15] For a further discussion of banking risks and risk management, see Chapter 2 of this Handbook.

A key banking risk is **funding liquidity risk**, which refers to the risk that the bank will not be able to raise sufficient cash to fund their operations.

The USA places a ceiling on the amount of money that the US Government can borrow. In other words, the US Government can borrow money as long as the debt does not exceed a ceiling (a limit) stated by law. To change the debt ceiling, Congress must pass a specific law (legislation) and the President must sign it into law. For example, the Congress increased the debt limit in February 2010 to $14.3 trillion and increased the debt ceiling again on July 31, 2011. The problem is that the US Government is spending more than its "revenues."[16]

Another increasingly important aspect of operational risk, certainly within the scope and objectives of this discussion, is the risk associated with the degree of comprehensive and timely access to accurate and detailed *data* – in all of its sources, forms, and uses – both throughout an institution and in the financial system as a whole. This has been discussed in Section 6.1.2 (The pervasive importance of data).

An additional component of operational risk as it applies to banking and other financial institutions is **model risk**. Given the importance and reliance that nearly all financial institutions place on analytical models of various types and uses for assessing risks into the future, it is particularly important that models be understood, vetted and calibrated.

Backtesting models to ensure that they produce results that are at least consistent and compatible with current conditions is extremely important. For example, due to their complexity, analytical models are often viewed as "black boxes" that emit "correct" results. Models are often accepted or relied upon without sufficient understanding or checks and balances. Also, as multiple models and methodologies are often employed in different parts (and levels) of an organization, it is important that the input assumptions and conditions of the models (along with the temporal evolution of scenarios under consideration) be synchronized in some fashion in order for the results of different models to be consistently combined or jointly evaluated to get a consolidated view of the institution.

There are three key non-Basel related risks. ***Business risk*** refers to uncertainty about the demand for banking products, the price that can be charged for those banking products or the cost of producing and delivering banking products. ***Reputation risk*** is another key risk in the retail sector that has taken on special importance due to the failure of many institutions to manage their mortgage related

[16] For example, in fiscal year 2011 the US Government estimated it would spend $3.8 trillion (in fiscal 2011) but expected to take in revenues of only $2.2 trillion, leaving a substantial "deficit." The USA reached an agreement on July 31, 2011 that there would be spending cuts (of around $900 billion over ten years) in exchange for increasing the debt limit (by $900 billion) but the political discussions around the debt ceiling and the need for the US Government to reduce spending has an indirect impact on interest rates.

risks carefully during the recent financial crises.[17] ***Strategic risk*** refers to the risk of significant investments for which there is a high uncertainty about success and profitability.

A key role of the risk manager is to make the risk across all banking products transparent. In the case of retail loans, the high predictability of retail credit losses means that the expected loss rate dominates retail risk. The risk of loss from commercial credit portfolios is dominated by the risk that losses will rise to some "unexpected" level. A bank with a portfolio that is diversified across regions and products has significantly less concentration risk than a portfolio that is confined to exposures in a particular region or product. Banks with broad diversified risks can make more reliable estimates of the percentage of the portfolio they "expect" to default in the future and the losses this might cause.

Another key feature of many banking portfolios is that a rise in risks is often signaled in advance by a change in the behavior of customers. For example, retail customers under financial pressure might fail to make a minimum payback on a credit card account. Warning signals like this are carefully monitored because they allow the bank to take pre-emptive action to reduce credit risk. The bank can alter its marketing strategies and customer acceptance rules to attract less risky customers or price in the risk by raising interest rates for certain kinds of customers to take into account the higher likelihood of default.

To sum up: given the high velocity and volume of transactions and events in the modern financial system – events which include the contracts and actions of customers and counterparties as well as macroeconomic conditions – it is increasingly important for senior management of financial institutions to have a very accurate and up-to-date view of the state of the institution based on current (internal and external) data. It is equally important that policies as well as analysis and modeling tools and methodologies used to measure risk are well understood and are synchronized with each other with regards to input assumptions and conditions.

6.2.5 The industry view

Banking has undergone dramatic changes that have revolutionized the industry over the past 25 to 30 years. The most prominent example of the industry trend on the retail side was in the mortgage market. We saw a dramatic evolution of combinations of products and financing available to borrowers, products that in turn created access to the housing market for an historically high proportion of the

[17] The dramatic decline in the market cap of JPMorgan Chase, after the sizable losses that occurred from trading positions came to light in May of 2012, is a good example of a reputational risk impact incident.

population. These combinations reached into the thousands when we consider different loan-to-values on top of the hundreds of other features.

The securitization and syndication of residential mortgage assets, pioneered by Government Sponsored Enterprises such as the Federal National Mortgage Association (FNMA, or "Fannie Mae") and Federal Home Loan Mortgage Corporation (FHLMC, or "Freddie Mac"), was subsequently vigorously pursued by investment banks and large global banks with capital markets facilities. The securitization process served to facilitate delivering capital to the industry and allowed for a very wide range of loan products to be offered to borrowers. This market expansion altered the behavior and competitive decisions of smaller regional and community banks. Smaller banks were able to sell product from their own balance sheet to intermediaries who in turn supplied product for the securitization vehicles.

The industry also evolved into a reliance on the use of increasingly complex credit scoring techniques. Further, the industry introduced systems that automated the approval process such as Automated Underwriting Systems (AUS). These changes were part of a series of building blocks that led to the development of complex mortgage products.

AUSs revolutionized the underwriting process with loans being approved in minutes and not hours or days. This greatly reduced the costs associated with approving a mortgage and increased the volume of loans. The AUS also allowed access to financing for a broader range of borrowers with decision models that relied on credit score, credit history, loan-to-value ratios, reserves and other factors. The AUS additionally created more consistency and control in standards of the loans the GSEs would purchase. Relying on the sole approval of the AUS created in part a lack of personal responsibility in the underwriter community. The AUS was seen as the final decision over the traditional underwriter and was deferred to in most cases.

We also saw a rapid decline in documentation standards. For example, we saw rising volumes of stated income (SI) as well as Stated Income/Stated Assets (SISA) mortgage loans. An SI loan is a specialized mortgage loan where the mortgage lender verifies employment (but not income). The SISA mortgage loan is a type of reduced documentation mortgage program that allows the borrower to state on the loan application what their income and assets are without verification by the lender (the source of the income is still verified).

Securitization is a big advantage for investors because a wide range of investors are able to invest in say a "senior class" of notes. A special purpose vehicle (SPV) or trust is set up. The structure requires active management of the loan portfolio. A qualified loan portfolio manager actively manages the portfolio within legal constraints. The bank originating the loans typically remains responsible for servicing the loans. The SPV is an efficient securitization structure because it allows the cash flows from a pool of loans to be pooled together and prioritized.

The impact of the increased infrastructure implementation costs required to meet the increased data reporting requirements to regulators will be harder to bear for the smaller banks and will be more easily absorbed by the larger banks. It is likely that industry groups will develop around sharing the costs of services and systems needed for increased data reporting compliance in order to mitigate the higher relative costs of these additional efforts on smaller institutions. This strategy would also work well in the cases where groups of banks are using common service bureaus for back-office processing of operational and accounting applications. The solutions offered by common service bureaus can be crafted once and repeatedly applied to multiple clients using the common data structures which the service bureau would use for all its business banking client firms.

6.2.6 The financial markets view

The asset-backed securities pipeline

As mentioned above, a notable aspect of contemporary banking is the securitization and syndication mechanisms that have been introduced not only to package and harness, but also to hedge and leverage, the cash flows of residential mortgages. Securitization is the process by which a set of cash flows from a retail portfolio, such as mortgage payments on a pool of mortgages, is transformed into the payouts of securities. First, the retail bank originates credit market assets (e.g., credit card receivables, auto loans, and home mortgages). The bank creates a security by pooling together many similar retail assets (e.g., home mortgages) whose aggregate income will provide the returns on the security. Finally, the security is sold to outside investors and the corresponding assets and liabilities are taken off the bank's books.

A catalyst for the development of mortgage securitization in the USA was the federal government's sponsorship of some key financial agencies (namely *Fannie Mae*, *Freddie Mac*, and *Ginnie Mae*). These agencies issue securities whose income is derived from pools of home mortgages originated by banks and other financial intermediaries. In order to qualify for inclusion in these pools, mortgages must meet various requirements in terms of structure and amount.

Many variants of the original mortgage-backed securities (MBSs) have appeared over the years. An asset-backed security (ABSs) is a term that applies to instruments based on a much broader array of assets than MBSs, (for example credit card receivables, auto loans, home equity loans, and leasing receivables). Commercial mortgage-backed securities (CMBSs) are frequently considered a different type of instrument altogether, because the risk characteristics and the degree of conformity are different from those associated with home mortgages. Collateralized mortgage obligations (CMOs) are a variation on the MBS approach that differs not in the nature of the underlying instruments, but in the structure of expected payments. CMO payments are divided into tranches, with the first tranche receiving the first set of payments and other tranches taking their turn. This structure makes the duration of the securities different and potentially easier to use for asset-liability management purposes.

Banks have an incentive to securitize assets since this provides an effective means for banks to deal with their funding problems. Generating mortgages and then securitizing them immediately obviates the need for ongoing funding for those assets. The money that investors pay upfront for securities can be lent out again to other customers.

A key aspect of securitization is that it calls for the retail bank to sell the cash flows to investors through the securitization process and therefore the retail bank gains a principal payment upfront. The securities might be sold to third parties or issued as tranched bonds in the public marketplace. The structure of the tranches, their rating and the spread paid over LIBOR depend on the credit quality of the assets securitized. Securitizations can take many forms in terms of their legal structure, the reliability of the underlying cash flows and the degree to which the retail bank sells off or retains the riskier tranches of cash flows. In some instances, the retail bank substantially shifts the risk of the portfolio to the investors and by this process reduces the risk associated with the portfolio. The bank gives up a portion of its income from the borrowers and is left with a profit margin that should compensate it for the initiation of the loans and for servicing them.

The mortgage-backed securities supply chain

The diagrams that follow (courtesy of the graphic visualizations and network data analysis of Valdis Krebs and Orgnet[18]), depict the relationships among the participants and institutions in the larger financial system. As shown in Figure 6.2, these participants include the original residential homeowner to the mortgage originator and further up the financial markets supply chain to the underwriters and capital markets brokers of the investment securities that were sold to financial investors.

It is worth mentioning that data visualization tools will be required to interact with and manage the extensive amount and complex structures associated with the products, instruments and portfolios in the contemporary financial industry. In addition to addressing data considerations, it is important to note that the right tools are required to tame the data beast. These include metadata dictionaries, extraction and translation mapping utilities and libraries, workflow sequencing and process automation and a wide variety of analytical applications that have the ability to integrate enterprise-wide scope with the ability to drill down to contractual details of products or portfolios. These include data visualization tools that interpret these structures and relationships, and some of the diagrams that follow serve to illustrate their use.

[18] www.orgnet.com

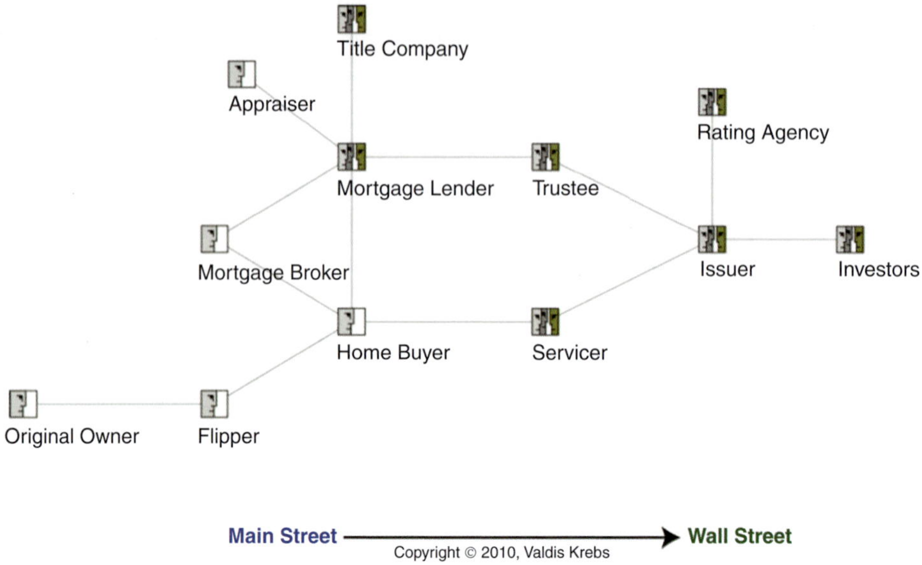

Figure 6.2 Relationships in the mortgage-backed securities supply chain (source: Valdis Krebs, Orgnet, used with permission).

There is an extensive set of data relationships in the mortgage industry supply chain pipeline, as shown in Figure 6.2. The provenance of data should be consistently maintained and digitally accessible throughout the financial product pipeline.

Figure 6.3 describes a particular example of the supply chain involving a residential mortgage that was included in the Goldman Sachs Abacus 2007-AC1 *synthetic CDO*.[19]

Figure 6.4 shows a detailed breakout of the failed Cleveland mortgages in one of the mortgage loan trusts (LBMLT 2006–4) that was part of the ABACUS 2007-AC1 synthetic CDO.

Figure 6.5 shows some similar foreclosed properties that were part of the Goldman Sachs Hudson 2006–1 CDO.[20]

The types of relationships among the various counterparties in a *synthetic CDO* (Collateralized Debt Obligation) are shown in Figure 6.6.[21]

Unlike a CDO that consists of pools of mortgages whose cash flows are sliced-and-diced in contractual tranches, the tranches of a synthetic CDO also contain CDS contracts on the corresponding tranches of an underlying CDO. This serves to illustrate how compound repackaging and multiple layers of referencing led to an explosion in the complexity of these vehicles. These products allowed for multiple

[19] See ABACUS (2007). [20] See Hudson (2006).
[21] `http://en.wikipedia.org/wiki/File:Synthetic_CDO_Diagram_-_FCIC.png`

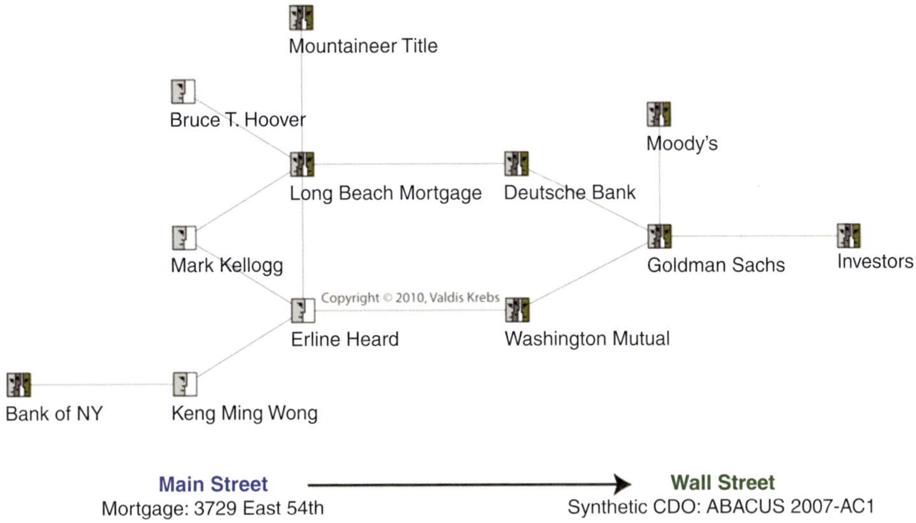

Figure 6.3 The ABACUS 2007-AC1 synthetic CDO supply chain (source: Valdis Krebs, Orgnet, used with permission).

investor participation on all sides of the vehicle. The fact that the institutional investors participating in the short unfunded positions were from a more rarified sector of the financial system than the bond holders, meant that there was significant interconnectedness among the institutional bankers, as seen in Figures 6.7 and 6.8.

6.2.7 The financial system view

Figure 6.9 depicts some of the major flows of high-volume data among entities in the financial system. New systemic regulation measures will themselves generate significant new flows of data throughout the financial system. In addition to the heightened standard to prevent market dislocations, additional data monitoring requirements in market oversight will come with development of new messaging facilities and standards for the now required central electronic clearing and depository recording of OTC derivatives trading. These market transaction data flows, coming primarily from global markets activities in securities and foreign exchange, would fall under the transaction monitoring desks of the regulatory agencies. The exchange depositories will be instrumental in providing counterparty relationship information for systemic regulators – but this will have to be done with the constructive participation of the banking industry private sector institutions.

Data flows from the "main street" side of the system, having to do with the markets and assets of the real economy that comprise consumer and business banking activities,

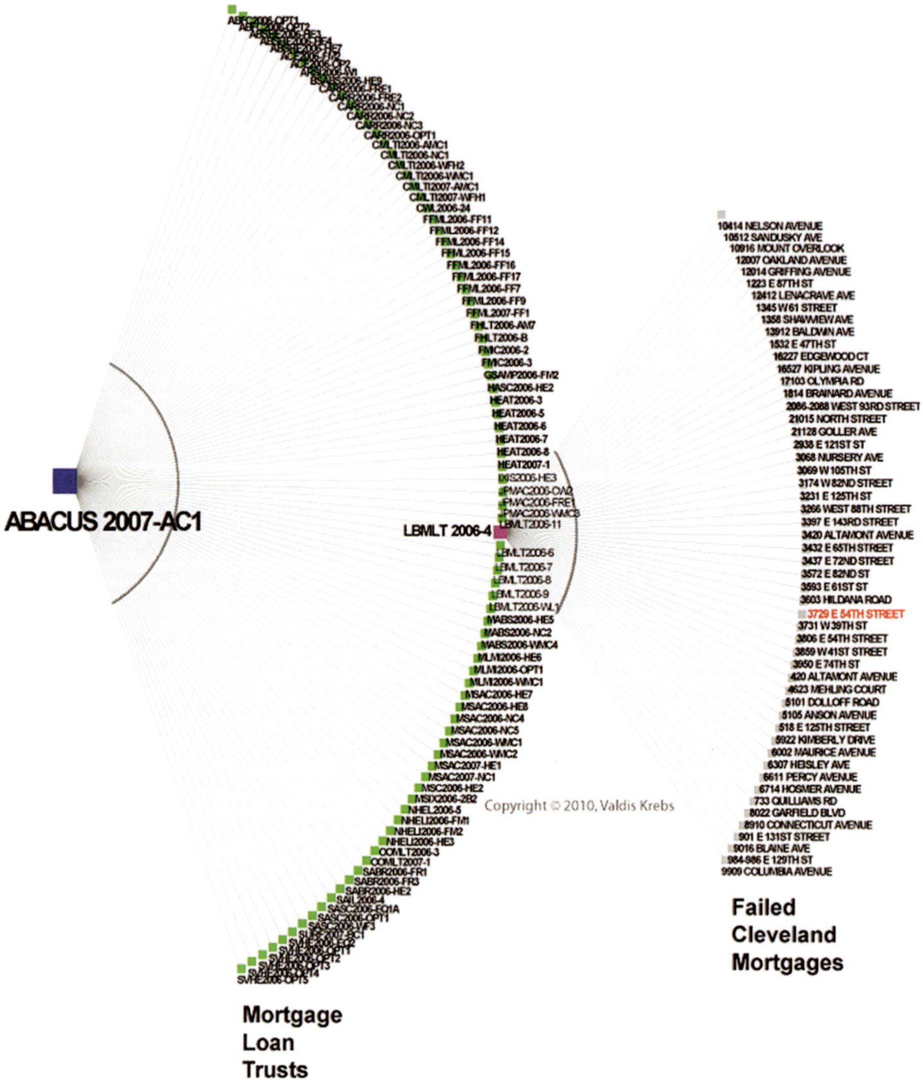

Figure 6.4 Composition of the ABACUS 2007–AC1 CDO (source: Valdis Krebs, Orgnet, used with permission).

including residential mortgages, fall more under the auspices of balance sheet supervision desks of regulatory agencies that are monitoring financial institutions. The increased frequency and volume of these prospective data flows will constitute a challenge for effective balance sheet monitoring at the level of the systemic regulator.

The broad mix of data flows in the financial market economy among primary classes of entities in the financial system can be legitimately and productively grouped into two large, and overlapping, camps. The different underlying

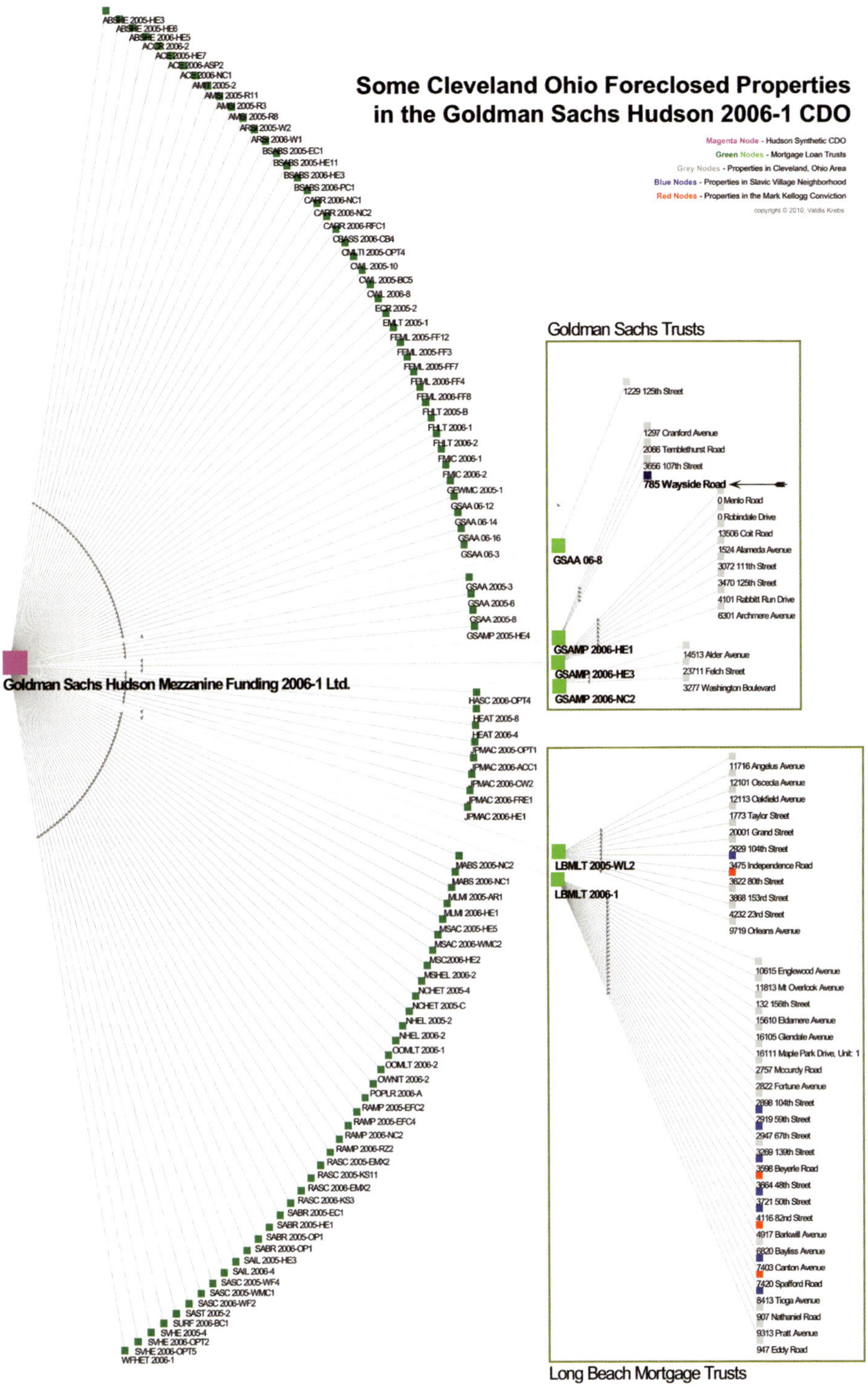

Figure 6.5 Goldman-Sachs Hudson Mezzanine Funding 2006–1 (source: Valdis Krebs, Orgnet, used with permission).

Synthetic CDO

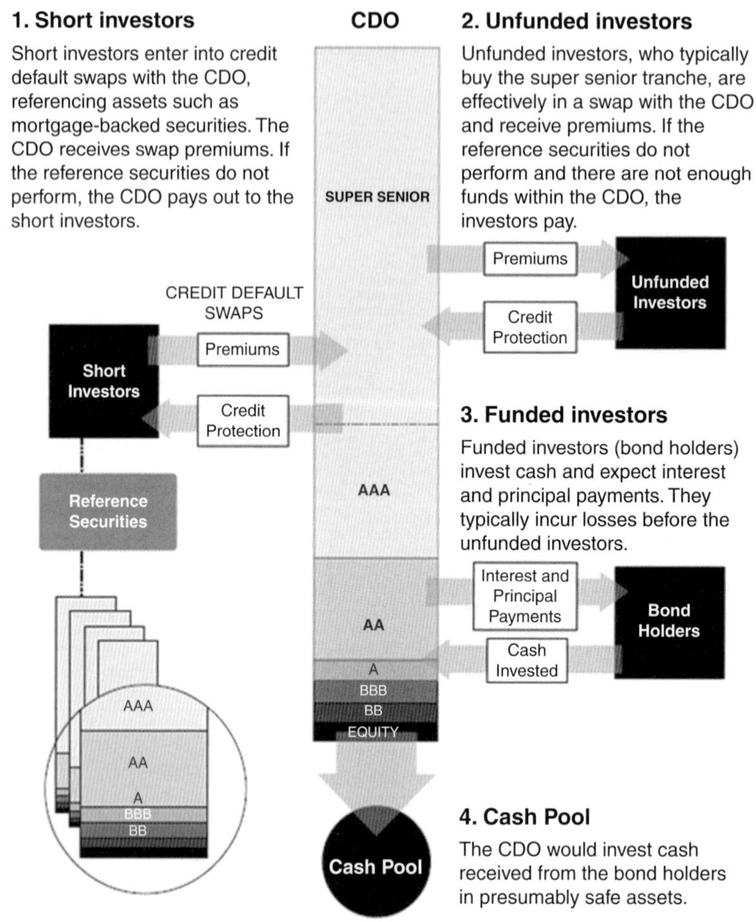

Source: The Financial Crisis Inquiry Report / FCIC

Figure 6.6 Multiple investors, counterparty relationships, and cash flows in a synthetic CDO.

characteristics of these camps have implications for how the continuous and sensible integration of the data is to be accomplished, both internally within each node in the system and as a network of nodes that need to be monitored as a system.

6.2.8 The global view

Banking product variations

Global financial capital markets comprise an extensive network of common financial instruments and products, particularly at the institutional buy-side or sell-side level, and in the wholesale markets among bank counterparties and central banks.

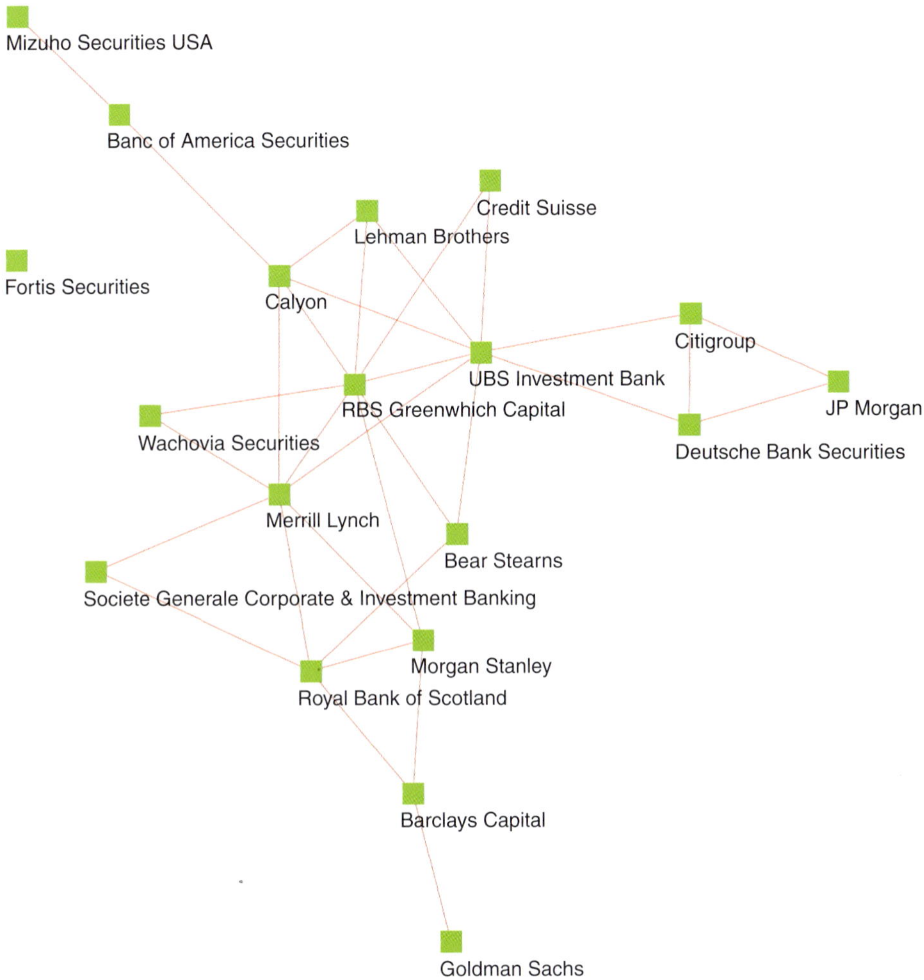

Figure 6.7 Circular CDO bank relationships (source: Valdis Krebs, Orgnet, used with permission).

However, when it comes to consumer or retail banking products and services at the local or domestic level within national borders, there are many variations in the modalities of financial products around the world. Some examples of these variations in products and behavior follow.

Japanese retail financial products In Japan, a system of twice-annual bonuses for employees is in place as part of the compensation structure. Japanese financial products are designed around these "dividend" like cash flows, since they occur en masse at the same points in the year and have an effect on the total system. For example, savings, loan payments and investment usually occur at the two bonus

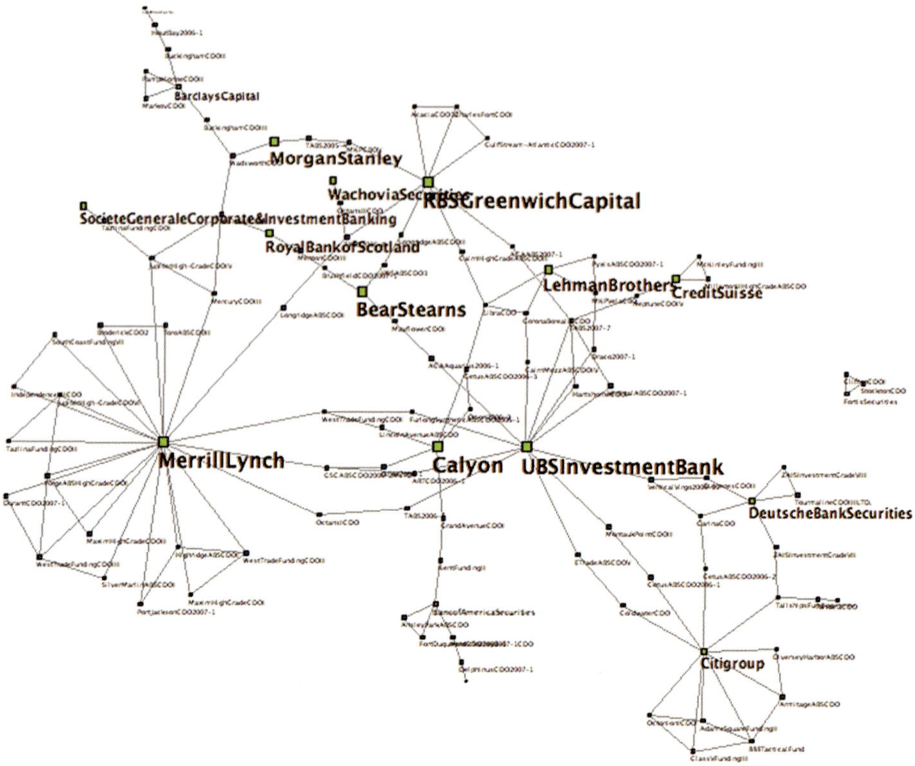

Figure 6.8 Cross-embedded CDO counterparties (source: Valdis Krebs, Orgnet, used with permission).

payment opportunities, as those cash flows represent something above the monthly budget that must be met the rest of the year.

Swiss managed-rate products The integration of the Swiss Franc in the Swiss economy – from the central bank to consumer interest rates for loans and, as a results, rents – is fairly profound. The Swiss banking rates are managed rates, meaning simply that they are set by the government much like the US Federal Reserve sets core rates based on policy review. But in Switzerland these rates are baked into nearly any time-weighted economic contract, be they "perpetual mortgages" (since monthly payments generally are interest only and principal reduction is rare) or simply rents paid to property owners. All rates reset every six months to whatever the prevailing managed rates are at the time. Loan rates and rents follow in turn. As a result of the prevailing semi-annual repricing frequency for nearly all products and the tight coupling of economic rates to the centrally managed rates, long-term fixed rate mortgages are rare, as is the need for an overly sophisticated duration-based asset liability management solution.

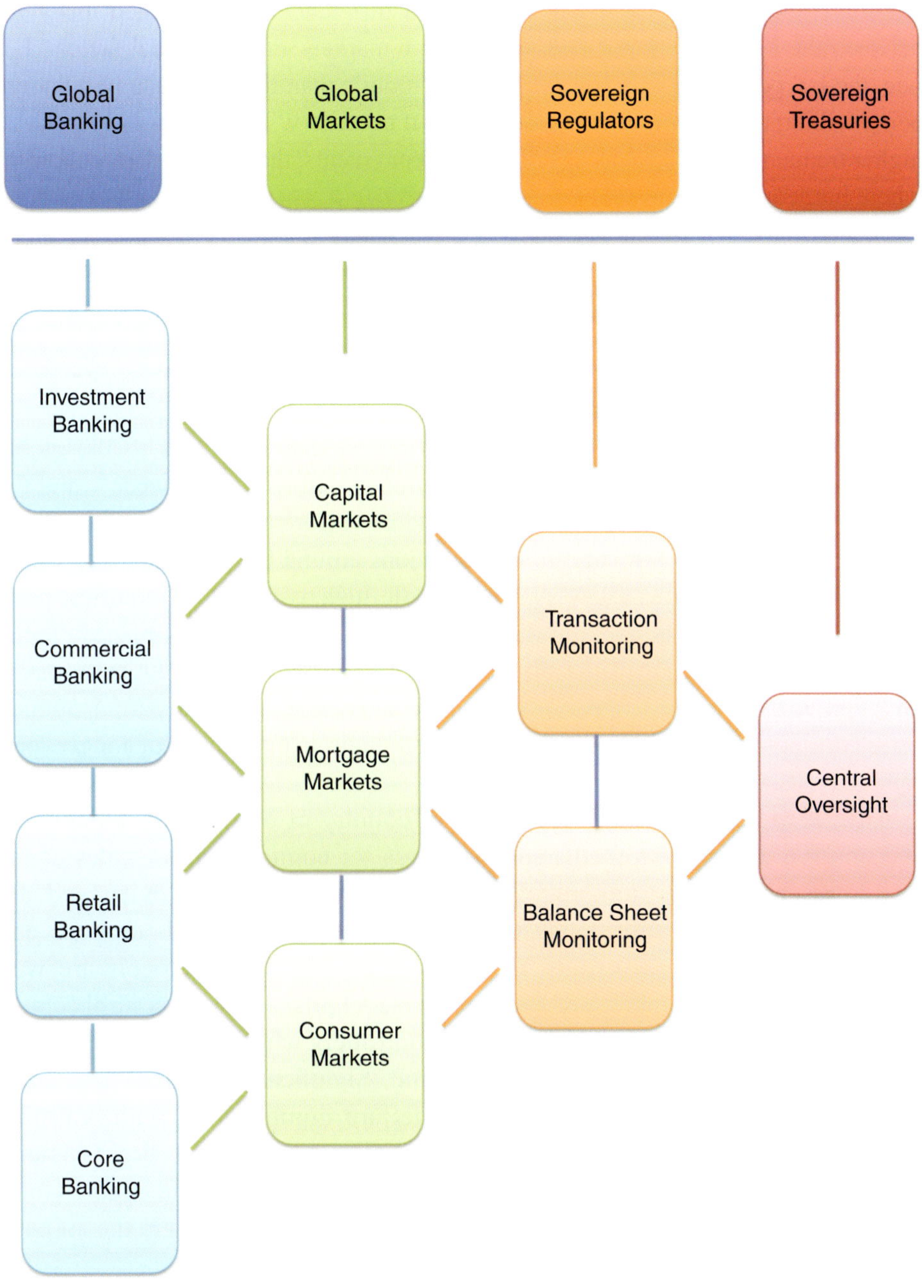

Figure 6.9 Data flows between major classes of financial system entities.

Refinancing behavior There is a significant difference in prepayment behavior between the US and European home mortgage markets, due to several factors. For one, the US population is traditionally more mobile than the European population, being willing to move many times in a career if necessary. This naturally produces a higher frequency of home sales, and hence prepayments, than is found in economies where homeowners relocate less frequently.

Second, in much of Europe there is a detectable reluctance to sell real estate property, since such property has often been in the family for multiple generations. Furthermore, owning real property is viewed as a desirable status, and selling it is to be avoided if possible.

Global operations and infrastructure

The global financial markets can be compared in many respects to the global aviation industry and the international scope of its informational needs – and the analogy can be instructive for the financial industry. Conventions and protocols for consistent air traffic control and flight operations around the globe are well established. Front office order-entry systems handle high volumes of inquiries, price quotes, and reservations booking transactions. Back-office systems manage logistics, operations, inventory allocation and planning for current and future commitments based upon contracts previously booked and ticketed. Skillful participation in petroleum commodity futures markets and strategic management of longer term capital assets are essential.

Airlines around the world, regardless of country of origin, have all adopted and observe common protocols and standards for communication, rules of the road, passenger safety and regulatory oversight. Procedures exist, and are invoked, for dealing with congestion, unpredictable weather disruptions and rare but highly stressful events or emergencies. Message protocols are well established for voice communication between cockpits, air traffic control centers that supervise regions, and airport tower controllers who handle traffic around airports, whether in the air or on the ground. Significant advances have been made with risk management and system failure minimization with real-time streaming of telemetry of all manner of statistics pertaining to flight history, fault reporting and sensor data monitoring mechanical devices and engine vibration – all using the Internet and performed whether the aircraft is in the air or in the hangar.

Other aspects of similarity between global financial markets and global aviation are: (a) the standardized use of the English language as the *lingua franca* among corresponding parties, and (b) the adoption of systems infrastructure and electronic standards that facilitate the application of information technology to monitoring and managing the industry.

Sovereign considerations

The upheaval in the US financial system that rippled through the global economy and drove most developed countries into recession revealed the degree of interconnectedness – and interdependence – of financial and economic activity throughout the world.

In the European Union, the rising cost of Greek sovereign debt created a difficult fiscal and political problem for the EU, despite the relatively small size of the Greek economy and its sovereign debt compared to the entire Eurozone. Spain is a much larger economy than Greece, and the crisis in Spain is one that impacts the banking sector directly because many real property market values are "underwater" – i.e., worth less than the amount of the mortgage loans funded by the banks.

Cross-border resolutions in times of financial or economic stress are far from straightforward, as they are always based on some form of agreement among sovereign states (some binding, some voluntary.) These relate to the handling of a variety of national and international legal situations, in particular those having to do with bankruptcies, transfers, and multinational assets and liabilities.

The global financial system is experiencing unusually severe gyrations linked to the effects of major subsystems on each other. Cooperation among regulators, central banks and the private sector is not a simple matter even in times of relative equilibrium. In times of heightened systemic risk it is all the more important to have mechanisms to foster cooperative and coordinated efforts to manage systemic risk actively. Those mechanisms require timely access to accurate and detailed data based on consistent data standards in order to be effective.

6.3 Foundations of banking data infrastructure

6.3.1 Data realms and domains

As suggested by Figure 6.10, there are many data domains and categories that overlap and intersect in the universe of economic relationships, financial positions, transactions, management decisions and business operations activities of financial service institutions. Some of the main category families and their shared domains of overlap (highlighted) are shown in Figure 6.11.

Figure 6.11 is a high-level, conceptual view that illustrates how the major collections of banking data have many overlapping categories and areas in which they interrelate.

The high-level data flow relationships among some of these domains are depicted in Figure 6.12, a conceptual illustration of financial data categories. The rough proximity in the grouping of domains is intended to signify a degree of interaction.

A conceptual distinction is made between operational (production) interactions, signified by the "Operational Bus," and the activity on an "Analytical Bus," where typically the main interactions are fundamentally on copies, extracts, or derivations of the production data.

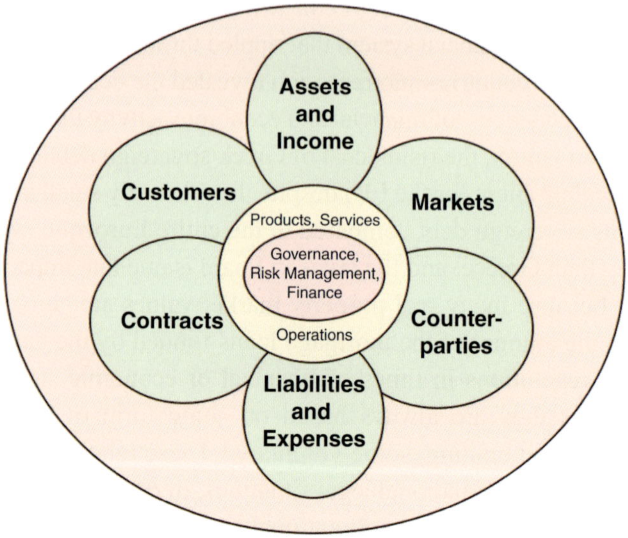

Figure 6.10 Overlapping data realms in banking.

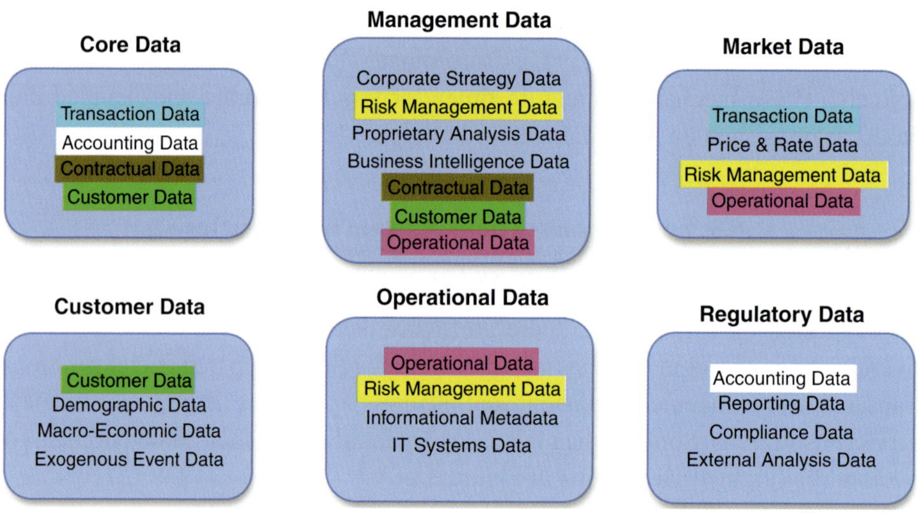

Figure 6.11 Overlapping categories and types of financial data.

A degree of replication of production data for analytical purposes is important in order to maximize transaction throughput (for productions systems) and analytical bandwidth (for decision support and risk management systems). It is important to minimize "infrastructure interference" between these two major modalities of information processing in financial institutions.

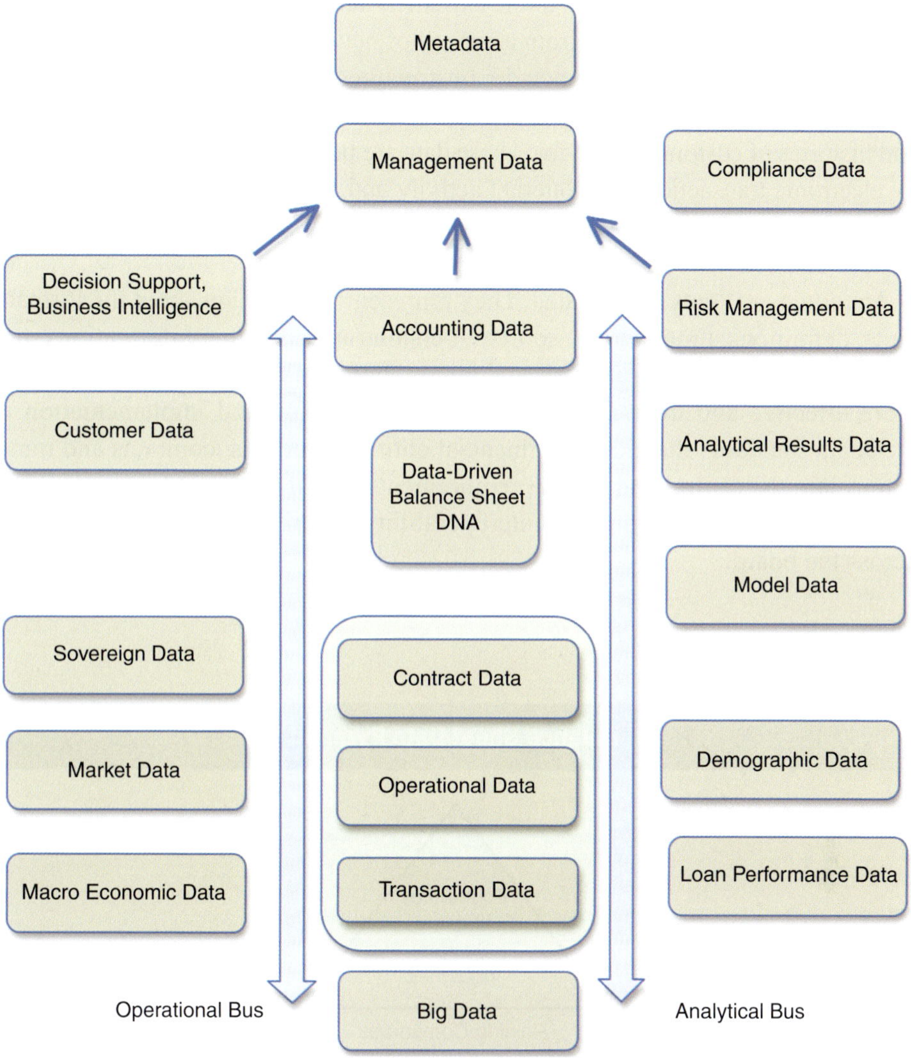

Figure 6.12 Financial data operational subdomains. This figure is not intended to be definitive or exhaustive and it is only a conceptual representation of data categories and relationships.

Some of these categories will be discussed below in more detail, including: accounting data, contract data, market data, customer data, operational data, risk management data, analytical results data, demographic data and macroeconomic data. Other categories (e.g., sovereign data, management data, compliance/reporting data, and model data) are discussed in the course of the narrative but not specifically elaborated.[22]

[22] "Sovereign Data"' is a reference to a conglomeration of data sources and types, including sovereign central banking, debt markets, national econometric data on industries and resources, geopolitical factors and other sovereign economy news.

Big Data is a term that refers to the extremely large volumes of a variety of data types generated and collected from a variety of new digital sources and processes. These new sources and types include: unstructured text, events, advertising campaigns, financial transactions, market tickers, media, web browser interactions, logs and histories of customer behavior. These data are being generated for the most part by electronic transactions and Internet activity and, depending on the industry and type of data, they benefit from technologies designed specifically to address their specific characteristics.

Metadata are "data about data." They represent information about the formats, units, definitions, time periods, sources, versions and any other information (such as indices) that can make analysis and interpretation of the underlying data faster, more effective and more reliable. Definition, adoption and implementation of metadata standards that describe financial entities, products, contracts and transactions are the central and critical requirements for all manner of improvements and efficiencies in data quality, interoperability and risk management practices across the board.

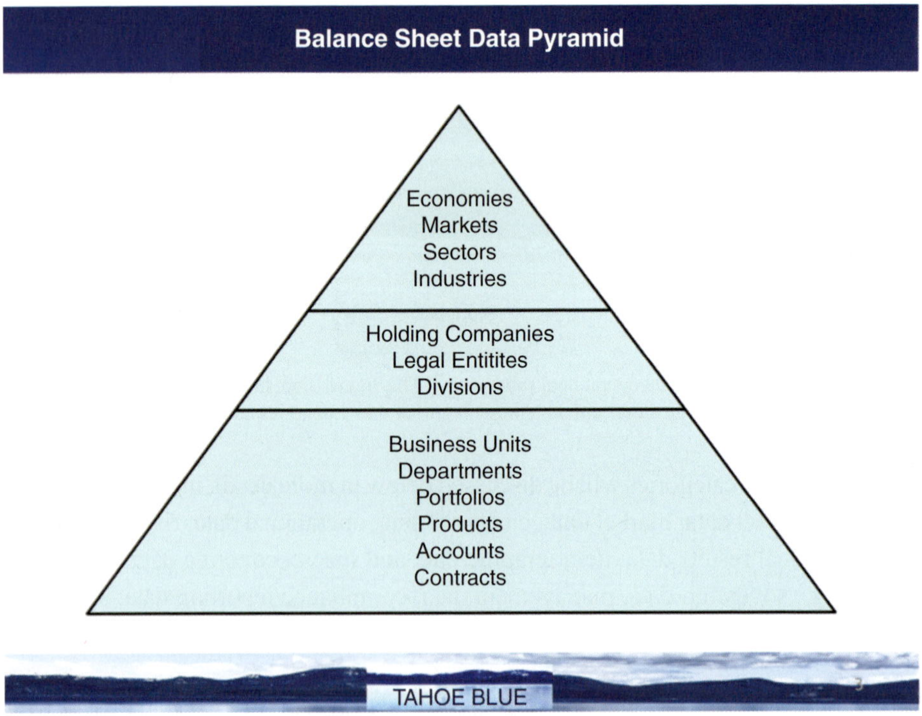

Figure 6.13 Accounting data pyramid.

Accounting data

Modern General Ledger (G/L) systems are far more sophisticated than traditional accounting systems in which the account record data elements consist essentially of some form of account number, line item description and an amount.[23]

In traditional accounting systems, the attributes and properties of a detailed account are associated with the account mainly by virtue of where the account "number," or key, is located within the overall schema of the complete Chart of Accounts. An "account number" was usually organized into multiple fixed-length fields in which the higher-order parts of the number were assigned to larger reporting categories and lower-order parts of the key were used to elaborate more specific subcategories or business units. The lowest level of the account number was used to assign the unique account to a specific item to be tracked at the lowest level of detail possible within the system, such as a retail checking account or mortgage servicing account.

The practice of assigning meaning or interpretation to an account by virtue of embedding ranges of codes or numbers into the account number ID field in a chart of accounts is by and large an outmoded practice. Doing so leads to innumerable problems and exceptions when, for example, a particular fixed range within the account key is exhausted and cannot accept additional accounts of the same type or subcategory.

Contemporary accounting systems, owing to their having been implemented using relational database management systems and not "hardwired" accounting applications, typically simply use a very large number as a unique account key. The informational attributes of a low-level account that are of interest are then associated with the unique account number for purposes of sorting and rolling the account balances and activity up into a variety of aggregate views, for example by customer, by business unit or by the chart-of-accounts hierarchy.

Derivation of decision support and business intelligence data pools The primary role of an accounting system is to do such things as create and process purchase orders, issue invoices and receive payments, post transactions to and update the balances of customer accounts. An accounting system also keeps track of budgets and the P/L of business units and ultimately produces summarized balance sheet and income statements for the entire business. The relational database approach, which attaches attributes to accounts without the need to overload the account key itself, makes it easier for accounting systems to interface with many

[23] Journal entries used to post transactions in double-entry accounting systems also include a second account number that is the account to be debited the same amount that is credited to the first account or vice versa, depending on the direction of the transaction.

other systems within the firm. These other systems and applications typically require the underlying data to be filtered, sorted and grouped based on a wide variety of other criteria. The criteria are themselves subject to change as a business evolves and new information requirements arise.

These systems include such applications as the following.

- Business Intelligence (BI) systems where data can be readily viewed and sorted by a number of different dimensions (e.g., customer, product, geography, demography, business line) in order to gain timely insights into up-to-the-minute performance of products or customers or business units without incurring the cost and delay of developing custom applications to perform the analysis.
- Decision Support systems are typically provisioned and used by departments or business lines. Within banks, however, these are also developed to meet the needs of the central treasury or executive committees (e.g., Asset/Liability Management applications). Such systems generally apply specific types of analysis or statistics for both tactical execution (e.g., fraud detection systems, transaction credit approval systems) as well as for strategic planning and risk management;
- Data mining applications typically utilize large collections of data and facts which have been amassed or accumulated through ongoing historical interaction with customers and markets. These data are used as sources for statistical analysis aimed at obtaining useful or hidden information about preferences, trends, or other correlated behavior that would inform product design, customer decisions and marketing campaigns.
- Full-scale Enterprise Resource Planning (ERP) systems allow management of workflow, capital investment, inventory, production and logistics and are used in manufacturing operations as well as informational industries and financial enterprises.
- Customer Relationship Management (CRM) systems.

The data interfaces between accounting systems and other systems and collections of data in an institution, whether the data flow into or out of the accounting system, are potentially quite numerous. Extraction of data from G/L systems for reporting or analysis purposes will naturally provide less information than may be required or desirable for many analytical and decision support purposes due to the general *reduction* or minimization of the number of detailed attributes that are supported when information is initially recorded in general purpose accounting systems.

The need for greater product parameter detail and aggregated data accuracy It is neither practical nor possible to use accounting system conventions or capabilities to try to associate very different numbers and types of parameters to G/L accounts, regardless of how flexible the accounting system software may be.

Even if it were theoretically possible, the interface between a bank's accounting system and the production system of record that handles a particular type of financial instrument or product is very limited.

For example, after an ARM loan is originated, the system of record production application is initialized with all of the specific terms of the customer contract. When the customer makes payments, those cash-flow payments are recorded in the system of record in order to update and maintain correct contractual outstanding balances for interest calculation and next-cycle payment processing purposes. The accounting system is also updated with the principal and interest component amounts in the payment (in order to be able to produce current balance sheet and income statements). But the accounting system has no specific details of the contractual parameters and conditions that drive the calculation and generation of expected payments and statements.

It is simply not possible to perform adequate risk analysis on the detailed assets and liabilities of financial institutions using conventional accounting systems – the information required to do so is simply absent, even if the lowest level customer accounts in the firm's accounting system are examined.

Complicating the situation further, data presented in G/L accounting system reports are, by necessity, aggregated. When low-level accounts and contracts are grouped and aggregated into broad product classes (for example, "15 year floating rate mortgages"), a large amount of information is lost.

When product balances are averaged together using broad categories and maturity ranges, contractual parameters necessary for projecting (or simulating) product cash flows become averaged together or, more often, simply ignored.

Applying linear risk-coefficient factors to aggregate product category balance numbers derived by adding too many disparate contracts together in order to determine capital requirements is an exercise based on "muddy" inputs, and will likely produce muddy results in return.

The need for more data dimensions and granularity Despite the ability of advanced accounting systems to assign multiple dimensions to account balances and transaction amounts, there are only so many dimensions that are suitable to be applied across the board. These types of dimensions, or attributes, do not include properties that are specific to one type of product or account but not another. The dimensional attributes that can be assigned to a one-size-fits-all accounting framework are truly a small proper subset of common dimensions, and represent a core set of data elements that would be required in a robust accounting system.

Common data elements do exist that, in addition to the account balance, might be applicable to nearly all types of account/product combinations. Depending on the capabilities of the accounting software, these data elements could be associated as

Table 6.2 *Core data elements and attributes common to all accounts and products*

Business unit	The organizational code of the business unit responsible for the account
Account type	The type of account
Account parent	The hierarchical parent account of the account
Product class	The general class of product (e.g., demand deposit, fixed-rate mortgage, credit card, commercial loan, certificate of deposit, etc.)
Currency	For multicurrency capable systems, the currency denomination of the account
Customer type	A category code for the type of customer (e.g., individual, small business, trust)
Customer location	Some level of location-based code, anywhere from country, to state, to major demographic market area, to zip code
Current balance	The current balance of the account
Original term/ tenor	The original tenor of the product type in the account, ranging from demand deposit to long term (30 or more years for mortgage loans)

attributes of G/L accounts for more robust reporting as additional dimensions associated with a unique account number or identifier. Attributes common to all types of account/product combinations are listed in Table 6.2.

Basic customer banking relationship

The products, assets and liabilities of the banking business are based on financial instruments that form the basis of contracts with customers. As such, a bank's assets and liabilities lend themselves to being expressed in some relatively straightforward relationships. These contractual customer relationships involve connections to the bank's accounting system, to some type of financial instrument contract (processed by production systems), and to a business unit that is responsible for managing the relationship between the customer and the bank. Figure 6.14 illustrates these basic relationships.

The *core account* in Figure 6.14 contains some of the data dimensions afforded by more flexible accounting systems. The core account is the node that connects accounting systems and production systems. Both accounting and production systems will likely share the data elements of the *core account*, in order for both systems to perform their functions and be synchronized in terms of both cash flows and balance sheet updates.

The accounts at the lowest levels of the G/L are internally "owned" by the business unit that manages the contractual relationship with the customer. The business units in turn roll up through the organizational hierarchy to the top level, represented as the legal entity of the bank.

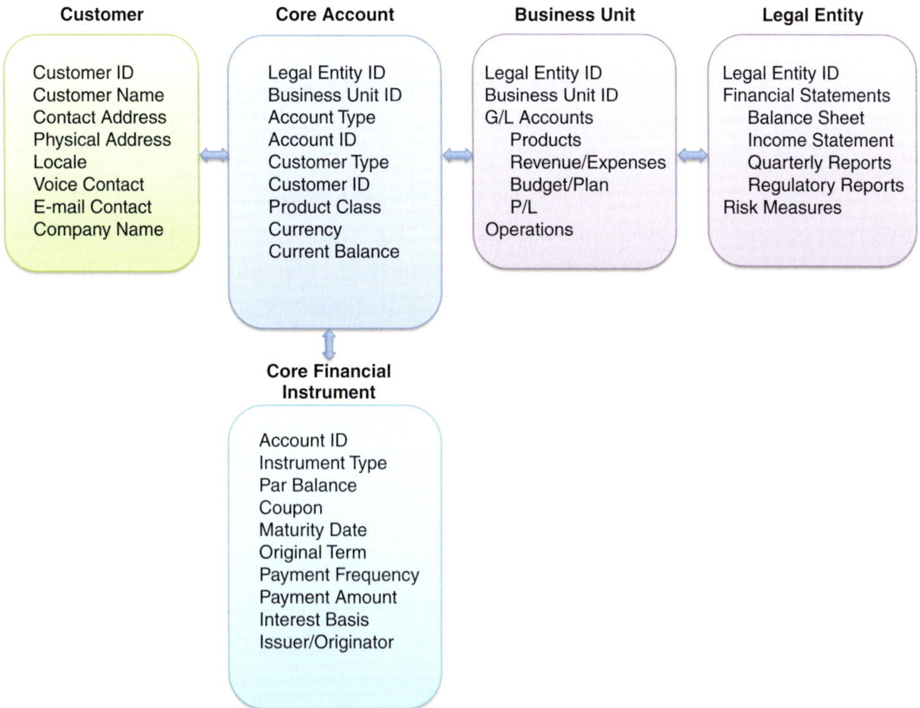

Figure 6.14 Basic customer banking relationship.

Contract data

Data requirements of different product instrument types The transactions and payments of the financial products of customer accounts are processed by production applications using the current state of account product data structures in the bank's systems of record. The *core financial instrument* in Figure 6.14 depicts the properties that are common to almost all financial products in a bank. These common properties will be used to construct a *core financial contract* (described below) for purposes of optimizing the performance and efficiency of physical data warehouse implementations.

Despite the core properties that are common to nearly all financial products, there are many financial instruments that utilize a wide variety of different configurations of required and optional terms in order to record and process customer account contracts and transactions under those contracts. Unlike general purpose accounting systems, the production applications in banks must correctly identify, maintain, access and process the specific configurations of detailed contractual parameters for the wide range of financial products that are stored in systems of record.

Instrument types are identifiers that define the particular sets and configuration of data requirements, structures and processing rules that make up the templates for the range of financial products around which each contract is constructed. In account servicing or other production applications, *instrument type* identifiers are used by operational systems to identify the data structure template of specific contractual terms and the logic rules governing the interpretation of those contractual product terms in order to process the contract going forward.[24]

Historically, different systems of record and processing applications have been deployed for different product types and the data formats and requirements of each system tended to be unique to the particular product class and line of business silo. This is still largely the case in legacy banking systems.

This fragmentation of contractual data processing has actually been magnified with the wave of mergers and acquisitions that have taken place in the banking sector in the past two decades. Establishing an enterprise data standard and production applications framework can be a major undertaking in and of itself. Migrating the data from the proprietary format of a recently acquired bank into the format and platform of the acquiring bank is not a process that lends itself readily to "cookie-cutter" efforts, since each acquired institution typically has unique business and system characteristics that distinguish it from previously acquired ones.

The incompatibility among different production systems that was commonplace in the era of mainframe fixed-record format production applications has been mitigated to some extent with the advent of the use of relational database management systems (RDBMS). Financial processing application software suites based on RDBMS platforms have been developed and marketed by major financial software solution vendors. Nonetheless, the use of independent – and incompatible – data formats in different production and operations applications within a bank remains a significant issue. This is not just true for the data quality control of each application system on its own. The diversity and lack of standardization of application systems of record for different product and contract categories is one of the significant barriers to effective management of the entire bank's portfolio of assets and liabilities – especially with regards to risk management practice.

Some account/contract types are quite straightforward with regards to the contractual terms or parameters required to process the account. As a case in point, the data elements needed to process a simple interest-bearing bond are the face (par) value, the coupon (interest rate), the payment frequency, and the maturity date.[25]

[24] It is important that field-level data rules for required and optional terms that an instrument type is intended to identify (e.g., for the servicing of contracts in production) be linked to the preparation and entry of data at the time that new contracts are created (e.g., in front office loan origination). The negative consequences of injecting incomplete or, worse, inaccurate data into the origination of new business cannot be overstated.

[25] Even in this simple case, it is also necessary to distinguish between annuities (where the principal is not returned at maturity) and principal-at-maturity bonds that pay interest – hence the importance of the instrument type attribute that identifies the specific business rule or algorithm for the contract.

On the other hand, the volume of data elements needed to originate and service an ARM is far greater. Ten to 20 parameters are required in order to process and issue a monthly statement that informs a homeowner what the next month's mortgage payment is to be, based upon the floating rate index that is tied to the contract. These parameters include: current balance, maturity date, starting interest rate, current interest rate, payment frequency, adjustable rate index, re-pricing frequency, spread over rate index, life cap, life floor, period cap, period floor, initial "teaser" rate, length of teaser rate period, prepayment penalty, etc.

Organizing the elements of financial instrument position structures It is critical to be able to examine and access all product contracts based on different financial instrument types across the full range of an institution's positions regardless of their type or structure. This requires a data modeling framework and corresponding information architecture that is up to the task and capable of representing and recording a wide variety of financial instrument types while at the same time lending itself to physical implementations capable of both high performance and the efficient use of resources.

Trading systems Information systems dedicated to the specific business and risk related needs of trading rooms must accommodate a wide variety of specific and potentially complex financial instruments. Products and positions based on financial instruments with more complex structure need to be evaluated quickly and frequently for the purpose of managing a trading book or portfolio. Systems dedicated for this purpose, which also apply sophisticated analytical valuation and pricing algorithms to positions in the portfolio, typically use data structures customized either by vendors or designed by in-house proprietary development teams in order to record the details of financial instrument contracts. Often, these data storage systems use *object-oriented database (OODB)* technologies that are able to define and record the product-specific nested structures of different instrument types.

This approach combines the accurate capture of financial instrument parameters with a tight and rapid coupling of the contract data with analytical tools. This approach works quite well for the specific needs and exacting demands and real-time execution constraints of the trading room. However, although well-suited for processing portfolios of similar types of positions and contracts, the dedicated nature of such applications poses challenges to obtaining more comprehensive views of financial positions and risk exposures across all product lines. In order to do so, it is often necessary to extract and map data from different trading systems into other systems that can integrate, summarize and report on the consolidated.

Balance sheet products: demand deposits and mortgage loans In the case of the bulk of the very large number of customer accounts on a bank's balance sheet, an approach is needed that will capture sufficient contractual detail of balance sheet products for those accounts. Products such as deposits and loans on the main balance sheet of the bank pose challenges due to the large number of accounts of these types in banks and the sheer volume and diversity of data. The challenge is further compounded because of the striking differences in the very simple structure of liabilities (e.g., demand deposits) on the one hand and the more complicated structure of assets (e.g., mortgage loans) on the other.

What is described in the discussion that follows is an approach that manages to capture the range of simple customer deposit account data, robust mortgage account data and more complex structured product data from the trading room. The approach *takes into account the physical storage and performance implications* and trade-offs that accompany the population of a data warehouse with data from a very large number of accounts whose contracts span a broad range of simple to complex financial instrument structures.

The data elements that are common to nearly all customer account product contracts in the entire bank can be identified. Figure 6.14 shows a base-level structure, the *core financial instrument* containing those elements. Included in this structure are such things as account ID, instrument type, account balance, interest rate, payment frequency and term (tenor) of the contract.

In Figure 6.14, the *core account* and the *core financial instrument* that is associated with a core account are shown as two different logical entities. By observing that a core customer account in a bank will almost always be associated with a core financial instrument, an optimization in which the core financial instrument is combined with the core account is possible. The result of this de-normalized combination is the *core financial contract* shown in Figure 6.15. The core financial contract structure is a physical implementation optimization that captures the information necessary to describe a large number of detailed accounts based on products such as deposit accounts or bonds. The relatively small size of this core structure means that, in large banks with tens of millions of customer accounts, the amount of space used to store the core structure is manageable.[26]

Next, an optional structure that is designed to accommodate the larger number of contractual parameters needed to express more detailed contractual terms and conditions for an ARM can be set forth. This structure, the *adjustable-rate extension*, is associated with the *core financial contract* structure as an optional extension for those cases requiring the more extensive set of ARM parameters. The *adjustable-rate extension* contains a complete set of financial instrument parameters, such that each

[26] This approach also makes it possible to efficiently access important information regarding the principal balance positions and interest rates of all accounts across the entire institution with a scan of a single data structure in a data warehouse that has been populated with data extracted and mapped from the various systems of record in the bank.

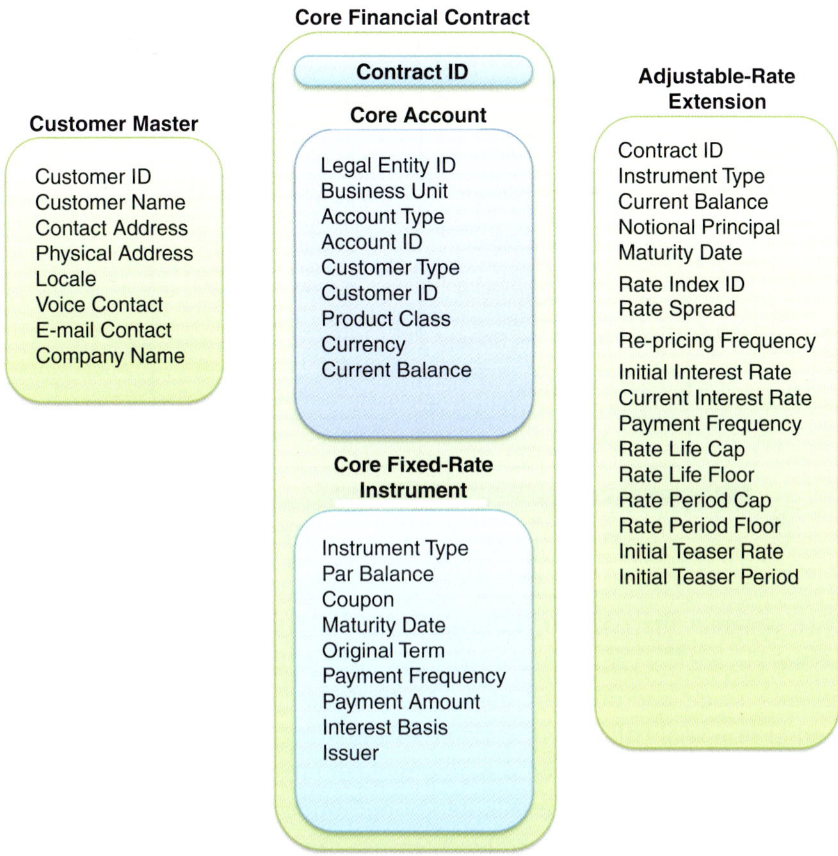

Figure 6.15 Core financial contract with adjustable-rate extensions.

adjustable-rate extension structure is not dependent on the core financial instrument structure to have a full set of instrument parameters. When an adjustable-rate extension leg is added, some of the parameters in the core financial instrument are redundant.[27] But by including a full set of adjustable-rate parameters, more than one adjustable-rate extension leg can be added. This makes it possible to express instrument types that have more than one adjustable or re-pricing type product leg (e.g., an interest rate basis swap).

Figure 6.15 depicts an illustrative example of a core financial contract structure approach that can be used in an account-level, enterprise-wide financial instrument data warehouse. It is a structure that allows for an optimal capture of a large number of basic banking customer accounts based on fundamental fixed interest rate products (e.g., deposits, bonds or fixed-rate mortgages).

[27] This is simply because the core financial instrument structure was designed to handle the large number of cases that did not require all of the parameters required by an adjustable-rate contract, i.e., it is an optimized subset of those parameters.

Market Data vs. Contracts

Market Data	Contractual Data
High Frequency	Low Frequency
Transaction Stream	Balance Sheet Change
Simple Structure	Complex Structure
Public/Transparent	Private/Opaque
Portfolios, Time Series	Customers, Counterparties
Valuation	Cash Flows
After the fact	Before the fact

Figure 6.16 Market data versus contract data.

If we separate the types of robust parameters and contractual terms that are necessary for ARMs (and other instruments whose cash flows are predicated on interest-rate resets and other contractual rules constraining changes to the interest rate and/or principal balances of variable-rate products) then it is possible for these kinds of contractual terms to be captured without creating an overly complex core record. Furthermore, there are technology and infrastructure benefits and advantages to keeping the core nodes of an enterprise model comparatively simple.

Multiple adjustable-rate extensions can be used to capture the additional terms required for product types with more than one adjustable rate re-pricing leg, such as interest rate basis swaps or floating-floating currency swaps. Central to such a system is a master Extension Table that, for each instrument type, defines all the extension data structures required for a complete description of that instrument.

Market data

Data from and about financial market transactions are fundamentally different from the core data in other domains in the real economy/financial sector space, such as contractual data or customer data. Figure 6.16 calls attention to some of these fundamental differences.[28]

[28] In Figure 6.16, "After the fact" and "Before the fact" are intended to connote the degree of "backward-looking" versus "forward-anticipating" states of market data compared to contractual (position) data. Market data (e.g., the prices and volumes associated with purchase and sales of equity shares) reflect decisions that have already been made and executed ("After the fact"). Data contained in active contracts and positions reflect information about the current state that will influence and inform future decisions about transactions that will affect the future state of the balance sheet and other financial measures of an institution, hence "Before the fact." Put another way, "fact" could be more precisely qualified as meaning "decision to take an action."

Market data is a very broad term to begin with, so some clarification is in order. After all, everything bought and sold is exchanged in a market of some kind, whether that market is the local grocery store, the local housing market or the art market. Electronically traded markets include foreign exchange, energy, stocks, bonds and commodities in addition to the more exotic instruments in the OTC derivatives markets.

Financial market data are assembled from a number of different types of data sources with different frequencies and interfaces. Much of these data are real-time feeds of trade price quotes and executions that are the serial record of the activity of financial market exchanges. Trades in markets for debt instruments (government or corporate bonds) that have coupon rates and prices (based upon the term or tenor of the instruments) are special types of market data.

The consolidated rates from the interest rate futures market for US Treasury securities establishes the interest rate term structure, or yield curve, for the so-called *risk-free rate*. The implied forward rates from the *risk-free yield curve* are used to determine discount factors that are then used to calculate the Net Present Value (NPV) of expected future cash flows. The yield curves for other financial products with term structure (such as interest rate basis swaps) have spreads on top of the risk-free yield curve that reflect the risk premiums of a particular product market as well as incremental risk spreads associated with the counterparties. The Credit Default Swap (CDS) markets are used extensively to express and reflect the financial debt markets' risk premiums associated with debt issues from all types of sources: municipal governments, corporations, and sovereign treasuries.

For banking, whose business is predicated on the spread between interest earned on assets and interest paid on liabilities, the so-called "*risk-free*" yield curve is fundamental.[29] In the USA, the risk-free yield curve is the term structure of interest rates paid on the wide range of US debt instruments, usually grouped in three categories: short, medium, and long term. Short-term is anywhere from overnight Fed Funds (drawn upon nightly for funding liquidity purposes in daily settlement of transactions and cash flows) up to 1 year (Treasury Bills). Medium-term is typically 1–10 years (Treasury Notes) and long-term 10–30 years (Treasury Bonds).

Financial market data differ from contract data in the sense that trades on exchanges reflect the history of decisions previously made. Contractual data, on the other hand – when combined with data about the counterparty or customer – enable a forward-looking horizon of analysis in which the future actions and

[29] Recent economic and financial events have to a degree called into question the notion of "risk-free" yield curves or indices. The credit ratings of many sovereign debt issuers (long assumed to be "risk-free") has been impacted by the exposure to rising public debt and declining tax revenues because of economic contractions. Also, the outcry over the apparent manipulation of the London Inter-Bank Offered Rate (LIBOR) by some of the participating LIBOR panel banks as a wholesale cost of borrowing index has complicated the concept of base measures of lowest-risk investment/funding measures.

decisions of the parties to the contracts are modeled and submitted to a wide range of scenario analyses.

Exchange-traded markets and instruments on the one hand and real economy balance sheet products on the other differ substantially in the volumes and frequencies of transactions as well as in the simplicity or complexity of their data structures. These differences in both structure and activity require information system architecture and data standards that are both wide and deep in order to digest, manage and analyze comprehensively the broad range and types of financial data products and events.

Customer data

Customer data in the banking industry can be viewed as data that are domiciled in or accessed from a combination of several overlapping and inter-related sources:

1. customer master file data that identify the responsible party (e.g., name and address) associated with a transactional or positional account and which are used in operational processing;
2. customer profile data contained within a bank's credit department systems that are used for evaluating the degree of risk exposure that the bank determines to be appropriate for a given customer; and
3. customer data that may exist outside of the institution but which the bank can access from a variety of external sources and which can be useful in both marketing, credit, and business development.

One of the long-standing data challenges with regards to customer data is that business lines in banks have traditionally operated in organizational silos with segregated data systems. This situation arose partly because of the difference in products and business data processing required by different divisions and partly out of mergers and acquisitions among banks which did not share similar production systems and which were not readily or effectively consolidated or migrated onto a single operational platform.

Operational data

Figure 6.17 illustrates, albeit in an oversimplified form, the segregation of data in traditional banking business line silos. The diagram following it (Figure 6.18) shows the rich nature of the data dependent interfaces in the operational cycles of financial firms.

Analytical results data

Many risk management analytical applications are built around scenario models that use current balance sheet contract information as input. In cases where forward-looking

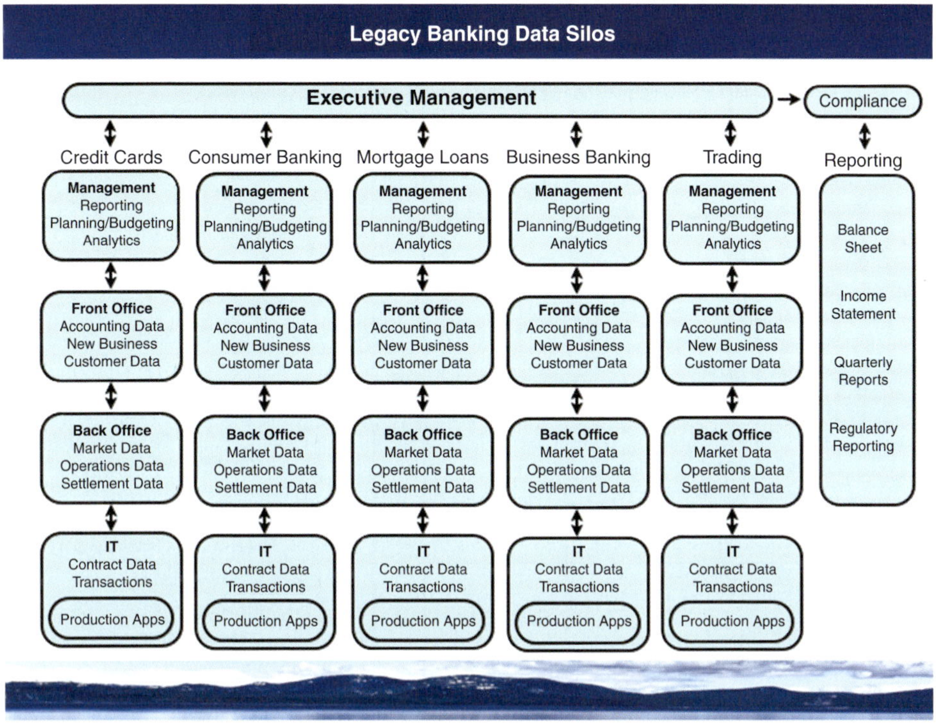

Figure 6.17 Legacy banking data silos.

stochastic cash flow Monte Carlo simulation models are used, the volume of results data that is produced can be significantly larger than that of the input data. This is principally due to the generation of a large number of scenario iterations in cases of forward-looking stochastic cash flow simulation models. The cycle of (1) sample, (2) process and (3) review and act upon results repeats itself, often on a daily basis, producing yet more results datasets. In addition to the need to manage these results datasets, this also gives rise to a need for the annotation and attribution of scenario assumptions corresponding to the data in the results data repository.

If results data from multiple analyses and model iterations are to be combined, there is a need for consistent contexts and synchronized scenario set parameters to be available and easily applied to independent model runs. Lack of such scenario consistency will destroy the ability to determine the volatility of aggregate results, since important aspects of the diversification across scenarios will be lost.

Organizing, analyzing, and summarizing large multidimensional results set distributions derived from multiple scenario loan-level detail cash flow analysis will require the appropriate tools and technologies to manage the scale of the data as well as support data mining and reuse of the data.

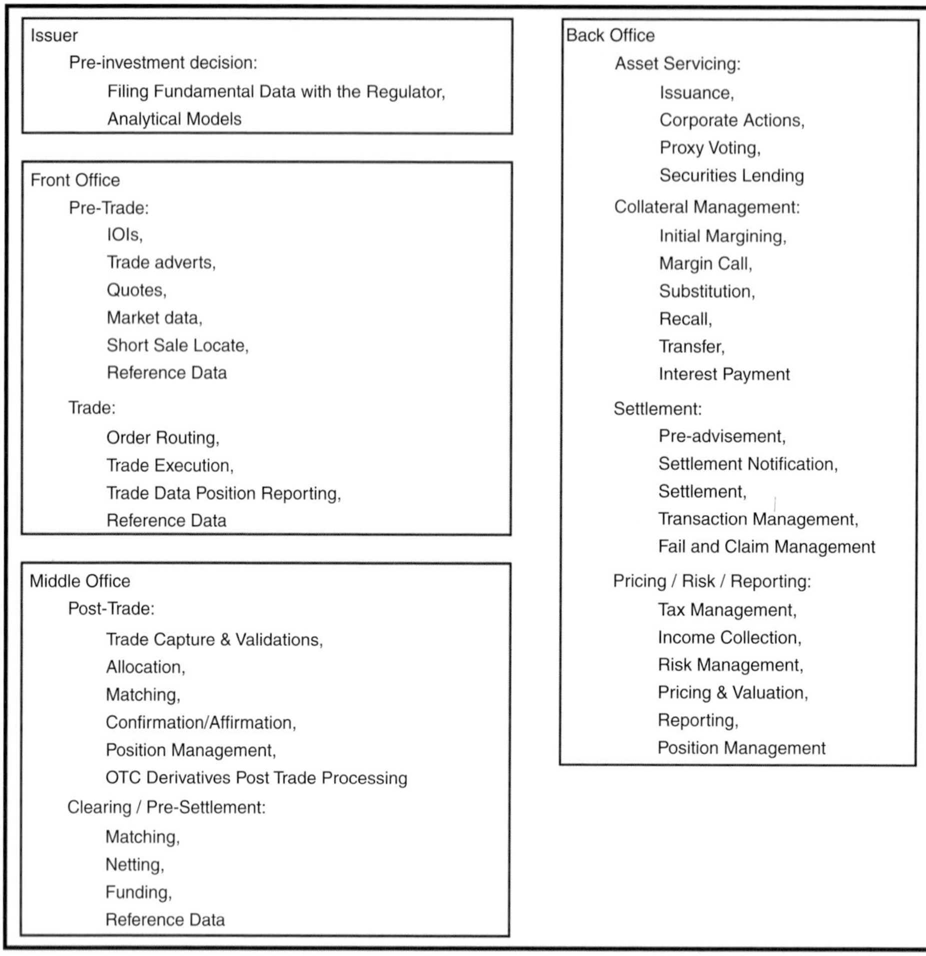

Figure 6.18 Operational product lifecycle data-intensive interfaces.

Availability and quality of customer and collateral data

In addition to specific terms and conditions of mortgage contracts and details of a particular customer's financial condition and credit quality, information regarding the quality of loan pools and demographic trends can be a valuable source of insight. This insight contributes to a comprehensive picture of the customers of banking institutions and how regional and demographic factors affect the context of those relationships.

Demographic and mortgage origination data

The Home Mortgage Disclosure Act (HMDA), which was enacted by Congress in 1975 and is implemented under the Federal Reserve Board's Regulation C, requires all banks, savings and loans, savings banks and credit unions with assets of more than $33 million and offices in metropolitan areas to report mortgage applications. These data, which are publicly available, are interesting and useful in a number of respects. For starters, the number of total applications and originations by year is a measure of the health of the overall home mortgage industry. Second, because the data are broken down by county and metropolitan area, they provide excellent insight into the ebb and flow of mortgage originations with a fairly detailed geographical focus.

The record layout of the information in the primary loan application record in the HMDA database is shown in Figure 6.19.

Other data categories

- **Reporting data**: data that are extracted from operations, financial data and management data (in the form of judgments expressed about business strategy, competitive conditions, areas of strength and weakness) for purposes of public or regulatory disclosure.
- **Compliance data** further extend the reporting data category with data that are required to be maintained for purposes of documenting regulatory compliance when a supervisory examination occurs.
- **Geopolitical/exogenous event data:** unpredictable, or uncontrollable, acts of man or nature; the Insurance Industry and actuarial data come into play here.
- **Operational risk data:** As part of an Operational Risk Management program, a potentially greater volume of operational incident data is collected, according to policy thresholds. These data are ongoing and need to be captured and assessed in order to provide effective feedback to the operational risk management program.
- **Sovereign risk data:** (related to geopolitical event data) reflect assessments of risks by sovereign entity and include econometric information regarding industrial sectors, public and private debt, natural resources, economic prospects and political factors.
- **Public data:** A tremendous volume of public and government-maintained financial data is available.
- **Loan performance data:** There are a number of qualified vendors who can provide detailed payment history, loan quality and performance analysis, usually as a service to a client institution and operating with the operational data of the client's loan portfolios.

6.3.2 Information architecture and data models

There are a number of levels of both data models and information architectures. Information models involve *metadata*, or data about data. They are commonly

HMDA Loan Application Register Format

Fields	Length	Type
As of Year	4	N
Respondent ID	10	AI
Agency Code	1	AI
Loan Type	1	N
Property Type	1	AI
Loan Purpose	1	N
Occupancy	1	N
Loan Amount (000s)	5	N
Preapproval	1	AI
Action Type	1	N
MSA/MD	5	AI
State Code	2	AI
Country Code	3	AI
Census Tract Number	7	AI
Applicant Ethnicity	1	AI
Co Applicant Ethnicity	1	AI
Applicant Race 1	1	AI
Applicant Race 2	1	AI
Applicant Race 3	1	AI
Applicant Race 4	1	AI
Applicant Race 5	1	AI
Co Applicant Race 1	1	AI
Co Applicant Race 2	1	AI
Co Applicant Race 3	1	AI
Co Applicant Race 4	1	AI
Co Applicant Race 5	1	AI
Applicant Sex	1	N
Co Applicant Sex	1	N
Applicant Income (000s)	4	AI
Purchaser Type	1	AI
Denial Reason 1	1	AI
Denial Reason 2	1	AI
Denial Reason 3	1	AI
Rate Spread	5	AI
HOEPA Status	1	AI
Lien Status	1	AI
Edit Status	1	AI
Sequence Number	7	AI
Population	8	AI
Minority Population %	6	AI
HUD Median Family Income	8	AI
Tract to MSA/MD Income %	6	AI
Number of Owner-occupied units	8	AI
Number of 1-to 4-Family units	8	AI
Application Date Prior 2004 Flag	1	N

Figure 6.19 Home Mortgage Disclosure Act (HMDA) loan application reporting record layout.

grouped into a number of levels (e.g., ontological, semantic, conceptual, logical and physical), ranging from "higher" levels that are more abstract and conceptual to "lower" levels more akin to detailed blueprints): semantic, conceptual, logical and physical.

At the upper end of that range, *ontological models* seek to identify discriminating descriptions and clarifications about and among things that exist, including the properties and relationships of "things." These discriminating attributes and properties that describe and differentiate classes of things are often translated into the logic structures and languages of dedicated ontology frameworks and knowledge bases.[30] *Physical models* are at the other end of the modeling spectrum, as they address engineering specifications for implementations.[31]

Taxonomies (and the hierarchies associated with taxonomies) are informational categorizations that are also associated with characterizing and organizing informational content and data models, but a taxonomy should not be considered as anything that could substitute for robust models and representations of the properties and inter-relationships of simple much less complex financial structures and relationships

Semantic models are not concerned with data issues per se but rather address the agreed-upon meaning and definition of elements in the model using descriptive language for a named property, or "thing." Usually accompanying a semantic model is a documented understanding of the behavior of elements acting on other elements.

There are semantic model standardization initiatives underway in support of efforts both to improve the efficiency and quality of enterprise financial risk management and to support industry-wide – and necessarily global – financial data standards. Semantic models are not concerned with implementation aspects, but rather with the clear definition of the operational meaning and interpretation of named things.

Conceptual models take on a high-level design of the groupings of informational elements, structures and processes that interact with each other, but the model is more for confirming human understanding of the system and its objectives than for explicit design of the data structures required. Conceptual models are the usual type of models that are developed in the course of business process requirements discovery and use case analyses in business development and information technology initiatives.

Logical data models precisely identify the entities and attributes that comprise the qualified domain of the model, enumerating and naming the entities and

[30] These ontology frameworks generally provide mechanisms to query, navigate and maintain ontologies. More advanced frameworks also allow testing assertions, theorem proving , ontology consistency checks. In the financial space, ontologies can be used to clarify the data requirements and dynamic behavior of classes of objects in the financial sector. For an example of a comprehensive ontology engine, framework and extensive knowledge base, see SUMO, developed by Adam Pease, `www.ontologyportal.org/`

[31] For example, encodings of data in physical media, be that how data is organized and stored persistently in file system or database structures, or how data are mapped and inserted into transactional or messaging protocols.

attributes and defining the data type for each attribute. Logical data models are informed by extensive use-case testing and subject matter expertise review in order to insure that the model and its structures fully capture all of the required data attributes (and business rules governing data attributes) needed to support a realized, functional implementation.

Physical data models translate the data requirements and properties expressed in the logical model into a particular implementation on an information technology hardware/software vendor system platform. A physical data model can generate the specific operations, procedures and data loads to create a functioning database instance of the logical data model.

Financial data integration

Financial data integration can be thought of as providing the data pathways and flows that are necessary initially to populate and then to maintain the ongoing implementation of a mission critical enterprise financial data management system.

If data are the blood of a financial enterprise, then data integration constitutes its circulatory system. Data integration involves the extracting, translating, associating, merging, constructing and loading of data from physical data sources into a data store based on a particular set of logical and physical models.

One of the key data integration missions in the enterprise is to bridge the data processing and application systems in production operations with the accounting system and the enterprise-wide financial data warehouse that supply data to the institution's decision support and management information systems.

The production application systems of record on the operations side of the house are typically not directly accessed by risk management applications on the analytical side. Instead, logical copies of the production systems of record data (contracts, transactions, account balances, etc.) are extracted, translated and loaded into a financial data warehouse.

This approach prevents the execution of computationally intensive analytical processes from degrading the performance and response times of operational systems. This is important because sluggish performance of those systems can disrupt service delivery and tarnish an institution's reputation with its customers.

6.3.3 Data management quality considerations

Table 6.3 lists some key considerations regarding data quality control stages in the data pipeline.

Table 6.3 *Key data quality control considerations*

Quality of data at the source
Correct identification of source data
Data provenance/collateral data
Quality of the data transformation between domains
Quality of the data transformation in aggregations
Quality of the data classifications procedures
Timeliness
Information architecture and data modeling completeness
Correct mapping to target model
Time and frequency considerations

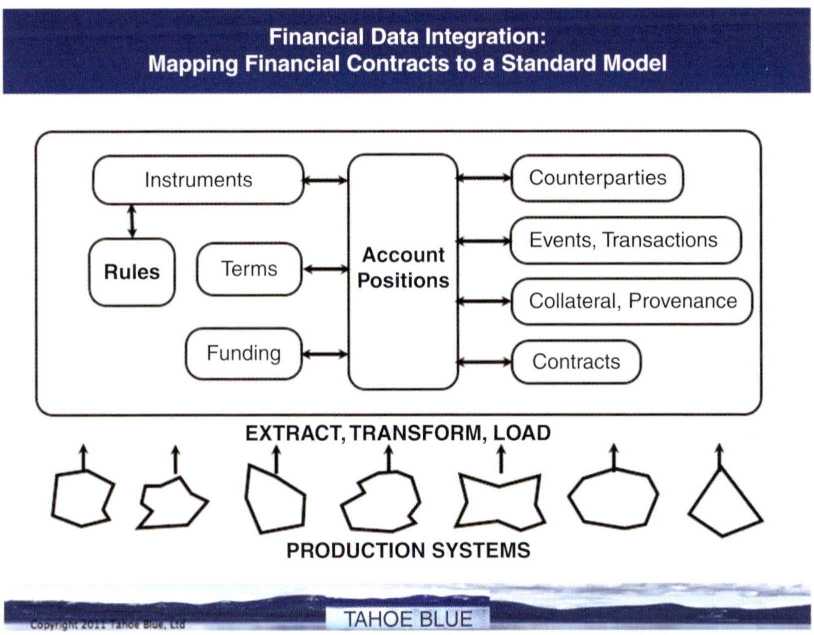

Figure 6.20 Mapping contractual data to a standard enterprise-wide model. This figure depicts the process by which contractual information from a wide variety of different financial products with diverse data structures and formats is extracted, translated, and loaded into a standard data model.

Artifacts, noise and loss of information One of the vulnerabilities of the data pipeline is that, once information is lost, for example, as a result of aggregation (as is usually the case with G/L data), then it is no longer possible to recover the lost information.

Systems of record production applications, on the other hand, naturally have sufficient detail. However, these details are often inconsistently

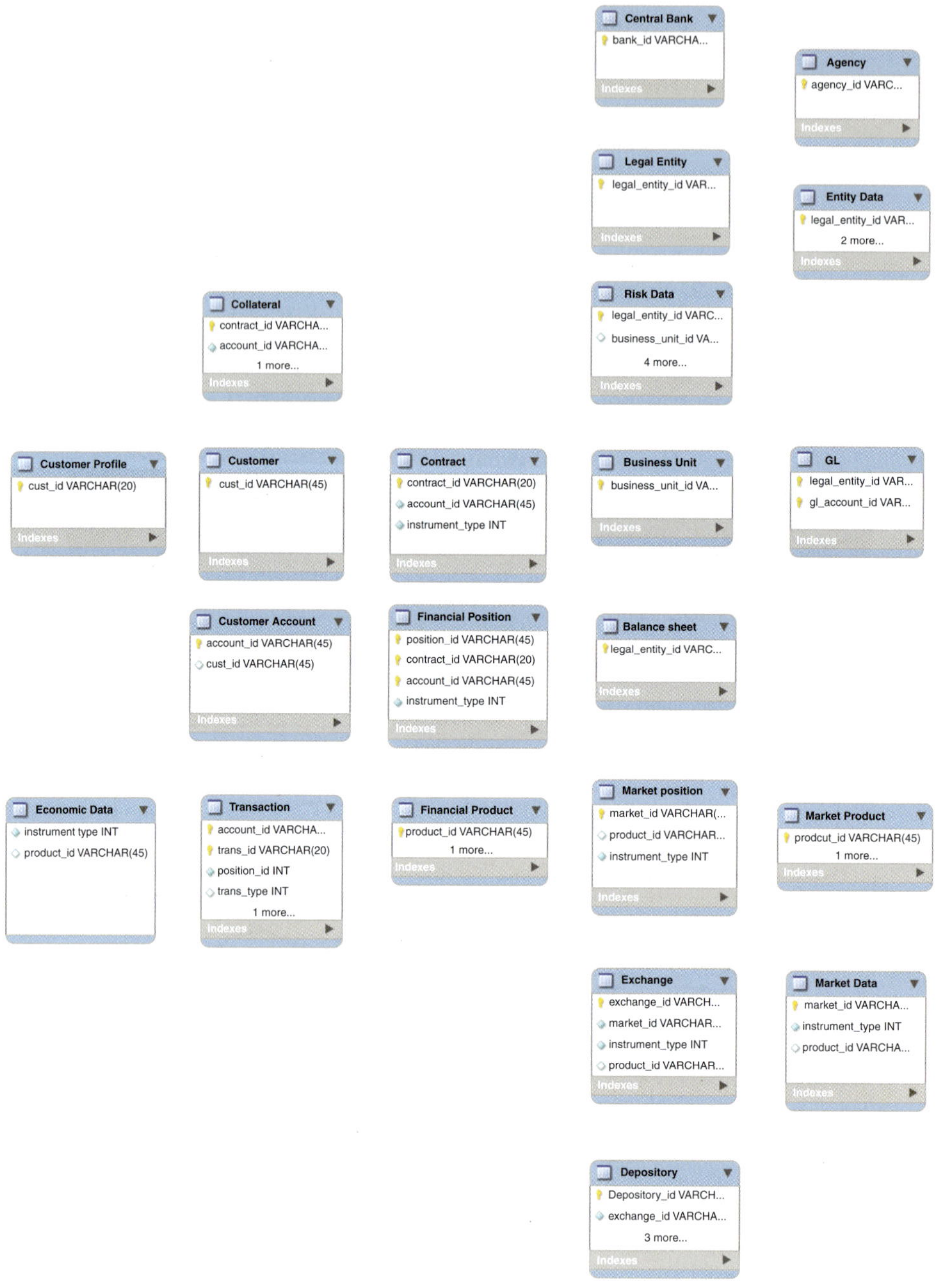

Figure 6.21 Related entities and object classes in the banking data space.

maintained across production systems. This typical lack of common data standards among production systems prevents the detailed contractual information from being accessed for purposes of enterprise financial data integration.

Loss of information via aggregation The most common and natural way for detailed information to be lost is via the process of data being aggregated, classified, and summarized as it moves up the organizational and reporting channels of an institution.

In order to retain as much contractual information as possible when aggregating balances over a variety of contracts and products of consistent structure but differing terms, one commonly used practice is to treat the parameter within the group as some kind of weighted average (usually where the amount of the contract balance serves as the weighting metric).

For example, if a \$100,000 10 year loan at 4% were combined with a \$400,000 10 year loan at 5%, the weighted-averaged rate would be 4.8 % for a \$500,000 10 year loan balance.

The example in Figure 6.22 illustrates how the choice of binning criteria (in this case 100 basis point versus 50 basis point thresholds applied to the coupon) can affect the resulting Weighted Average Coupon (WAC) and, more importantly, the total *error characteristic* resulting from different binning criteria – in the course of WAC-based aggregations.

Table 6.4 *Comparison of general ledger accounting systems and production systems of record*

System	Standardized	Level of detail
General ledger/accounting	Yes	Too aggregated
Systems of record production	No	Too detailed

(a)

One WAC pool of 4 loans, in a *single 100-basis point coupon bin*

	Loan Balance	Loan Rate	Par Interest	WAC Interest	WAC-Par Interest	*Distance* |ABS|
1	400000	4.900%	19600	17840	–1760	1760
2	600000	4.800%	28800	26760	–2040	2040
3	400000	4.200%	16800	17840	1040	1040
4	600000	4.000%	24000	26760	2760	2760
	2000000	**4.460%**	89200	89200		
		WAC				**7600**
	2000000			**89200**		**8.520%**

Figure 6.22 (a) Example of Weighted-Average Coupon (WAC) calculation on a pool of four loans with 100 basis-point bin. (b) Example of Weighted-Average Coupon (WAC) calculation of the same loans in two pools using 50 basis-point bins.

(b)

Two WAC pools of the same loans, in a *single 50-basis point coupon bins*

	Loan Balance	Loan Rate	Par Interest	WAC Interest	WAC-Par Interest	*Distance* |ABS|
1	400000	4.900%	19600	19360	−240	240
2	600000	4.800%	28800	29040	240	240
	1000000	**4.840%** WAC	48400	48400		
3	400000	4.200%	16800	16320	−480	480
4	600000	4.000%	24000	24480	480	480
	1000000	**4.080%** WAC	40800	40800		**1440**
	2000000			**89200**		**1.614%**

Figure 6.22 (cont.)

In the examples in Figure 6.22 there are four simple fixed-rate loans listed, for which two different aggregation calculations are demonstrated using all four of the loans.

The first case, shown in Figure 6.22(a), uses a binning criterion that combines loans with interest coupons in 100 basis-point-sized bins. Observe that all four of the loans end up being weighted-averaged together into a single loan vector in the amount of $2 million and with a WAC of 4.46 %, as highlighted.

The second case, depicted in Figure 6.22(b), uses a binning criterion that aggregates loans with interest coupons that are in 50 basis-point-sized bins. Note that the four loans end up being weighted averaged into two separate vectors consisting of two loans each, with respective WACs of 4.84% and 4.08%, again highlighted.[32]

[32] The weighted-average coupon (WAC) of a pool of loans is determined by weighting the interest rate of each loan in the pool by the percentage of the loan amount to the total loan amount of all of the loans in the pool. For a pool of *n* loans with interest rates R1, R2, . . . R*n* and loan principal balances P1, P2, . . .P*n*, the WAC can be derived using the following expression:

WAC = (R1*P1 + R2*P2 + . . . + R*n**P*n*) / (P1 + P2 + . . . + P*n*).

Similarly, the Weighted-Average Maturity (WAM), where the term to maturity in months of each loan in the pool is M1, M2, . . . M*n*, is obtained by:

WAM = (M1*P1 + M2*P2 + . . . + M*n**P*n*) / (P1 + P2 + . . . + P*n*).

Weighted-average contractual terms and parameters of a portfolio of financial products (e.g., a pool of mortgages) would be derived in similar fashion. When aggregating financial products using weighted-averaging based on the principal amount to derive the weighted-averaged contractual terms of the pool, it is very important to limit the members of the weighted-averaged pool to contracts whose terms and parameters have a relatively close affinity to each other.

In other words, the maximum difference in the value of contractual parameters among the members of the pool should be within a tolerance that is sufficiently small so that the behavior of the financial vector that results from the weighted-averaged aggregation process is appropriately representative (as a proxy) of the behavior of the individual contracts under similar forward-looking scenarios.

The total loan principal balance in both cases is, of course, the same, as is the total interest that would be generated from (a) the original loans and their original coupons, (b) the single-vector case, and (c) the case with two separate 50-basis-point vectors – as it should be.

Although the total principal balance and the *total* interest generated in both cases is identical, the results are not equivalent. It is instructive to compare (1) the interest that each loan would produce using its original coupon with (2) the interest that each loan principal balance would produce using the WAC computed for the bin into which it falls, for both binning cases.

The difference between the interest amount generated by applying the contractual coupon of each loan versus the interest derived by applying the common WAC to each loan is calculated in the column labeled "WAC-Par Interest." In the final column, this difference is converted to a simple linear *distance* by taking its absolute value, and this distance is summed for all loans in both cases.

In the first case, the error function total is 7600. In the second case, the error function total is 1440. If we now compare the relative magnitude of the difference in total loan pool interest generated in each of the two cases to the total interest generated (in all cases) of 89200, a characteristic that is representative of the "goodness of fit" of each strategy relative to the original underlying loan contracts can be derived (i.e., as a simple ratio). In the single-vector, 100-basis-point binning case, that ratio is 7600/89200, or 8.52%, as highlighted. In the lower case, where 50-basis-point binning produced two vectors, the ratio is 1440/89200, or 1.614% (highlighted).

Even though both weighted-averaged binning criteria produced identical total principal balance and total interest amounts, the 100-basis-point binning criteria produces a total error characteristic that is five times the total error characteristic of the 50-basis-point binning case.

Financial contract vector data compression One could think of the process of aggregating financial contracts in order to reduce the total number of financial contract vectors that need to be analyzed as similar to digital image compression algorithms (e.g., like JPEG). Digital image compression has the objective of reducing the storage size of digital image data while preserving a specified level of quality and detail of the original image as *determined by the requirements of the subsequent use of the compressed data*. (The end-uses of such visual data run the gamut from postage-stamp-sized "thumbnail" images and lower-resolution images for web publishing to high-resolution images required for photographic printing).

Although there are "lossless" compression algorithms (from whose output the original data can be restored without any loss of information), most compression algorithms are designed to achieve a range of higher compression ratios with a corresponding range of information loss as part of the design trade-off.

Producing a smaller number of aggregated contractual vectors (for example, to help reduce computational work load, produce extracts for input to ancillary applications, or reduce the amount of data for human review and examination) can be viewed as financial contractual data compression that will likely involve degrees of information loss depending on how the aggregation is conducted.

In the case of the two simple examples of aggregation illustrated above in Figure 6.22, the weighted averaged loan vector produced by the first case results in more information loss than the second case. The result of the first case could be thought as being more "fuzzy," or "pixelated," than the loan vectors produced in the second case. It is important to quantify and understand the impact and sensitivity on the quality of risk measurement analysis that degrees of reductions in the resolution and information in contractual vectors can cause.

Consider the analysis or simulation of the prepayment behavior and outcomes under a variety of widely different interest rate scenarios using the two aggregation examples described above. In interest rate scenarios with either up or down transitions between 4% and 5%, the prepayment incentives of the single vector case with a WAC of 4.46% will differ markedly from the incentives of the vectors with WACs of 4.84% and 4.08% under identical scenarios. More importantly, the behavior of the vectors produced by the 50-basis-point binning criterion will be more similar to the behavior of the actual underlying loans than will result from the same scenario analysis applied to the 100-basis-point criterion vector.

The interest coupon of a financial instrument (such as a mortgage) is only one of many parameters, contractual terms, market demographics and customer characteristics that can distinguish and differentiate one instance of a mortgage loan from another. One can extend this exercise of "goodness of fit" for analytical purposes to any number of the many dimensions that rightfully should be taken into account to determine optimal levels of dis-aggregation when combining financial contracts into collections and groups in multifactorial data aggregation stages.

For example, weighted-averaged maturity (WAM) is an important product of a contractual financial vector aggregation process. It is not difficult to see how using binning criteria that are too wide applied to the term (or tenor) of financial contracts will produce significant increases in the error characteristic of the compressed dataset produced by the rollup.

This is particularly true in the case of a parameter such as the maturity date of a financial contract, owing to (a) the central role that *time-weighted rate of return* plays in the analysis of contracts and the calculation of Net Present Value (NPV) using discount factors precisely applied to the size and timing of future cash flows, and (b) the importance of clearly anticipating the timing and extent of positive and negative cash flows on both sides of the balance sheet in order to manage funding liquidity risks and the spreads between interest-earning and interest-bearing commitments.

The error-distance measures of a weighted-averaged contractual parameter dimension can be summed and the characteristics of metrics across parametric dimensions can be calculated. In this way, a multidimensional view of the "goodness of fit" of various data aggregation strategies can be derived and instituted.

Decisions about the appropriateness of data aggregation binning criteria can be framed as a trade-off between the minimization and materiality of these statistical error characteristics versus the size and number of vectors in the aggregated dataset that would be produced and the computational complexity involved to subsequently process or analyze those vectors. More importantly, this methodology for data quality analysis can be automated and embedded directly into the aggregation processes.

It is important to note that a key reason to diligently strive for aggregation with minimal information loss (and optimal "goodness of fit") is the fact that financial data produced in an aggregation stage – particularly stages that are closest to the level of individual account contracts – will be used as inputs to successive stages for subsequent analysis, reporting and further aggregation.

It is also important to note that the consequences of suboptimal aggregation strategy may change over time. Ideally a periodic examination and quantification of the size of errors introduced by the prevailing aggregation strategy is advisable.

Loss of information via classification Classification and categorizations are necessary in order to present manageable information to executive management. However, when classifications that are too broad are introduced too early in the data pipeline, as opposed to being applied simply at the point of presentation using higher quality source data, information loss and data distortion can occur. Classifying 30-year mortgages into three categories representing maturity in one of three decades, for example, is a categorization by time interval that lumps too many products together that are likely to behave differently. *The granularity of classification categories (and their hierarchies) has a direct impact on the goodness of fit of aggregations that incorporate such enumerated categories.*

Information architecture and data modeling completeness

In addition to capturing details, a good enterprise financial data model implementation will also allow for the data to be viewed from a variety of perspectives. For example:

- Transactional/Exchange perspective
- Contractual/Balance Sheet perspective
- Liquidity/Solvency perspective
- Cash Flow/Accounting perspective.

Selection of evaluation criteria for credit quality determination Table 6.5 lists several evaluation criteria and types of data properties and characteristics that are relevant and important to include when determining commercial credit quality assessments.

The expanding scope of financial industry data standards

Much of this chapter has focused on data that are internal to a financial institution. Data that are external to a firm or that are involved in a firm's interaction with external entities generally fall into one of a number of large groupings, shown in Table 6.6.

Not surprisingly, data standards for different facets of the financial industry have been in existence for quite some time, driven largely by the need to have common formats for participation in financial markets, exchanges, and clearing organizations. Examples of the broad standards categories already in commercial use are: the Depository Trust and Clearing Corporation (DTCC) using Committee on Uniform Security Identification Procedures (CUSIP), retail credit and debit card transactions, Electronic Funds Transfer (EFT) using Society for Worldwide Interbank Financial Telecommunication (SWIFT) and Automated Clearing House (ACH) protocols and other settlement and clearing operations between financial institutions.

However, these data standards have been fragmented to the extent that they address specific or more specialized aspects of banking operations and financial transactions. For example, in the equities markets alone, there are any number of ways that traded securities and the companies with which they are associated are identified, often depending upon the type of data reporting service and the particular markets that are involved.

In the last decade, efforts at data standards have gone beyond reference data (e.g., for public companies and market securities identifiers), and are being extended to the structure and attributes of financial products and accounting data.

In the area of transaction protocols and messaging, one such example is the Financial products Markup Language (FpML), an example of which is shown in Figure 6.23.[33] FpML defines a taxonomy and structure of financial derivative products

[33] FpML started as a private initiative in 1997 on the part of JP Morgan (now JPMorgan Chase) and, with the participation of PricewaterhouseCoopers, a draft definition was proposed in 1999 for the purpose of standardizing the format of financial derivatives traded among firms in the OTC derivatives markets. An

Table 6.5 *Evaluation criteria for credit quality determination*

Industry risk
Competitive position
Management and corporate strategy
Operating performance
Capital adequacy
Investments
Largest obligors
Financial flexibility
Enterprise risk management
Liquidity
Leverage
Sovereign risk

Table 6.6 *Data external to a firm*

Reference data	Data that identify and/or describe an organization, location and its products
Market data	Data such as price, rate, volume data emanating from financial market transactions
Operational data	Data included in electronic transactions and messages between counterparties, exchanges, clearing organizations, and depositories
Financial and business data	Data that are publically available for investors and regulators, augmented by additional confidential data to regulators and/or rating agencies
Corporate data	Data related to corporate actions, changes in corporate structure or equity, mergers and acquisitions
Risk data	Data included in, resulting from, or supportive of analysis and identification of risk exposures and contingencies

using the eXtensible Markup Language (XML) standards. and includes structural definitions not only for derivatives, but also for the underlying financial instruments and assets to which financial derivatives contracts must necessarily refer. Residential mortgages and the pools and tranches of mortgages which have become securitized and which are referenced in derivatives such as Credit Default Swaps (CDS) are among the types of underlying assets and securities that can be expressed in FpML.

industry consortium, FpML.org, was formed and, in 2001, the International Swaps and Derivatives Association (ISDA) announced the intention of integrating this effort into the ISDA organization. Standardization efforts extending coverage to derivatives of all types is ongoing.

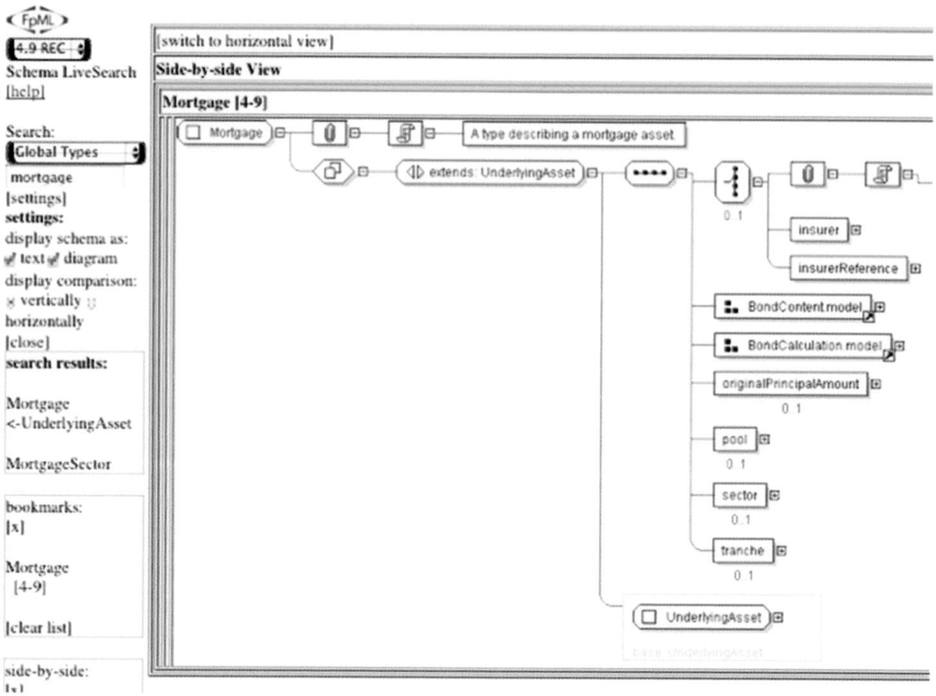

Figure 6.23 FpML Schema LiveSearch representation of a mortgage loan (`www.fpml.org/tools/schemalivesearch/index.php`).

Data standardization efforts

In order to manage a large portfolio of risk effectively – be it at the level of a large bank, a national regulator or the global economy – it is obviously imperative to have data that can flow electronically through the system at least as easily as cash flows themselves are moving. In the cases where a variety of standards have grown up in different territories or markets and have yet to be reconciled, translating between standards ought to be straightforward as well.

The first step that the Office of Financial Research (OFR) took towards the goal of arriving at national, and hopefully international, data standards was to launch the work to establish a standard Legal Entity Identifier (LEI), the purpose of which is to consistently identify a legal entity in all financial transactions.

Legal entity identifier The identification by the OFR of the need for an LEI standard was the first step in implementing data standards in the USA that will support more accurate assessments of institutional as well as systemic risks as stipulated by the Dodd–Frank Act. Subsequent to the announcement of this initiative by the OFR, the G20 took up the task to further define the LEI as an international standards initiative to be developed at a global level by the Financial Stability Board (FSB) at the Bank of International Settlements (BIS).

The FSB LEI Implementation Group (IG) recognized that this ambitious undertaking needed broad-based input and support from the private sector in order to help identify business requirements, data standards, operational procedures and system architecture for a federated implementation of the global LEI system.

In mid-2012, the FSB LEI Implementation Group convened the Private Sector Preparatory Group (PSPG) by inviting representatives from four principal financial industry sectors to form work groups to tackle business process, data standards and requirements analysis of the future LEI system.[34] These four sectors are:

- Advisory, consulting, academia,
- End-users of the LEI (this includes non-financial companies as well, such as agribusiness),
- Financial data and infrastructure service providers, and
- Standards organizations and industry associations.

A proposed ISO draft standard for the LEI (ISO 17742) was developed. ISO 17442 specifies the physical size of the LEI (18 alpha-numeric characters followed by 2 check digits), as well as a minimum list of reference data attributes that are specified in order to uniquely identify a legal entity.

ISO 17442 (and its 2012 revision) does not specify an implementation of the LEI (i.e., how the LEI is to be structured, allocated, managed, issued and disseminated). However, the FSB LEI Implementation Group, in the course of reviewing alternative proposals for how to manage the issuance of LEIs by a federation of LEI registrars, decided on an LEI structure which the LEI address space is partitioned using a 4-character, high-order prefix in order to assign exclusive and unique partitions to individual LEI registrars to use to allocate and issue LEIs to legal entities that request them.

Reference data associated with the LEI include:

- The official name of the legal entity as recorded in the official registers.
- The address of the headquarters of that legal entity or the address of the fund manager.
- The country of incorporation.

LEIs are to be issued to all financial market participants, including

- Private businesses,
- Public corporations,

[34] The authors were actively involved in the work of the PSPG.

- Sovereign treasuries,
- Trusts,
- Joint ventures or
- Other legal entities (such as special purpose vehicles created for embodying structured financial instruments).

It is the reference data of each legal entity associated with an LEI that actually uniquely identify the legal entity, not the LEI itself. The LEI contains no information about the entity, and must be submitted to the LEI system in order to access the public reference data defining the identity of the legal entity associated with the LEI key.

LEI system as a global federation of LEI registrars Adopting a global LEI system is an important first step in a planned major reform of the global financial system's reporting and systemic risk monitoring infrastructure. Focused efforts are underway at the FSB to create the seminal LEI Foundation that will anchor and oversee the process of authorizing and accrediting additional LEI registrars in a variety of jurisdictional and market communities in order to stand up a global LEI system and make the global LEI standard viable in an international context comprised of many independent sovereign countries.

One consequence of promoting the LEI as a global standard is that organizational and operational solutions that might have been workable in the context of a particular domestic implementation must be thought through carefully in a global context in order to eventually meet the needs and requirements of international sovereign states and regulations.

The raising of the LEI to the level of being developed as an international standard by the G20 has been followed by concerted efforts on the design and implementation of a globally federated LEI public utility based on a federated of LEI registrars (Local Operating Units, or LOUs). The LEI Regulatory Oversight Committee (ROC), established in January of 2013 (see `http://leiroc.org`), was tasked with the responsibility to stand up the LEI system.

Relationships among legal entities The core reference data are intended to also include data that record facts pertaining to such things as the ownership structure of the legal entity, or, more generally speaking, relationships *between* legal entities. A flexible data structure consisting of a collection of tuples defining a number of different possible ownership relationships between other entities and a particular legal entity would be the more extensible choice, in lieu of a few single fixed data elements that attempts to record an ultimate and/or an immediate "parent" entity of an organization.

The need for a simple, expandable "fact" structure is because there are many inter-entity legal relationship situations where multiple parties are involved, each having some stake, responsibility or degree of control in a particular set of legal inter-relationships. Legal relationship configurations can range from the simplest case of a subsidiary that is 100% owned by a parent entity to the more complex multilateral arrangements involved in special purpose vehicles (SPVs) such as Collateralized Debt Obligations (CDOs).

A simple data structure that expresses relationships like "LEx *owns* <23%> of LEa", and "LEy owns <42%> of LEa" is a better approach than allocating a fixed number of fields in the master profile reference data record of the legal entity. A simple, "relationship fact" data structure can record any number of relationship facts in a list, one entry for each fact about ownership relationships that may apply, and it is inherently more flexible than assigning fixed attributes in a single master data record to record relationship data.

When the LEI is subsequently used as the identifier of entities in market transactions and balance sheet reporting, regulators for the first time will be able to assemble a network graph of the number and size of counterparty exposures at the systemic level, similar to the counterparty interdependencies for financial institutions shown in Figures 6.7 and 6.8, above. The introduction of an LEI standard will make this possible for the first time by regulatory agencies.

Next steps in financial data standardization efforts The development of the LEI is an important foundational step in establishing financial data standards in the financial industry that, for the first time, are global in scope. A global, unique identifier of financial market participants is needed before the financial transactions and positions of financial market participants can themselves be identified and associated with the entities that own or transact them for such purposes as systemic risk analysis.

In addition to tagging financial transactions and products with LEIs in order to identify the counterparties to financial transactions, identifying the types and instances of financial transactions and products will also be required. Efforts to classify, name and identify financial products and transactions have been evolving for some time. In addition to refining and elaborating the description and enumeration of the data elements and attributes that distinguish financial object classes and instances from one other., the root-level classes and characteristics of financial objects are now being developed and articulated from an ontological and a semantic perspective.

It should be noted that future efforts to standardize and identify financial objects for purposes of systemic risk management are not meant to suggest that such standardization efforts must be extended into the internal information infrastructure implementations within financial institutions in order to be effective at the systemic level. The initial rationale for establishing common standards for financial data structures is predicated on the need for such standardization to support the reporting of the activities and inventory of financial market participants to regulators and the public. However, financial firms would be well advised to seek to realize the benefits of such standardization efforts for internal financial analysis and risk management purposes as well.

Just as with the LEI providing the unique global identifier for information about a specific legal entity, instances of the product and transaction superclasses will be associated with unique global identifiers as well. Some of the categories of financial industry objects that will be classified and identified follow.

Financial products owned (FPO)

Definition: Financial Assets(Liabilities) on the balance sheet or, in some cases, assets/liabilities that may not be on the balance sheet but over which the entity has either some control, responsibility to manage, or recourse. Examples of the latter would be such things as collateral, custodial accounts, repurchase agreements, or derivative contracts that "restructure" the balance sheet. Each instance of FPOs would be identified with a *Financial Product Identifier* (FPI).

The financial properties and attributes of balance sheet positions would be recorded in FPO data structures for which each instance would be associated with a different FPI. FPO data structures would also record the LEI of the owner entity, as well as the *Financial Instrument Type* (FIT). The FIT is an important property of a financial product, as it is essentially a super-class reference to a template of contractual rules and processing logic that determine the behavior of the products (contracts).

Financial transaction exchanges (FTE)

Definition: Exchanges of assets, investments and financial products. Examples include:

1. Securities traded on exchanges, i.e., purchases and sales,
2. Payments made between counterparties made under the terms of a governing contract (e.g., loan payments, or bond dividends),
3. Movements or transfers of assets (including cash, e.g., deposits and withdrawals).

These *Financial Transaction Exchanges* are defined in the narrower sense of a transaction as being the exchanging of a product, commodity or cash flow at a point in time. (Transactions in the larger sense used to connote the execution of business contracts that memorialize broader relationships or agreements between parties is not the scope of the term intended here.) Financial Exchange Transactions would be associated with a Financial Transaction Identifier (FTI)

Regulators are calling for the introduction of more complete contractual and data standards to be defined and agreed upon. Regulators are generally striving to obtain collaboration and input from the private sector to arrive collectively at a consensus solution under the auspices of regulator review, modification, endorsement and subsequent support at sovereign levels like the G20.

6.3.4 Data quality benchmarks

The Enterprise Data Management Council (EDM Council) is an industry association whose members come from the departments responsible for data management in financial organizations. The EDM Council, in conjunction with Carnegie Mellon University, have proposed a Data Maturity Model, consisting of a range of criteria to use in benchmarking and analyzing how well an organization is meeting and managing its data assets. An example is shown in Figure 6.24.

The EDM Council is also active in developing ontological models and taxonomies for reference data in the financial industry. The universe of financial

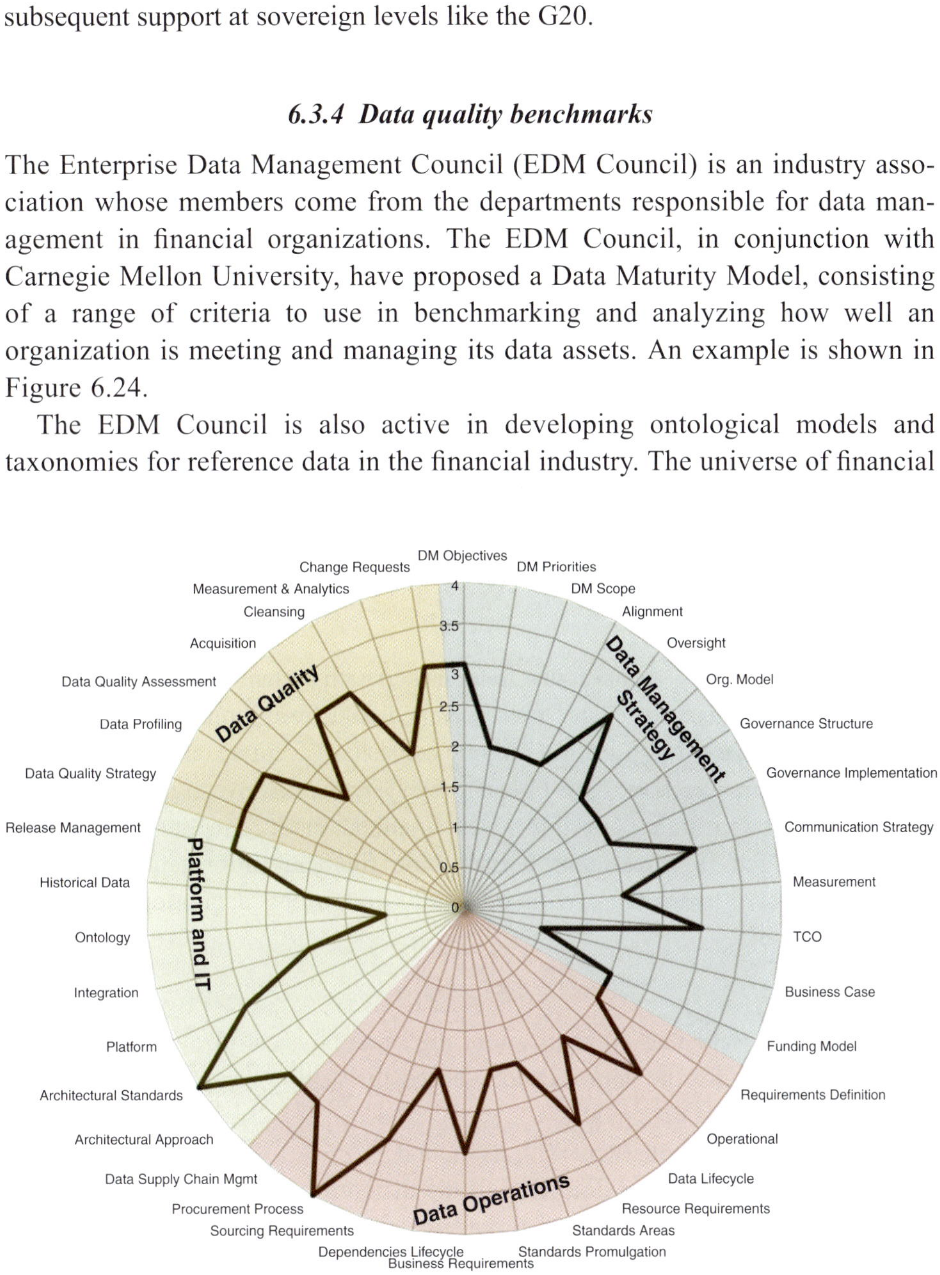

Figure 6.24 EDM Council data maturity model evaluation benchmark framework for data quality (`http://edmcouncil.org/PDFs/20110630.DMM.Core.Model.2.1.pdf`).

instruments is also the subject of semantic descriptions on the part of the EDM Council.

6.4 Financial risk information management planning

In Section 6.3, *Foundations of banking data infrastructure*, we looked at the supply-chain of data both in the capital markets and in the real economy, and the transmission of data across common boundaries and interfaces. We also addressed the problems that both firms and regulators have historically had in collecting and managing such data, as well as the challenges of performing effective risk management as a result of data and informational challenges.

In this section, we look at what to do from a business and infrastructure perspective to anticipate and meet these challenges.

6.4.1 Rationale and assessment

An overall strategy and road map forward could include the following considerations.

First

- Regulatory agencies – in particular those tasked with systemic risk management – will ultimately need to receive more detailed data from thousands of financial institutions. These data must not only be consistent, accurate, and timely – they must also be manageable, digestible and by systemic risk oversight.
- In order for this to be possible, each financial institution must previously have managed to implement a firm-wide financial information and risk management infrastructure with internally consistent, accurate, timely and digestible data.

Second

- It is not necessary that all firms adopt the same physical or even logical data model in order for the approach to succeed. It is only necessary that a firm's enterprise financial data model
  - be workable and consistent with itself, and
  - be able to capture and domicile all the different types of data that are part of the appropriate realms of financial data applicable to the business of the firm.
- If firms have tended to their own gardens and successfully achieved their own internally consistent, enterprise-wide financial risk management implementations, it will be comparatively straightforward to establish maps to transform

firm-wide realms to a target data model and API interface standards that organizations like the Office of Financial Research (OFR) could ultimately either adopt or stipulate.

Third

- In order to ensure workability of the process, it will be important to contain the following potentially substantial costs:
  - construction and development costs born by firms to achieve an effective financial risk management system internal to the firm, and
  - ongoing costs to perform the additional and recurring work of mapping from the internal data of the firm to reporting structures and formats adopted by regulators.
- Maximizing shared consistency (and precision) in the definition of each firm's data standards is highly desirable. This should be accomplished, if possible, at the outset of the standards process in order to minimize the number of differing implementations that would require re-mapping due to definitional differences at the time of construction.
- Opportunities for leveraging shared standards among collaborating firms include reference data, financial instrument structure representation and financial information data conventions.
- Other situations where a collaborative approach would clearly be cost/beneficial include cases where mid-sized banks use service bureaus for back office data processing, loan servicing, trade execution, and access to financial market data.
  - Due to the commonality of service bureau applications and data formats, work done to establish enterprise data standards and mappings for one banking firm can be applied to other banking clients of the service bureau.
  - In this way, redundant duplication of work on the part of each bank could be greatly reduced, and collaboration of the banks would produce joint economic value and cost savings.
  - These savings could be significant for small to midsize banks, for which the fixed costs of ultimately responding to heightened compliance and data reporting requirements to central regulators constitute a much higher percentage of their earnings than would be the case for much larger firms.

Fourth

- There clearly will be a period of transition as the financial system evolves toward a more global system that incorporates improved data standards that support more effective financial risk management in individual firms and the financial industry as a systemic whole. The question is how long that period of transition is likely to be: 2 years? 5 years? 10 years?

- The strategy must address how both firms and regulators should try to proceed in this transition – so that rules and regulations do not all change at once, but would preferably be staged in phases.
- An important objective should be to avoid having the data highways and technical resources tied up in extensive, complicated and ongoing construction and modification of institutions' data infrastructures.

Fifth

- It should be possible to begin to address, in parallel and ahead of the curve, the question of what an optimal data interface between financial institutions and regulators would, could, or should look like. Much work on this front can be accomplished even before the standardization of reference data, exchange-traded instrument structures and balance sheet financial contracts is accomplished. The more complete standardization of these substantial data domains in the global financial sector, beginning with the process of establishing a standard for implementing a global Legal Entity Identifier (LEI), will very likely require more time than the Dodd–Frank Act rather optimistically stipulates.
- Similarly, efforts to harmonize such efforts in the USA with those of risk management oversight rules and regulations at a broader global level (e.g., on the part of the G20 as undertaken by the Bank of International Settlements (BIS) and its Financial Stability Board (FSB)) will also take some time to coordinate and implement. This would be true in more stable times, but the many stresses and crises that are still gripping the financial industry at the time of this writing further complicate the ability to make rapid progress in this regard.

Figure 6.25 depicts how a data interface between financial institutions and industry regulators could be set forth. Reporting and data gathering activities among government financial oversight and regulatory agencies can be consolidated under the auspices of systemic risk oversight organizations and regulators.

Consensus on a simple but extensible specification of a financial data model would allow financial institutions to each stand up a "shuttle," or "docking" database API to access common information exchange interfaces between the compliance and risk management functions in the financial sector on the one hand, and systemic regulators such as the OFR on the other.

Such a "docking" database containing data would be populated by a financial institution and could in fact be subsequently electronically accessed by the OFR Data Center. The data model for the "docking" database would be standardized as part of the new data reporting requirements for the OFR so as to be compatible with the data model of the OFR data center as it pertains to that part of the OFR Data

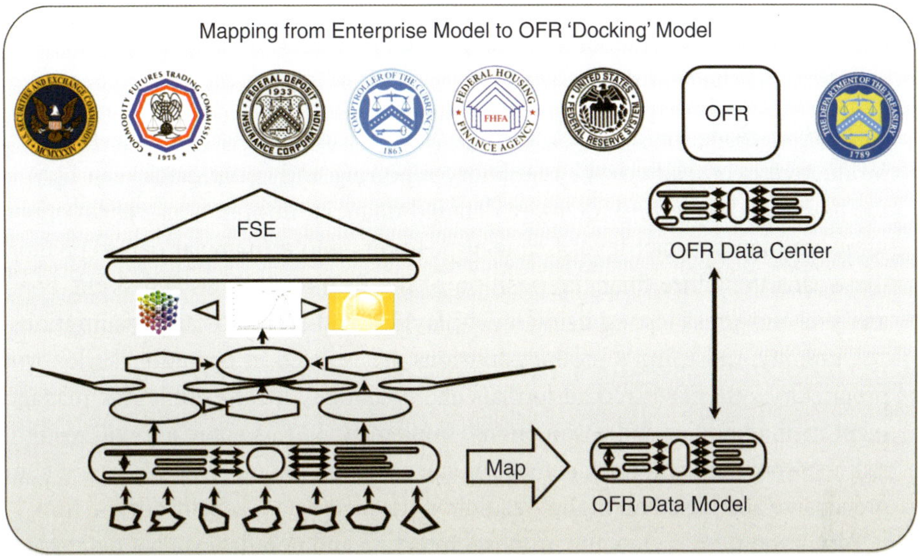

Figure 6.25 Mapping from the enterprise model to an OFR "docking model."

Center model tasked with collecting more detailed contractual and counterparty information from each institution.

6.4.2 The joint economic path forward for both the private and public sectors

Managing (even measuring) systemic financial risk with comprehensive cash flow analysis and dynamic simulation at more precise levels of detail and frequency is a significant departure from most traditional macroprudential risk management and supervision practices.

The results of more detailed contractual and counterparty analyses offer better insights into the risk dimensions of organizations and the financial system than do large-brush composite risk measures applied to balance sheet product categories from the top down. Applying risk category coefficients to aggregated balances of major product lines may simplify the determination of financial risk and reduce the burdens of compliance and regulatory reporting. However, in practice this approach can generate a constant flow of new reporting demands as the issues that concern regulators shift over time and the data needed to examine them change.

If accounting methodologies aimed generally at showing a longer-term view of a firm's "trailing average" (or smoothed-out) behavior are applied to risk management, then artifacts can arise that obscure or mask risk. The one-time events that accounting practices seek to footnote should in all likelihood be headline topics for risk analysis and stress testing.

Some fundamental premises of how best to respond, as a firm and as an industry, to expected regulatory requests for more detailed contractual and counterparty information, are the following.

1. ***Individual firms*** should seek to turn what is traditionally viewed as a non-productive overhead cost of regulatory reporting and compliance – in light of the new mandate to provide more detailed financial information – as an opportunity to map contract-level financial positions into a financial data repository that spans the entire financial position of the firm and allows across-the-board analysis and stress testing using level-playing-field scenarios and assumptions.

   Currently, most firm's product divisions are isolated in operational silos with proprietary systems-of-record formats and potentially incompatible risk management methodologies and assumptions. Implementing this approach will result in risk measurement and management practices which are more timely, more comprehensive and based on higher-resolution detailed data – a 'win' for the firm.

   Firms should not view the proposal to create and populate such a database as an attempt to jack up their entire financial operations and insert a new "ground floor," nor to be a replacement for internal data warehousing initiatives. Rather, the model is to allow an interface database to be populated with appropriate mappings from existing systems within the firm.

   A database compiled and extracted from legacy production systems, although more contractually detailed than traditional G/L reporting system, is not "mission critical." As part of a decision support solutions architecture, such a database provides a key, productive platform for more robust risk management.
2. ***The industry as a whole***, in conjunction with regulators, should strive where possible to agree on common data standards to represent low-level financial positions and contracts such that each firm does not create its own proprietary version of such a data model for compliance and regulatory reporting. Proprietary data models for each financial institution would require further re-mapping and translation (and likely incompatibility) at the level of systemic oversight.

   Having the financial industry and the public regulators agree on common standards, with requisite standardized reference data, will be a "win" for the firm as well as the public good. It will greatly reduce the cost and complexity of making sense of more detailed financial information for purposes of analysis at the systemic level in the OFR.
3. ***By implementing a form of distributed reporting repository***, each institution can make the database available on a secure basis not only to regulators but also to internal staff as well as vendors who can supply value-added reporting and analysis tools predicated on the standard model. This is yet a third win for

economic efficiencies that would make available better risk management practices to a wider range of institutions who otherwise would not choose or be able to develop such tools.

The increased regulatory requirements mandated by the recent passage of the Dodd–Frank Act are most likely not a welcome development for financial institutions. Furthermore, the challenges to fulfill the specific functions of the OFR as delineated therein would be formidable even with full cooperation from the industry. It is clearly in the best interests of the financial industry and the public if the projects can be pursued in a manner that will produce substantial long-term benefits to offset the additional costs incurred. Given that the work needs to be done, we also need to take into consideration the time and effort expended.[35]

6.4.3 Staging phases and risk management information infrastructure guidelines

Successful implementations of information system solutions will have broad and critical impact on the future operation and viability of a firm when completed. In addition, management of the demand placed on scarce resources of the firm during the design, development, testing and deployment phases will require careful planning.

The tables that follow present a succession of five major stages that constitute a plan for a phased implementation of public sector and private sector responses to evolving risk management infrastructure requirements. These major strategic stages (which require skillful handling and planning on the part of executive management) are:

1. Establish internal and external standards
2. Implement enterprise data integration
3. Upgrade risk management analysis methodologies
4. Improve enterprise governance and risk management culture
5. Evolve systemic risk management and regulatory oversight.

The process commences with the establishment of common data standards, followed by the population of a firm-wide database of financial contracts with a data integration methodology incorporating those standards. A comprehensive framework for analytical tools, components and applications can then be built out and assembled using the firm-wide financial database as a consistent foundation.

[35] See Braswell (2010b).

The combination of powerful, modular analytics supplied with standardized, detailed data can then be pressed into service for dynamic enterprise risk management. Subsequently, the benefits of improved firm-wide risk management can be leveraged and applied to the difficult but important tasks of systemic risk management.

In addition to developing plans for these strategic stages, a cost/benefits analysis and an estimation of the ROI of the investment in risk management improvements and efforts at the firm level should be undertaken.

Standards	**1. Establish internal and external standards**
1.1	Establish a joint strategic task force, within the firm, with participation from corporate management, business lines, subject matter experts and IT resources.
1.2	Authorize the task force to oversee the planning, design, development, testing and deployment of the desired firm-wide financial data warehouse, and give it the backing of the C-suite.
1.3	Participate in, and track, initiatives on the part of the industry and regulators to standardize reference data, counterparty identifiers, financial instrument models and taxonomies, and regulatory reporting initiatives that anticipate and meet new rules and requirements.
1.4	Adopt a set of financial data standards and definitions that can comprehensively describe and identify all entities, financial instruments and contracts of the institution.
1.5	Contemporaneous with internal standardization efforts, work strategically with communities of industry peers and regulatory bodies with the goal of identifying and defining external standards and workable interfaces.

Data integration	**2. Implement enterprise data integration**
2.1	Accomplish the accurate, detailed capture of all contractual terms, parameters and provisions from operational systems of record – all contractual terms and parameters required to process, evaluate, exercise or simulate contractual behavior going forward.
2.2	Populate a financial data warehouse that provides a common and consistent snapshot source of enterprise-wide data to all analyses and models which draw upon the financial instrument warehouse.
2.3	In any use of aggregation designed to reduce the number of low-level accounts (or financial "vectors") to be analyzed in demanding computational contexts, do so in a manner that (1) ensures data quality and resolution with sufficient levels of *disaggregation* and (2) preserves parametric detail across all performance dimensions such that the contractual and behavioral accuracy of the aggregated vectors have a very high "goodness of fit" with the source data. (See the section on *Financial contract vector data compression*, above, in Section 6.3.3.)

Risk analytics	**3. Upgrade risk management analysis methodologies**
3.1	Simulate forward-looking contractual cash flows under a large number of scenarios (backward-looking correlations do not provide an adequate compass for the future).
3.2	Incorporate dynamic submodels that address contractual contingencies such as prepayment, non-performance, default and embedded-option exercise.
3.3	Incorporate dynamic submodels tasked with modeling management strategy with respect to funding, pricing, origination and product design.
3.4	Generate GAAP and accrual-based accounting transactions in concert with and in response to all simulated cash-flow events.
3.5	Integrate stress testing scenarios within and among the scope of analysis scenarios

Enterprise risk management	**4. Improve enterprise governance and risk management culture**
4.1	Simulate the change in the balance sheet going forward over a reasonable horizon (1–2 years).
4.2	Generate internal risk management analyses and reports for strategic and tactical decision support.
4.3	Incorporate relative improvements in risk measures with business unit P/L recognition and incentives.
4.4	Focus on corporate culture and data quality values as part of operational risk management.
4.5	Map key risk scenario data to downstream regulatory reporting requirements.

Systemic risk oversight	**5. Evolve systemic risk management and regulatory oversight**
5.1	Engage in industry-wide efforts to establish a private-public financial reporting interface standard in cooperation with national and global regulatory bodies.
5.2	Identify ways that collectively funded and commissioned solutions development in the industry can be utilized to reduce cost of implementation.
5.3	Anticipate extensive use of emerging Internet-enabled information and technology trends.
5.4	Provide flexible global regulatory standards with architectural frameworks that support federated organizational and operational implementations in order to facilitate widespread adoption by sovereign states and to provide incentives for commercial uptake.

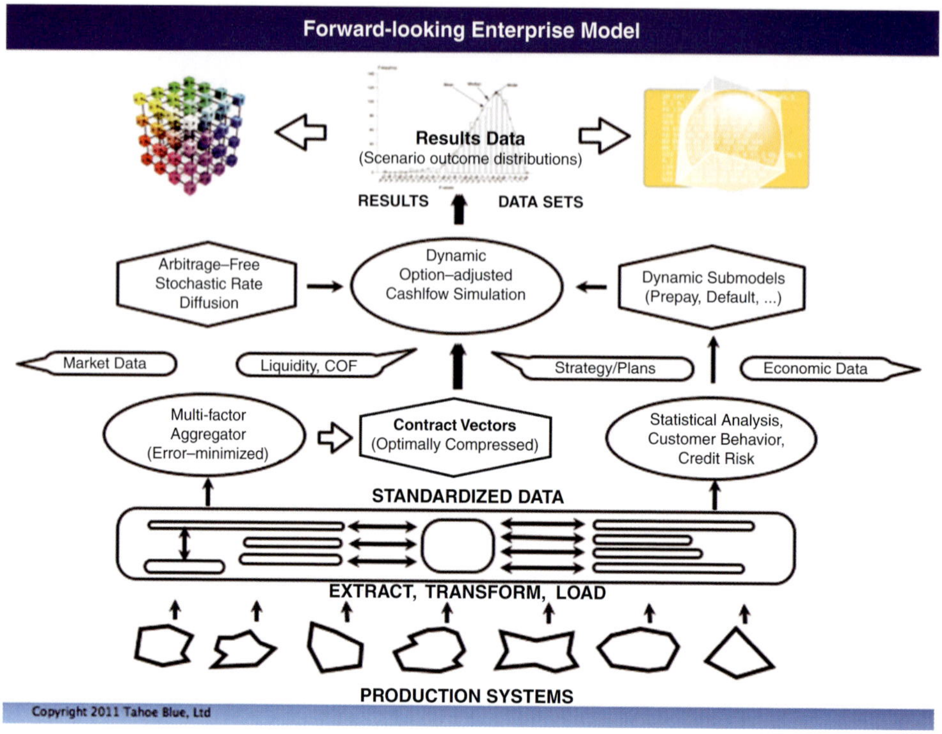

Figure 6.26 Forward-looking enterprise simulation model.

6.4.4 Forward-looking dynamic enterprise model

Figure 6.26 shows an integrated architecture for forward-looking enterprise risk analysis and management. An enterprise financial data warehouse at the instrument or contract level has been populated from production systems of record, and is the source of balance sheet data in the model.

This balance sheet detail is compressed into a dataset of *contract vectors* using multifactor optimal aggregation techniques that minimize the informational error between source and compressed datasets within data confidence intervals specified by the Chief Risk Officer. (See the important discussion of the loss of information via aggregation and financial contract vector data compression in Section 6.3.3, *Data Management Quality Considerations*.)

The purpose of performing degrees of aggregation on financial data is generally two-fold:

1. Reduce the number of *contract vectors* to be analyzed in order to make the computationally intensive nature of *dynamic, option-adjusted cash flow simulation models* more manageable;

2. Reduce the amount and scale of information presented to management in order to facilitate the cognitive understanding of the financial health of the firm and enable informed human judgment and decision support regarding the slope and curvature (first and second derivatives) of the change of financial health of the firm in the areas of the firm's greatest material exposure to risk.

It is generally better to perform aggregation of information needed to meet the objectives of the second step after minimal amounts of aggregation are used to accomplish the objectives of the first step.

Furthermore, even when information is presented to senior management in the form of dashboards that highlight and summarize risk across the enterprise, it is critical that drilling down into underlying levels of more detailed data – if necessary back to the source – be not just "possible," but easy, intuitive and immediate.

For purposes of incorporating interest-rate risk into the analysis, an *arbitrage-free, stochastic interest rate diffusion*[36] is integrated with the simulation of contractual instrument behavior across the resulting distribution of interest-rate scenarios.

The instrument-level simulation engine incorporates *dynamic submodels* that are used to introduce scenario-dependent modifications of outcomes that occur as a result of a number of factors (e.g., prepayment, default, multicurrency, macroeconomic trends, and financial stresses).

Depending on the degree of robustness of the analysis and the computational bandwidth that is available, these dynamic submodels can themselves incorporate stochastic variables for multiple scenario distributions of the behavior of the submodels. Incorporating multiple stochastic variables in analyses of this kind can greatly increase the amount of computation required, but prepayment, credit default and multicurrency submodels do lend themselves to this approach.

One of the benefits of using a simulation engine for risk analysis is that stress tests can be incorporated as scenarios that a simulation engine can evaluate, thereby bridging stress-testing analysis and analysis of more continuous deviations from current state under one roof.

Customer behavior submodels with multiple parametric input dimensions are calibrated in part using statistical analysis and current customer transaction histories, and backtested.

Results datasets (such as the collections of cash flows from multiscenario simulations) are collected in high-performance, RAM-resident containers that can be immediately viewed or quickly saved for subsequent recall and analysis.

[36] A large set of future possible evolutions of interest rates in a pseudo-random probability distribution which is consistent with the implied forward rates of the current yield curve.

6.4.5 Standards for consistent enterprise risk data

One of the problems with the variety of analytical approaches and methodologies that are deployed in departmental silos is that, regardless of how appropriate a particular tool or application may be for analyzing a specific portfolio, it is often difficult to integrate the results of different departmental solutions at the level of the entire firm in order to arrive at a comprehensive and consistent view of the total risk profile of the firm.

"Dashboards" can effectively list and summarize the risk analyses, metrics, and assessments of various product and business line silos, but managing the net risks of the firm does not necessarily follow readily. This is for a number of reasons.

First of all, a single customer may appear across multiple business lines and products. Departmental analyses may arrive at different results should the total exposure of a customer be known as opposed to only the exposure which may be segregated by different product types or parts of the organization.

Second, to the extent that results of some risk measurement methodologies are expressed in terms of relative ratios or probabilities, as opposed to risk exposures or capital requirements, then it is not possible to aggregate that type of risk metric in a total risk assessment rollup.

More importantly, if different analytical applications do not share common and consistent starting assumptions, input parameters and equivalent forward-looking scenario evolutions, then the results of disparate analytical applications cannot be aligned and reconciled, and hence cannot be merged with the knowledge that the different applications were responding to shared conditions and assumptions.

Risk management tools and procedures in different departments and lines of business should (1) share a common and consistent balance sheet snapshot as staged in a standardized data warehouse of all of the contractual data in the firm, and (2) perform their analyses using a common, shared set of scenario assumptions.

Fungible analytics

Many departmental or independent risk analysis solutions provide summary risk measures (such as loss expectations, risk capital requirements, or Value at Risk calculations). It is often difficult to combine (or consolidate) risk measures such as ratios, point estimates, or statistical approximations that are obtained by applying factors like correlation or risk capital coefficients to aggregate portfolio balances.

It would be preferable if the primary output of multiple-scenario or stress test risk analysis were distributions of cash flows which would arise using a consistent set of a large number of scenarios applied to each subset of the financial product tree analyzed.

If there is consistency in model assumptions and scenario evolutions, then cash flows and accounting events produced from simulations of contract terms and customer behavior can be combined and are capable of being merged. By so doing, a very robust and flexible view of the distribution of outcomes of the entire institution across all scenarios and stressed environments can be achieved.

6.4.6 Capability checklist

A checklist for the capabilities and functionality that are recommended components of a comprehensive financial enterprise risk management methodology follows, organized and grouped into several capability subcategories.

- Foundations of the methodology
- Dynamic submodel integration
- Distributions of cash flow outcomes and dynamic balance sheet evolution
- Dynamic scenario analysis
- Distributions of stochastic scenario simulation outcomes
- Institutional snapshot datamart
- Counterparty risk network analysis.

Taken together with the staging phases and risk management information infrastructure guidelines in Section 6.4.3 above, this checklist lists more of the capabilities of effective and integrated risk management methodology for both financial institutions as well as systemic regulators.

1. Foundations of the methodology
   1.1 Financial enterprise data model
   1.2 Data extraction and mapping tools
   1.3 Optimal compression on financial positional vectors
   1.4 Forward-looking cash flow simulation
   1.5 Stochastic scenario assumptions consistent across the board
   1.6 Option-adjusted cash flow analysis

2. Dynamic submodel integration
   2.1 Consumer credit risk
   2.2 Commercial credit risk
   2.3 Prepayment behavior
   2.4 Macroeconomic conditions
   2.5 Corporate business strategy modeling (product pricing, marketing, investment)
   2.6 Treasury, funding and liquidity

3. Distributions of cash flow outcomes and dynamic balance sheet evolution
   3.1 Provides complementary and consistent means to integrate total risk across lines of business
   3.2 Allows accounting treatments of cash flows to be projected to the income statement and balance sheet
   3.3 Produces realistic simulations of changes in the balance sheet and income statement of the bank going forward
   3.4 Supports the integration of stressed scenarios
4. Dynamic scenario analysis
   4.1 Consistent application of scenario assumptions
   4.2 Rational continuity between "normal" and stressed scenarios
5. Distributions of stochastic scenario simulation outcomes
   5.1 By product
   5.2 By customer/counterparty
   5.3 By scenario
6. Institutional snapshot datamart
   6.1 Useful for internal analysis
   6.2 Reporting interface for regulators
   6.3 Optimized level data detail/aggregation
   6.4 Rollup into multidimensional OLAP and business intelligence
7. Counterparty risk network analysis
   7.1 Incorporate business hierarchy legal entity exposure
   7.2 Key systemic risk analytical domain
   7.3 Calculate net systemic exposures to markets, currencies, multinationals, and sovereigns

6.4.7 Conclusion: time to overhaul the timing chain of the economic engine

At the outset, we stated that the rapid changes in banking, the broader financial system and the global economy demand more immediate and actionable risk management capabilities derived from comprehensive and timely access to accurate and detailed data. The widening gap in the rapidly accelerating volume, speed and complexity of financial information flow – relative to the current linearly growing ability of data and analytical infrastructure to help humans keep pace with and manage this growing gap – is introducing new forms of incremental risk and volatility in the financial system.

We have recently witnessed the severity of the consequences that were produced when the financial timing chain of the economic engine malfunctioned at high RPM and caused the cylinders of the economic engine to seize up and stall. Five years after the onset of the breakdown, the global economy is still struggling in a state of low torque as major developed economies, laden with debt, sluggishly seek to gain traction and get back on the road to growth and prosperity. Despite the fuel of liquid cash that sits in the tanks of many corporate treasuries, the stewards of the real economy in the private sector have been reluctant or unable to apply more gas to the engine and advance the spark, and a degree of tactical and strategic uncertainty lingers.

The challenges that face the financial industry and the global economy are considerable at the present time. A more effective response to meet these challenges on the part of consumers, financial institutions, corporations and small businesses, regulators, sovereign governments and central banks is sorely needed.

The deceleration and plateauing of the velocity of the real economy at the current economic trough in which we find ourselves has certainly served to raise dramatically the level of awareness and concern about the current state of affairs and what it bodes for the future. A degree of urgency exists on the part of the public as they clamor for the mechanics and engineers to rectify what are perceived as design flaws in the overhead cams and valves of the engine. More than a simple tune-up, individual consumers and government regulators are likely to agree that significant steps to overhaul the financial timing chain are required in order to avoid the economic engine "throwing a rod" again in the future.

The time to do this overhaul is certainly now, while the perceived need to analyze and fix any design flaws or quality control issues is at its highest. If too much time elapses, the willingness to disrupt an economic acceleration in the future will wane. If nothing is done by the private and public sectors, working productively together, to address these issues effectively, the odds of debilitating crashes occurring in the next cycle are significantly increased rather than reduced.

It is crucial to deliver more accurate and timely instrumentation data to the onboard analytical computers in order to provide accurate and timely information to the dashboards of senior financial management as well as their regulators. The financial industry would be well served to learn from global industries such as the airline and retail industries that have paid careful attention to improving the transparency, accuracy and timeliness of data with a combination of data standards, robust infrastructure resources and corporate culture that understands the importance of effective data management.

The priority that the airline industry places on safety, coupled with the industry's ample servings of rigorous discipline and global protocols and data standards, has produced a long-running global system with an enviable track record of very high overall reliability and, notably, the absence of crashes in a very demanding and inherently risky industry.

The retail industry invested in global supply chain, manufacturer, product and transaction data standards over 40 years ago. Furthermore, the retail manufacturing industry adopted a global federated infrastructure to curate and maintain the utilization and application of the standards among a large and diverse international member community. Not only were these data standards and technologies beneficial to all parties involved, but the standards that made accurate electronic point-of-sale scanning of retail products possible also gave rise to further technological and process innovations that in turn yielded multiple additional benefits and economic efficiencies.

The financial industry does possess some NASCAR- and Formula One-like abilities to harness information technology for high-frequency trading, portfolio analysis, global credit card transactions, electronic funds transfer and the processing of billions of accounts and financial products on a daily and even on a real-time basis for consumers and businesses around the world. Banking really is at the nexus of the real economy. The automobile racing industry, however, has the somewhat rueful distinction that rarely does a racing competition occur without some form of dramatic crash. The financial industry seems to have inherited this same gene, even if it tends to manifest itself every decade or so instead of every 500 laps around the oval.

The accelerating gap between the amount of financial data and information that is growing exponentially in the global economy compared to the ability of risk management practices at the firm, industry, and systemic levels to keep up with the pace of this acceleration requires a concerted response on the part of the private and public sector in order to address the fundamental data issues that face the industry. We are at a juncture in the history of the financial industry where global standards and rules are being pursued by government regulators and central banks in hopes of achieving strategic improvement in the transparency and manageability of financial risks. It is important for the banks and other financial institutions to step up and play a concerted role in improving the quality of their own data management methodologies and risk management practices, as well as to work constructively with financial regulators who seek improved monitoring and manage of financial enterprises and systemic risk.

We naturally welcome any and all suggestions and comments as to how this process can be improved, including how to establish better metrics for identifying, calibrating or benchmarking the quality of financial risk information and data as part of the creative response to address these challenges.

Bibliography

ABACUS 2007-AC1 CDO, www.math.nyu.edu/faculty/avellane/ABACUS.pdf

Accounting for Financial Instruments Constituent Outreach Summary May 2010, Financial Accounting Standards Board, www.fasb.org/cs/BlobServer?blobcol=urldata&blobtable=MungoBlobs&blobkey=id&blobwhere=1175820818351&blobheader=application%2Fpdf

Black, Lamont and Lieu Hazelwood, 2012, *The Effect of TARP on Bank Risk-Taking*, Board of Governors of the Federal Reserve System International Finance Discussion IFDP 1043, March, www.federalreserve.gov/pubs/ifdp/2012/1043/ifdp1043.pdf

Bloomberg, www.bloomberg.com/news/2012-04-27/economy-in-u-s-expands-at-2-2-annual-rate-less-than-forecast.html

Board of Governors of the Federal Reserve System, Troubled Asset Relief Program (TARP) Information, www.federalreserve.gov/bankinforeg/tarpinfo.htm

Braswell, Jefferson, 2010a, *The Data Challenge for Systemic Risk Management*, Presentation to FDIC/PRMIA, November, http://tahoeblue.wordpress.com/2010/11/23/systemic-risk-data-challenge-and-the-office-of-financial-research-ofr/

Braswell, Jefferson, 2010b, *A Unique Opportunity for a Win-Win in Financial Risk Management*, http://tahoeblue.wordpress.com/2010/11/25/a-unique-opportunity-for-a-win-win-in-financial-risk-management/

Braswell, Jefferson, 2012, *Recommendation for LEI Issuance via Authorized Federated LEI Registrars*, http://tahoeblue.files.wordpress.com/2012/04/recommendation-for-lei-issuance-via-federated-lei-registrars-using-registrar-ids2.pdf

Braswell, Jefferson, 2012, *Predictive Analytics: Past, Present and Future*, Bringing the Power of Today's Advanced Predictive Analytics to the Practice of Risk Management, ERM Symposium 2012, http://cas.confex.com/cas/erm12/webprogram/Session4933.html

Crouhy, M., D. Galai and R. Mark, 2000, A comparative analysis of current credit risk models, *Journal of Banking and Finance*, 24, January.

Crouhy, M., D. Galai and R. Mark, 2000, *Risk Management*, McGraw-Hill.

Data Management Maturity Model, EDM Council, http://edmcouncil.org/PDFs/20110630.DMM.Core.Model.2.1.pdf

FCIC, *Financial Crisis Inquiry Report*, Synthetic CDO diagram, http://en.wikipedia.org/wiki/Synthetic_CDO

Financial Accounting Standards Board, *Summary of Statement No. 115*, www.fasb.org/summary/stsum115.shtml

Financial Accounting Standards Board, *Summary of Statement No. 157*, www.fasb.org/summary/stsum157.shtml

Hudson Mezzanine Funding 2006–1, http://fcic-static.law.stanford.edu/cdn_media/fcic-docs/2006-10-00_Hudson%20Mezzanine%20Funding%202006-1_CDO%20Term%20Sheet.pdf

International Convergence of Capital Measurement and Capital Standards – Revised Framework Comprehensive Version, June 2006, Basel Committee on Banking Supervision, www.bis.org/publ/bcbs128.pdf

Krebs, Valdis, *Circular CDOs: Contagion in the Financial Industry*, http://orgnet.com/cdo.html

Krebs, Valdis, *Connecting the Dots in the Mortgage Meltdown*, http://orgnet.com/meltdown.html

Krishna, Dilip and Robert Mark, *Risk Information Management for Complex Financial Products*, Teradata White Paper, www.teradata.com/article.aspx?id=4773

Mark, Robert, 1994, A credit grade risk migration utility (CGRMU), in: Robert A. Klein and Jess Lederman (eds.), *The Handbook of Derivatives and Synthetics*, Irwin Professional Publishing, ch. 17.

Mark, Robert, 1994, Option's risk measurement units, in: Robert A. Klein and Jess Lederman (eds.), *The Handbook of Derivatives and Synthetics*, Irwin Professional Publishing, ch. 20.

Mark, Robert, *Operational Risk: Where is the Value?* Laurent Birade, SAS White Paper, http://blackdiamondrisk.com/images/uploads/files/wp_4323.pdf

Mark, Robert, *Risk Adjusted Return on Capital: Getting the Data Infrastructure Right!* www.teradata.com/resources/white-papers/Risk-Adjusted-Return-on-Capital-eb45899/

Mark, Robert, *Six Guidelines for Constructing Stress Tests*, www.sas.com/knowledge-exchange/risk/regulatory-compliance/six-guidelines-for-constructing-stress-tests/

Mark, Robert, 2000, The World According to Robert Mark, interview, *Derivatives Strategy*, September, www.derivativesstrategy.com/magazine/archive/2000/0900qa.asp

Observations on Developments in Risk Appetite Frameworks and IT Infrastructure, Senior Supervisors Group, www.newyorkfed.org/newsevents/news/banking/2010/an101223.pdf

Spinner, Karen, 1998, CDOs under fire, *Derivatives Strategy*, November, www.derivativesstrategy.com/magazine/archive/1998/1198fea1f753.asp

Tabb, Alexander, 2011, *Feeding the Options Beast: Big Data in the US Options Space*, Tabb Group, www.tabbgroup.com/PublicationDetail.aspx?PublicationID=994

Wall Street Journal, 2011, A look inside the Fed's balance sheet, interactive graphic, http://blogs.wsj.com/economics/2011/06/21/a-look-inside-the-feds-balance-sheet-11/tab/interactive/

7

Capital markets data

Martijn Groot

7.1 Introduction

This chapter gives a high-level overview of capital markets data. We will survey the different types of capital markets data and provide different angles for categorizing them, from the perspective of the instrument or the trade lifecycle. We follow this with a discussion on the data generation process and how that is changing. After this we discuss different metadata aspects – such as quality measures on data, business process context information that specifies the intended usage, as well as information that is typically inferred from capital markets data. We conclude with some observations on developments in capital markets data, how they are sourced and distributed.

Capital markets data comprise generic information about financial products (terms, prices), financial markets participants (issuers, guarantors, trading counterparties, customers, investors), operational and transactional data (master agreements, settlement instructions) and opinions (equity research, ratings). They also include participant specific information including trades, positions/holdings, portfolios, books, hierarchies, collateral and specific proprietary valuation methodologies, trading strategies and algorithms.

To lend some order to financial products, market practitioners, trade associations (e.g., ISDA), standards organizations (e.g., FpML, ISO, FPL) and infrastructure providers (e.g., SWIFT) have created different classification schemes and descriptive standards. These range from financial product classification and identification standards to complete data models to describe terms and conditions of financial products.

When it comes to categorizing capital markets data, we can for example look at the types of content that play a role in the transaction lifecycle. If we split up the trade cycle into pre-trade, trade and post-trade phases, we can map specific data requirements and information focus per function, as shown in the table.

Handbook of Financial Data and Risk Information, ed. Margarita S. Brose, Mark D. Flood, Dilip Krishna and Bill Nichols. Published by Cambridge University Press. © Cambridge University Press 2014.

	Function	Data requirements
Pre-Trade	• Quantitative modeling • Broker/trader discussions • Management approval • Account management	• Research, Time series, Reference data • Counterparties, Costs • Limits, P&L histories • Accounts, Counterparties
Trade	• Execution – Order management – Order routing • Margin and securities lending • Account management	• Execution – Parameters of models and algorithms, Brokers, Execution strategies – Reference, Counterparties • Inventory, Availability, Margins • Accounts, Counterparties
Post-Trade	• Clearing and settlement • Corporate action processing • Risk management • Compliance management • Account management • Fund management	• Transactions, Standing settlement instructions, Accounts • Positions, Corporate actions accounts • Counterparties • Positions, Prices, Risk parameters • Positions, Regulations • Accounts, Counterparties, Positions, Prices

The table shows that different types of content will be required in the pre-trade (idea generation, price discovery process), trade (execution) and post-trade (settlement, asset servicing, revaluation, risk reporting) phases of the lifecycle of a transaction.

Another way to look at capital markets data is by how they are sourced and by taking a process-oriented view. For example, capital markets data could be split into the following four categories:

- publicly available information, e.g., EDGAR filings;
- commercially sourced information, e.g., real-time data feed from an exchange;
- data exchanged peer to peer with trading counterparties, e.g., a master agreement governing OTC derivative trades or transaction details;
- proprietary information produced within a company, e.g., portfolio composition, proprietary trading strategy.

Yet another common way of looking at capital markets data is by looking at where they sit in the organization: front, middle or back office. This would lead to categories such as risk or middle office data (such as exposures, limit information, Value at Risk numbers) and front office data (e.g., trading books, real-time market data).

In the discussion below we will divide the world of capital markets data into the following high-level categories by looking at the data type and by high-level functional usage:

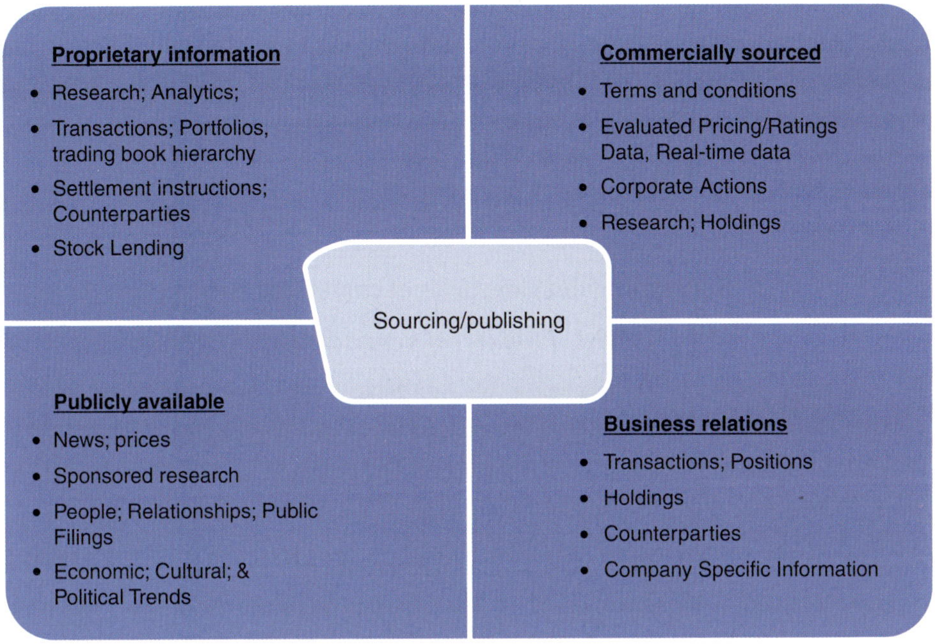

Figure 7.1 Different sourcing and publishing options for different categories.

Product data -Securitymaster -Corporate actions -Pricing data	**Entitydata** -Counterparty data -Corporate structure -Customer data
Trade information -Transactions -Holdings/portfolios/trading books -Trade support information	**Research data** -Analyst opinions -Ratings

Figure 7.2 Schematic overview of capital market data categories.

- product data (reference data, i.e., terms and conditions);
- product data (quotation information but also various forms of derived data, e.g., risk and performance metrics, curves, surfaces, smiles, correlation matrices);
- trade information (e.g., trades, payments, positions/holdings, portfolios, books, hierarchies);
- entity information (legal structure, collateral information, counterparty exposure, client information);
- analyst information including ratings, benchmarks, indices, analyst information.

We should note that capital markets data distribution is a large business in its own right. Estimates of spend are in excess of 20B$ per year. (Burton-Taylor International Consulting LLC). Most of the spend is on real-time data that determine when, what and where to trade, but a lot is spent on the collection, aggregation and cleansing of data which are needed to support the post-trade process and for internal risk, investor, shareholder and regulatory reporting.

7.2 Survey of main categories of capital markets data

7.2.1 Product reference data

In discussing product reference data, we first have to talk about financial product classification. After this we discuss the various descriptive (reference data) elements that can be associated with financial products, as well as corporate actions that can affect these terms and conditions.

Product classification

A basic distinction within financial products is that between securities, which are instruments that represent ownership such as stocks and bonds, versus derivatives and commodities which are either bilateral contracts or represent physical goods such as copper, oil and soy. Another basic distinction is that between over-the-counter (OTC) and exchange traded products. OTC products are essentially contracts: any two trading parties can create a new OTC product when doing a transaction. A security is a claim on a corporation: either an ownership interest (equity) or a loan (bond). Exchange traded derivatives are standardized derivative contracts created by the exchange. Some product examples are provided in the table; note that some OTC derivatives become standardized and move to an exchange traded environment.

	Securities	Derivatives
OTC	Often corporate bonds, illiquid bonds, sometimes large block trades in equity off exchange	FX options, interest rate swaps, credit derivatives
Exchange	Bonds, equities, Exchange Traded Funds (ETFs), some funds	Exchange traded options and futures

FX is a special category. It is traded over the counter but many trades are centrally cleared through CLS Bank.[1]

[1] CLS originally stood for "Continuous Linked Settlement," which is a good description of the institution's role. It was designed to be open round-the-clock to overcome the potential settlement risk posed by different currency markets operating on very different time windows throughout the global day. This often meant remitting one side of an FX trade many hours before receiving the corresponding funds in the second currency. Failure of a counterparty during this timing mismatch period could lead to a loss many times larger than the net pre-settlement gain or loss on the trade. The classic example of this was the failure of the German Herstatt Bank in June 1974 in the middle of the European trading day.

The variety and pace of product innovation, combined with the often unique needs of investors for which products are tailor made, is such that any list that claims to be exhaustive will be out of date very quickly. One of the more commonly used public classification schemes is that of the Classification of Financial Instruments (CFI codes, ISO standard 10962). The CFI classification has been designed to group financial products in a uniform and consistent way. It includes codes for financial product types to facilitate electronic processing. The first character of the CFI code indicates the category, and is the highest level of classification. These categories are:

- Equities (E)
- Debt instruments (D)
- Entitlements (Rights) (R)
- Options (O)
- Futures (F)
- Others/Miscellaneous (M).

The CFI code consists of six letters in total. As can be seen, the first letter indicates the type of instrument. The second letter signifies a finer subclassification, and the remaining letters show various attributes of the security. **Equity** and **Debt** instruments are also collectively labelled "Securities." Because debt instruments are often tailored to conform to certain funding or investment requirements, there can be an abundance of terms in these instruments that determine the cash flows, the redemption profile and the conditions on the cash flows.

Whereas equities represent ownership rights and debt holders own loans that are obligations of other entities, for example corporations or countries, **Rights** include Warrants which are instruments that give the owner the right to purchase a certain amount of, for example, a commodity, a currency or a financial instrument during a certain time for certain strike price. Warrants are often attached to other financial instruments (e.g., bonds). Options are similar to warrants in that they are rights. **Options** give the holder the right to buy (call option) or sell (put option) a certain amount of an underlying instrument at a certain price during a certain time. For example, the option to buy Euros at $1.40 during a certain time period or the option to sell Dow Chemical shares at $37 by December 2014. **Futures** are the obligation (not the right) to buy or sell something at a certain price by a certain date (expiration date). The **Miscellaneous** category includes two main groups. One consists of referential instruments that are often used in other instruments such as options and futures. These referential instruments cover currencies, interest rates, indices and commodities. The second group includes a mix of other asset types not yet covered such as real estate and insurance policies.

Here are a few examples of complete CFI codes.

- For a debt instrument, an example would be D-W-F-S-F-B; in this case D stands for debt, W means "with warrants attached," the F stands for "fixed rate," the S denotes that it is secured, the F means that the maturity is fixed and the B means that it is a bearer instrument.
- For an option, an example would be O-P-E-C-C-S; O for option, P for put, E for European exercise style, then C to identify the underlying (currency), C for the delivery method (cash delivery) and S for standardized.

The CFI code is used around the world and many countries started to allocate CFI codes during the 1990s. However, similar to the coverage of other financial services ISO standards such as the International Securities Identification Number (ISIN) code, the use of the CFI code is skewed towards exchange traded products. This could be due to more automated processing for exchange traded products or could be due to the relation between exchanges and Central Securities Depositories (CSDs) and numbering agencies in many countries. In any case, within the CFI classification, many derivative instrument types are grouped together under the "miscellaneous" category.

An alternative way to think about types of financial products is through that of a *stacked* model, moving from underlying physical goods owned by legal entities, to securities that represent claims or ownership on those entities, to derivatives with indices and currencies as running cross vertical pillars. Derivatives are often grouped into their underlying product type, for example,

- Interest rate derivatives
- Credit derivatives
- FX derivatives
- Commodity derivatives
- Equity derivatives
- More esoteric derivatives such as weather derivatives and derivatives on mortality rates for life insurers.

A common standard and classification scheme to encode and represent terms of OTC derivatives in a standard way is the Financial products Markup Language (FpML).

Yet another useful way to categorize financial instruments – in terms of processing as well as for risk measurement – is by looking at the frequency, type and level of uncertainty in an instrument's *cash flows*.

We could look at the incoming and outgoing cash flows along two criteria: the certainty as to when they will occur and the certainty as to the amount. Typically, the purchase of a **security** is associated with one initial outflow against a future cash flow stream representing for example dividends, coupon payments and the redemption in case of a bond. The only occasions when there are subsequent cash payments arise as a result of corporate actions when for example a rights issue is used to purchase additional stock or when they are needed for fees or services on the securities such as custody charges.

In many derivative contracts such as futures, forward rate agreements and interest rate swaps, there will normally be no initial cash flow but a periodic settlement of the balance of the contract instead. In the table we show examples of instruments according to the various natures of the cash flows associated with the instruments.

Cash flow size/type	Example instruments	Cash flow timing	Variance depends on
Fixed by contract	Deposits, bonds, repurchase agreements	Fixed, e.g., annually or semi-annually or one-off (zero bond); frequency also determined by market conventions	Credit default, settlement errors
Fixed by contract	Puttable bond	At holder's discretion	Market conditions
Fixed by contract	Callable bond	At issuer's discretion	Market conditions
Fixed, some flexibility in contract	Complex bonds where the coupon is dependent on other factors, e.g., business performance, price of commodities	Fixed	Contract-specific terms
Variance according to dividend policy	Common stock	Fixed	Company performance, dividend policy change
Depending on price differential of asset with regard to benchmark at one point in time	Options, warrants, swaps	Fixed	Performance of reference or of underlying asset
Depending on price differential of asset with regard to benchmark averaged out over time	Asian options, exotic options	Fixed	Performance of reference or of underlying asset
Bandwidth within contractual minimum and maximum	Retail structured products and collars	Fixed	Performance of reference or of underlying asset
Depending on price movement of a group of underlying products	Basket products, first-to-default, worst case, passport options	According to contractual resettlement	Performance of reference or of underlying asset

Cash flow size/type	Example instruments	Cash flow timing	Variance depends on
Zero upfront, then periodic, direction depending on market	Interest rate swap, FRA	Periodic, e.g., annually, semi-annually or quarterly	Market dependent
Zero upfront	Future	Initial margin payment, after that potentially daily maintenance margin depending on market movement	Market dependent

The frequency of the cash flows can vary from daily (e.g., to keep up the maintenance margin for exchange traded futures and options) to once in 30 years for a 30 year zero coupon bond or perhaps never at all, for example when a purchased option expires out of the money.

The cash flow frequency has major consequences for the work needed to process and monitor the instrument after the trade in the back and middle office.

If you hold a bond or a deposit you can book the accrued interest as revenue. If you hold a more complex instrument you have to start making assumptions as to **when** and indeed **if** a cash flow will appear. Also, in the case that cash flows could go out on a daily basis, for example keeping up maintenance margin at a clearing house or posting new collateral, the work entails processing the exchange's margin call on a daily basis. An assessment of the likelihood of and probability distribution of the size of future cashflows is part of the pricing and risk management of the product. Reflecting their primary purpose of risk transfer, many OTC derivative products are cash settled periodically instead of the underlying being delivered. Calculating the exact amounts due hinges on accurate administration of all the terms and conditions of the transaction.

Note that regulators would also classify financial products according to their risk bracket, with both a market stability and an investor protection angle in mind. For example, certain types of financial products (often corresponding to minimum level investments) may also be sold to specific, for example informed, investors. From a risk and solvency perspective, products such as bonds were classified into different risk brackets based on the perceived default risk category of the issuer.

Apart from the public domain classifications, many content providers have their own product classification, typically reflected in the way their products are structured and sold. An example of this is the Bloomberg yellow key classification corresponding to the high level classification within the Bloomberg database and the Bloomberg Open Symbology instrument identification.

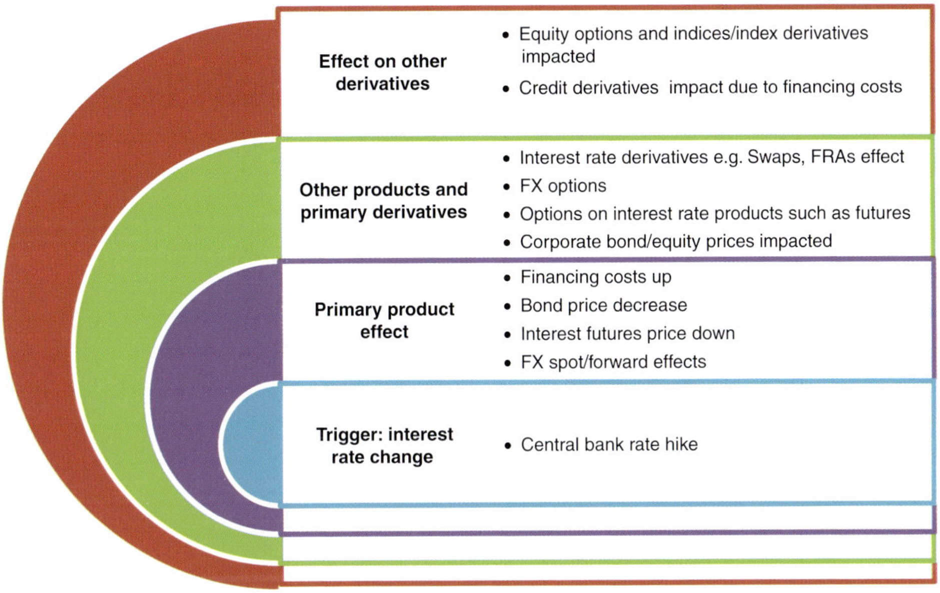

Figure 7.3 Interrelations between different product types due to shared risk factors: some kind of spider web. Ripples through multiple directly or indirectly related financial products.

Note that financial product classifications should not distract from the fact that financial markets are highly interrelated and the behavior of different product types is often driven by shared risk drivers. Also, effects on one product can percolate through to other, at first sight dissimilar, products. An interest rate change for example will have effects throughout different product categories (see for example Figure 7.3).

Security master

Security master describes the terms and conditions of securities. Terms and conditions basically define the frequency, size and variability of a financial product's cash flows.

The starting point with financial product reference data is identification. Through standards such as ISIN (ISO 6166) combined with place of trade, for example, it is possible to identify exchange traded products unambiguously. It gets trickier when we get to OTC derivatives. Typically these are identified by listing a number of descriptive elements such as product type, tenor and currency.

Reference data range from a few descriptive elements on a simple bond (e.g., currency, issuer identification, maturity date, coupon rate, payment frequency, interest calculation method/daycount basis) to identify a security to a vast quantity of terms for structured products, for example.

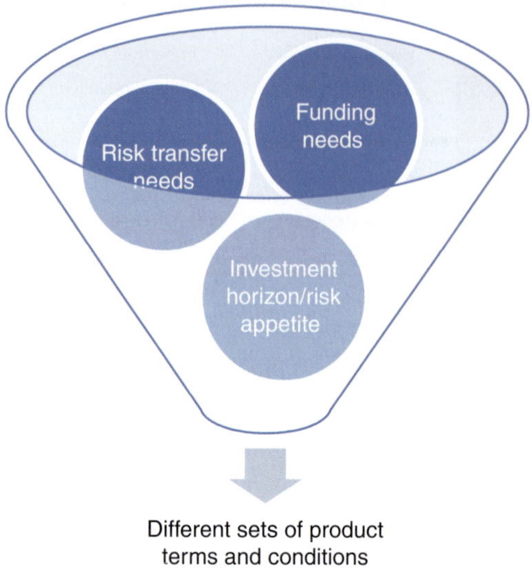

Figure 7.4 Different sets of product terms and conditions.

The number of terms potentially needed to describe securities can be enormous. Many data vendors that offer larger aggregated products around a security master have data dictionaries that contain many thousands of different fields. Financial products are created to address and bridge funding, investment and risk transfer needs. To cater for and to optimize the balance between different investor types, risk categories and funding profiles, a large number of financial product terms has been created. Especially in structured products, large numbers of parameters can be required to describe the structure fully and accurately. Various initiatives have been launched to give greater uniformity in reporting, for example the Asset Backed Security reporting templates from the European Central Bank. Bankers are sometimes compared to "cocktail bartenders" that have to come up with interesting new mixes and products to keep their audience engaged. On top of the funding, risk transfer and investment needs, financial products have also been created to arbitrage between different fiscal and regulatory regimes.

Some content vendors specialize in providing complete sets of structured product information, whereas some of the larger data aggregators only provide summary information. This is due to the variety of financial products and terms that arises mostly due to different profiles of investors and structures and different regulations and fiscal regimes to which products are tuned.

Corporate actions

Within reference data, corporate actions data play a special role in that they may change the terms and conditions of a security. Corporate actions are initiated by the issuer of a financial product and range from relatively straightforward and regular events such as for example cash dividends and annual meeting announcements, to very complex one-off events in the case of mergers, for example. Corporate actions processing has long been a matter of manual work and there remains a core of corporate action types which is very difficult to automate.

Because the details and terms of a corporate action are at the discretion of a company, they cannot be fully captured in advance by any fixed data model or prescriptive standard; some narrative element will necessarily remain. Nevertheless, to facilitate and streamline their processing, corporate actions have been classified into different types in the ISO 15022 standard using four letter acronyms. Examples include

- DLST = delisted
- CAPG = capital gains distribution
- REDM = final maturity
- INTR = interest payment.

The different perspectives, depending on their role, that financial institutions and departments have on corporate actions are also reflected in capital markets data. There are pure corporate actions feeds as well as larger content products that contain corporate action information as ancillary to maintaining the security master. A special category of content products is that of the integrated corporate actions dataset that aims to bring order to this difficult data space. Some custodians or very large investors have been able to offer products in this space that compete with the pure aggregators. This is possible because they are in a unique position and can leverage/capitalize on the information they already need to run their own business.

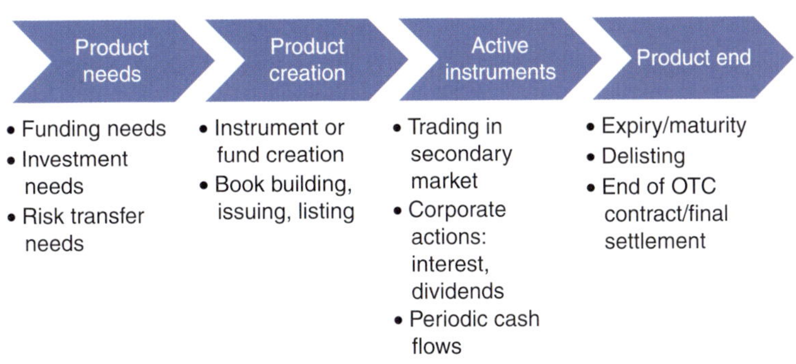

Figure 7.5 High level instrument lifecycle.

7.2.2 *Product data: pricing and time series information*

In contrast to security master information, with pricing and time series data we mean every piece of capital markets information we expect to change very frequently due to market circumstances.

Pricing data or time series data cover pricing information that can be directly observed from the market as well as various forms of inferred or calculated time series data such as risk factors. Large volumes of time series data are created through analytics and financial models taking other time series as input.

The data elements provided in a quote differ depending on the product type. For exchange traded products where trading takes place through a well-defined process using fixed hours and a central order book, there will be official "high," "low," "open," "close" and "volume" information as well as level 2 or order book data showing the current best bid and ask and information on the depth (liquidity) of the market. For OTC markets where quotes are assembled through brokers, there will often be bid and ask quote elements only, which could be just indicative rather than firm (actionable) quotes.

Also, **what** is quoted will depend on the specific product. For equities, it is typically the price to buy or sell a share whereas for a bond it will be a percentage of the nominal value or a yield. This means that for example a BID ASK pair could refer to a **price** or to a **rate** or to a **yield** or to **volatility** or to a **correlation** or to a **spread** to give just a few examples. Below we provide some examples of different quotation types.

- Price – the price of an asset, e.g., equities, rights.
- Rate – the rate that is traded, e.g., deposits, bankers acceptances.
- Fixing – a fixing done by a committee or a group of banks, e.g., London gold fixing or LIBOR, or the price for very illiquid equities.
- Correlation – correlation between two assets or an asset and a benchmark that is traded.
- Yield – the yield that is traded, e.g., government bonds.
- Volatility – volatility that is traded, e.g., swaptions and FX options.
- Index – the value of an index.
- Spread – the traded difference between two other series, e.g., basis swaps, FX forwards, credit derivatives. Spread can come in different varieties as well including:
  - Price spread
  - Credit spread
  - Yield spread
  - Rate spread
  - Correlation spread
  - Volatility spread.

Note that apart from the currency of the product, for commodity products, for example, it is also critical to know the unit of measurement for which a price is quoted. We need to include it in the gold price for example, which is usually quoted in USD per ounce, whereas silver can be quoted per kilogram. Oil is normally quoted in USD per barrel whereas natural gas could be quoted in for example USD/mmBTU or in CAD/gigajoule.

In the case of options or futures contracts, the **size** of the underlying needs to be taken into account as well. For example, is this an option on a 100 shares or is there a deviating contract size. When it comes to futures, we need to include the contract terms because it needs to be clear whether we are talking about a money market future on 1M USD or on a deviating amount.

Apart from directly observable prices, additional time series information is typically inferred from these quotes. This includes constructs such as bond curves, zero curves, commodity forward curves, volatility smiles and correlation matrices. These are used both to serve **risk factors** to calculate the risk in a trading book or portfolio and to **price** new products. For example, the prices of deposits and FRAs, money market futures and swaps, or bonds, could all be fed into a bootstrapping algorithm to create a zero curve. The quality of the inferred data hinges on the quality of the input datasets.

Time series analysis is a discipline in its own right and is used for the backtesting of trading strategies and risk analysis or for example to infer patterns used for new trading strategies. This can be both at the ultra high frequency tick data level to come to new algorithmic trading strategies and at the longer term level to find seasonality effects, for example. Other patterns can include patterns due to fiscal calendar, macroeconomic cycle, product correlations, investor psychology and business-cycle based demand/scarcity applicable for example to commodity products.

One common perspective on pricing data is to classify them by their frequency and latency. Often this distinction is also made in content licensing agreements with real-time data being expensive and delayed data often being freely available. We can, for example, distinguish between real-time, snapshots, historical and end of day (EOD) data.

Note that the concept of time series is not limited to prices, order books and volumes. For example, banks have to keep histories of internal credit ratings and operational losses as part of new regulatory requirements under the Basel II accord. This allows for a more detailed quantification of credit and operational risk. In general, there is more and more focus on data retention, so in a way everything becomes a time series.

With regards to frequency, we can classify time series data roughly as follows.

- **Real-time** or streaming data refers to a direct quote and price stream from a broker or execution venue. Regulation on both sides of the Atlantic (REG NSM in the USA and MiFID in the European Union) has aimed to make this pre-trade information more transparent to help investors determine on what trading venue they can get the best price for equity trades, for example. The motivation behind this is investor protection: to make sure the investor gets the best available price. The added complexity of this is that in the case that an instrument trades on various execution venues, an institution may need to consolidate different quote streams and order books internally to get the full liquidity picture.

- **Snapshots**. For OTC markets there is no official close price, therefore the institution that has to revalue a position has to take a snapshot at a certain cut-off point. Consistency in time for all instruments in the portfolio is important. Sometimes an institution takes streaming data from an interval around the desired cut-off point and takes the median or the average price to prevent a freak outlier or rogue price at the snap time from distorting the picture. Sometimes an institution takes for example the 4:00 pm price from a respected market-maker in that particular instrument. Sometimes a combination of both is used, for example by taking the median price of a set of trusted market-makers approved by risk control. Note that for revaluation, snapshots could also be proxies or theoretical prices. Fair value accounting standards such as FASB157 and IAS39 classify assets to be revalued into different categories. Apart from liquid (level 1) instruments, there are less liquid or illiquid (levels 2, 3 respectively) instruments that will have to be revalued using a model price or a proxy based on a comparable instrument.
- **Historical data**. These can be historical end of day (official close or consistent snapshot) data or historical tick data. Historical end of day data can be used for risk management purposes and also to calculate the historical volatility needed to price derivatives. They are also used for the backtesting of strategies that rely on the microstructure of the market and they can be used to study market impact and to measure liquidity and as a source for trading algorithms. Often between one and seven years of historical data are used for risk purposes in banks, but in the insurance space the history taken can be much longer than that.
- **Other frequencies**. Time series data that are not available on a daily basis include macroeconomic data. Numbers such as unemployment, inflation and GDP are typically released on a monthly, quarterly or yearly basis. Financial statements can also be considered as time series data and are typically made available on a quarterly, semi-annual or annual basis.

When a portfolio needs to be revalued, some assets are easier to value than others. In the case of illiquid instruments, "mark to model" is sometimes the only option. There are different organizational set-ups for this. The quotes can come from an independent middle office function that collects them from trustworthy sources and signs off on any models that may be used. Alternatively, the quotes to revalue them can come from the traders, and the middle office compares those against for example a neutral source from the market.

Driven by investor scrutiny and regulation (such as Dodd–Frank in the USA and the AIFM directive for hedge funds in the EU), there is an increasing focus on an independent valuation function, whether an internal department separate from portfolio management or a wholly separate business to which asset managers can outsource this middle office function. For scalability reasons, it is often outsourced to an independent provider. Keeping track of the sometimes enormous array of terms and conditions of complex products can cause operational complications and sometimes hidden risks for wealth managers, custodians and investors.

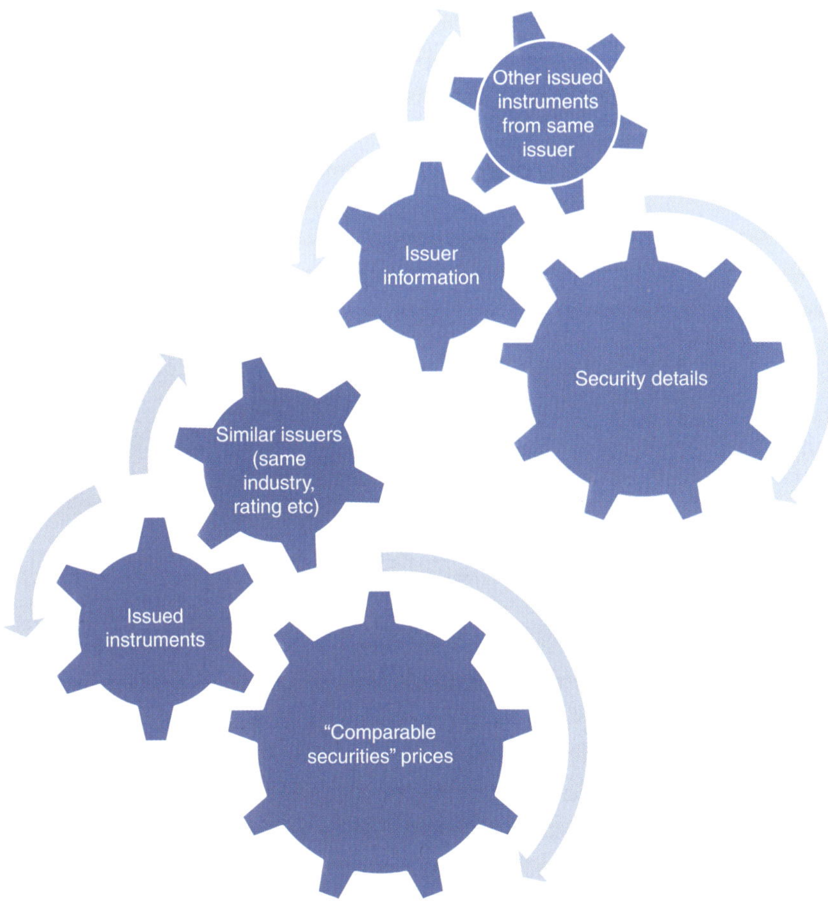

Figure 7.6 Interlocking data elements: connections can only be made through common and unambiguous instrument and issuer identification.

Valuing of complex or illiquid products is difficult and often labor intensive. Proxy valuations are usually done by finding and using the prices for comparable products. The issue can be to find the right comparable products and also – when there are multiple pricing sources – to keep track of multiple prices for the same instrument. You will likely get different candidates for a "comparable" instrument by selecting different comparison criteria such as rating class, maturity band or industry sector. Also, different products should be linked to, for example, relevant indices, comparable securities or main price drivers to check or proxy the price and price movement. This can mean extensive data modeling and linking of different financial products and risk drivers so that adequate comparables can be taken in a consistent and repeatable way.

7.2.3 Transactional information

In this category of capital markets data we discuss trades, payments and positions/ holdings, portfolios, books and hierarchies. We also highlight what we term "trade support" data which include settlement instructions and master agreements.

Transactional data

From a data categorization perspective, we can make a distinction into the following transactional information sets.

- Transactions. Any exchange of a financial product between different parties. Typically the data elements of a transaction would minimally cover the product traded, the amount, the price at which it is traded and some instructions as to how the trade will be effected, for example through settlement instructions. There are multiple transaction reporting obligations depending on the circumstances:
  - banks have to report large cash transactions, both for fiscal/tax reasons and to detect money laundering;
  - banks have to report securities transactions, e.g., for MiFID.

In some cases there have been trade registries where all trades in certain asset classes are kept, for example, FINRA for US corporate bonds and increasingly through derivatives trade repositories.

The volume of transactions differs by type of player. Some trading operations generate tens of thousands of transactions a day (especially in, e.g., algorithmic trading). Depending on the strategy, some hedge funds may only make a few strategic bets a year. Also, Market makers trade all the time but, broadly speaking, maintain a flat position. In general the trend has been for more frequent trading, the volume of orders has risen significantly in equity markets but the average order size has gone down. This also reflects a lower cost per trade.

There are several transaction types. Typically a financial product is exchanged against cash (e.g., buying X shares of a company's stock or selling 5000 USD nominal value of US Treasuries). In many cases, however, cash is exchanged against cash (in FX trading, e.g., buy 10M EUR with 14M USD) or different cash flows are swapped (e.g., through an interest rate swap where a fixed cash flow stream is swapped against a variable cash flow stream).

- Holdings or positions. The total amount a player (or portfolio) has in any given instrument, e.g., 10,000 shares of Citibank or 50B EUR of Greek government bonds. Transactions update holdings and the tracking of how and when holdings change and at what price is investment accounting. Financial services firms organize their holdings in different ways. A logical collection of positions could be called a portfolio or book in a bank, whereas an investment manager would usually speak about portfolios. A portfolio could correspond to a mutual fund or a specific mandate from an institutional investor. A bank would talk about books for specific investment horizons or product classes. For example, it would talk about the "global currency options book." A common distinction is between a "trading book" which is marked to market and a "banking book" which could record loans at nominal value. Another set of portfolios can be sets of collateral per client.

Trade support data

Under trade support data, we include information needed to facilitate the transaction process and to run the daily operations of a financial firm. There is some overlap here with company information, for example, contract data and data for the settlement of transactions. We also include business holiday calendars.

Contract information refers to the legal framework trading parties put in place and includes agreement on settlement terms. The settlement risk can be reduced by agreement between two parties to have all their dealings in a certain product type (e.g., credit derivatives) governed by the terms of an ISDA master agreement. Under a master agreement, confirmations of specific trades fall under the provisions of the master agreement. This will, for example, imply a *netting agreement*, the cash transfers between two parties will only involve the balance of all the sums due for all the trades under the master agreement. Contract data also include the administration of which products can be sold to which customers. This includes both exposure limits to cap the risk of the institution, but also the set of "eligible" products for different customers, for example retail clients, high net worth individuals, professional counterparties and so on.

Increasingly, financial institutions have to take proper care that they only offer products appropriate to, and commensurate with, a client's risk appetite and investment objectives. In the case of professional counterparties, contract information can also include details on "best execution" agreements that stipulate for instance the eligible execution venues for a transaction plus other execution criteria.

Holiday calendar data contain the business day conventions of a country or region, for example listing the typical working days, Monday to Friday, or Sunday to Thursday, the bank holidays when trades cannot be settled; they can also include trading hours of exchanges. Holiday calendars can be organized by a geographical indication such as country, region or city but normally by country or by trading center such as the exchange. Major providers of holiday information include Swapsmonitor (`www.financialcalendar.com/`) and CoppClark (`www.coppclark.com/`).

Settlement and payment instructions data indicate how to rout payments or where to settle certain products for certain counterparties. This category of information is typically pooled from the market players and includes the following subcategories:

- commercial payment data, which include the SWIFT BIC (bank identification codes) directory and the CLS bank directory for settlement of foreign exchange trades;
- on securities' settlement instructions, there are several "poolers" of account information such as Omgeo.

7.2.4 Entity information

Apart from transaction and product information, capital markets data also include a category of descriptive information on entities, whether government bodies, corporations or individuals. An institution could need this information for various reasons.

- For a corporate banking customer, to whom loans are made, for example, financial information would be needed such as balance sheet, profit and loss statement and annual reports as well as corporate structure and comparative information on the company's industry classification. There should also be links to any financial product (e.g., loans) transacted with this entity and what can be sold to whom (e.g., signoff authority levels in the client). Information can also include external and internal credit ratings.
- For a retail banking customer, information such as credit scores will be kept plus an overview of products sold and cross-selling opportunities.
- For a trading counterparty, documentation would be kept on any master agreements that have been made that govern for example the trading of interest rate swaps and credit default swaps with this party as well as the type and amount of collateral that needs to be pledged. Account information and standing settlement instructions would also be kept.

Descriptive information includes basic data such as name and address, relations to other entities (e.g., parent or subsidiary relations), identification, credit rating, industry sector or customer category and information as to what can be sold to whom.

It often also includes identifiers should they exist and information on the company's business, shareholders and management. Some entity data products specialize in the cross-referencing between various identifiers and showing the linkage between issuers and issued instruments. This can support client identification within a bank and is needed for credit risk to link issuers to issues. Unlike the case of financial instruments where ISIN is a global standard, there is no corresponding standard for legal entities. Although the BIC code from SWIFT is often used, this is an identifier for a branch and serves to identify the endpoint of a trade rather than a legal entity. In addition, there is an ISO standard 16362 called the IGI – the Issuer and Guarantor Identifier (IGI) – and a more recent global legal entity identification (LEI) standard (ISO standard 17442). Following the financial crisis, there is increased regulatory focus on the unique identification of legal entities, and the LEI will be rolled out globally.

A separate set of content products provides details on corporate structure, showing a holding company with all its wholly owned or majority owned subsidiaries. Good quality information on who owns whom, who owns what and who is ultimately responsible for any obligation incurred somewhere in the legal hierarchy tree is very important.

Product and entity data are closely related and the boundaries are sometimes straddled. For example, a larger corporate can be a customer of a bank, for example for financing needs or for M&A advisory work. At the same time, it can be a counterparty of the bank in trading FX or interest products and the bank itself could trade the company's equity or credit default swaps on the company. Indeed, in the case of a credit default swap it could be said that (the credit of) the entity has become the product.

7.2.5 Research information

Within the category of research data we include all information used to form an opinion on the value of a certain product. This includes the categories of fundamentals, estimates and macroeconomic data, but it also includes credit ratings, indices and news.

Fundamentals, estimates, macroeconomic data

Earnings estimates and fundamentals are important data categories. Periodically public companies must file their financial statements such as income and cash flow statements and balance sheets (e.g., into the EDGAR database). Information on past performance is therefore in the public domain and is disclosed in different ways.

Private companies also have to file some information and the level of granularity depends on the jurisdiction.

News

News is provided by agencies which can be commercial wire services, corporations (e.g., Thomson Reuters), and cooperatives such as newspapers that pool their content. Financial news on for example macroeconomic data, key hires, financial results, patents, client wins, lawsuits and so on is actionable information and a major driver of prices. Other companies provide add-on services to make the classification, processing and distribution of news content easier. The analysis and mining of news and social media is a separate area. Because news and social media normally contain *unstructured data* not conforming to any predefined layout or data model, it is harder to process this news and to make automated inferences from it. Nevertheless, there are software packages that analyze news feeds and social media updates to gauge the prevailing sentiment on for example a specific market or company. Conversely, some news has started to be tagged and marked up with certain keywords, allowing easier manipulation, routing and processing.

Rating information

Ratings are assessments of creditworthiness of a retail or corporate client, of an issuer, of a financial instrument such as a bond or structured product or of a country.

As the pricing of credit is the core business of a bank, it could be asked why it is left to an agency? The answer could be that an agency rating is an independent assessment and therefore valuable as a second opinion. An analogy is the revaluing of an investment portfolio where a company will normally not use the trader's own price but an independent price instead. However, agencies fulfil other roles as well. Their credit stamp on an instrument can make the difference between an instrument or structure being eligible for investment by many pension funds and mutual funds, since many of these can only invest in investment grade instruments while others may specialize in junk bond markets. Thus an official credit rating from a certified rating agency can be a necessary condition for market adoption.

Ratings are either provided by research agencies through behaviorial, financial and statistical analysis, or are created within a financial institution (internal rating). We can distinguish between different types of ratings:

- credit ratings (external or internal)
- mutual fund ratings
- commercial ratings
- retail credit scores.

Several large rating agencies such as Moody's, Standard and Poor's, Dominion and Fitch dominate this market. In the wake of the financial crisis, rating agencies will be more closely supervised and new rating agencies with different business models or focusing on different geographies may enter the market. In the EU, rating agency supervision will be done by the European Securities and Markets Authority (ESMA). In general, there have been calls for more insight into the working processes and the methodologies used by rating agencies.

Indices

An index may not be strictly research information. Rather it can be seen as a "market dashboard" that summarizes the direction of a specific market, product or geography. Note that indices are also a financial product and are included in the CFI standard in the Miscellaneous category.

The most well-known indices are the equity or property indices that make it to the daily news, but there are also indices in different product areas such as interest rate and credit. Indices are typically created by condensing information on a large number of variables (the index constituents such as the stocks in the S&P 500 index) into a single number via a calculation procedure that gives a weighted average with weights based, for example, on market capitalization in the case of equity indices.

Indices are very valuable bits of data real estate because they are often very well known and either serve as the benchmark for investment products or are the basis for an investment product. Many of the popular Exchange-traded Fund (ETF) products are designed to track a certain index. Indices carry considerable brand value and investors may pressure asset managers to use certain indices.

More and more index families and topical indices have been created (for example based on securities with a certain market cap, with certain dividend characteristics, based on certain industries or geographies) and there is a fine line between an index and an investment strategy.

7.3 Data generation

We reviewed the various categories of capital markets data. But how does all this information come about? In this section we describe different ways in which capital markets data are produced. We discuss various ways in which new information is originated including:

- initial product creation ("issuance"),
- financial markets where products are traded,
- research services, and
- regulatory data.

Product creation

New financial products are created all the time. This can be done in different ways.

- Companies can go public or list on new exchanges, leading to new listings. After this, through corporate actions, these products can change or, for example, merge due to a corporate takeover.
- Corporations, governments or some supranational bodies can issue debt to meet their funding needs leading to new bonds which eventually mature.
- Financial services companies can create new products targeted at investors. This could range from new mutual funds or warrants for the retail market to very specific new structured investment products targeting specific private banking client segments.
- Exchanges can create for example new futures series or option series on new equities.
- Financial services firms that trade peer to peer may invent a new bespoke derivative structure, for example an investment bank coming up with a structure specifically designed for an asset management client or insurance client. To be extreme, every OTC trade, every opinion, every quote produces new financial data.

In many cases, new financial products need a document outlining the investment proposal, the prospectus. Demands on what information should be minimally in the prospectus differ by jurisdiction, but typically regulation aims to ensure potential investors get a good idea of the risks and returns associated with the investment. Purely descriptive text can be ambiguous and data standards such as XBRL can lead to a more uniform format for prospectuses and should help to ensure that all the required disclosure information is included.

Information on new financial products that are commonly available is typically included in data products from the major aggregators such as Six Financial Information, Bloomberg, WM Daten, Thomson Reuters and Interactive Data. Information on new investment products is often distributed through, for example, marketing campaigns via fund supermarkets.

Financial markets

Financial markets are where buyers and sellers meet to trade products, thereby creating pricing information and transactions. There are different ways to look at the financial markets. Typical classifications of markets include the following.

- *Primary* markets (where new issues are sold to investors) or *secondary* markets (where existing issues are traded). For example, there is a primary market for US government bonds with agents placing the bonds with investors. This concept of primary versus secondary primarily holds for securities and is not really applicable for derivatives, which are essentially bilateral contracts and are created from scratch whenever a transaction takes place.

- *Over the counter* (OTC) and *exchange* markets. The difference is that exchanges are organized places of trade and OTC markets are a set of participants trading in a peer to peer model and generating corresponding data flows. Exchanges trade standardized products and typically act as a clearinghouse or central counterparty that acts as the buyer for every seller and the seller for every buyer, thus removing the settlement risk, consolidating the pre-settlement risk across multiple trading partners into the exchange itself. By contrast, in OTC trades there is direct exposure to the credit risk of each individual counterparty. Frequently, larger stocks are listed on multiple exchanges and there would be a distinction between the exchange of *primary* listing and other listings.
- Classification by instrument type gives rise to terminology such as foreign exchange markets, commodity markets, credit markets, interest rate markets, equity markets.
- Classification by client type, for example, the wholesale market, the private banking market, the retail market.
- Classification based on *market model*, i.e., the procedure that is used to arrive at transaction prices. These can be fixings, periodic auctions, continuous auctions and so on.

The formerly clear-cut division between OTC and exchange markets is blurring. There are a number of reasons for this.

- In OTC markets we also see product standardization as product volumes increase. Industry bodies such as ISDA are driving standardization of terms and legal frameworks. In some cases, regulatory pressure has led to the requirement to report transactions to a central transaction warehouse, thereby increasing transparency.
- Exchanges are creating new products in the areas of credit and foreign exchange which were historically the domain of brokers. Exchanges are doing this as part of their business development, to remain competitive and also to make these kinds of products available to the retail investor.
- The central counterparty model of a clearing house which was always inextricably linked to an exchange is also being introduced into OTC markets. Examples include the CLS (Continuous Linked Settlement) Bank which is owned by a large number of banks and provides intraday settlement of foreign exchange transactions. Clearing services by for example LCH.Clearnet also provide Central Counterparty services for OTC products such as interest rate swaps and repos.
- Regulatory changes have eroded the monopoly positions that exchanges enjoyed in some jurisdictions, allowing trades to be executed elsewhere. Due to ongoing

consolidation in the financial services industry, often the liquidity of a certain product is concentrated in less than a dozen players. These players can form consortia themselves and pool their liquidity off exchange.

Other information

Research services such as ratings agencies continuously provide updates to their opinions, increase their coverage or refine their models, methodologies and information products with additional information. Part of the input into their credit models includes the balance sheet, income statement and cash flow statement filings to regulators which become public information. Banks have to supply information on their risk exposures and risk profiles to regulators, for example in the context of the Basel accords.

Additionally, enormous amounts of information are exchanged on a peer to peer basis between financial services firms including information on:

- open OTC trades, for example, payments/settlements on interest rate swaps,
- valuation statements for funds by fund servicing companies that provide pricing services, for example,
- margin calls, when the value of the collateral posted against an exposure has become too low and the holder of the collateral demands that the counterparty top up the collateral.

In Figure 7.7 we present some example information flows between different capital markets participants.

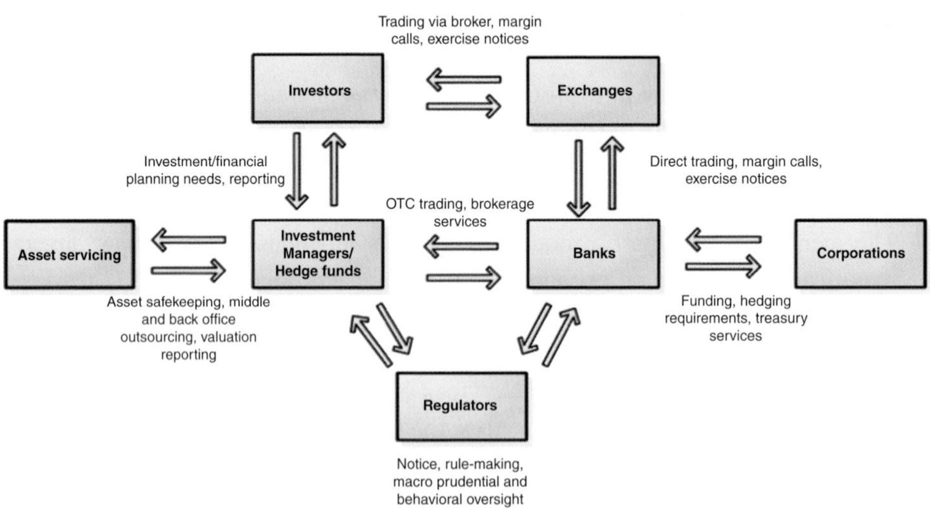

Figure 7.7 Schematic overview of information streams between some capital markets participants.

7.4 Metadata aspects

So far, we have looked at the various capital markets data categories. However, on top of the information itself, *context* in the form of data ownership, confidence level as to its accuracy and audit information is required to process and interpret this information adequately. Such business process context documentation helps users make the best use of the underlying information and includes definitions of the data, specifications as to intended usage, including what usage is permitted or prohibited, where and how the data are stored, naming conventions and what values the information may take. In short, it tells users where the data are, how they can access them and what the boundary conditions are in terms of information content as well as usage.

A common understanding of this business process context is a prerequisite to any discussion about information quality. For example, any data quality measure is only possible when there is a common understanding of how the data are to be used. Typically, asking people how to define and measure quality can cause a lot of hazy looks and tends to trigger lengthy philosophical and often heated discussions. In this section, we list a number of factors pertaining to different elements of quality. To a large extent, quality depends on the needs of the people at the receiving end. The dimensions on which data quality is measured will vary from department to department and from data type to data type.

Metadata – data about data – is just as important as the data itself. We summarize some of these metadata aspects in Figure 7.8.

Although *accuracy* and *timeliness* are always important, the *relative* importance of quality aspects differs based on the perspective – different users and roles would have different demands on both data and metadata. For example, when the credit market grew very quickly in the early 2000s, for smooth trade processing it became important to identify unambiguously the underlying reference entity in credit default swaps. This led to an identification scheme and also made CDS quote data very valuable and sought after.

After the bankruptcy of Lehman Brothers, with its legal structure composed of well over 2000 entities, legal entity data, their identification and the ability to roll them up to the ultimate parent company became critically important and were the subject of intense regulatory focus. Algorithmic trading led to an arms race in terms of ultra-low latency data.

In Figure 7.9 we show the relative importance of key data elements and metadata criteria along the lifecycle of a transaction, from trade idea to external reporting.

To illustrate these different quality considerations, we provide a number of examples below. Note that this is by no means an exhaustive list.

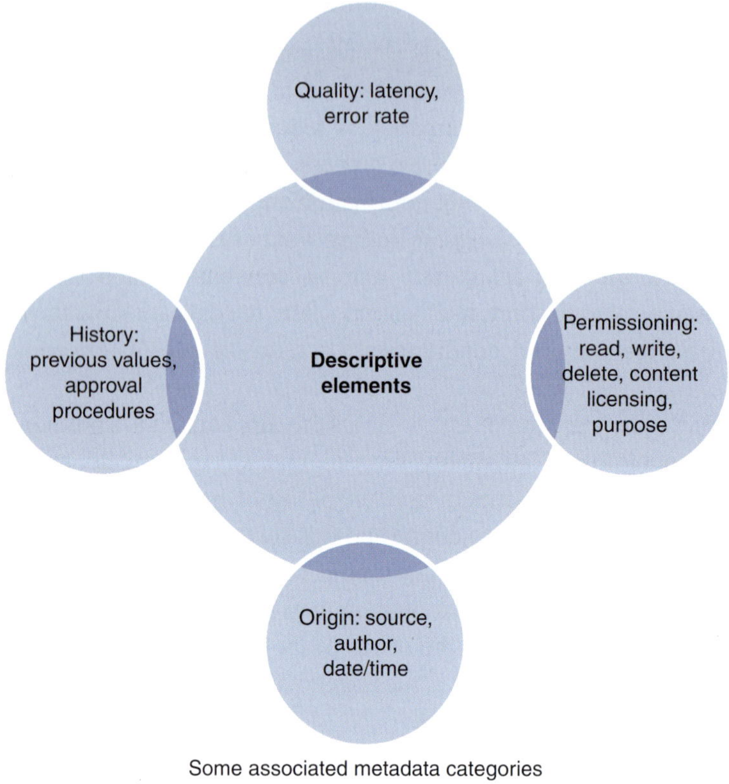

Figure 7.8 Different metadata categories providing business process context around the data themselves.

- *Speed* or the time delay aspect of data. When can you act on them? What is the decay factor in their value? Often a trading opportunity exists for a microsecond. Information must get to the right place at the right time to service a client, to set up a new account, to reply to an indication of interest and so on. In all cases an overly long delay can lead to loss of business.
- *Accuracy.* In some cases you need to be 100% sure of the accuracy of your data, for example the precise spelling or precise coding. It is necessary to know the full legal name underlying a CDS from an authoritative data source; an inaccuracy here can undermine the validity of the transaction.
- *Consistency.* Consistency refers both to data and to business processes. We should have internal consistency, for example if you have different trading books for risk purposes you want to use the same exchange rates. From a process perspective, are you treating your various clients consistently or some more favorably than others? Institutions will want to present one voice to the external world, not just from a corporate marketing perspective. Globally active, diversified financial services institutions can have many product lines that all serve the same customer.

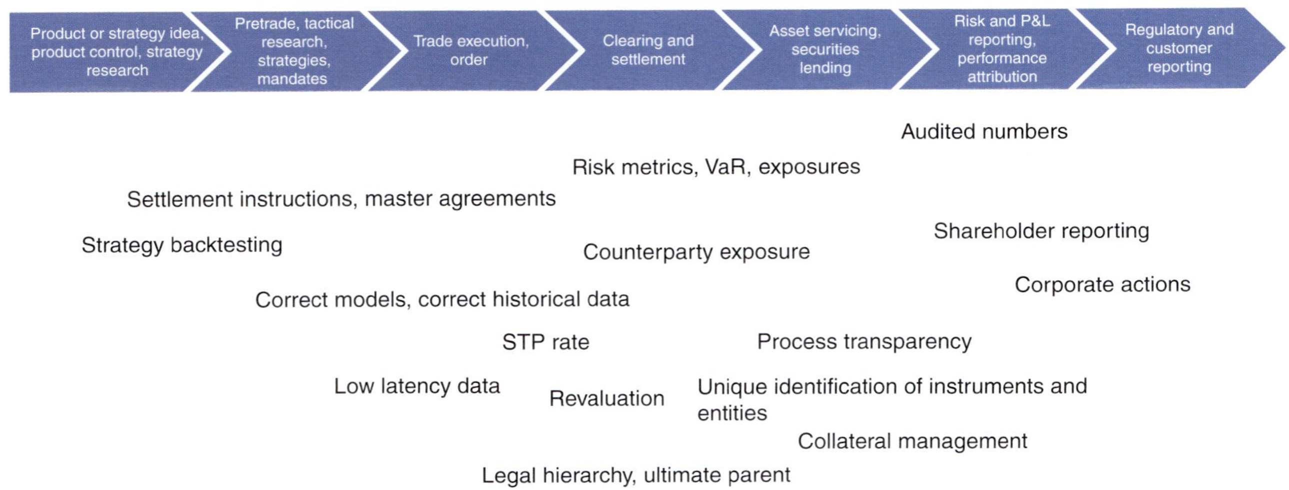

Figure 7.9 Various information and metric types along the transaction flow.

Apart from that, organizations need to be on the same page with regards to prioritization of information needs. Different ideas about what is most important lead to confusion. An important measure is *interconnectness*. Do organizations link different data sources together or do they allow "information islands" to exist? Although data are not a wasting asset, there is a maintenance cost associated with them, and links between data categories help point out stale information much more easily.

- *Authentication*. Has the source of the information confirmed the validity? Is a quote indicative or actionable? Has the trade been confirmed? Are there liability ramifications if the price, settlement instruction provided or other information prove to be wrong? Is there a "twilighting" process in place to reaffirm validity periodically?
- *Transparency*. To what extent is the complete data lineage clear? For example, what information was used at what point around the decision to take on a new client? For example, in the context of best execution reporting requirements, an institution needs to know what elements went into a price used to revaluate a position.
- *Synchronicity and order*. Is information for regulatory reporting preserved in the right order, even if a long sequence of transactions and quotes took place at the microsecond level? When the May 2010 "flash crash" was investigated, regulators had to zoom in on the complete transaction log for the affected stocks. Also, if you have to piece together a report and rely on cut-and-paste information and queries from various repositories, chances are that your information has not been obtained at the same time and may be out of synch immediately.
- *Completeness*. Is the full picture on an instrument available? For example, you can have 25 covenants in a loan or a bond, and you could have 100% accurate but nonetheless incomplete information if data are only available for 24 of these. Another example of completeness can come from legal entity data where you will want complete information on the legal structure, the guarantees and the liabilities of an entity. Missing one type of information can have major ramifications. It is much easier to start the entity data provision with the on boarding process of a customer, which maybe requires 10 to 15 fields. The compliance function will need more information on audit and documentation. In some cases, firms can have different quality level markers and there can be different levels of completeness of information depending on the function, for example in the case of information on a legal entity, it could be:
  - *research ready*, a subset of information is present,
  - *compliance ready*, a potentially different subset of information is present,
  - *trading ready*, the complete set of attributes is available and has been checked.
- *Relevance*. Is information filtered in useful ways? Users do not want to be cluttered with useless information and want the useful bits extracted out of the

torrent. Ways to address this could be to filter out the relevant updates only, to alert users only on a real update, to prioritize or filter information offered by *held securities*, by the top 25 exposures, through filtering news feeds on key words or to condense it by sentiment indicators, to filter information by credit rating, by price volatility, or by complex product exposure. A credit rating itself is also a summary measure.

- *Control and maintenance*. Macros and spreadsheets have put strong IT tools in the hands of business users but without a management framework, for example for version control, they can grow out of control and lead to a maintenance nightmare. Also, too much power and operational risk can be concentrated in whoever understands the 80MB spreadsheet.
- *Accessibility*. Do data reside on local desktops? Can people find the data they require within a reasonable time frame? Can everybody who needs the data access them? Are access permissions, Chinese walls and separations of duties in line with content licensing terms, conflict of interest rules (for example between research and advisory functions, between corporate finance and sales and trading) and data protection legislation. In particular, data protection and privacy laws should be areas of attention when a firm outsources or off-shores the collection or processing of customer information. Are the data (or access to them) leaving the building, leaving the country or leaving the organization? Each of these acts may or may not be allowed. How easily does the information lend itself to automated retrieval and processing? Often, information may be residing *somewhere* in the institution, but there are often no yellow pages in place directing you to the relevant department. Which people have *backdoors* allowing them direct and privileged access to data stores, for example database administrators, IT staff and management. If the number of people with backdoor access is fairly large, the controls put in place for other users will become meaningless. Another aspect of accessibility is availability. Can you always access the data? Is there an uptime target such as four nines or five nines?
- *Usage restrictions*. Is the organization abiding by the content licensing terms in the contracts and by for example Chinese wall policies stipulated by regulators. Are privacy laws and policies being obeyed?
- *Security*. This is important both in terms of abiding by content licensing terms (operational risk) and also as the flipside of accessibility. Is unauthorized access prevented? How vulnerable are the (customer) data to unauthorized external access? Is information being encrypted? Are the standards of the institution with regard to security level of information (from public to confidential and shades in between) being adhered to?
- *Service and support* around data. If you want to change something, how flexible is the process and what is the turnaround time? If you have an issue

with the data (query, error or enhancement) what is the turnaround time in addressing such issues from your vendor or from your internal data management team? It can be difficult to incorporate *new* data elements that do not fit an existing data schema into a firm's information infrastructure. What are the effects, both on the loss of time to market as well as on the robustness of the information infrastructure, when the data framework is "tweaked" to accommodate new data? Are the new data modified to shoehorn them into the existing schema?

Whether a user relies on the data for revenue generating capabilities, cost controls or to assure compliance can lead to different perspectives on quality. Various subsets of these quality criteria are more relevant depending on the type of data (pricing or static, equity or commodity, counterparty or corporate action). For example, information quality aspects associated with the quality of order execution would include measures on effective spread, rate of price movement (either in absolute terms or versus a benchmark), fill rate of the order and speed in turning it around. There is no single right answer to the quality question. When it comes to defining standard data services, different users will want to embed different combinations of the factors above in their service level agreements and assign them different weights.

To take the concept of metadata one step further, one could envision "self-aware" data elements which would entail associating information both with a set of metadata elements and with related information. This could turn pieces of information into Lego bricks that could easily and effortlessly be combined into different reports and larger constructs. A dataset on security master agreements should always have hooks into, for example, the issuer and other associated entities (e.g., guarantor, underwriter). It should hook into corporate events and should contain metadata such as "where and by whom it can be traded." Every financial services firm would want more complex, more flexible and more frequent ways to view, analyze, manipulate and adapt their business data. This tagging of information with links into other categories can vastly increase the self-correcting capability of the data infrastructure. So far, in identification schemes such as ISIN codes, check digits are included, but the idea could be applied much more broadly to make logical inferences on data gaps and suggest candidate new values.

Various XML standards are pushing information in this direction through adding a descriptive layer and tagging data with metadata elements. In the end this could turn data into information nuggets that know the answer to questions such as: Who am I? Where do I come from? How do I relate to others? Where can I go? Carrying the information and context to answer these questions reliably would truly transform data into information assets.

7.5 Conclusions and challenges

Not only is the volume of capital markets data increasing, more is going to be demanded of it. Shareholders of banks, customers of investment managers and regulators policing the market for both financial stability and consumer protection all want more dependable and timelier information on a more detailed level than ever before.

While reporting needs go up, financial markets themselves are continuously changing as well. New asset classes continue to be created and, while standards steadily chip away at information disorder, it is still often very difficult to share financial market data uniquely and unambiguously across market participants and systems.

Finally, the number of sources of capital markets data also continues to grow rapidly. New trading venues appear that publish data; data vendors in new geographies or specializing in new asset classes are created. Apart from the traditional aggregators and analyst firms, unstructured data from social media and newsfeeds are also gleaned for news and parsed to gauge market sentiment. Perhaps a useful analogy to conclude the chapter is to draw a parallel between financial markets information infrastructure and the energy markets.

The energy space has been moving to a "smart grid" which is about detailing and tracking the consumption of energy, but also allowing previously passive consumers to become contributors to the grid via decentrally generated energy.

Something similar has been going on in the data space. In the past you had a few large central places where trade data were generated – exchanges. From there on it was centrally pushed out. Different developments over the last decades have contributed to a move in the direction of a data "grid."

- OTC products were traded peer to peer, so new products were essentially created decentrally.
- Part of the order flow which used to go to the exchanges is crossed internally, by systematic internalizers turning banks into mini exchanges.
- Exchanges have been consolidating but new trading venues have been set up as well.
- Finally, more and more information is shared peer to peer between financial service firms, leading to a complex web of interactions and information flows. This is due both to outsourcing of middle and back office functions from one financial services firm to another and to a recognition that it makes sense to collaborate on standards and information sharing.

Collaboration and sharing can make sense for the types of data that essentially represent the plumbing and the infrastructure of markets. These data need to be correct and everyone individually can spend a great deal of time and money assuring such accuracy, but such data are generally of a non-competitive nature and offer little competitive advantage. For such data it may make commercial sense, both individually and collectively, to create a single master source as a more efficient solution, lowering costs for all parties concerned while also creating a more robust information foundation for the capital markets.

In information management, there is always the question of data standards. Whatever is commonly used across different user pools could be standardized. Whatever is proprietary and part of the "special sauce" of a business, a strategy, you want to protect and will retain bespoke elements. Even in this last case, however, standards help to put in place the foundation on which you can more solidly construct your proprietary information. Perhaps an analogy with musical scales is a useful one. Music is standardized onto musical scales, but these scales provide the foundation for your own unique improvisation.

8

Capital markets reference data

Marc Alvarez

To seasoned veterans of the market data world, the phrase "capital markets reference data" at first sight may appear somewhat contradictory. Indeed, the notion of the world of "capital markets" brings to mind bright flashing screens displaying prices in real time as transactions occur on the world's securities exchanges or over the counter between firms and investors. In fact, to most professionals in the space, the concept of "reference data" comes as an afterthought (and not usually a pleasant one at that).

Such an assumption is entirely misplaced. Not only do "reference data" contribute significant value to users of capital markets information services, they represent mandatory content required for those services to work. Furthermore, in today's rapidly globalizing economy, it is in the realm of reference data that a significant portion of investment and research is taking place as this is where the opportunities are being uncovered via the development of new trading strategies and improved operations performance.

In *Market Data Explained: A Practical Guide to Global Capital Markets Information* (Alvarez, 2007a), the general market and reference data landscape is laid out as a logical framework aimed at providing an introductory level of navigation through some relatively complex content. It serves to demonstrate that with a few well thought out fundamental rules, the universe of data content (or at least a significant portion of it) deployed in the capital markets on a daily basis can be classified in a single, coherent structure. In turn this approach allows for applications to share a common source of content, such as a database, thereby benefiting the broader organization by ensuring consistency and eliminating redundancy.

This chapter looks in detail at a subset of the "reference data" universe and focuses on a set of fundamental rules and associated content needed to make "capital markets data" a consistent service to today's enterprise. It identifies those elements

Handbook of Financial Data and Risk Information, ed. Margarita S. Brose, Mark D. Flood, Dilip Krishna and Bill Nichols. Published by Cambridge University Press. © Cambridge University Press 2014.

within the reference data universe that are effectively mandatory and must be considered as part of the scope of all applications that consume capital markets data.

8.1 Reference data defined

For several decades, the capital markets industry has survived quite well with the following definition of "reference data":

> Reference Data encompass all the content not reflected in real-time data.

The obvious weakness of this definition is that it is based on exclusion and negation. In fact, it really does not define much at all, since it lacks any sort of precision.

It is this lack of precision that can, and does, create some serious pitfalls in the design and functionality of business applications. Symptoms of the disease include duplication of content from multiple sources, leading to inconsistency, and even downright conflicts, in actual values used by the business. In a time of less

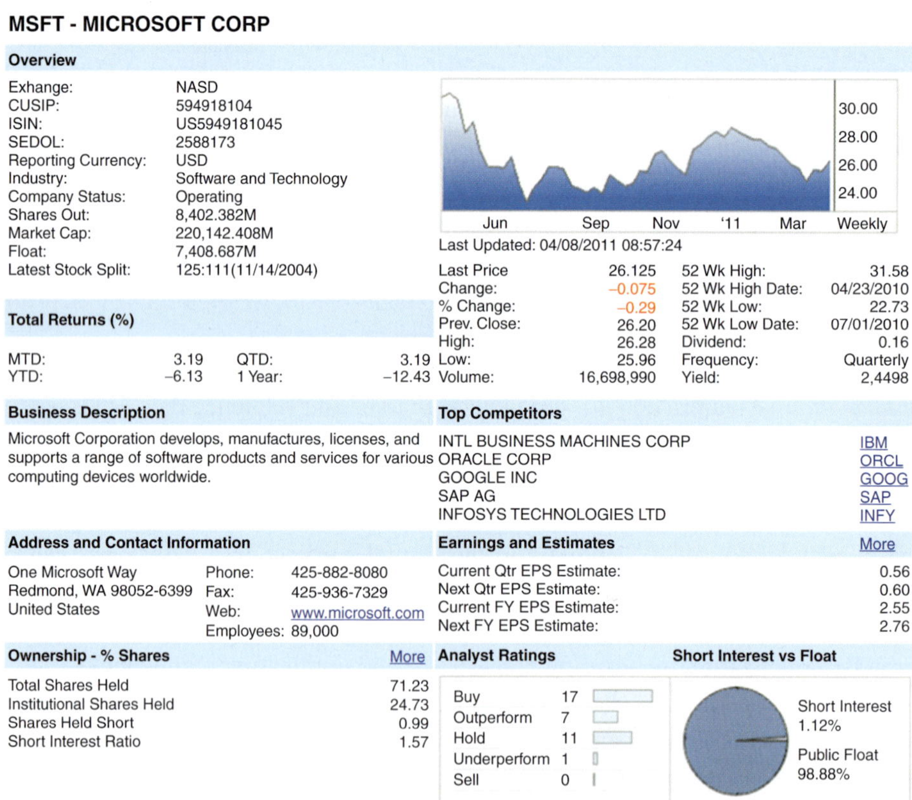

Figure 8.1 Desktop trading application (courtesy of Interactive Data Corporation, Market-Q).

regulatory and shareholder oversight, such risks may have been manageable. In today's highly globalized and data optimized industry, however, such situations can have seriously damaging consequences.

Futhermore, this definition by exclusion is just plain wrong. Take a look at Figure 8.1.

This screen shot provides a basic instrument record sourced from Interactive Data's popular *Market-Q* desktop investment research and trading application.[1] The screen consists of a collection of both reference and real-time data fields. The overview section and the pricing pane show a pretty standard application of blending the two data types. According to the definition above, the reference and real-time data fields can be classified as shown in the table.

Data attribute	Definition	Update frequency
Reference data		
Ticker symbol	A textual identifier issued by an exchange to identify uniquely a security. In this case, the ticker MSFT refers to Microsoft Corporation common shares.	At setup, updated for changes as and when announced.
CUSIP	An industry coding scheme used to identify a security uniquely, independently from any on exchange or other trading venue. The term is an acronym of Committee on Uniform Security Identification Procedures. The CUSIP distribution system is owned by the American Bankers Association and is operated by Standard and Poor's.	At setup, updated for changes as and when announced.
ISIN	An alphanumeric security identification standard (ISO 6166) issued by the International Standards Organization (ISO). The term is an acronym of International Security Identification Number.	At setup, updated for changes as and when announced.
SEDOL	A security identification issued by the London Stock Exchange. The term is an acronym of Stock Exchange Daily Official Listing, and is unique to the combination of a security and the market in which it is quoted.	At setup, updated for changes as and when announced.

[1] Courtesy of Interactive Data Corporation.

Data attribute	Definition	Update frequency
Reporting currency	The national currency unit (e.g., USD, Euro, CAD etc.) associated with the pricing values contained in the display.	At setup, updated for changes as and when announced.
Industry classification	Identification of the industrial grouping to which the issuing organization is assigned. There are multiple industry classification schemes in use such as ISIC (United Nations), NAICS (USA, Canada, Mexico statistics authorities) and NACE.	At setup, updates in response to reclassification arising from corporate actions and/or other business events.
Company status	An indication of the activity of the issuer in the market. Examples include operating, defunct, bankruptcy etc.	As soon as new value is quoted in the market.
Shares outstanding	The nominal number of the securities issued to the market.	As soon as new value is quoted in the market.
Market capitalization	In calculated value that represents the total value of the tradable shares of a publicly traded company; it is equal to the share price times the number of shares outstanding.	Typically daily at market open. However, in theory this value changes with every new price quoted in the market.
Float	Refers to the number of shares of a public company's stock held in public hands.	At setup, typically updated monthly or quarterly.
Latest stock split	The most recent calendar date on which a share split affecting the security took effect.	Irregular, in response to corporate actions announcements.
Real-time data		
Last price	The previous value at which the security traded in the market.	Following every price quote.
% Change	The net change, expressed in percentage terms.	Following every price quote.
Previous close	The value recorded at the end of the previous day for the security.	Once per day.
High	The maximum value recorded for the security during the trading day.	As a new maximum value is realized during the course of the trading day.
Low	The minimum value recorded for the security during the trading day.	As a new minimum value is realized during the course of the trading day.
Volume	The total number of individual securities transacted in the market during a given trading day.	Following every price quote.

As the table shows, this screen contains both "reference data" and "real-time data" elements in the context of what, by general consensus, would be deemed a "real-time" application. Indeed, any change to any of the "reference data" attributes (a change to the dividend amount say) would very likely result in a significant change to the "real-time" attributes. For example, such a change could occur as a result of restructuring of the underlying instrument which, in turn, could trigger the issuing of a new CUSIP and ISIN. As a result, there is a tight relationship between "reference" and "real-time" data.

This is an important characteristic of reference data – they are not distinguished based on frequency of update but rather the use to which the consuming application puts the content. To a lesser extent this is also true of pricing values (the "real-time" data attributes presented above) but it is central to the universe of reference data.

An alternative definition put forward for the term "reference data" is the following (Alvarez, 2007a, p. 21):

> Reference Data embody a collection of facts that describe an issuer, a financial instrument or a market quote and their associated interrelationships.

This definition is provided as part of a broader logical framework that looks at the world of capital markets data content. When looked at in the context of the application illustrated above, it refers to a very precise subset of reference data which, in turn, are augmented by additional business data content.

As such, this approach also falls short of the goal of a commonly accepted definition because it relies on a specific context. It is closer to providing a meaningful frame of reference but it misses the wider universe.

Somewhere between these two extremes lies a common, easily understood, definition. First, start with the phrase "reference data" itself. As it suggests, this captures the characteristic that the content "refers to" or "has references to" something else. In technology terms this suggests that the data are "referential" – that is they provide a means of linking from one item to another.

A key element of any definition, therefore, needs to incorporate this concept of linkage. For the capital markets, this characteristic means that "reference data" are at the core of achieving unique identification. As the second definition suggests, the key entities that need to be identified uniquely and precisely are organizations, instruments and individual listings of securities. For technology applications, this suggests that reference data include at a minimum the *primary*[2] and *foreign*[3] keys to the universe of content.

[2] In relational database design, a primary key is a value that can uniquely identify a single row in a table, see `http://en.wikipedia.org/wiki/Primary_key`

[3] In relational databases a foreign key is a field in a relational table that matches a candidate key of another table. Typically, foreign keys can be used to cross reference tables, see `http://en.wikipedia.org/wiki/Foreign_key`

Unique identification in and of itself is useful, but it falls short as a self-contained definition. Since different users of the content may use different means of recording such unique definitions, it is possible, and indeed occurs every day in the capital markets, that two separate users may use the same value (a number, for example) to identify two completely different items (such as a security).

The latter case is all too familiar to users of reference data. It manifests itself in the array of coding schemes in use for a variety of data content including (but most certainly not limited to):

- Issuing organizations,
- Securities,
- Market listings,
- Markets,
- Asset classification schemes, and
- Ratings schemes.

One of the biggest frustrations of working with reference data is this diversity and complexity. While past industry efforts have attempted to standardize unique identification (for example, the ISIN code for securities), no final resolution has yet been defined or adopted by the industry for all data items (although some high profile efforts are underway such as the Legal Entity Identifier (LEI) initiative). Worse, not only does the diversity persist, attempts to deploy such standards have historically been done in a piecemeal fashion – no comprehensive program has yet been attempted to identify the universe of content and deploy agreed standards as a cohesive set.

Ironically, it is this aspect of reference data that is fundamental to all market participants. The interchange of content between parties requires a common means of identification. As a result, each party to a financial transaction must identify, that is *refer to*, the means they are using to identify the particular item of exchange (a security in this example). Failure to do so risks a mismatch in the transaction, requiring an expensive manual intervention in order to correct the error. Given the enormous volume of transactions carried out in the capital markets every day, this represents a significant amount of expenditure which is passed on to investors via service fees and other costs.

To minimize this cost, firms spend significant amounts of time and other costly resources to perform confirmation and reconciliation of transactions. The obvious solution would be to ensure that all firms refer to a single list of data attributes. While this is a simple, common sense concept, the reality is that no such standard exists or is likely to exist any time soon.

Therefore users of reference data content need to focus attention on precise identification of securities, their issuers and associated trades. In addition, in order to support this objective, the definition further requires a corresponding subset of qualifying attributes to identify precisely the entity of interest. To this end, this chapter works on the basis of the following definition of "reference data":

> Within the global capital markets context, "reference data" refers to the subset of data attributes used to identify and describe uniquely an organization, a security or a listing on a market.

This definition points to a critical working hypothesis for capital markets applications – by focusing on a subset of core reference data attributes and their relationships, the balance of the information set can be treated largely as descriptive data content associated with one organization, instrument or listing item. Making these data available for use by multiple business applications introduces the benefits of simplicity of navigation, thereby reducing the time required to develop and deploy new business applications to support growth and better manage costs.

8.2 A working taxonomy

The need for easy to follow navigation is a critical factor as it underlies the ability of a firm to reuse the same content across multiple business applications. By making use of the consistent subset of reference data across multiple applications and/or lines of business, firms can achieve significant benefits including:

- **Operating efficiencies** – eliminating costly reconciliation and confirmation while at the same time broadening the deployment of automation,
- **Error reduction** – reducing the number of failed trades and other mismatches and the associated cost and complexity,
- **Faster time to market** – improving the availability of new applications to service business functions by supporting reuse of data content as an infrastructural asset,
- **Regulatory and audit compliance** – consistent terminology and nomenclature applied across a firm's operations allow for improved coherence in reporting and accounting functions.

It is this fundamental subset of the reference data universe that is essential to deploying reference data as a strategic asset across the enterprise. This observation suggests that the reference data universe can be divided into two logical categories:

- **Core reference data** – the minimum mandatory elements required to identify uniquely an organization, the instruments it issues and the markets where the instrument is quoted. Within this definition core reference data are made up of two fundamental subcomponents:

- attributes that identify uniquely and describe sufficiently individual organizations, instruments and market listings,
- commonly accepted industry identification schemes and coding standards that correspond to each organization, instrument or market listing,
- **Business data content** – the universe of attributes that collectively describe an organization, instrument or market listing for the purposes of individual business functions. Common examples include:
  - Security Master – the data that describe a security in detail along with associated terms and conditions,
  - Corporate Actions – data that describe events affecting a security either as directly announced by the issuer or at arm's length,
  - Pricing – data that record the value of an instrument as quoted in a market (such as an exchange) or by other means (such as evaluations) at a given time.

The key difference between these two subsets is really a question of scale. As the working definition implies, the core reference data universe is typically a relatively small universe of content, while business data content represents the whole universe of the capital markets – a domain of content that is known to be complex and very diverse.

Secondly, it is vital to recognize that business data content relies on the logic and structure of core reference data. As the following sections show, without a solid structure, navigating the universe of business data content can be very frustrating and historically has led to duplication of data within firms in a bid to streamline access to data content – a situation which has become increasingly evident and alarming to the industry.

Since duplication of common data (quite frequently from different sources within a single firm) can easily lead to different values, mismatches can occur within a firm, and are also a risk when exchanging data – via a trade confirmation for example – with counterparties. As a result, data need to be verified between participants in a transaction and, where mismatches occur, reconciliation must be conducted. This is a known to be an expensive burden on the industry.

The universe of business data varies widely and includes categories of content such as the following (Alvarez, 2007a).

- **Securities descriptive data** – attributes that describe the security and its features in detail. These data are typically relatively static and are not updated with predictable frequency.
- **Trading data** – data that capture information associated with the purchase and sale of the security between counterparties. For some security types, such as equities, futures and options, trading data can have a very high frequency on the

order of thousands of transactions per day. For other security types, such as bonds, trading information can be very sparse as the instrument is seldom exchanged.

- **Ratings** – assessments of the risk associated with owning the security as published by third party agencies. Ratings are issued for both individual securities as well as their issuers.
- **Relations** – linkages between securities, issuers and listings are common. Examples include the ownership hierarchy between an issuer and its subsidiary and the list of instruments used in calculating an index.
- **Terms and conditions** – detailed information on the rights and responsibilities associated with a particular security
- **Corporate fundamental data** – data that record the business performance of an organization over time. Examples include balance sheets, income statements and cash flow statements.

Indeed, this list of information categories can be whatever a firm needs to meet its business requirements. It is the identification and navigation provided by the core reference data that are fundamental to successful application of the content at an enterprise level.

The key goal is to achieve and maintain a comprehensive dataset that is aligned across all market participants. Not only does this serve to minimize the risks of mismatches, it also supports a more streamlined flow of data within an individual organization, improving its overall operating efficiency.

8.3 Core reference data

Applying reference data successfully to support business applications requires an understanding of two key dimensions:

- the relationships that are implied by the use of reference data within the capital markets, and
- the content that makes up the core reference dataset.

The combination of relationships and content form the backbone of a navigation framework through the complex domain of reference data. More importantly, as the following sections demonstrate, by establishing a coherent and managed core of reference data, this navigation can be tailored to meet the needs and working methods of any firm and avoids the difficulties of imposing a single fixed view of capital markets content across a firm's operations. Getting the core reference data right, then, is the key to unlocking the power of consistent reference data across the enterprise.

8.3.1 Understanding relationships

As mentioned earlier, reference data are by definition relational in nature. This is especially true of the core reference data subset. It follows, then, that understanding the nature of the relationships associated with this subset is fundamental to developing best practices to apply to business functions.

No reference data relationship is more fundamental than the statement:

> Organizations can issue one or more instruments which are traded on one or more markets by participants in those markets.

Presented as a relational diagram, this takes the form shown in Figure 8.2.

As Figure 8.2 illustrates, the universe of core reference data can be physically separated into separate groups. In data modeling terms, each of these groups is

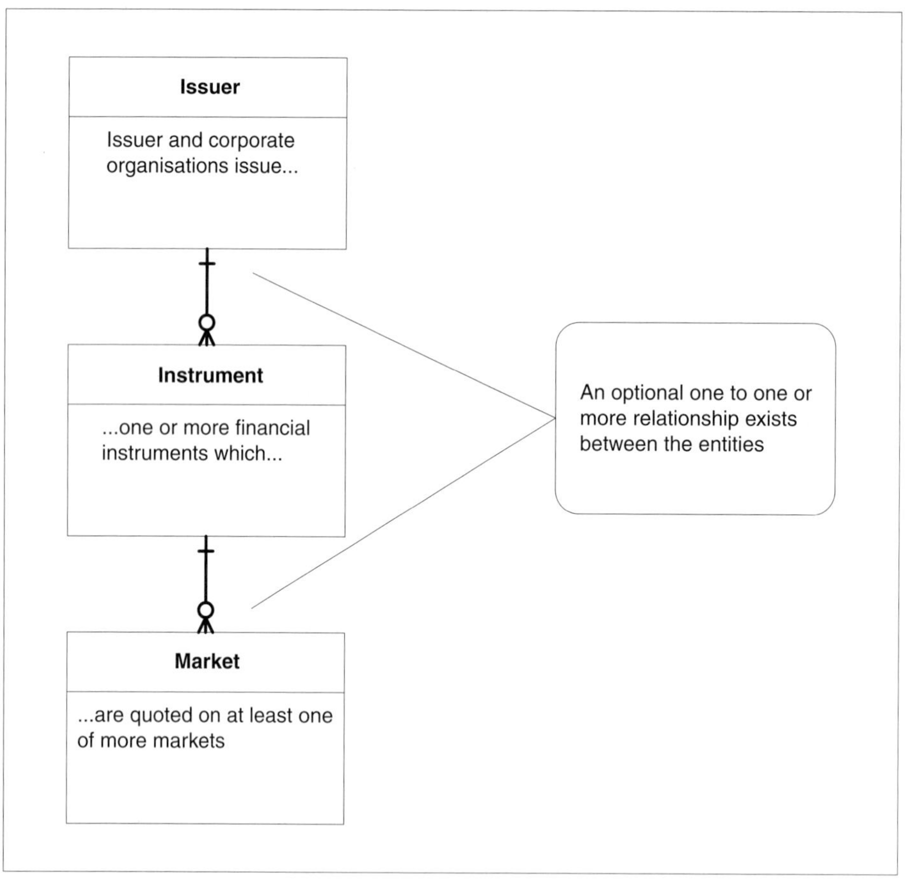

Figure 8.2 Core reference data relationship (Alvarez, 2007a, p. 23).

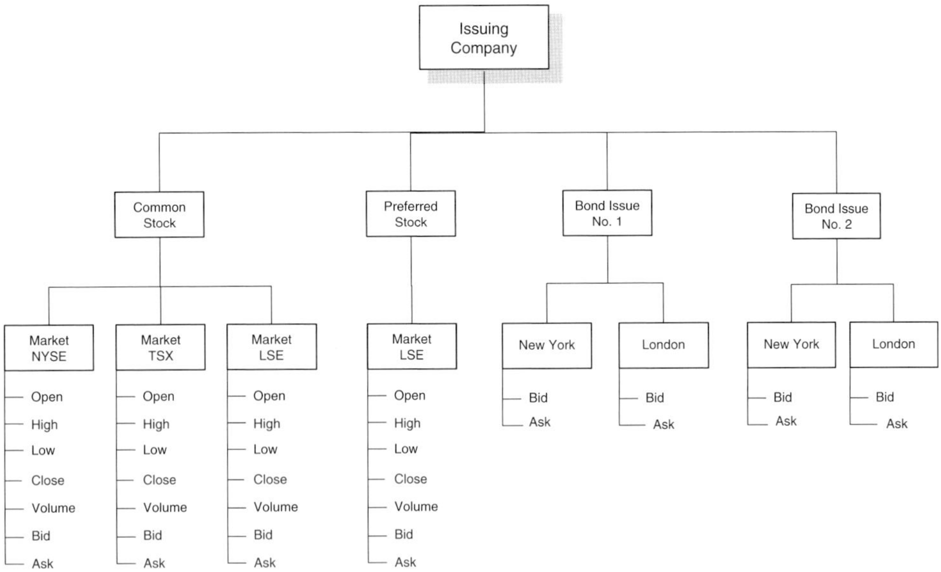

Figure 8.3 Example of a hierarchical relationship.

referred to as an "entity." In database terms, each entity can be thought of as a table, with each table consisting of a series of fields (or "attributes") unique to that table.

Each of the Issuer, Instrument, and Market entities is linked by a defined relationship (identified by the crow's foot symbol). In this case, the relationship between the entities is hierarchical as there can be zero or more records for each record higher in the relationship. An example is shown in Figure 8.3.

Separating out the content in this way is fundamental to understanding and navigating the content. Not only does it make it very easy to identify where an attribute relates to the content, it also serves as a blueprint to building out physical implementations to store and manage this content. The important point to recognize is that the logical representation (the data model) does not need to be the same as the physical implementation (the database).

In fact, one of the most important points to recognize in working with reference data is that there are many different applications for the content. Therefore, multiple varying physical implementations can and must be supported across the enterprise in order to support active business functions. Examples of differences include:

- local database copies tailored to meet specific business functions,
- XML data formats used to support web applications,
- Excel spreadsheets used to analyze content.

While each physical implementation can be expected to vary in order to meet business requirements, the goal should be to make sure each shares the same underlying logical model. In practical terms this means simply maintaining a linkage between the name of the entity/attribute combination of the logical model (analogous to a table/field combination in a database) and the individual data item used in any particular application (a process typically referred to as "mapping").

Enforcing this discipline serves to maintain a record of what data are used where on a consistent basis, since all items delivered to support a specific application can be reconciled back to a single logical item name. Of greater benefit, though, is that this represents a relatively small set of data which can be easily managed yet ensures consistent use of content across the organization.

Of course many business applications derive or transform source data content in order to meet specific needs. This can be as simple as changing a name (for example renaming "bid price" to "bid" to support easier display) or performing arithmetic or other operations. This need to support transformation represents one of the thorniest problems in the reference data business. Not only are reference data required across all applications, but it can be an absolutely nightmare to reconcile the underlying data with data to which a firm subscribes. However, by establishing a basic logical inventory of the content and maintaining the mapping between the source and the application, a fundamental level of accounting can be achieved. Such accounting opens up significant benefits as it allows for identification of commonly used data items which, in turn, can start to serve as the basis for eliminating duplicate data sources (which as discussed above often carry different values) and improving consistency across the firm.

Unfortunately, no single such blueprint (or data model) is commonly applied in the industry today. Consequently, each firm attempts to implement their own view of capital markets content. While this usually proves adequate within the firm, it tends to be at the root of problems that occur when data interchange takes place between firms, as it is common for two counterparties to make reference to individual data items with different underlying interpretation. In turn this may cause settlement and other operational risk, affecting the operating efficiency of both firms.

In most cases the source to application mapping is buried in software code, configuration files or database tables. These tend to be very difficult to read and next to impossible to govern in a consistent manner. Consequently, many firms simply give up trying to compile a definitive mapping capability or, worse, simply implement a new version to support a new application. Ultimately that leads to further diversity in data content used across the organization even though the content is already in use elsewhere.

It is all too easy to conclude that such imperfections in the interchange of data content can be catastrophic. The reality, however, is that the additional cost these inefficiencies imply are readily managed by capital market firms, and efforts are underway via regulatory and standards agencies to address some of the more egregious failures. Such efforts may not lead to a perfect solution that can be applied to all businesses, but they are an indication of the industry's collective efforts to minimize the costs and risks associated with the imprecision that comes from working with inconsistent reference data.

8.3.2 Core reference data content

The content that makes up each of the core reference data entities introduced earlier is an interesting topic. While the basic structure is well understood and widely used across the industry, there is considerable diversity in selecting precisely what data are stored at each level of the Issuer-to- Instrument-to-Market hierarchy. In turn this situation can be further complicated by the varying content needs of each individual consuming application.

It is all too easy to conclude that, as a result, significant portions of the content (if not all of it) need to be replicated for each consuming application. However, the opposite is actually the case – by implementing a consistent core reference data subset and enforcing its use across the organization, much of the diversity and duplication of content can successfully be supported. In turn this opens up the possibility of rationalizing business data content over time, leading to reductions in redundancy and related inefficiencies.

There are three principal components to consider when setting up the core reference data:

- **Primary keys** – the choice of identifier codes that uniquely identify each individual issuer, instrument and market listing,
- **Basic descriptive data** – a small selection of additional descriptive content that fully qualifies a specific record,
- **Cross reference codes** – support for the ability to link commonly used industry identifiers (such as exchange ticker symbols, ISIN codes and SEDOL codes).

A key success factor in managing reference data is to view each of these components as mandatory attributes. Unless all three components are adequately satisfied, the identification of individual records of interest should be considered suspect. Figure 8.4 presents a conceptual design for the relationship between these content sets and how they can be applied to support capital markets business requirements.

Primary keys

From the perspective of supporting reference data requirements as an internal service, this characteristic raises some interesting issues. By far the most important

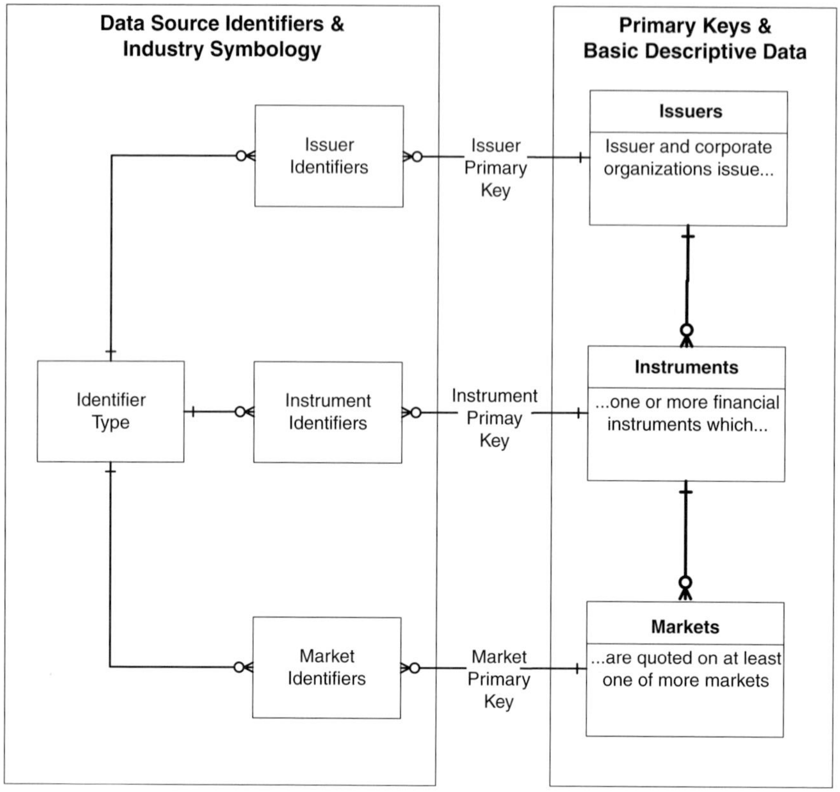

Figure 8.4 Core reference data.

is the question of what values to use for primary keys[4] to identify records at each of the issuer, instrument, and market levels. Two options most commonly considered are:

- make use of industry standard coding schemes,
- deploy an internally defined, system generated set of values.

There are arguments for and against both approaches. At the level of individual business applications, users are likely most familiar with various industry code types such as ticker symbols, CUSIP, SEDOL or ISIN codes. However, at the enterprise level, such coding schemes are known to fail to support the global universe of data content – conflicts and mismatches are common.

[4] In relational database design, a primary key is a value used to identify uniquely each row in a table. It comprises a single column or a set of columns, see `http://en.wikipedia.org/wiki/Primary_key`

Ensuring uniqueness of records at the issuer, instrument and market levels is a fundamental requirement for using reference and market data. In fact, poor indexing of content leads to record and attribute duplication which can easily result in business problems requiring confirmation, reconciliation and subsequent correction.

The commonly adopted approach to addressing these problems is to apply a primary key value to each of the Organization, Instrument and Market entities. Alternatively, common use of industry provided codes can also be adopted.

There are costs and benefits to both approaches. Implementing a primary key requires additional effort and management. A system needs to be put into place that creates and manages primary key values through the lifecycle of the underlying data records. On the plus side, this lets the data user manage the identification process explicitly, avoiding reliance on externally supplied values.

Using industry standard coding schemes offers the benefit of simplicity of implementation, but relinquishes control to external agents. Known problems with this approach include the risk of time latency (changes may not be available prior to required usage) as well as poor linkage to other coding schemes.

A further distinction between the two approaches is that when industry code values (such as ISIN etc.) are used, any changes that are received must be detected and fanned out across the entire database and distributed to consuming applications. Depending on the volume and complexity of data in the application, this can result in very high management overheads as changes need to be cascaded across a wide universe. In contrast, the use of a primary key allows for changes to be made in one place as a simple cross reference (see below).

For a single user application such as a spreadsheet, this is probably not a very big issue. For an enterprise class database though, the difference in operating overheads and associated risk is significant. As such, use of a primary key is likely the better approach as it supports scaling up a single use case to the enterprise. The cost of implementing and managing an internal primary key is easily outweighed by the flexibility and reuse of data content from a single source across the organization.

Core descriptive data

By themselves primary key values are of limited use. In order to make the core reference data content useful, a core set of descriptive data is called for in order to qualify each individual record fully. The choice of data content is relatively arbitrary, but should include the basic information to support the definition of a primary key, thereby ensuring the uniqueness required by financial applications.

The table lays out some of the basic attributes that are typically considered within the core descriptive dataset. These are intended as a general recommendation only, as they are known to provide the basic qualification and definition needed to support the application of primary keys to reference data.

Issuer descriptive data items	Instrument descriptive data items	Market descriptive data items
• Issuer primary name • Issuer type • Industry classification • Organization status (e.g., operating, defunct, bankrupt etc.) • Country of headquarters	• Instrument primary name • Instrument type • Issue date • Issue status • Trading status • Primary market of trading • Currency of issue	• Market identification/code • Currency code • Trading status • Primary name • Listing date

The key success factor here is to avoid combining those attributes required for unique identification with the broader universe of business data content. This approach eliminates the need to duplicate the core reference data content as part of each data application. Rather, by using a single shared source for the core reference data, all applications benefit from the same identification and definition of the issuer, instrument or market listing of interest. Keeping the core reference dataset as small as possible makes this a manageable task without involving the need to maintain and support the broader universe of content.

Cross referencing

Probably the source of the most frustration and failure in the use of reference data is the area of cross referencing. As no single identification scheme exists to identify individual issuers, instruments and market listings, the capital markets industry must support multiple coding schemes across their various lines of business. In turn, data suppliers have introduced their own, proprietary coding schemes which have achieved varying levels of success and adoption.

In order for any application to make use of business data content it must first define the means by which to identify individual records of interest. Furthermore, in order to enrich data content by appending it to other data content (by aligning a pricing table to related corporate action announcements for example) the records must be reconciled and confirmed to relate to the same record. Therefore, by definition all applications must include the notion of matching underlying keys to data content in order to integrate a final dataset.

A good approach to supporting this need is to segregate all cross reference records as part of the core reference data subtype as illustrated in Figure 8.3 above. The first point to note is that cross reference codes apply at all three of the issuer, instrument and market levels, as shown in the table below. This is an indicative list only and is not meant to be a complete definition.

Issuer code types	Instrument code types	Market level code types
• Dunn & Bradstreet DUNS numbers • CABRE numbers • Legal Entity Identifier (LEI – currently being defined)	• CUSIP • ISIN • VALOREN	• Exchange ticker symbols • SEDOL • QUICK • Reuters Instrument Code (RIC)

Simply put, cross referencing represents a necessary tax to users of market and reference data content. It provides the means by which the user's frame of reference (using a ticker symbol for example as part of a trading application) can be linked to other applications in the firm (such as ISIN codes as part of the clearing and settlement process). By extension it represents the means by which data can be requested from various sources (whether an internal database or an external vendor) and in turn matched in order to be integrated into a single view of content required to support individual applications.

That may sound a little silly and inefficient, but the fact is that this process is common to all participants in the capital markets industry. While efforts at standardization have been made in the past (to varying degrees of success) no single global set of identifier standards is currently in place. Consequently, the need for cross referencing functionality will be with the industry for some time to come.

As Figure 8.3 shows, this functionality can be easily supported within the core reference data concept. By implementing a primary key strategy each of these code types can then be linked to any other, whether at the same level of the hierarchy or not. An example is shown in Figure 8.5.

Figure 8.5 shows that this functionality allows for seemingly different datasets to be merged into a single, consolidated record with content from both sets. Note here that the ISIN code is used by both sets of data as a means of identifying an individual security. The outcome is structured to include the primary key of the data service, thereby allowing the data in the user portfolio to be referenced by the internal data service (as long as there is a match on the ISIN codes).

In database terms, this logical linkage is referred to as a "join." Where the values of the primary keys match, the records can be linked and in turn accessed (requiring what is called a "select"). It is vital to note here the intended use of the primary key value. It is included in this example to provide a linkage between the cross reference codes contained in the two datasets. This allows the access to the business data content to be dynamic – not only can the cross reference codes change without impacting the business data content, but the primary key values themselves can be changed across the database if necessary.

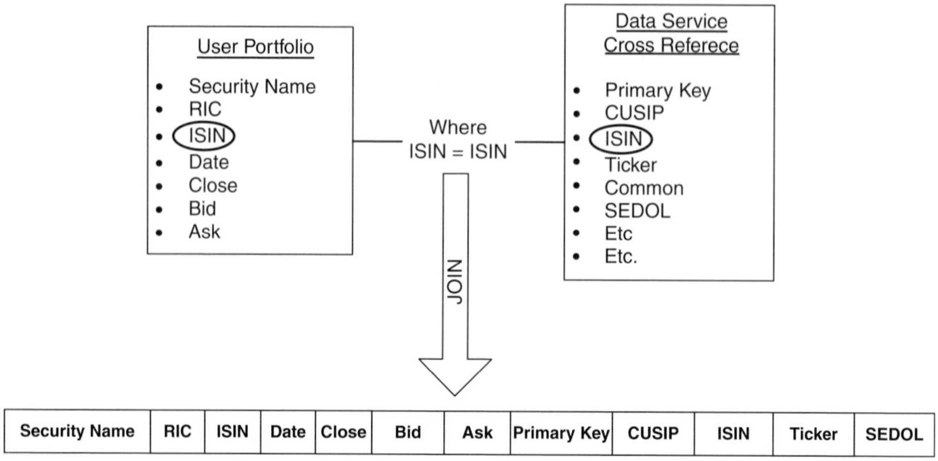

Figure 8.5 Cross reference JOINs.

The use of the primary and foreign key values used to perform these joins should be clearly understood. The logic can only work if the key values are dynamic – they should never be stored within an application or other data store. In this way they can be relied upon to provide the most up to date linkage between the records of interest when they are accessed. This insulates users and applications from problems that arise in changes to issuer, instrument and market listing codes – these can and do happen all the time. By referring to the primary key value generated objectively by a system, the vast majority of these problems can be avoided.

Therein lies a second objective for deploying reference data content – core reference data content should be managed in a dynamic manner. The content can and does change in response to various business events such as corporate actions announcements. Symbols can and do change (some with alarming frequency). So it is essential to cater to the data by acquiring and processing these changes on a real-time basis in order to ensure consistent navigation to the business data content. Nothing frustrates business users more than being unable to navigate to or retrieve a previously accessed record as a result of having to wait for an end of day update to take place.

8.4 Business data content

The biggest irony in dealing with reference data content is that when implemented successfully, its presence and utility can be pretty much taken for granted. The real value of information services is not the infrastructure that supports navigation and ensures uniqueness, but rather the actual business data that are needed by consuming business applications.

Thus far, the discussion has focused on the practice of segregating a core subset of reference data from the broader content with the hypothesis that using this approach supports enterprise scale use of data content. Figure 8.6 aligns the recommended core reference data type with some example business datasets commonly used within firms (note that this is intended as an indicative list only).

The key point to recognize here is that each of the business data categories can exist independently from the core reference data content. The primary keys and their cross references to industry identification schemes make it relatively straightforward to perform joins with specific business data content as and when required.

Figure 8.6 puts forward a conceptualized view of this navigation. In this example the primary keys contained in the core reference data subset provide the means of linking to the universe of business data content (grouped loosely here into the categories of Security Master, Corporate Actions and Pricing). As long as the primary key values are used to identify the underlying business data content records they can be accessed by resolving for the records of interest from the core reference dataset.

An alternative approach (and one more common in the industry) is to use the cross referencing capabilities independently. Legacy systems that uniquely identify records using industry coding schemes can be reconciled with other systems inside the firm. In turn, this serves to compile the master list of records that the firm needs to manage on behalf of the business users.

This points to the conclusion that cross referencing and symbology management are not optional in the design and deployment of data services to the enterprise. It is a core function that requires attention when setting up the service to achieve the goal of deploying a platform that can support multiple use cases at the same time using a single and consistent set of content.

8.5 The reference data lifecycle

So far this chapter has focused on the importance of implementing a subset of core reference data in a manner that is extensible and planned to cater for growth across an enterprise. While such a structure serves as the foundation for a broader capital markets reference data strategy, it is just as important to pay attention to the lifecycle associated with core reference data themselves.

Core reference data elements – and the cross reference symbology subset in particular – differ from other forms of data such as pricing and descriptive content. From the section above, it is easy to see that taking account of this lifecycle is essential in order to ensure that users have easy access to the business data content.

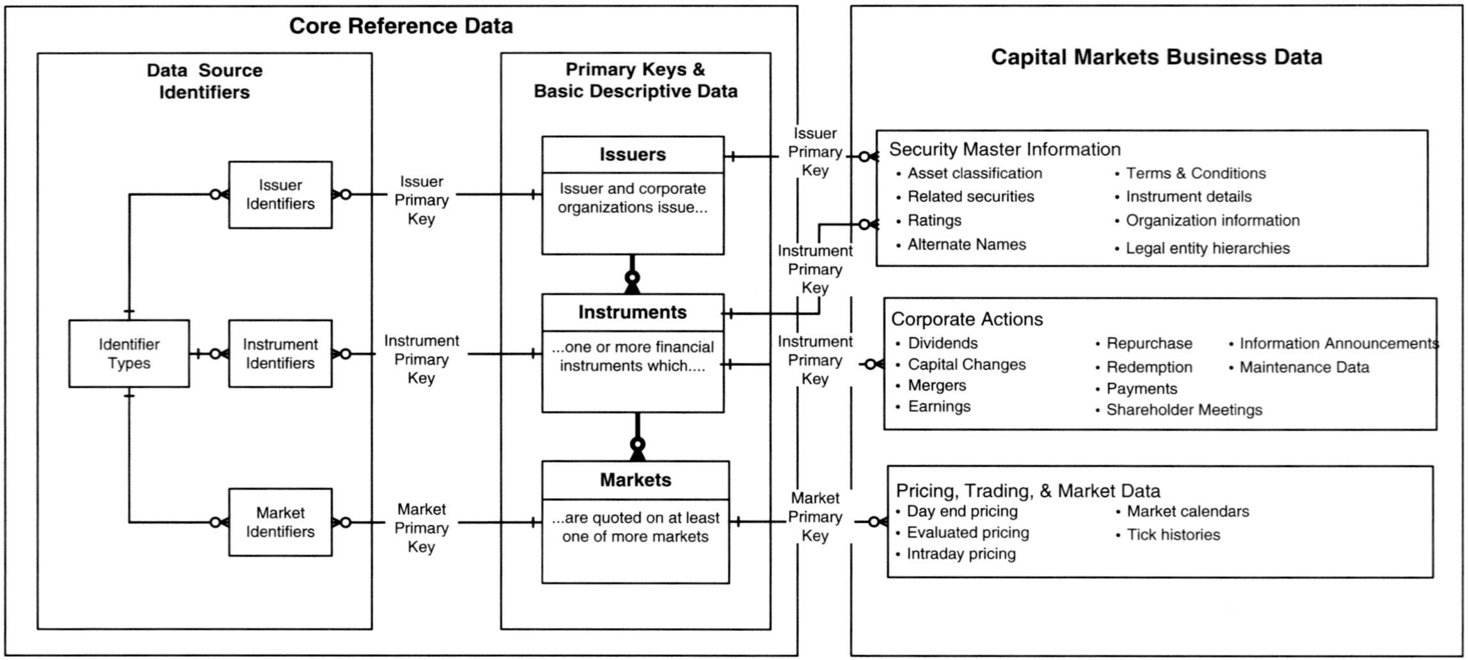

Figure 8.6 Core reference data and business data.

If a symbol change is missed, for example, it is very likely that a user will receive an error message if they submit a request to the data service based on a newly issued identifier. It is entirely possible that business data still reside within the service (such as a database) but they cannot be accessed unless the primary key has maintained a link between the old and new symbols. This is a frustrating and all too frequently encountered situation when working with reference data.

This example highlights some of the core characteristics of the reference data lifecycle. Three of the most important to consider are the following.

- **Core reference data are mandatory** – since the core reference data are the means by which navigation to actual business content is achieved, they therefore must be viewed as mandatory.
- **Reference data are not static** – the content is dynamic, it can and does change. Some of those changes can have direct impact on the ability of the information service to operate, while others can have a more subtle impact.
- **Symbology is self-referring** – industry supplied identifiers contain both explicit and implicit relationships among themselves that need to be considered

Each is explored in detail below. On their own, none of these characteristics necessarily represents difficult logic or workflow. When taken together, however, they represent the makings of a relatively complex and dynamic system, requiring careful design in order to maintain high service levels for business users.

8.5.1 Mandatory

The first thing to recognize is that regardless of the general approach, all business applications require a minimum subset of reference data. From the perspective of designing a database or an information system, this points to constraining a subset of attributes to be mandatory. They must be populated in order for any other content to be stored or made available to users.

By far the most commonly encountered aspect of mandatory data is cross referencing of industry symbology. Consuming applications must recognize that all access to business data requires the use of some form of identification. Furthermore, as discussed above, cross referencing between symbol types is an inexact science and subject to the lifecycle of the data content as made available by suppliers such as data vendor firms.

By enforcing the population of cross references (i.e., as mandatory data attributes) the system ensures that users can access the content using a standard and well-recognized means of identification (such as a ticker symbol or CUSIP code). In effect, the business data content can only be accessed if the mandatory field is populated.

Symbology cross reference is one example, however, most applications require a minimum set of mandatory data content to round out the identification of an issuer, an instrument or a market quote. Typically these fields will vary among firms and applications. For securities, typical fields to consider as mandatory include:

- at least one valid industry instrument level identifier (e.g., CUSIP, ISIN etc.),
- some identification of the source of the record (e.g., Interactive Data, NYSE etc.),
- a security name,
- an asset classification (e.g., Debt, Equity, Future, Option etc.),
- a status indicator (active, inactive etc.).

The list will likely vary by asset type, but at a minimum the selection of mandatory attributes sets the basis for issuing the primary key for the instrument. Together, the combination of the primary key and the mandatory attributes provides the minimum set of data to identify the instrument uniquely and ensure that users have access and can navigate to the associated business data. This relationship is illustrated in Figure 8.6 above.

The need for unique identification is stressed here as it is fundamental to establishing the quality of service that can be provided to business users. Failure to ensure unique identification can easily lead to multiple sets of potentially conflicting data returned to the user, requiring costly research and reconciliation to remedy. An old saying captures the problem well: "A person with one watch always knows what time it is. A person with two watches never does."

8.5.2 Dynamic

Core reference data can and do change, sometimes frequently and with serious impact on business applications and users. In order to maintain navigation to the underlying business data content, these changes need to be processed as soon as possible.

Take for example the relatively common situation of a ticker symbol changing from value "A" to "B." Failure to apply these changes will lead to one of three scenarios.

- A user submitting the new symbol "B" could find that no data are returned from the database. If the cross reference to the primary key has not been updated, then the value "B" will not yet be in the database and therefore cannot link to any content.
- A user submitting the previous value of "A" may find that content for a different security is returned if the update process has supplied new content to the database.
- No data are returned in response to a request for either "A" or "B" if the symbol change logic has not been implemented properly.

All three of these scenarios point to a service outage commonly referred to as "stale data." In particular, if the cross reference symbology is not maintained on a timely basis, then it is entirely possible for a transaction to take place between two counterparties referring to different securities. This situation will cause the trade to fail as the clearing process will ultimately detect the mismatch. However, this represents a costly situation that can be avoided through the disciplined application of reference data management.

It is worth thinking about the scale of the problem. Any application that purports to support global markets is looking at approximately 10 million instrument records.[5] If, conservatively, even 1%[6] of these records change in a given year, that results in 100,000 transactions to be processed. At 225 business days per year, this means that, on average, approximately 450 changes can be expected each business day.

While this may not seem a major number on the surface, if there are multiple systems that require the updates, the amount of effort to support the dynamic nature of cross reference codes needs to be correspondingly multiplied. The scale of the problem accelerates very quickly, further complicating the task of managing the service to business applications.

Secondly, experience shows that such changes are not distributed evenly over the course of the year. In fact, the single biggest cause of such changes tends to be exchanges modifying their symbology. In these situations, literally thousands of symbols need to be updated as quickly as possible and with the added need to be complete before they take effect (or else the underlying business data will not be accessible).

There are a variety of methods for handling such a situation, but the most recent trend is a move to adopt a more event based approach to managing the core reference data universe. Where previously these data were typically updated as part of a daily batch process at the end of the trading day or on an overnight basis, business applications are moving to a more dynamic, intraday approach. This is part of a wider industry trend to move to improved timeliness of all reference data content. Technologies such as real-time feed management and data streaming all provide the opportunity to make reference data more a truly mission critical asset for the firm. The trend here is inescapable – reference data are rapidly becoming real-time data (Alvarez, 2007b).

8.5.3 Self-referring

Another important characteristic of cross reference symbols and other core reference data attributes is that, by definition, relationships to other data content are involved. Take the following example.

[5] This is roughly the number of unique instrument level records maintained by Interactive Data in 2012.
[6] Based, again, on the Interactive Data experience.

- An issuer (IBM, for example) announces a restructuring resulting from an acquisition (*Algorithmics*, a private company). This change is deemed a material event, requiring a change to the CUSIP code assigned to the issuer.
- When the CUSIP code changes, by definition the corresponding ISIN code will change (since the root of the ISIN contains the CUSIP value for North American securities, as per ISO standard 6166). In this case there is an actual, transaction level link between the two code types.
- Equally, the act of a primary clearing code very likely will correlate closely to other code changes (such as the SEDOL or perhaps a ticker symbol) as well as other key fields (such as the instrument name). Note that this is a correlation only and not a causal link. Not all symbol changes will necessarily affect another code type.

It is plain to see that where a database or other reference data application makes use of industry symbols as the primary key, then these changes would need to be applied to all records making up the service. This represents a fairly complex maintenance task which likely places constraints on the efficiency and performance of the application.

Secondly, it should also be noted that the act of the change itself needs to be recorded. All business data previously linked to the original CUSIP code need to be linked to the replacement value. Maintaining the history of these code changes supports the ability to look at data content on a historical basis, providing an audit trail that can be used to support the business.

The example quoted here is relatively straightforward but it is important to identify these characteristics as part of the design of reference data applications. Making the right logic available is essential to ensuring that business users avoid experiencing either service outages (bad enough) or incorrect behavior, leading to potential losses or other unacceptable situations.

8.6 Key reference data success factors

The primary goal of this chapter has been to identify the fundamental categorization of reference data required to deploy an enterprise class information service. For the most part, any system that services information requirements in a firm implements at least some of these aspects – there are, after all, only so many ways to skin the market data cat.

It is important, though, to take a step back and recognize that in a rapidly globalizing and changing landscape of data content, technology and user requirements, certain factors need to be included as part of the design of an information service if it is to scale to meet these needs. Some of the most important to take into consideration include the following.

- **Completeness** – the data framework needs to cover the whole landscape of all the businesses in which the firm is engaged. It further needs a logical separation of mandatory data attributes and structures from descriptive and supporting attributes. This distinction can be a bit tricky and in many cases will depend on the needs of the consuming business application. So it is important to establish a logical model that anticipates growth into new asset classes and data types over time.

- **Navigation** – this is often overlooked by users of reference data, but the simple and unpleasant fact is that, for most users, simply being able to navigate to and select items of interest as and when they are needed is just as important as characteristics like timeliness and frequency. In order to ensure rapid adoption and active use, the information service must include the concept of an easy to follow roadmap for business users.
- **Put the reference into "reference data"** – by definition reference data are referential. Yet many people do not recognize this implicit characteristic, leading inevitably to an implementation with limited prospects for reuse across the broader enterprise. In turn this tends to lead to redundant purchases and implementations of precisely the same content and unnecessary technology. Understanding and communicating these fundamental relationships as part of the information service can be vital to avoiding the risks of duplication and the resulting downstream problems.
- **Reference data are dynamic** – not only are reference data fundamental to the use of underlying business data, they also can change with alarming frequency. Given its importance to navigation and access, such changes can be catastrophic. Do not underestimate the complexity of the reference data lifecycle. It is vital to initialize the universe fully and to maintain its consistency on as timely a basis as possible.
- **Discipline and management are critical** – success will depend on establishing and maintaining a disciplined approach to the management of reference data. This applies not just in the application of technology, but also in the skills and experience of staff assigned to manage reference data and of users within the organization.

This is by no means meant to be an exhaustive list. However, it does provide the basics to defining an information architecture with an eye to servicing the full enterprise. Adopting these principles as part of the baseline design of an information service will go a long way to generating faster time to market for new applications and improved return on investment.

Bibliography

Alvarez, Marc, 2007a, *Market Data Explained: A Practical Guide to Global Capital Markets Information*, Butterworth Heinemann.

Alvarez, Marc, 2007b, Reference Data Realtime – The Coming Revolution in Data Management, *Inside Market Data*, February.

9

Risk management data and information for improved insight

Margarita S. Brose, Mark D. Flood and David M. Rowe

Herein we provide an introduction and overview to the field of risk management for interested non-specialists, with a particular focus on the data and information requirements.[1] Financial institutions produce enormous amounts of data. The challenge for risk managers is finding those data most useful for the task at hand, which might involve analyzing market, credit, operational, liquidity or systemic risk. All data are not equal, and their interpretation has been the subject of much debate, especially after the recent financial crisis. Some have asked whether data provide guidance or distraction in understanding financial risks. This chapter first discusses some fundamental distinctions to keep in mind relative to all risk indicators used when actually managing risk. It closes with a brief tour of some of the core risk modeling tools in current practice, with an emphasis on their data and information needs.

9.1 Data versus information

9.1.1 The value and pitfalls of "actual" data

The modern world, characterized by pervasive computing power and massive electronic storage, is awash in data. As far back as 2007, Google was reportedly maintaining 100 exabytes of data (100 EB = 1×10^{20} bytes = 1 billion GB), largely to index the far more voluminous material on the Internet. Overall global storage capacity is measured in zettabytes (1 ZB = 1×10^{21} bytes; see Hilbert and López, 2012). In their raw and unorganized form, however, the data that surround us do not constitute information. While the massive amounts of data within our reach have the potential to change the way we think (e.g., Hey et al., 2009), the effect is not magic. Taken alone, raw data do

[1] Book-length overviews of financial risk management include recent textbooks by Hull (2012a) and Matz and Neu (2011). McNeil et al. (2005) cover advanced topics.

Handbook of Financial Data and Risk Information, ed. Margarita S. Brose, Mark D. Flood, Dilip Krishna and Bill Nichols. Published by Cambridge University Press. © Cambridge University Press 2014.

not condition our judgment or help us make better informed decisions. That requires a coherent process of selection, organization and interpretation to transform raw data into useful information that can shape and improve our judgment.[2]

An additional important distinction is between observed and derived data. Some data are directly observed in nature or in social interactions. For example, the last price each day at which a company's common shares traded on the New York Stock Exchange is directly observable, crystallized as a published transaction price. On the other hand, the daily value for the London Interbank Offer Rate (LIBOR) is not a directly observed transaction value but an average of unaudited estimates submitted in a daily survey of 18 banks conducted by the British Bankers' Association which calculates the inter-quartile mean (the mean of the middle 50% of the reported values); see Economist (2012).[3]

Other data are the result of far more complex processes of observation, interpretation and estimation. The figures for real gross domestic product (GDP) are the end result of an extensive process of compiling and interpreting reports from a wide variety of sources, including industry associations, regulatory authorities, government treasuries and company filings as well as public and private surveys. Many components of the initial GDP release combine preliminary reports for two months of a quarter with estimates based on other indicators for the third month. GDP figures are then subject to change as inputs are revised and as additional data replace initial estimates. GDP can be revised many years after the fact when the figures are benchmarked to detailed data from the Census of Manufactures that are only available every five years; see Fixler and Weadock (2006). Finally, many macroeconomic series are seasonally adjusted to compensate for regular patterns, such as the bulge in temporary hiring in the retail sector for the Christmas shopping season. The vital point is that not all data are created equal. Some are simple observations based on an objective event or measurement. Many, however, result from complex processes involving varying degrees of judgment, transformation and estimation.

Even confining attention to objective or "raw" data that require little or no judgment to create, the scope of the dataset can be frustratingly incomplete. For example, the interest rate yield curve might be "bootstrapped" from prices for any one of several different instruments, including forward rate agreements, interest-rate swap contracts,

[2] See Choo (2005) for a general discussion of the relationships between data, information and knowledge in the context of organizations. At a more practical level, SSG (2010) identified leading practices and recommendations on how financial firms should implement technology infrastructures for capturing risk data and aggregating them into actionable knowledge. IIF (2011) expands on this with additional implementation recommendations.

[3] In 2012, a major scandal erupted over evidence that employees from some of the 18 survey banks had colluded to manipulate the value of the reported LIBOR rate. Even absent outright fraud, the survey method proved to be problematic in the midst of a significant liquidity crisis that severely depressed activity in the interbank lending market. One consequence of this scandal was that the British Bankers' Association agreed to surrender oversight of the LIBOR determination process to UK bank regulators. This transfer of authority is planned to occur in early 2014. If nothing else, this scandal reinforces the relevance of understanding the distinction between observed and derived data.

or interest-rate futures contracts (Bianchetti, 2009). In the latter case, the market offers quarterly contracts expiring at fixed future dates every three months. These future expiry dates roll forward one day at a time until the nearest contract matures, at which point the designation of "nearest contract" abruptly jumps forward three months to the next closest active contract. Thus, the futures market only offers actual observed interest rates for four fixed calendar dates per year.

The obvious answer is to interpolate rates for all the intermediate dates to allow publication of a consistent yield curve, but this requires some convention for the interpolation. When rates are trending higher (or, less commonly, lower) as maturities lengthen, the method chosen is not likely to make much difference. When the curve changes direction across maturities, however, for example first rising and then falling, the interpolation methodology can have a greater impact. Lastly, rates in the cash treasury market may vary somewhat from a theoretical yield curve based on the futures market due to temporary liquidity differences across markets. The choice of benchmark is usually based on the degree of liquidity in each market. Depending on the particular purpose under consideration, the standard choice for the benchmark may or may not be most appropriate.

Data vary enormously in the degree of analysis and judgment needed to produce them. Understanding these processes can be as important as knowing the data themselves. This knowledge provides a basis for judging the range of uncertainty to impute to individual values and the types of analysis for which the data are suited.

9.1.2 The two kinds of "risk"

Our lack of information about unknown future events falls into two fundamentally different categories. The first category can be characterized as "distributional" in nature, meaning that periodic observations are individually unknowable before the fact but that collectively they appear to be the product of a stable random process. "Stable random process" simply means that there is a mechanism generating the observations and it exhibits a stable mean, dispersion, skewness (lack of symmetry), etc.[4] The practical consequences are that any two large samples of random events from the same mechanism will resemble one another; the larger the sample size, the smaller the discrepancy between samples. The mechanism is not perfectly known, but it is measurable in this sense.

The second category refers to situations where the mechanism is not measurable, either because observations are too rare (or even unique), or because the mechanism itself is unstable, perhaps because human agents are in the mix disrupting the

[4] A formal definition involves a discussion of issues of stationarity and ergodicity that is beyond our scope here. For details, see Sheppard (2012) or Tsay (2010) and the references therein.

process. Such rare events – especially if economically significant – are sometimes called "Black Swans" – plausible creatures but quite surprising when first encountered. Knight (1921) assigned the name "risk" to the first type of unknown event: measurable randomness. Black Swan events fall into what Knight called "uncertainty." Simply put, "Knightian risk" can be analyzed in a distributional framework and "Knightian uncertainty" cannot. "Risk" in this sense is characterized by repeated realizations of random events generated by a stable stochastic mechanism, or, at least, one with a high degree of stochastic inertia (i.e., where the structure of the randomness changes only slowly over time).[5]

Value at risk (VaR) is one of the best known measures of Knightian risk, and a great deal of criticism has been leveled at its use.[6] In truth, VaR is representative of distributional analysis in general. In the early 1990s, when the concept was introduced, VaR was the first reasonably effective means for communicating risk implications between traders and general managers. It was soon incorporated into regulatory bank capital standards. Unfortunately, hype helped create a false sense of security among senior managers and watchdogs. VaR was sometimes referred to as a "worst case loss," which it clearly is not. The "expected minimum twice-a-year loss" would be more accurate, and might naturally provoke the question, "how bad could the loss be on those two days?"

In contrast, "uncertain" events are typically rare or non-recurring. In a social or economic context, such events can depend on the infuriatingly mercurial influence of human emotion. Shifts in mass psychology can be sudden and unexpected. It is problematic – even counterproductive – to force a distributional apparatus into the analysis of such non-recurring events. Tinkering with distributional assumptions, for example applying extreme value theory, may improve the performance of a model in a back test, but there is an irreducible core of unpredictable uncertainty that will never yield to classic statistical analysis. Risk and uncertainty require fundamentally different evaluation approaches. Evaluating Black Swan uncertainty demands a different mental framework and different training and experience than is required for the conduct of distributional analysis. Conversely, risk is also present in financial markets, and statistical tools such as VaR have an important role.

The profession of financial risk management is currently dominated by applied statistical analysis. In reaction, Bootle (2011, p. 101) argues that risk managers should

[5] Coleman (2012) provides an overview of the risk/uncertainty distinction. A similar distinction appears in the engineering literature between "aleatoric uncertainty" (Knightian risk) and "epistemic uncertainty" (Knightian uncertainty); see Der Kiureghian and Ditlevsen (2009). Risk in the Knightian sense was the basic subject of Bernstein's (1996) book, *Against the Gods: The Remarkable Story of Risk*. Punning on this, *The Economist*'s (2010) review of risk management's failures during the global financial crisis was titled "The gods strike back." Throughout this chapter, we use the unqualified term, "risk" to refer to both Knightian risk and uncertainty; we reserve "Knightian risk" and "uncertainty" to indicate the specific subtypes.

[6] See, for example, Rowe (2009), Bookstaber (2009), Daníelsson (2002) or Johnson and Kwak (2009). VaR is described in greater detail in Section 9.2.1.

study less mathematics and more history. The statistical approach is exemplified by market risk analysis (see Section 9.2.1 below), which has the statistical luxury of many millions of regularly occurring price observations for many thousands of securities. While supply and demand change from day to day (or microsecond to microsecond), the institutional structure governing transactions typically does not. This allows risk analysts to search for patterns and clues in a relatively high-frequency, low-dimensional space; distributional methods were designed for problems of this type. At the opposite pole, Knightian uncertainty describes the problem of low-frequency events in a high-dimensional space. This is exemplified by operational risk (see Section 9.2.4 below), which must address problems as diverse as embezzlement, hurricanes and data-entry typos. Although the Basel approach to operational risk views it through a largely statistical lens, managing operational issues well ultimately requires a more abstract, less data-centric skill set. This includes tools such as business analysis and process design, internal audit, systems engineering, legal liability analysis, and organizational architecture. The information to support these tools is less structured and frequently resists precise quantification. For example, inputs from country risk officers, industry analysts and macroeconomists should be integrated into regular deliberations about risk. Addressing both risk and uncertainty will require executives and policymakers to abandon the comfortable idea that risk can be reduced to summary statistics like VaR. Only those organizations that are prepared to grapple with risk and uncertainty in all their messy multidimensionality will have a reasonable chance to avoid the worst effects of the next crisis. This is relevant at the aggregate level as well, where authorities must formulate policies to address systemic risk (see Chapters 13 and 14 of this Handbook).

9.1.3 Comparative static versus sequential scenario analysis

Given the limitations of distributional analysis in the face of uncertainty, stress testing is another obvious alternative. For example, scenario analysis is at the heart of the Federal Reserve's Comprehensive Capital Analysis and Review (CCAR) process for banks with assets over \$50 billion.[7] Such exercises have their own limitations. As currently structured, they are fundamentally exercises in comparative static analysis. The overall vulnerability is a combination of the portfolio's sensitivities to various factors, multiplied by potential changes to those factors. A bank's financial position at a point in time is evaluated assuming that market conditions experience a large, *instantaneous* change. A commonly applied heuristic

[7] Read more about the Federal Reserve's microprudential supervisory activities in Chapter 11 of this Handbook. The new European Banking Authority (EBA) is conducting similar stress tests in Europe; see Chapter 13. Banks and securities firms also conduct stress tests as part of their internal risk management; see CGFS (2005).

is that stress scenarios should be, "exceptional but plausible" (e.g., CGFS, 2005, pp. 3–4). Despite this artificiality, instantaneous stress scenarios allow detailed analysis of the impact of hypothetical changes in market conditions. This can reveal facts about risk exposures embedded in a firm's portfolio that traditional accounting or VaR analysis cannot.

Unfortunately, the instantaneous character of the analysis often limits the range of market changes that are plausible even as extreme scenarios. For example, a financial crisis may unfold gradually as a loosely connected set of related events, where damaging changes in one period trigger further problems in subsequent periods. In practical terms, introducing temporal dynamics expands the range of possible circumstances exponentially. As a result, dynamic scenarios cannot be analyzed in the same level of detail as a comparative static scenario. Nevertheless, reasoning through such dynamic scenarios should allow for consideration of certain changes that might otherwise be dismissed as implausible.

For example, the "volatility paradox" that characterized the mid-2000s – low market volatility as risks were building – made it implausible to propose a stress scenario involving a 30% to 40% drop in home prices in the USA.[8] On the other hand, a scenario involving a 2% to 5% drop in home prices over 12 to 18 months should have been plausible; this had occurred as recently as the early 1990s. In playing this scenario out over time, many subprime borrowers would have exhausted their teaser rates and been unable to sell at a profit or to refinance on affordable terms. This would have triggered defaults and foreclosures, increasing distressed home sales by lenders and further depressing prices. It is such interconnected risks and self-reinforcing feedback loops that can create full-blown systemic crises. Because such loosely connected and overlapping causal relationships unfold over time rather than all at once, they are difficult to catch with comparative static risk analysis.

Even if the initial exercise falls well short of the future reality, it is likely to sensitize people to the sequence of events as they begin to unfold. Such advance brainstorming can be instrumental in preparing an organization to take defensive action sooner than it would have done otherwise. This requires more than simply digging deeper into distributional analyses such as VaR. It involves a holistic perspective, including dynamic analysis to capture potential feedback loops. Neither data nor information alone will improve decisions unless the analysis fits the problem. This requires detailed structural data on positions sufficient to simulate prices under a wide range of market conditions both currently and in the future. The

[8] On the volatility paradox, see Brunnermeier et al. (2012, p. 25). A prominent example of the assertion of implausibility was the so-called "Anti-Bubble Reports" produced by the National Association of Realtors (NAR) prior to the crisis, for example, NAR (2005).

new Y-14 data collections mandated by the Federal Reserve to support the CCAR process represent a major step in this direction; see Chapter 11. Data that fail to illuminate uncertainty are ill suited to measuring and managing dynamic scenarios – including crucial aspects of systemic risk.[9]

9.2 Risk modeling tools

Financial institutions must manage all types of risks and uncertainty, which differ widely in their source, modalities, statistical structure, and in business lines they affect. Essentially any business activity has risks attached, each of which typically has recognizable characteristics that make it amenable to measurement and management. For example, there are political risks, inflation risks, reputational risks, legal risks, and systemic risks to name a few; the full gamut is beyond the scope of this chapter. We focus instead on four areas among the most central to financial institutions (and therefore among the most developed analytically): market, credit, operational and liquidity risk.[10] In each case, we consider the particular risk-analysis toolkit in the context of a particular portfolio type or business line. This pairing of a specific example is not, however, an indication that the given risk type occurs only in that context.

There is a strong emphasis in these toolkits on statistical methodologies. This is a simple reflection of the state of the art. In the case of market risk, which depends on market prices for its inputs, this emphasis is natural. Prices are typically readily available as repeated observations from a stable, well-understood process. In the case of credit risk, which depends on Knightian uncertainty about borrower creditworthiness, or operational risk, which depends on uncertainty from many sources, the advantages of a distributional approach are less obvious. Yet, here too the statistical methodologies have made deep inroads. For example, the broad expansion over the past two decades of loan securitizations and credit derivatives has had the effect of converting the Knightian uncertainty of loan underwriting into the market risk of traded securities.[11] Managers should pick the best tool(s) for the job, staying alert to the nuances of the problem at hand.

9.2.1 Market risk: the example of equity portfolios

Market risk refers to the changes in the value of a portfolio induced by fluctuations in market prices, including stock prices, bond prices and interest rates, exchange

[9] See Castrén et al. (2009) and the references therein, particularly Section 15.3.2, for a discussion of the data needs to address contagion and the feedback dynamic in the context of systemic stress testing.

[10] The recent financial crisis highlights the crucial importance of systemic risk, which is covered in greater detail in Chapter 4 of this Handbook, and in a recent survey by Bisias et al. (2012). Chapter 3 discusses the organizational process of portfolio risk monitoring and risk governance.

[11] Flood et al. (2012) provide an overview of some of the data and information challenges implied by this conversion.

rates and commodity prices. A myriad of external influences affect prices, including man-made and natural events, and business and governmental policies. For example, business-specific events may result from internal decisions (the successful or unsuccessful launch of a new product), or events (the explosion on British Petroleum's Deepwater Horizon drilling rig in 2010). These "business fundamentals" typically affect the prices investors are willing pay for a stock, currency or commodity.

Fluctuations in market prices are seemingly random, partly for the simple reason that they occur outside of the control of the individual portfolio manager. However, there is a more fundamental and powerful argument that predictable patterns in market prices should not exist in unfettered markets, as buyers (sellers) bid up (down) the price of under-valued (over-valued) securities. By construction, the predictable component disappears, and what remain in equilibrium are the price movements that market participants could not predict. This basic notion is the core of the efficient markets hypothesis.[12] Vast quantities of price data are available for thousands of securities, in some cases at time resolutions down to the microsecond or less, making the apparently random fluctuations amenable to sophisticated statistical analyses. In the case of equity portfolios, problems are simplified by the fact that trading typically occurs on organized stock exchanges, electronic communication networks (ECNs), and alternative trading systems (ATSs) which provide pre-trade price transparency in the form of bid-ask spreads and limit-order books and post-trade transparency via published transaction prices.[13]

For statistical analysis and risk management, price series are typically converted to returns. This conversion normalizes price changes to facilitate comparisons across stocks and over time. The most relevant calculation for equities is the "total" return, which includes both price changes and dividend payments.[14] Continuous compounding is the typical convention, as this comports readily with the continuous-time models used in derivative markets (see Taylor, 2005, ch. 2; Daníelsson, 2011, ch. 1). Financial market returns, including equities markets, are characterized by three general and important statistical patterns (Taylor, 2005, ch. 4). First, the distribution of returns is non-normal, exhibiting a relatively (compared

[12] Malkiel (2012) presents the efficient markets hypothesis in detail. There is, however, a devil in those details. See Lo and MacKinlay (2012) for a discussion of some of the complications that afflict the simple random-walk model of prices.

[13] Details of trading systems are beyond our scope; see Schmidt (2011). See Muhlerin et al. (1991) for an insightful history of the stock market tickers in the USA, resulting in exchanges' formal property rights in the price data. While the technology of exchanges has evolved, the basic principle of post-trade transparency remains the same.

[14] Ordinary cash dividends are typically calculated separately and added to the price. However, there are myriad other corporate actions to accommodate, such as stock splits, stock dividends, reverse stock splits, etc. These are typically factored in via adjustments to the price; see CRSP (2000). Even tracking a stock over time through mergers, liquidations, spin-offs, exchange listings, etc. can be a challenge, making tickers and CUSIP codes unstable; the CRSP PERMNO is generally a more stable identifier, where available.

to the normal) high central peak representing lots of small returns together with relatively heavy tails filled with large returns.[15] The "extra" observations near zero and at the extremes are balanced out by a relative dearth of medium-sized returns. Second, consistent with the market efficiency hypothesis, there is essentially no correlation in returns from day to day (or over longer time lags). Third, there is correlation in variance from day to day. This is the so-called volatility clustering effect (commonly referred to as ARCH or GARCH effects), that high-volatility days tend to be followed by additional high-volatility days, and similarly for low-volatility days.[16]

Volatility modeling and clustering

Volatility – especially its forward-looking estimate – is the central issue for market risk management. There are two broad approaches to extracting volatility estimates from returns data: historical and implied. Remarkably, while the level of returns is very difficult to predict from historical data due to market efficiency forces, the expected volatility is easier to forecast due to the volatility clustering effect.[17] As the frequency of trading has increased over time and data have become available, the estimation of historical volatility has moved to intraday time scales, where it is commonly known as *realized volatility* (see McAleer and Madeiros, 2008; Heber et al., 2009).

Implied volatility exploits the fact that the value of put and call options is positively related to volatility. Except for volatility, all of the other inputs to the basic Black–Scholes pricing model are unambiguously observable. Conditioning on these other inputs (i.e., strike price, time to expiry, risk-free interest rate and underlying price), the formula for option value becomes an invertible function of the volatility, and one can back out the model-implied volatility estimate for a given market price for the option.[18] Because they are based on current (i.e., forward-looking) option prices, implied volatilities have an advantage over historical

[15] The technical term for this peakedness is "leptokurtosis." The central limit theorem establishes that the distribution of the mean of a sample of independent observations from a stable data-generating process with finite variance will converge to a normal distribution as the sample grows large, which is not what we observe here. The problem lies in the instability of the process, driven primarily by ever-shifting variance.

[16] ARCH and GARCH are acronyms for "(generalized) autoregressive conditional heteroskedasticity." The two abbreviations are used largely interchangeably. There are many different implementations of the time-series econometric models for estimating GARCH patterns; see Bollerslev (2010) and Engle (2001) for further details. Other (i.e., non-GARCH) stochastic volatility models are also available, along with simplified exponentially weighted moving average (EWMA) techniques popularized by RiskMetrics (see Daníelsson, 2011, ch. 2; and Alexander, 2008, ch. II.3).

[17] Poon and Granger (2003, 2005), and Andersen et al. (2007) provide extensive surveys of the literature on historical volatility estimation and its use in forecasting. GARCH models are the workhorses; see Brownlees et al. (2011).

[18] See Hull (2012b, ch. 14) for further details. The Black–Scholes model is also of the risk measures known as the the "Greeks," which represent derivatives of formula with respect to its various inputs; Hull (2012a, ch. 7; 2012b, ch. 18).

volatilities in adapting to abrupt shifts between high-volatility and low-volatility regimes. However, implied volatilities have complications of their own, many of which are due to the assumption of a particular pricing model as the correct rule for translating between option prices and implied volatilities. For example, an implied volatility surface calculated via the standard Black–Scholes model for a range of strike prices and expiry dates typically reveals a "volatility smile," whereby out-of-the-money strikes exhibit higher implied volatilities than at-the-money contracts (Hull, 2012b, ch. 19). Clearly there are limitations to the theory. The main shortcoming of Black–Scholes is the suspect notion that a single volatility statistic (standard deviation) can fully capture the dispersion of returns in a world where those returns exhibit much larger tails than the normal distribution can accommodate.

Value at risk

In practice, one of the most popular tools for capturing the dispersion of returns is value at risk (VaR).[19] VaR is defined as the dollar loss threshold that will not be exceeded α percent of the time (e.g., 99%) for a given holding period (e.g., ten days). It is well documented, broadly understood, and widely used for allocating capital both for regulatory and internal risk management purposes. Unlike simple volatility measures, it is clearly focused on the tails of the returns distribution. It is also widely applicable, not just to equities, but to bonds, derivatives, commodities and other asset classes. Unfortunately, VaR has shortcomings as well. Foremost, it says nothing about what might happen in the $(1 - \alpha)$ percent of cases where extreme events do occur. It is unchanged by large risk exposures that occur sufficiently infrequently to fall beyond the VaR threshold, and the ubiquity of VaR as a tool for setting risk limits thus encourages traders to take on large tail risks (Rowe, 2009). Some common implementations of VaR, such as the "variance-covariance" method and the "square root of time" rule implicitly and unrealistically assume normally distributed returns arriving independently from day to day with no jumps in the process; these are recipes for underestimating tail risk (BCBS, 2011b). An alternative is historical simulation VaR (HS VaR), which in its most common form builds a hypothetical profit/loss (P&L) time series from daily data for the securities in the current portfolio.[20] HS VaR is simply the relevant percentile observation from the histogram of this distribution. HS VaR methodologies avoid the need to make specific distributional assumptions, but are susceptible to regime shifts: low-volatility episodes in the estimation window will reduce VaRs, and can thus create unwarranted complacency (Pritsker, 2006). This is the essence of the volatility

[19] The myriad details of VaR calculations are well covered in Alexander (2009) and Jorion (2006). See also Chapters 1 and 2 of this Handbook.

[20] Frésard et al. (2011) document that typical applications of P&L data tend to contaminate loss data with fees and commissions, leading to systematic underestimates of tail exposures.

paradox described above. The Basel rules (BCBS, 2005) require a minimum of one year of daily data be used to estimate VaR, but this was not long enough to see beyond the protracted episode of low stock-market volatility from late 2003 to early 2007. In the aftermath of the 2007–2009 crisis, regulators are reconsidering the reliance on VaR adopted under the 1996 Basel market risk amendments (BCBS, 2012b). "Coherent" risk measures, such as expected shortfall (i.e., the expected loss, conditional on being in the tail) have gained in popularity (Daníelsson, 2011, ch. 4). Expected shortfall addresses the problem of ignoring loss exposures in the tail, but it is still susceptible to improper distributional assumptions and misleading estimation windows. Stress testing is another alternative to examining a portfolio's response to tail events (Alexander and Sheedy, 2008; CGFS, 2005). If undertaken deliberately, it can avoid some of the pitfalls of purely statistical approaches, and thus form a useful complement to VaR.

Case study: limitations of value at risk

On April 18, 2008, UBS released *Shareholder Report on UBS's Write-Downs*, with the purpose of documenting the history and causes of "losses incurred in relation to structuring, trading and investment activities in mortgage and asset-backed securities, in particular with respect to securities referencing US 'Subprime' residential mortgages."[21] The discussion centers on "AMPS" (UBS, 2008, p. 14):

> Amplified Mortgage Portfolio ('AMPS') Super Seniors: these were Super Senior positions where the risk of loss was initially hedged through the purchase of protection on a proportion of the nominal position (typically between 2% and 4% though sometimes more). This level of hedging was based on statistical analyses of historical price movements that indicated that such protection was sufficient to protect UBS from any losses on the position. Much of the AMPS protection has now been exhausted, leaving UBS exposed to write-downs on losses to the extent they exceed the protection purchased. As at the end of 2007, losses on these trades contributed approximately 63% of total Super Senior losses.

Essentially it had simply been *assumed* that a 2% to 4% change in value was the most that could possibly be anticipated. This appeared to be justified "based on statistical analyses of historical price movements." Clearly the thought process involved was confined to a narrow distributional mindset rather than incorporating broader structural trends and realities. The following passage describes some alternate analysis that was being performed at this time. One particular statement drives home the essential point that being hedged for VaR purposes does not imply protection from major systemic upheavals (UBS, 2008, pp. 30–31, italicized text added):

[21] This is a remarkably candid disclosure. While it is somewhat technical in places, most of the key points are accessible to general readers. Reviewing it is a valuable exercise for anyone wanting to gain deeper insight into widespread technical and organizational realities that gave rise to the 2007–2009 crisis.

> Once hedged, . . . the Super Senior positions were VaR and Stress Testing neutral (i.e., because they were treated as fully hedged, the Super Senior positions were netted to zero and therefore did not utilize VaR and Stress limits). The CDO desk considered a Super Senior [*position*] hedged with 2% or more of AMPS protection to be fully hedged. In several MRC [*Market Risk Committee*] reports, the long and short positions were netted, and the inventory of Super Seniors was not shown, or was unclear. For AMPS trades, the zero VaR assumption subsequently proved to be incorrect as only a portion of the exposure was hedged . . ., although it was believed at the time that such protection was sufficient.

The following passage describes some alternate analysis that was being performed at this time (UBS, 2008, p. 38):

> The historical time series used to drive VaR and Stress are based on five years of data, whereby the data was sourced from a period of relatively positive growth. Regular work being performed during the relevant period focused on confirming the efficacy of existing scenarios based on broad based economic developments and historical events. When updates to methodologies were presented to Group and IB Senior Management, hindsight suggests that these updates did not attribute adequate weight to the significant growth in the US housing market and especially the Subprime market. The Market Risk function did not develop scenarios that were based on more fundamental attributes of the US housing market.

Unfortunately this effort did not carry the weight it deserved in shaping senor management's view of the Black Swan exposure of UBS's subprime mortgage business. Overconfidence in the view that the positions were hedged allowed distributional analysis to dominate and crowd out broader and more structural forms of risk analysis. Reassuring indications based on the analysis of "risk" were deemed to be broadly conclusive and blinded management to the equally important need for analysis of Knightian "uncertainty."

Portfolio risk and covariance estimation

A core result of modern portfolio theory (MPT) is that it is misleading to consider a single firm's returns in isolation, because much of the idiosyncratic variation for an individual equity will be averaged away when the position is held in a diversified portfolio (see Elton et al., 2009). As a result, the covariance matrix of returns (or its normalized counterpart, the correlation matrix), plays a central role. The volatility, measured as standard deviation of returns, for a portfolio is a simple function of the covariance matrix and the fractions of the portfolio invested in each stock. Small (or negative) correlations increase diversification benefits, making idiosyncratic stocks especially attractive for diversified portfolios.

Covariance estimation also usefully illustrates many of the challenges of working with actual financial data. The covariance matrix is a theoretical ideal; in practice, real data are messy and uncooperative, and a number of complications can arise (see Alexander, 2008, ch. II.3 for an overview of the basics). For example, in a world of tens of thousands of traded equities, a full covariance matrix may have tens of millions of entries (covariance parameters) to estimate. To keep the problem manageable, it is common to impose some structure to reduce the number of parameters, for example assuming that all correlations come from shared exposures to a limited set of underlying factors, instead of allowing every stock to correlate separately with every other (Briner and Connor, 2008). Another problem is that many popular multivariate techniques require special restrictions to guarantee that the estimated covariance matrix is positive semidefinite.[22] When working with high-frequency data, institutional effects (so-called "microstructure noise") such as "bid-ask bounce" can contaminate the estimates, and non-simultaneous observations can bias correlation estimates downward (Andersen et al., 2011).

In estimating correlations for risk management purposes, it is vital to remain mindful that it is the *forecast* that matters. To paraphrase Kierkegaard, market risk is understood backward, but must be lived forward. Thus, the most important empirical fact in this context is the presence of intertemporal patterns in correlations. Some of these patterns are extensions of the intertemporal clustering of (univariate) volatility described above. Naive historical estimates of a stationary covariance matrix are inappropriate when time-series patterns can be exploited to improve the forecast, and a range of econometric techniques exist for multivariate GARCH estimates of the full correlation matrix (Andersen et al., 2007). Although these models can capture intertemporal clustering, they are challenged by abrupt changes in the statistical regime, such as occurred with the Lehman failure in 2008. Such crisis turning points are very difficult to predict with statistical models, and typically bring with them large shocks and sharp increases in correlations (Ang and Chen, 2002). This can effectively eliminate diversification and invalidate supposed "hedges" precisely when they are needed most (Daníelsson, 2002).[23]

22 Positive semidefiniteness (or lack thereof) is a technical characteristic of the covariance matrix that is important for a range of applications, including regression analysis and portfolio optimization. For example, without positive semidefiniteness of the covariance of returns, the estimated or forecast variance of a portfolio's returns could (nonsensically) become negative. In theory, the covariance matrix must be positive definite by construction, as is also the case for the standard textbook estimator of the unconditional sample covariance matrix. However, more sophisticated techniques, including many popular multivariate GARCH models, do not automatically produce positive semidefinite estimators; see Andersen et al. (2007).

23 Statistical estimation of these events is plagued by scarcity of data. Econometric approaches to the so-called "tail dependence" phenomenon of simultaneous large moves in multiple markets include extreme value theory (McNeil, 1999) and copula modeling (Alexander, 2008, ch. II.6). However, because the data are scarce and the stakes are high, traditional statistical methods are usually supplemented by some form of stress testing (CGFS, 2005; Daníelsson, 2011, ch. 8; Alexander and Sheedy, 2008).

Structural models of correlation are not merely expedients for estimation of large systems; they also typically have useful interpretations in terms of other markets and fundamentals, thus facilitating hedging strategies as well as the understanding and communication of risk exposures. The most basic and best known of these structural models posits that there is a broad-based "market" factor, R_M, responsible for all co-movements in returns, and that any other variability, ε_i, (which may be significant) is firm specific:

$$R_i = \alpha_i + \beta_i R_M + \varepsilon_i. \tag{9.1}$$

The market factor here is typically proxied by a standard market index, such as the S&P 500 (S&P, 2012b), with the parameters α_i and β_i estimated by regression analysis. The firm-specific error term, ε_i, typically is neither hedged nor priced because it can be diversified away. The parameter β_i captures the stock's sensitivity to the general market factor. This linear sensitivity might be hedged as appropriate, for example via an offsetting position in stock index futures. Given N stocks, notice that the estimation of a full $N \times N$ covariance matrix effectively collapses here to the estimation of the $2N$ α and β parameters (along with estimates of means and variances for $R_M + \varepsilon_i$); when N is large, this parsimony is enormous.

The so-called "market model" in equation (9.1) is similar to the famous capital asset pricing model (CAPM) (Elton et al., 2009). Both models relate individual stock returns via a sensitivity parameter (usually labeled "beta") to the return on a market portfolio or market factor.[24] The CAPM differs in two key respects: (a) it is an equilibrium model derived from first principles as to how the market *should* behave if the assumptions are correct; and (b) the equilibrium result – the "security market line" – is stated in expectations, and in terms of the market return premium over a risk-free interest rate, R_f:

$$\mathbf{E}(R_i) - R_f = \beta_i[\mathbf{E}(R_M) - R_f]. \tag{9.2}$$

Risk managers prefer the simpler and more flexible market model (9.1), as risks must still be managed whether the market is in equilibrium or not. The market model is also easily extended to incorporate other factors; a more general form of (9.1) is:

$$R_i = \alpha_i + \beta_{i1} R_M + \beta_{i2} R_S + \Sigma_j \beta_{i3j} F_j + \varepsilon_i, \tag{9.3}$$

[24] The CAPM earned Harry Markowitz, Merton Miller and William Sharpe a Nobel Memorial Prize in 1990. The CAPM can be nested as a special case of the arbitrage pricing theory (APT), also an equilibrium model. The CAPM and APT have been the subject of thousands of books and papers over the years. See Rubinstein (2006) for an annotated sampling of these models within the larger context of investment theory. The CAPM is also the basis for the Sharpe ratio, which divides a portfolio's excess return over the risk-free rate by the portfolio's volatility to get a simple risk-adjusted performance measure. Note that this and other measures that assume a stationary portfolio strategy (e.g., the information ratio) are susceptible to manipulation; see Goetzmann et al. (2007).

where now there are many factors, including the market return, R_M, the overall return on the relevant industry sector, R_S, and any other factors, F_j, that are deemed appropriate, such as "momentum" or firm size. Again, the sensitivities are linear, which facilitates hedging, and provides a natural economic interpretation, which facilitates risk reporting and strategy development.[25]

9.2.2 Liquidity risk: the example of fixed income portfolios

Simply put, liquidity risk is the risk that market liquidity will not be available when you need it. Despite its prevalence in discussions of markets, "liquidity" remains multifaceted and often vaguely defined, largely because it can manifest itself in many different ways depending on institutional realities and the task at hand. As with other types of risk, adequate governance is crucial for successful implementation (see Matz, 2011; Matz and Neu, 2011). We focus here on measurement issues.

A common distinction is between "market liquidity" and "funding liquidity" (Brunnermeier and Pedersen, 2009), which hinges on whether one is trying to retain or unload risky assets. Market liquidity – sometimes called microstructure liquidity or market depth – refers to the ability to sell an asset for cash with a minimal impact on price. It can be measured, for example, by the price concession the seller must accept to effect an immediate sale. In contrast, funding liquidity refers to the ease of finding lenders or investors willing to finance ongoing portfolio holdings. This is the typical assignment of a bank treasurer, who must daily replace the bank's maturing short-term funding with new deposits.[26] Monetary authorities sometimes use the term "global liquidity" (CGFS, 2011), referring to the overall availability of funding from both public and private sources, reflecting the fact that liquidity is a phenomenon that emerges in the aggregate. Despite the apparent ambiguity, all usages of the term refer in some way to the timely availability of counterparties for financial transactions.

Fixed-income portfolios and the term structure of interest rates

We use the example of fixed-income portfolios to illustrate the basic measurement techniques for liquidity risk. "Fixed income" refers to the fact that principal and interest payments are contractually defined. Familiar examples are corporate bonds, government debt, mortgages, commercial loans, defined-benefit pensions and most bank assets and liabilities.[27] The fixed commitment eliminates a key source of

[25] Alexander (2008, ch. II.1) provides a good summary of factor models.

[26] Bank funding activities usually occur in the interbank money markets, where interbank deposits are only one of many funding vehicles available. See Cook and Laroche (1993) or Stigum and Crescenzi (2007).

[27] The payment obligations may be complicated by floating rates, teaser periods, option clauses, and other issues beyond our scope. See Fabozzi and Mann (2011) for additional details on fixed-income portfolios. See Amihud et al. (2005) for a survey of liquidity models applied in a variety of markets, including fixed income.

uncertainty, placing the focus instead on the exogenous fluctuations in the interest rates used to discount these predefined cash flows.

The behavior of interest rates is an intensely intricate topic largely outside our scope here. In brief, interest rate dynamics are dominated by five factors.

1. There is the possibility that a borrower will fail to meet his contractual obligations. Debt contracts are typically laden with detailed subordination rules and covenants to handle these eventualities. (We discuss credit risk in the next section.)
2. According to the Fisher equation and the Expectations Hypothesis, interest rates will compensate lenders for expected inflation, in addition to providing a profit from putting their funds at risk (Mehra, 2006).
3. Loans of similar maturity are close substitutes, and supply and demand will force them to have similar yields to maturity. This natural consistency among discount rates across the maturity spectrum gives rise to the *term structure* of interest rates (James and Webber, 2000; Fisher, 2001; Campbell, 1995).
4. The ordinary randomness that characterizes all markets also affects interest rates. Given the consistency constraints across maturities implied by the term structure, this implies an interesting statistical problem, and many econometric models have been developed to address it (Brigo and Mercurio, 2006).
5. Monetary policy is typically implemented via direct manipulation of the short-term (overnight) interbank rate by the central bank. Adjustments to the short rate propagate to affect rates for longer maturities via market expectations of future short-rate policies (Fisher, 2004; Dai and Singleton, 2003).

An important implication is that the investment performance and risk profiles of fixed-income portfolios depend on the term to maturity and promised interest rate. These idiosyncrasies tend to make fixed-income markets more fragmented and less liquid than for securities without fixed payment promises. A simple example of illiquidity is the difference in prices between on-the-run and off-the-run US Treasury securities of similar maturity (Krishnamurthy, 2002; Fleming, 2003; Amihud et al., 2005). On-the-runs are those involved in the current weekly auction of new bonds. The auction attracts attention and investors, and on-the-runs trade for slightly more than comparable off-the-run issues. This liquidity premium is greater for corporate bonds, which trade in more specialized niche markets depending on issuer credit risk, seniority rules, covenants, embedded options, etc. The problem is exacerbated further for mortgages, commercial loans and other contracts with specific collateral, which tends to be highly idiosyncratic. Indeed, a central attraction of the recent growth in securitization is that it provides a channel for banks to sell off credit risk exposures (i.e., mortgage loans) concentrated in their own local origination markets.

Pricing data are a key resource for measuring market liquidity, as investors demand a liquidity premium (i.e., a lower price and therefore a higher expected return) to hold assets that will likely be more difficult to unload. As the on/off-the-run comparison suggests, the liquidity premium often emerges in a spread between two aptly chosen price or rate series. Such comparisons can be fraught with complexity, however, as price differentials can capture several effects simultaneously. For example, the so-called TED spread between yields on US Treasuries and Eurodollar deposits attracted considerable attention during the 2007–2009 financial crisis as a metric for liquidity strains in the banking and shadow banking sectors.[28] However, it is difficult to disentangle liquidity and credit premiums when both are present, especially in the context of Knightian uncertainty about the underlying risks (Brunnermeier, 2009). Investors can be skittish because they are uncertain about the creditworthiness of the banks, or because they suspect their fellow investors are skittish – or, more likely, both at once.

Bid-ask spreads and market depth

In many over-the-counter (OTC) markets, the predominant liquidity providers are market-makers, traders who commit to provide price quotes upon request for both buy and sell orders. Market-making requires significant capitalization because it exposes the dealer to potentially large order flows. Many markets do not have designated market-makers, however, while others combine some form of dealing with other mechanisms for handling incoming orders. Market microstructure is the study of these institutional details (Harris, 2002; Hasbrouck, 2007), which can have significant implications for market liquidity and its measurement. A simple example of a microstructure measure for liquidity is the bid-ask spread. Competing bidders and offerors jockey to attract transactions. With quotes arranged by price priority, buy (sell) order flow satisfies the most aggressive offerors (bidders) first, and the best ask (bid) price will temporarily rise (fall) as the next most attractive offeror (bidder) is exposed. A narrow (wide) spread is therefore symptomatic of a liquid (illiquid) market, as depicted in Figure 9.1. Fleming (2003), for example, finds that market-makers' reticence as reflected in the bid-ask spread between buy and sell quotes for Treasuries is a more timely and accurate gauge of liquidity than a range of other metrics. In dealer markets, such as those for corporate bonds, Treasuries and mortgage-backed securities, quote information is dispersed among the individual dealers, making the bid-ask spread difficult to observe directly. Roll's (1984) technique estimates the bid-ask spread from reported transaction prices rather than

[28] The TED spread originated in the early 1980s at the Chicago Mercantile Exchange, where it tracked the pricing differential between the 3-month T-bill futures and 3-month Eurodollar futures contracts, which traded in neighboring pits (McCauley, 2001). It is often proxied via direct comparison of the underlying (i.e., non-futures) yields on 3-month T-Bills and 3-month LIBOR.

directly from bid or offer quotations. Assuming that buy and sell orders arrive independently, observed transaction prices will "bounce" at random between quoted bid and ask prices. For example, a purchase order followed by a sale followed by another purchase will create a high-low-high pattern in transaction prices. On average, the magnitude of this negative serial correlation in transaction price changes can be extracted to infer approximately the width of the market-wide spread. For example, Bao et al. (2011) apply this measure to the corporate bond markets, finding significant explanatory power.

Liquidity can also manifest itself in the overall level of prices. Securities perceived as difficult to unload will be less attractive to investors. In equilibrium, this liquidity risk should affect the overall valuation of securities, and it does. Corporate bond prices, for example, are often expressed as a yield spread relative to Treasuries, since the latter are (relatively) highly liquid, and free of default risk, covenants or option clauses. The price thus embeds a number of valuation effects in a single yield premium. Chen et al. (2007) consider several techniques for disentangling the liquidity premium, including bid-ask spreads, the percentage of zero returns in daily time series (evidence of no trading activity), and an econometric model that imputes the liquidity threshold that information-based trades must overcome to be profitable. They collect bid-ask spreads manually from a Bloomberg terminal. Price information is taken from Datastream, which provides average daily prices across all market-makers for a given bond.

Given the intricate nature of the problem, it can be helpful where possible to apply additional information, beyond prices alone. Statistics on trading volume and order flow are an important source, because liquidity is a relative notion. "Market depth" refers to the number of potential buyers or sellers willing and able to take the other side of a trade. Market depth need not be infinite to create a liquid market, but merely adequate relative to the net order flow it must absorb. Figure 9.1 depicts a market with asymmetric depth on the bid and ask sides. A rough gauge of market depth is trading volume. Kyle's lambda exploits this fact; it is typically estimated via a linear regression of daily price change on daily net trading volume. Net trading volume, in turn, is the net of external buy orders less sell orders, and is often proxied as daily volume signed by the direction of the price change.[29]

Exchanges publish daily statistics on trading volumes in their individual securities. On the other hand, trading volumes in OTC markets are often available only through regulatory or voluntary industry surveys (e.g., FRB-NY, 2012a, 2012b; SIFMA, 2012; ICMA, 2012; BIS, 2010). Derivatives exchanges additionally typically publish open-interest volumes describing the number of contracts outstanding

[29] Kyle (1985) develops the original measure. Kyle and Obizhaeva (2011, eq. 23), suggest a simple alternative specification that estimates price impact at higher frequencies and that they argue is stable across markets.

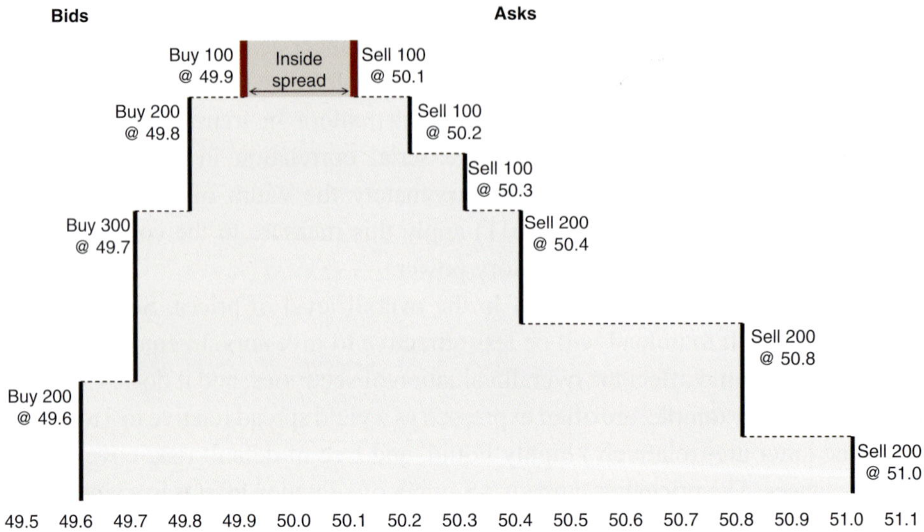

Figure 9.1 Market depth in a simple limit order book. The chart depicts the widening of the bid-ask spread for a limit-order book arranged according to price priority. The most aggressive bid and offer prices rise to the top of the book, producing a relatively tight spread. As buy and sell order flow arrives, these transactions eat through successive layers of the book, revealing wider and wider spreads.

at the daily close of trading. Care must be taken in interpreting transaction volumes, however, as high-frequency algorithmic trading comes to dominate markets. The "algos" generate large volumes, but are usually very thinly capitalized. As the "flash crash" of May 2010 demonstrated, their ability to provide true liquidity in the face of significant one-sided order flow is limited; see CFTC-SEC (2010).

Funding liquidity and asset-liability management

Reliance on the markets to provide liquidity in sufficient depth is often neither necessary nor wise. Tirole (2011) calls this a finance-as-you-go strategy and emphasizes various forms of hedging and reserving as an alternative. These activities occur at the level of the firm, where "funding liquidity" becomes the dominant concern. Funding liquidity is the ability to finance increases in assets and meet obligations as they come due (BCBS, 2008). For example, liquidity concerns may arise from short-term cash needs: an individual portfolio manager may face a margin call due to ordinary price fluctuations. A financial firm in this context is like a fund of funds, in which one business line's cash needs can be balanced against another's cash surplus. Cash is paid, received, and managed at the level of the firm. Bank runs are another example of a funding liquidity shock. Runs are rare but spectacular events (see Gorton, 2010).

Managing liquidity trade-offs at this level is the traditional purview of asset-liability management (ALM) and the firm's treasury group (Choudhry, 2007). An asset-liability committee (ALCO) defines the scope of ALM, sets overall risk appetites (including but not limited to liquidity risk), and coordinates risk and cash needs across the various portfolios and business lines in the firm. The ALCO typically uses an internal funds transfer pricing (FTP) system to aggregate risk exposures across the firm. For clarity and fairness, the FTP should be based on market prices wherever possible, although add-ons and adjustments for position risks are commonplace.

For liquidity purposes, ALM is organized around a set of daily cash-flow projections, sometimes called "cash-flow ladders" that help identify impending cash shortfalls that require funding. Structural patterns in the cash-flow ladders create risk-management challenges. For example, unlike mutual funds and securitization trusts which act as simple or tranched pass-throughs for portfolio income, banks typically have short-term obligations to depositors and bondholders that are not tied directly to the longer-term holdings on the asset side of the balance sheet. In some cases, an individual asset position will be match-funded with a funding liability that mirrors the risk exposure of the asset. Frequently, however, such "micro-hedging" does not apply, and the bank exhibits both duration gaps and repricing gaps (Saunders and Cornett, 2010). A repricing gap calculation measures the extent to which maturing short-maturity liabilities along the ladder will be replaced with equivalents at dearer (or cheaper) interest rates relative to the cash flows on the original positions.

In addition to the timing of cash flows, the sensitivity of position and portfolio values to interest rates is crucial. If sensitivities can be measured accurately, then the portfolio can likely be hedged, sharply reducing the chance of a forced liquidation in unfavorable market conditions. Two common sensitivity measures are duration and DV01. Duration comes from the basic valuation formula for fixed-income instruments as the discounted value of promised cash flows, using risk-adjusted interest rates for the relevant maturities. The simplest case, Macaulay duration, assumes a flat yield curve (i.e., the same interest rate at all maturities) and calculates the first derivative of a bond's value with respect to the constant discount rate.[30] This calculation requires only the current interest rate along with the contractual terms and conditions for the portfolio positions. An alternative approach measures effective duration and effective convexity via direct observation of market interest rate changes and corresponding market valuations (or marks to model) for the portfolio. DV01, also called basis point value (BPV) or delta, is a similar first-order sensitivity,

[30] Expressed as a percentage of current portfolio value, this derivative is measured in units of years; hence the name, "duration." More sophisticated calculations involve the second derivative (the portfolio "convexity"), or derivatives relative to individual discount rates along the maturity spectrum (key-rate durations). See Saunders and Cornett (2010) for further details.

measuring the dollar change in position value for a one basis point parallel shift in the yield curve (Choudhry, 2007, ch. 5).

An important convenience of linear sensitivity measures is ease of aggregation; for instance, the Macaulay duration for a bond portfolio is simply the price-weighted average of the individual bond durations. Linear sensitivities also facilitate hedging. For example, a variety of duration-matching strategies such as "barbells" can hedge the interest-rate sensitivity of a portfolio. The most important limitation of linear sensitivities is that they are local approximations. As the relevant change in interest rates grows, the accuracy of the sensitivity measure will deteriorate. A variety of refinements exists, including convexity, the second-order counterpart to duration. Convexity approximates the curvature of the bond valuation function to parallel yield curve shifts. It too has its limitations; for example, the embedded prepayment option in most residential mortgages creates a negative convexity – if rates fall too low, the value of the mortgage stops increasing at an increasing rate, because borrowers are likely to prepay and refinance to a new mortgage at a lower rate instead.

Duration gap is a supplement to repricing gap for tracking the funding liquidity exposure of an institution as a whole. (Because it measures sensitivities to market interest rates, duration gap is also a market risk metric.) Duration gap is the weighted difference between the duration of assets and duration of liabilities, normalized by total assets so that the gap is measured in years:

$$D_{gap} = [AD_A - LD_L]/A. \tag{9.4}$$

Pension funds, for example, typically have a negative duration gap, since their liabilities are dominated by interest-rate sensitive annuities while their assets are more of mixed bag of fixed-income and equity investments. Banks, on the other hand, will typically have a significantly positive duration gap due to their maturity transformation service that converts short-term deposits into long-term loans. For banks, then, the gap provides a rough gauge of the aggregate unhedged interest-rate risk in the balance sheet. If the gap is judged to be too large, one possible remedial action is to enter into a "balance sheet hedge," often in the form of an interest-rate swap, to reduce the exposure.

Hedging is not a panacea. It can help address liquidity risk, but will succeed only if the firm has sufficient reserves to ride out an episode of illiquidity while keeping the hedge in place. As a contrived example of the challenge, consider a "laddered" portfolio containing equal amounts of bonds at each annual maturity between 1–5 years, delta-hedged with barbell funding consisting of 30-day paper and 10-year debt.[31] The

[31] Statistical arbitrage ("stat arb") in the hedge fund sector is a less contrived example with the same basic characteristic: a strategy that is sure to pay off in the long run, if only the portfolio can weather a series of short runs to get there.

funding amounts can be calibrated to reduce the duration gap to zero, but the short-term funding must roll over after 30 days. If funding liquidity is unavailable at that moment, the hedge collapses simultaneously with the need to sell assets to pay off the maturing short-term debt.[32] Reserving (sometimes called "hoarding"), is a more direct alternative to hedging for liquidity risk management. It involves setting aside a reserve of ultra-liquid assets to defend against a possible funding crisis. For instance, this was institutionalized for banks long ago in the form of regulatory reserve requirements for demand and time deposits. Establishing a source of long-term funding can have a similar effect. Pre-arranging a line of credit from a bank or other lender can be seen as a form of contingent hoarding, whereby the cash reserve is created only at the moment it is needed.

Reserving or hoarding has received renewed regulatory impetus in Basel III (BCBS, 2010) in the form of the liquidity coverage ratio (LCR) and net stable funding ratio (NSFR). Moreover, recent consultative guidance (BCBS, 2012a) highlights a number of intraday measures of funding liquidity for banks:

1. Daily maximum liquidity requirement
2. Available intraday liquidity
3. Total payments
4. Time-specific and other critical obligations
5. Value of customer payments made on behalf of financial institution customers
6. Intraday credit lines extended to financial institution customers
7. Timing of intraday payments
8. Intraday throughput.

The need for intradaily monitoring in part reflects the increased pace of activity for large dealer banks. Chapter 11 of this Handbook contains further discussion on banking regulation, including the Basel rules.

Global liqudity

We close this section by noting that aggregate funding liquidity is a central concern of monetary policy (CGFS, 2011; Goodhart, 2008). An important hazard is the possibility that coordinated liquidity surprises across financial institutions can create a crisis in which the banking system as a whole does not have sufficient reserves – so-called "inside" or private liquidity – to pay off all maturing claims, even if existing reserves are optimally reallocated via the interbank funding markets. Depending on the depth of the shortfall, banks may be forced to sell assets at fire-sale prices (Shleifer and Vishny, 2011). Fire sales can create a feedback loop if they reduce the mark-to-market value of collateral held under margin agreements, or by

[32] Tirole (2011) calls this the "double whammy" problem.

introducing Knightian uncertainty about the true value of assets or collateral. Under such conditions, the central bank should increase money available to the system – "outside" or "official" liquidity, for example via the discount window – to compensate for the private shortfall (see Tirole, 2011, for a discussion of inside and outside liquidity). In the 2008 crisis, the Lehman failure provoked a withdrawal of inside liquidity from the shadow banking sector as investors suddenly became more skeptical of the credit quality of much of the collateral available. The problem was exacerbated by the fact that shadow banks did not have immediate access to central bank liquidity facilities such as lender-of-last-resort services. CGFS (2011, section 3) highlights price-based metrics such as short-term interest rates, and quantity-based measures such as monetary and credit aggregates to measure global liquidity. Domanski et al. (2011) offer a number of specific examples.

9.2.3 Credit risk: the example of debt portfolios

Where liquidity risk is about a portfolio's relationship with an abstract and anonymous market, credit risk concerns the relationship with specific, named counterparties. Credit risk is the probability of an unexpected change in the credit quality – including default – of a contractual counterparty. Credit risk can arise in any contract with a significant financial component. The central information challenge is that "creditworthiness" is a very abstract notion with very concrete implications. A wide range of observable attributes of deals and counterparties seem to correlate with the underlying credit quality, leading to a diversity of modeling approaches and data requirements.[33]

Another important consideration is the risk management context. A loan officer might be interested in the creditworthiness of an individual auto loan applicant or the credit quality of a specific mortgage; attributes of the borrower and Knightian uncertainty will loom large. At the level of the business line or firm, such idiosyncrasies recede and more general economic factors such as the business cycle predominate. For regulators, who look across many firms, the focus is less on general profitability and more on extreme loss events, especially on shocks that might affect many firms simultaneously. These differences in context and objectives affect the details of credit risk measurement.

The data used for measuring, monitoring and managing credit risk present four crucial challenges. First is the scarcity of information on credit quality. Defaults, in particular, tend to be rare events. The translation between concrete observables (e.g.,

[33] We only scratch the surface here; for a more extensive overview at an introductory level, see Baesens and van Gestel (2009). Duffie and Singleton (2003), Bielecki and Rutkowski (2002), Jarrow (2009), McNeil et al. (2005, especially ch. 8 and ch. 9), and Lando (2004) all provide advanced coverage of the technical details.

published financial statements or consumer payment histories) and abstract notions of creditworthiness (e.g., probability of default) is complicated and imprecise. Analysts should typically calibrate models to specific contractual details, borrower cohorts and economic environments, thus slicing the available data even more finely. Second, information is not freely and equally available. Borrowers have a natural incentive to disguise facts that might reveal unflattering aspects of their credit quality, creating an informational asymmetry between borrowers and lenders. This incentive tends to increase as creditworthiness deteriorates; at the extreme, it can create a market failure (i.e., credit rationing) as lenders withdraw in the face of uncertainty. Third, consistent with the rarity of defaults, credit loss distributions are highly skewed. Over time, debt portfolios tend to generate many small gains offset by occasional large losses. This raises the stakes for credit risk modeling enormously; small errors in estimation may threaten profitability or even solvency. Fourth, credit risks may be correlated across borrowers, for example, due to shared exposure to the general economy. This creates a statistical phenomenon of default dependence, whereby loan quality tends to deteriorate simultaneously across the portfolio. Default dependence increases the skewness of the portfolio loss distribution, further raising the stakes.

Loan underwriting and portfolio management

Credit quality is ultimately tied up with the idiosyncrasies of individual counterparties. A first step in credit risk measurement, then, involves reducing this nuanced picture to a set of measurable categories that can be applied consistently across a broad range of exposures. Already, the picture is complicated, as the appropriate categories depend upon the market involved and the purpose for the measurement. The Basel II guidance for banks (BCBS, 2006, paragraph 452, p. 100), for example, defines the crucial condition of "default" to have occurred if either or both of these events have taken place.

- The bank considers that the obligor is unlikely to pay its credit obligations to the banking group in full, without recourse to actions such as realizing security.
- The obligor is past due more than 90 days on any material credit obligation to the banking group.

Note that the presence of adequate collateral does not negate the occurrence of default under these rules. Meanwhile, the International Swaps Dealers' Association (ISDA, 2003, Article IV) defines six "credit events":

- Bankruptcy
- Failure to pay
- Restructuring

- Obligation acceleration
- Obligation default, and
- Repudiation/moratorium.

Because these events are used to define payoff triggers for the ISDA (2002) standardized credit derivative master agreements, there is a sharp focus on clear, objective measurability of each event.[34]

Case study: mis-hedging a credit portfolio

In *The Big Short*, Lewis (2011, pp. 200–219) describes how even a trader who anticipated the crash in the subprime market could lose billions by being too clever by half. By April 2006, Howie Hubler, the head of Morgan Stanley's asset-backed bond trading, had concluded that the subprime mortgage market was headed for a major correction. To leverage this view he had purchased $2 billion in credit default swap (CDS) protection on the mezzanine tranches of subprime CDOs. Hubler believed that eventually he would realize a profit equal to a substantial portion of this $2 billion position when the underlying securities collapsed in value. Unfortunately, this position was costing Morgan Stanley $200 million a year in fees, or about 10% of the asset-backed bond unit's annual profit target. Even the firmest subprime skeptics were not sure just when the collapse would occur.

To relieve the pressure on his running costs, Hubler decided to sell some subprime protection to generate offsetting fee income. Rather than sell protection on the mezzanine tranches which would have effectively canceled the cherished short position, Hubler sold CDS protection on a series of AAA-rated senior tranches in a handful of massive trades with Goldman Sachs, Deutsche Bank and others. However, because these AAA protection fees were much smaller (per notional amount guaranteed) than the mezzanine protection fees he was paying, he had to sell AAA protection on roughly ten times the $2 billion face amount of his original bet to offset the cost of the mezzanine protection. In the end (Lewis, 2011, p. 206), "Hubler had sold credit default swaps on roughly 16 billion dollars' worth of triple-A tranches of CDOs. . . . Between September 2006 and January 2007, the highest-status bond trader inside Morgan Stanley had, for all practical purposes, purchased $16 billion in triple-A CDOs, composed entirely of triple B-rated subprime mortgage bonds, which became valueless when the underlying pools of subprime loans experienced losses of roughly 8 percent."

When the reckoning finally arrived, Hubler's purchased mezzanine protection paid off, but the correction was so severe that the sold AAA protection paid off even more: in late 2007 Morgan Stanley reported a $9.2 billion trading loss. CEO John Mack

[34] However, even here there are complications. Within the FpML schema specification (FPML, 2011), for example, the creditEvent element defines a substitution group with six elements corresponding to the six ISDA types. However, the separate creditEvents element lists 14 possible child elements, adding eight subtypes to the six main ISDA types: failureToPayPrincipal, failureToPayInterest, distressedRatingsDowngrade, maturityExtension, writedown, impliedWritedown, defaultRequirement, and creditEventNotice.

explained to analysts that certain "hedges" against subprime mortgage risk "didn't perform adequately in extraordinary market conditions of late October and November."

The moral of the story is that one should always be skeptical of claims that "It's hedged." Hedges that track well for small movements in prices or rates may be inadequate for larger moves and especially as protection against a major market upheaval. In Morgan Stanley's case, the "hedge" greatly amplified losses in a systemic crisis, illustrating how Black Swan risk is fundamentally different from the more commonplace risks amenable to statistical tools such as VaR that have become so popular in the past two decades.[35]

The definitions above describe key triggering events in the credit process after the fact; equally important are ex ante characterizations of creditworthiness. For corporate borrowers, this typically begins with some form of internal or external credit rating. Treacy and Carey (2000) and OCC (2001) describe the implementation of internal ratings systems in the USA. A basic distinction is whether general borrower creditworthiness (e.g., probability of default, or PD) is estimated separately from the risk of individual contracts (e.g., loss given default, or LGD), or whether the two factors are reduced to a single combined effect. Table 9.1 (adapted from Treacy and Carey, 2000, Table 1) depicts the basic structure of a two-dimensional framework of rating grades, where PD interacts with a fixed LGD figure of 25% to give combined expected loss (EL) percentages in the final column. Rating grades 7–9 in this example are considered "problem assets" deserving more intensive monitoring. Note that both PD and LGD use the same nine categories, as is typical in two-dimensional frameworks. LGD is held constant in this example, but this need not be the case.

In Treacy and Carey's (2000) study, the majority of large banks (60%) measured combined, one-dimensional effects. The finalization of the Basel II guidelines, however, provided new impetus for internal ratings, and two-dimensional frameworks in particular. The internal risk-based (IRB) rules under BCBS (2006, paragraphs 394–437) identify the Pillar I requirements for internal ratings systems at large banks. Schuermann (2005) provides a useful overview of the Basel formulation. The centerpiece of the framework for credit risk is expected loss formula based on probability of default (PD), loss given default (LGD), and exposure at default (EAD):

$$\mathrm{EL} = \mathrm{PD} \times \mathrm{LGD} \times \mathrm{EAD}. \tag{9.5}$$

Note that this construction implicitly assumes that the three factors are statistically independent. This is a strong assumption (Rösch and Scheule, 2006); it may be

[35] This case study summary draws heavily on Rowe (2012).

Table 9.1 *Example of internal ratings buckets*

	Internal category	PD (%)	LGD (%)	EL (%)
1	Extremely low risk	0.00	25	0.000
2	Low risk	0.10	25	0.025
3	Moderate risk	0.30	25	0.075
4	Average risk	1.00	25	0.250
5	Acceptable risk	3.00	25	0.750
6	Borderline risk	6.00	25	1.500
7	Special cases	20.00	25	5.000
8	Substandard	60.00	25	15.000
9	Extremely high risk	100.00	25	25.000

a useful expedient for a regulatory capital standard that is applied uniformly across firms, but may be inappropriate for internal risk management. Allowable PD estimation techniques under Basel II are (BCBS, 2006, paragraph 462): internal default experience, such as a credit scorecard; mapping to external rating grades; or statistical default prediction models.

Credit ratings

The most common sources of external rating grades for corporate debt are the commercial rating agencies (Langohr and Langohr, 2009). The SEC has certified 10 firms as Nationally Recognized Statistical Rating Agencies (NRSROs; see SEC, 2012), but the two most important for traded corporate bonds are Moody's (2009) and Standard and Poor's (S&P, 2012a). Traditional scoring systems are based on straightforward logit/probit regression or – as with the original Altman (1968) Z-score – linear discriminant analysis. Most vendor models, in contrast, are proprietary. Although the two main rating agencies label their ratings buckets similarly, the PD estimates for "corresponding" categories do not line up perfectly across the two agencies (Livingston et al., 2010). Since the mid-1990s, both Moody's and Standard and Poor's have also rated syndicated corporate loans, with a positive net effect on debt issuance (Sufi, 2009). Standardized ratings buckets facilitate the monitoring and modeling of ratings transitions, such as downgrades, for rated securities.

For retail credits such as mortgages and credit cards, external ratings typically come in the form of credit reports and credit scores from one or more of the three dominant providers: Experian, Transunion and Equifax (Avery et al., 2003). Credit reports contain four broad classes of data (Hunt, 2006):

- Identifying information – name, address, Social Security number, etc.
- Credit information – bank accounts, credit cards, etc., including payment histories

- Public records – bankruptcy filings, tax liens, criminal convictions, etc.
- Inquiry count – number of times the consumer has requested his score.

Factors that could encourage legally sanctioned lending bias (race, national origin, religion, gender or marital status) are excluded. Many banks combine credit report data with other information and analysis to create internal credit scores for loan applicants. All three vendors also calculate some customized variant of the FICO score originally developed by Fair, Isaac Corp. The core methodology is proprietary. FICO scores range between 300 and 850, with the vast majority falling between 600 and 800. Among other uses, FICO scores are a crucial input into automated underwriting systems used by mortgage originators (Gates et al., 2002). Increasing reliance on automated, data-driven credit analysis has produced unintended consequences as loan applicants learned to game the system (Foust and Pressman, 2008).

Statistical credit risk models for PD and LGD

Researchers and practitioners have devoted significant effort to modeling both PD and LGD. This literature is extensive, and the details are largely beyond our scope. The basic approach involves designing and fitting a statistical model that will allow the user to extrapolate historical data patterns into the future to estimate distributions of future defaults and losses (Jarrow, 2011). These distributions in turn support risk management, origination decisions, the allocation of capital, and the pricing of credit-risky instruments including loans, bonds and credit derivatives. Credit risk typically involves low-probability events, since managing the legal and operational consequences of a default is usually a costly and difficult process. If default is likely ex ante, the contract should probably be avoided altogether (a.k.a. credit rationing).

There is a remarkable diversity of statistical modeling approaches for credit risk. Ultimately, this is a reflection of the facts that: (a) there are many classes of obligations, each with its idiosyncrasies; and (b) defaults are rare, making pattern detection difficult. Relatively simple approaches include the actuarial or "mortality" techniques of the Credit Suisse's (since discontinued) CreditRisk+ model and the VaR-based CreditMetrics model (RiskMetrics Group, 2007). The former is an example of a two-state (default versus no default) model; the latter tracks transitions across multiple ratings grades as a Markov process. Another common distinction is between so-called structural and reduced-form models (see Hilscher et al., 2008; Giesecke, 2004). Structural models derive from Merton's (1974) application of Black–Scholes analysis of corporate debt as a put option on the assets of the firm. These models, such as the one maintained by Moody's Analytics/KMV, focus on the liabilities structure of the firm and measure a distance-to-default process over time. Where equity prices are available, the estimates can be updated in real time. In other cases, such as retail loan portfolios, the structural model must be modified to estimate the distance to default from other data (Allen et al., 2004). Reduced-form PD models, such as Kamakura, estimate a jump-to-default process without specifying a structural cause such as decline in asset values. Instead, an abstract default intensity process (e.g., a Poisson process) allows for unnamed triggering events to arrive at random. Although there is a debate over appropriate horizons for PDs, standard industry and regulatory practice is to estimate PD over one year (BCBS, 2006, paragraph 447).

Credit risk modeling is also required for LGDs. This is typically reduced to the equivalent problem of modeling of the recovery rate (RR), since LGD = 1 − RR. Again, a detailed survey is beyond our scope (for an overview, see Altman et al., 2004). Peter (2006) provides a valuable overview of the practical data and systems requirements to support LGD estimation and reporting. Modeling recovery rates involves important institutional considerations, including collateral (which may be general or specific), loan guarantees, bond covenants, bankruptcy costs, seniority, margining etc. (Altman et al., 2005; Moody's, 2011). Credit enhancements such as seniority, collateral and guarantees can reduce LGDs significantly (Schuermann, 2005).

The business cycle is important for both PDs and LGDs. Both are strongly correlated with macroeconomic expansions and contractions. As with market risk, backtesting of credit risk models is critical, and the estimation/testing windows should include prior downturns. Indeed, Basel II requires the use of downturn LGD estimates, making them positively correlated with corresponding PD estimates (Bruche and González-Aguado, 2010). From a capital policy perspective, this correlation matters for another reason: if minimum capital standards are set based on credit risk exposures at a point in time (PIT), then greater (lesser) leverage will be tolerated as the economy expands (contracts), encouraging the boom (bust). The procyclical nature of such a capital regime would exacerbate financial instability, leading regulators to prefer "through-the-cycle" (TTC) capital rules (Kashyap and Stein, 2004; Gordy and Howells, 2006; Gonzalez et al., 2004; Rowe, 2002, 2003a, 2003b, 2003c, 2003d). Aside from capital policy, there are other concerns. Because PDs and LGDs tend to move together through the business cycle, the interaction exacerbates credit problems as the two factors reinforce each other in economic downturns, producing fatter tails in the overall loss distribution. Figure 9.2, taken from Mora (2012, p. 20), depicts three portfolio loss distributions with identical expected loss (30%), but with three different 99th-percentile credit VaR thresholds: 65% loss for the (simplistic) normal distribution, 69% loss for a beta distribution with the same volatility as the normal, and 82% for the second beta distribution. This increasing skewness and credit VaR is a typical pattern that results from increasing correlation between PD and LGD for loans in the portfolio.

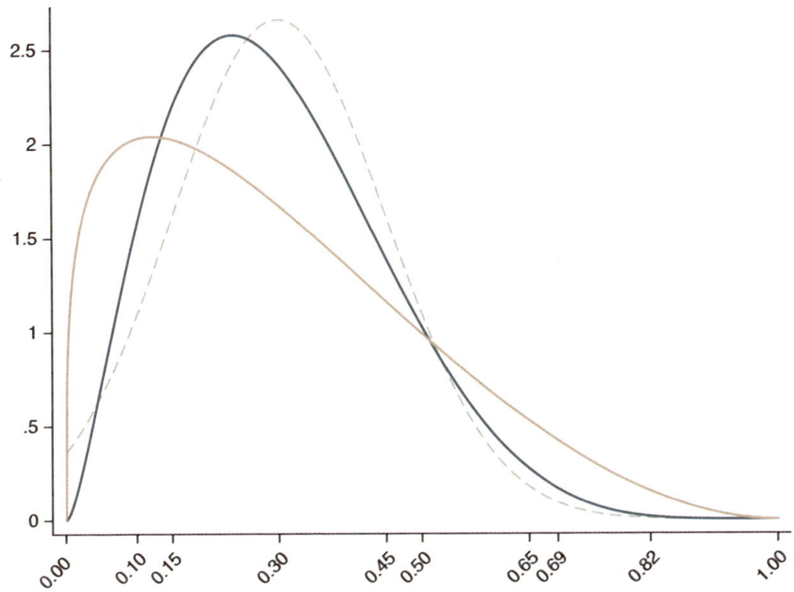

Figure 9.2 Portfolio loss distributions.

Portfolio credit risk

New concerns emerge at the level of the credit portfolio. Similar to the effect of cyclical correlations that raise risk exposures in economic downturns, cross-sectional correlations can do the same within a portfolio at any given time. The general problem is one of diversification: the likelihood that one or more credits in the portfolio will default or be downgraded simultaneously.[36] Such coordinated credit events might arise through common exposure to a single underlying factor, such as an industrial sector or national economy, or via contagion from a failing firm to the credit quality of a counterparty. The Basel (BCBS, 2006) guidelines, for example, posit a single underlying common factor and assign correlations to specific asset classes. Concentrated regional economic exposures were an important motivation for the securitization of mortgage loans, for example.

Capturing the benefits of diversification typically involves estimating a multivariate distribution. For corporate credits, traded equity prices might be used to estimate a correlation matrix for firms' asset returns, which can then be used in a

[36] Simple VaR risk measures tend to break down in this context, because they are not "sub-additive." A sub-additive risk measure rewards diversification of credit risk: applied to two portfolios after they have been merged, a sub-additive measure will be no greater than the sum of their risk measures before they were merged (Hull, 2006). It is easy to create realistic portfolios for which the simple VaR of the combination exceeds the sum of the individual VaRs.

structural default model.[37] More generally, because defaults are extreme events, interest centers on what happens in the multivariate tail. Unfortunately, the workhorse normal distribution tends to under-predict extreme financial events, and this shortcoming is compounded in a portfolio context. "Tail dependence" is the tendency for a multivariate distribution to exhibit extreme outcomes in several dimensions simultaneously, such as correlated defaults. Unlike some other distributions such as the multivariate *t*, the multivariate normal distribution does not allow for tail dependence.

It is frequently difficult to find the data necessary to estimate portfolio credit risk models. Defaults are rare events, and joint defaults are much rarer still. Copula modeling attempts to address this paucity of data by estimating univariate marginal tails, and then combining them into a joint multivariate distribution via the copula function. However, this cannot manufacture the missing information; rather it requires an assumption (often arbitrary) of the appropriate copula function. Securitization can partially address the information shortage by diversifying away certain obvious risk factors, such as the identity of the originating institution.[38] Securitization introduces new challenges, however. For example, most mortgage securitization liabilities are tranched into seniority buckets, and many are re-pooled and structured as CDOs. Such structuring tends to isolate finer slices of the default distribution, making estimation still more difficult (Coval et al., 2009). Lastly, the calculations are typically complicated further by the presence of third-party credit support. For mortgages, this credit enhancement might occur at the individual mortgage, for example via private mortgage insurance, FHA guarantees, etc. Alternatively or additionally, it might occur further up the securitization chain, via credit default swaps (CDS) or other credit derivatives (see Giesecke, 2009; Taksler, 2008; Schönbucher, 2003; and Mengle, 2007).

9.2.4 Operational risk: the example of hedge funds

Operational risk is a wide-ranging category, encompassing hazards as diverse as fraud, software bugs and natural disasters. Roughly speaking, there are two general approaches to operational risk measurement: one focuses on processes, while the other focuses on outcomes. These two general philosophies both have strengths and weaknesses, and they are not mutually exclusive. We offer examples of two frameworks that illustrate the two ends of this continuum, although neither is confined to one extreme. The Basel guidance, with its focus on capital requirements, tends to

[37] Chernih et al. (2006) review the literature. Chaplin (2010, ch. 6) presents a practical estimation procedure using rating transitions rather than asset returns.

[38] Ashcraft and Schuermann, (2008), and Ashcraft et al. (2010) describe the securitization process in greater detail.

emphasize the realized outcomes and the estimation of a loss distribution from operational events. The framework of the Committee of Sponsoring Organizations of the Treadway Commission (COSO) emphasizes operational processes and the need for well-designed controls; it is a largely non-statistical approach.

Hedge funds and their quantitative performance and risk measures

We consider operational risk in the context of hedge funds (where it is a major cause of fund failures) because this uninsured and lightly regulated industry is not constrained to use a particular risk measurement approach, but funds must nonetheless demonstrate the maturity of their operational processes to external investors to earn their trust.

Hedge funds run the gamut from global trading firms to small family-run funds. They provide alternatives to traditional investing and advertise returns to make the standard 20% performance fees palatable (Stowell, 2010). The financial crisis caused regulators and investors alike to review the risks inherent in investing in and running a hedge fund.

The Dodd–Frank Act and hedge fund reporting

The Dodd–Frank Wall Street Reform and Consumer Protection Act of 2010 (Dodd–Frank) made numerous changes to the registration, reporting and record-keeping provisions of the Investment Advisors Act of 1940 (40 Act), specifically targeted at hedge funds. Effective July 2011, advisors to most private equity and hedge funds must now register with the Securities and Exchange Commission (SEC), with some exceptions.[39] While previously hedge fund advisors could voluntarily register with the SEC, Dodd–Frank mandated that the disclosure principles of the 40 Act would apply to hedge funds like all other publicly managed funds, giving potential investors the ability to assess the risks and rewards of investing more effectively.

Dodd–Frank also directed that the SEC be able to obtain information from registered investment advisors that would assist in the assessment of systemic risk by the Financial Stability Oversight Council (FSOC). In joint rulemaking by the Commodity Futures Trading Commission (CFTC) and the SEC, a new Form PF is required to be filed with the Commission by registered hedge fund and private fund advisors, and to the CFTC by certain commodity pool operators and commodity trading advisors; see CFTC-SEC (2011). Form PF is designed to provide the FSOC with information that will enable the Council to assess systemic risk, but it also results in the provision of information to investors that had heretofore not been required to be made public.

While the rule-making pursuant to Dodd–Frank will provide greater transparency by providing investors with data upon which to base investing decisions, the new data do not completely erase the uncertainty of investing in a hedge fund. The financial crisis pointed out that understanding and evaluating operational risks were just as important to

[39] SEC (2011); see also Champ (2012). Other countries have or will be enacting similar legislation.

hedge fund investors as understanding historical returns. There is an uncertainty implicit in the still largely unregulated financial entity that is a hedge fund, and prudent investors might also do their own due diligence to improve their understanding of the level of uncertainty before they commit their funds to a hedge fund manager.

Most disclosures regarding performance by hedge funds prior to the rule-making under Dodd–Frank had been voluntary, generally provided to vendors or fund advisors with the aim of marketing the funds to investors. The new Form PF requires investment advisors to provide specific data, including assets under management; the size, leverage and performance of all private funds subject to the reporting requirements; and basic information regarding the hedge funds. However, the rules do not require advisors to change their existing procedures for calculating and reporting fund performance; see CFTC-SEC (2011, end of p. 71144). Investors will have more information available following implementation of the Dodd–Frank rules, but the risks and uncertainty in investing in hedge funds and private funds remain.

The CFA Institute (formerly the Association of Investment Management and Research, AIMR) has promulgated voluntary Global Investment Performance Standards (GIPS) to provide investment managers and their clients with a well-defined framework for reporting investment results (CFA Institute, 2012).[40] A key concept of GIPS is that firms should present performance for composites that include all fee-paying discretionary accounts managed for a given investment strategy or objective. This quantitative approach is designed to avoid selection bias – the practice of including only accounts with good returns – assuring investors that the manager is presenting comprehensive picture of fund performance.

Hedge funds use risk measures just like any other investment fund to manage their investing returns (Lo, 2010). Examples of quantitative risk and performance measures include the following.

- Volatility (standard deviation) of returns measured in percentage terms, usually provided on an annual basis.
- VaR, usually at a 5% threshold.
- Downside Capture, which measures how correlated a fund is to a market when the market declines. The lower the downside capture, the better the fund preserves wealth during market downturns.
- Drawdown, the percentage drop in cumulative return from the most recent high-water mark.

[40] An international non-profit organization for investment practitioners and educators, AIMR was founded in 1990 from a merger of the Financial Analysts Federation and the Institute of Chartered Financial Analysts. It renamed itself the CFA Institute in 2004.

- Leverage, which magnifies both returns and risks and can force the fund to sell assets at steep discounts to cover margin calls.
- Alpha, an indicator of a manager's contribution to performance, calculated as the fund-specific intercept from a regression of fund returns on market controls in a factor model.
- Performance ratios, such as the traditional Sharpe, Treynor and information ratios, but also more specialized measures such as the Sortino, Stutzer and Omega ratios; see Cogneau and Hübner (2009).

These quantitative measures provide a basis for performance management and guidance on asset allocation and risk management.

Most hedge funds provide their performance results to data vendors such as Lipper TASS (2012), Hedge Fund Research (HFR, 2012) or Dow Jones Credit Suisse (2012, formerly CS/Tremont), which publish manager results and calculated indexes. But data vendors must balance access to data against the possibility of selection bias, since the very nature of the data request filters the funds that qualify or will choose to participate (Lo, 2010, ch. 2). Although industry standards such as GIPS exist, they are voluntary, creating the possibility of incomplete information in reported results. For example, there is the possibility of survivorship bias in the reporting of returns, since poor performers tend to cease operations or stop reporting, leaving behind an unrepresentative population of better-performing active funds (Getmansky et al., 2004). Lipper TASS, for example, maintains a "graveyard" database of funds that have fallen out of the active sample under one of seven attrition categories. One hopes that the more comprehensive regulatory disclosures under Form PF will alleviate some of the biases in voluntary reporting.

The typical hedge fund investor is sophisticated; indeed, most hedge funds solicit investments from professional "fund-of-fund" managers, in addition to high-net worth individuals. Due to the relatively light regulation of hedge funds, concerns especially about "Madoff risk" loom large (Gup, 2012), and fund-of-fund managers engage in a program of operational due diligence before investing (Scharfman, 2008). These investigations cover not only a manager's investment returns but also the internal controls and organizational soundness of the fund where their monies are invested. Operational risk looms large, in addition to market and credit events.

Basel II framework: operational risk capital measurement

In its updated "International Convergence of Capital Measurement and Capital Standards: A Revised Framework," the Basel Committee on Banking Supervision (The Committee) defines operational risk as the risk of loss "due to failed or inadequate processes, people, or systems, or from external events. The definition includes legal risk but excludes strategic and reputational risk" (BCBS, 2006, p. 144; see also Chernobai et al., 2007, ch. 3). In the USA and other large countries, the framework has been incorporated into official regulation for banking firms (see Chapter 10 of this Handbook). The revised framework includes global standards for the management of operational risk, and requires that banks set aside risk capital to cover it. These requirements formalize an approach for measuring and mitigating the risk of operational loss. Because they are part of a minimum capital rule, there is special emphasis on more accurately predicting the probability and impact of loss events.

The framework provides three methods to comply with the requirement to set aside operational risk capital: (1) the Basic Indicator Approach, (2) the Standardized Approach, and (3) the Advanced Measurement Approaches. It was expected that banks would move up the spectrum of approaches as they developed their operational risk systems and practices (BCBS, 2006, p. 144). As the name indicates, the Basic Indicator method is the most simplistic of the approaches. It directs that banks hold capital for operational risk equal to the average over the previous three years of a fixed percentage of positive annual gross income. This approach is easy to implement and easy to manage, but does not give the firm any flexibility in assessing its operational risk capital and has limited responsiveness to a firm's specific needs and characteristics. No qualifying criteria are required to use this method (BCBS, 2006, p. 145).

The Standardized Approach, the next level up, splits the institution into its business units and business lines, and the capital charge is sum of the specified percentage (the so-called "beta") of each business line's average annual gross income over previous three years (BCBS, 2006, p. 146; see also Table 9.2). Banks must demonstrate they have effective risk management and controls in place, including active involvement from a board of directors and senior management, an operational risk system that is "conceptually sound" and sufficient resources in each of the business lines as well as in the risk and control areas ("qualifying criteria"). The bank must also demonstrate a risk management framework and regular audits of the operational risk management process and key inputs (BCBS, 2006, p. 148). Internationally active banks have additional criteria if they follow this approach, including the need to demonstrate the ability to collect, analyze and report loss data across all business lines.

The Advanced Measurement Approaches (AMA) are meant to be utilized by institutions that have developed operational risk measurement systems and have the resources and expertise to utilize the qualitative and quantitative criteria set out by the framework. Under AMA, the institution calculates its own risk capital requirement, using the risk measure generated by the bank's internal operational risk management system. A bank must receive approval from its regulators to implement the AMA (BCBS, 2006, p. 149). The AMA standard has distinct quantitative requirements for complying banks, including the use of an operational risk measurement system that meets the supervisory soundness requirements set out by The Committee, the calculation of regulatory capital using the sum of Expected Loss (EL) and Unexpected Loss (UL), and the use of internal and relevant external loss data, scenario analysis, and business environment and internal control factors as inputs to the capital risk analysis (BCBS, 2006, pp. 150–154).[41]

European banks were the first to be required to comply with the standards of Basel II, starting in 2006, but the Federal Reserve Bank and other US regulators have since adopted similar operational risk capital standards for their member banks (FRB, 2008). Hedge funds are not required to assess or quantify this type of risk, nor to set aside capital to manage this particular risk of loss. But hedge fund investors are increasingly asking questions of their potential investment managers with regard to the types of operational risks, how the firms are prepared to handle them, and what controls are in place to mitigate them.

As part of the Advanced Measurement Approach under Basel II, banks must collect internal loss event data as an input to the capital calculation. Institutions must use a consistent methodology for tracking the data, for example, establishing a monetary threshold at which events are reportable. Banks typically track the monetary amount of loss (or gain), characteristics of the event (including systems and departments involved), and categorize the loss consistent with the Basel risk categories (see Table 9.3).

Banks must also use external events in their risk analysis (BCBS, 2006, p. 153). Most large financial institutions contribute their event loss data to a consortium called ORX, which manages one of the largest databases of operational risk loss data in the industry. The Fitch Group also has created a large database of publicly disclosed operational risk loss events and publishes the data for the benefit of its customers (Chernobai et al., 2011, p. 1685). As part of the Advanced Measurement Approach under Basel II, firms must collect both internal and external data to support their analysis of operational risk capital and ORX and Fitch provide their clients with external data for that analysis.

ORX provides a common reporting standard for its members, which is developed and updated by the consortium's Definitions Working Group. The effort to provide a standard assists regulators as they review and evaluate the banks in a consistent manner. But these standards are only as good as the quality of their application by the reporting members; selection bias in reporting events is a possibility, as is the probability that not all operational risk events are reported internally, and so not captured and reported by the members.

[41] One variant of the AMA method is the Loss Distribution Approach, in which the bank uses its internal data to estimate two probability distribution functions for each business line and risk type. Chernobai et al. (2007) cover this approach in detail.

Table 9.2 *Standardized approach – business line mapping*

Level 1	Beta factors	Level 2	Activity groups
Corporate finance	18%	Corporate finance Municipal/government finance Merchant banking Advisory services	Mergers and acquisitions, underwriting, privatizations, securitization, research, debt (government, high yield), equity, syndications, IPO, secondary private placements
Trading and sales	18%	Sales Market making Proprietary positions Treasury	Fixed income, equity, foreign exchanges, commodities, credit, funding, own position securities, lending and repos, brokerage, debt, prime brokerage
Retail banking	12%	Retail banking Private banking Card services	Retail lending and deposits, banking services, trust and estates Private lending and deposits, banking services, trust and estates, investment advice Merchant/commercial/corporate cards, private labels and retail
Commercial banking	15%	Commercial banking	Project finance, real estate, export finance, trade finance, factoring, leasing, lending, guarantees, bills of exchange
Payment and settlement	18%	External clients	Payments and collections, funds transfer, clearing and settlement
Agency services	15%	Custody Corporate agency Corporate trust	Escrow, depository receipts, securities lending (customers) Corporate actions Issuer and paying agents
Asset		management Discretionary fund management Non-discretionary fund management	12% Pooled, segregated, retail, institutional, closed, open, private equity Pooled, segregated, retail, institutional, closed, open
Retail brokerage	12%	Retail brokerage	Execution and full service

The Basel categories for risk (see Table 9.3) are

- internal and external fraud,
- employment practices and workplace safety,
- clients, products and business practices,
- damage to physical assets,
- business disruptions and failures, and
- execution, delivery and process management.

Table 9.3 *Basel risk categories*

Basel risk category	Examples
Internal fraud	Misappropriation of assets, tax evasion, intentional mismarking of positions
External fraud	Theft of information, hacking damage, third-party theft and forgery
Employment practices and workplace safety	Discrimination, workers compensation, employee health and safety
Clients, products and business practices	Market manipulation, antitrust, improper trade, product defects, fiduciary breaches, account churning
Damage to physical assets	Natural disasters, terrorism, vandalism
Business disruptions and failures	Utility disruptions, software failures, hardware failures
Execution, delivery and process management	Data entry errors, accounting errors, failed mandatory reporting, negligent loss of client assets

The most visible operational risk events for financial services companies in current times are rogue trader events, but events with no publicity, such as system outages, also can easily affect a firm's bottom line.[42]

Quantitative measures and statistical models for operational risk

Firms measuring operational risk capital using the AMA use data collected from internal (realized losses) and external events, as well as other indicators, and feed statistical models to calculate their risk capital. The common methods used today by the financial industry to calculate operational risk capital include: the loss distribution approach (LDA), the scenario based approach, and methods based on extreme value theory (EVT), or a hybrid of all three (Medova and Berg-Yuen, 2008; Chernobai et al., 2007). The calculation of operational risk capital is left to the discretion of the bank, although the methods used are subject to regulatory scrutiny. Researchers have reviewed the various modeling approaches and reached various conclusions about the sufficiency of the quantitative analysis dictated by Basel II. Cope et al. (2009) argue, for example, that the Loss Distribution Approach (LDA) can be overly sensitive to a small set of low-frequency, high-severity losses, further increasing the data sufficiency requirements and increasing the volatility and uncertainty of the capital estimates (Cope, et al., 2009, p. 3). Medova and Berg-Yuen (2008, p. 1) argue that the Extreme Risk Capital Model provides a suitable view of a bank's operational risk profile for risk management purposes.

[42] Examples of banks with massive rogue trading losses include Barings Bank (1995), Société Générale (2008) and UBS (2011). See Gapper (2011).

Basel II provides guidance to banks on the measures that they should incorporate in their analysis of operational risk, using data obtained from operational risk events. Firms use a variety of modeling methods, including Pareto, Poisson and binomial to calculate loss distributions using data gathered (Chernobai et al., 2007, chs. 5–7). Statistical measurement provides a guide for normal risk, but operational risk events of concern are fat tail, or Black Swan, events, like the terrorist attack on September 11 or the losses incurred by a rogue trader like Nick Leeson, in which enormous sums are lost. When such events occur, institutions using AMA must increase their estimates for fraud loss and reevaluate whether these types of events are really one in 100 year events, or should be considered a more frequent possibility.

COSO framework: internal controls

Basel II mandates operational risk management approaches for banks as part of a regulatory capital framework. This implies a need for an even-handed framework that can be used to compare a wide range of banks of various sizes and business models in multiple jurisdictions – hence its standard and comprehensive lists of business lines and risk types. It also implies a need to map risk exposures into loss probabilities for capital calculations. The COSO approach emerged instead from an internal audit tradition, and this heritage shows in its framework. The COSO emerged from the Treadway Commission, a private-sector accounting expert group formed in the 1980s in response to regulatory concerns about bribery and accounting fraud. After the initial Treadway report (COSO, 1987), COSO developed a benchmark framework against which companies and organizations could evaluate internal control systems for the management of risk.[43] In a similar fashion, the US Sarbanes–Oxley Act (SOX), requires officers of SEC-registrant companies to attest that they have a strong internal control framework that supports the integrity of the firm's financial statements. The Sarbanes–Oxley Act of 2002 was passed in the wake of the Enron scandal and applies to all public companies (and therefore is not required for hedge funds).

The COSO framework is organized into the "COSO cube," the cross product of business objectives, business lines and risk components. There are four categories of business objectives (excerpted here from COSO, 2004):

- Strategic – high-level goals, aligned with and supporting its mission
- Operations – effective and efficient use of its resources
- Reporting – reliability of reporting
- Compliance – compliance with applicable laws and regulations.

[43] COSO (1992) was the first version of the framework, and it has been updated repeatedly since. COSO (2004) expands the scope to an enterprise risk management framework. COSO (2009) is the most recent official version, while efforts progress on the next update.

There are eight risk components (again from COSO, 2004):

- Internal environment – The internal environment encompasses the tone of an organization, and sets the basis for how risk is viewed and addressed by an entity's people, including risk management philosophy and risk appetite, integrity and ethical values, and the environment in which they operate.
- Objective setting – Objectives must exist before management can identify potential events affecting their achievement. Enterprise risk management ensures that management has in place a process to set objectives and that the chosen objectives support and align with the entity's mission and are consistent with its risk appetite.
- Event identification – Internal and external events affecting achievement of an entity's objectives must be identified, distinguishing between risks and opportunities. Opportunities are channeled back to management's strategy or objective-setting processes.
- Risk assessment – Risks are analyzed, considering likelihood and impact, as a basis for determining how they should be managed. Risks are assessed on an inherent and a residual basis.
- Risk response – Management selects risk responses – avoiding, accepting, reducing or sharing risk – developing a set of actions to align risks with the entity's risk tolerances and risk appetite.
- Control activities – Policies and procedures are established and implemented to help ensure the risk responses are effectively carried out.
- Information and communication – Relevant information is identified, captured and communicated in a form and time frame that enable people to carry out their responsibilities. Effective communication also occurs in a broader sense, flowing down, across, and up the entity.
- Monitoring – The entirety of enterprise risk management is monitored and modifications made as necessary. Monitoring is accomplished through ongoing management activities, separate evaluations, or both.

The framework at this level is quite abstract, which makes it possible to tailor the rules to a range of firms in a variety of industries. Befitting the generality of the framework, it is lacking any specific guidance here of the sort seen in the Basel rules on loss event types, loss distributions or business lines. There is a clear emphasis on goals (risk appetite, objectives), awareness (event identification, risk assessment, monitoring), and reaction (control activities, risk response, communication). COSO has been criticized as too complicated (Shaw, 2006). It would be facile to conclude that a framework designed by large, for-profit accounting consultancies would be intricate and challenging to implement. However, the same complaint has been levelled against the SOX and Basel requirements. Another possible conclusion is simply that modeling, monitoring and controlling operational processes in a large organization is hard work.

Hedge fund due diligence

The foregoing outlines two very different approaches to operational risk monitoring, reflecting two distinct traditions. The Basel process comes from the data- and probability-centric context of international bank capital regulation. It provides detailed measurement strategies specific to finance. The COSO process comes from the controls- and accountability-centric context of internal auditing. It is much less focused on the statistical distribution of loss events, and more focused on processes and how they are managed. This distinction between operational risk viewpoints is similar to the contrast that Brown (2012) draws between "bottom-up" (e.g., Basel) approaches and "top-down" approaches based on evidence of legal and regulatory problems in operational due-diligence reports (see also Brown et al., 2008).

In this context, hedge funds represent an interesting accidental experiment. While neither framework is a requirement for hedge funds (nor is SOX), they nonetheless face a very immediate operational risk challenge. As they compete for funds against each other and against well-capitalized, regulated, and (relatively) transparent banks, they must regularly convince skeptical investors that their funds will be safely kept. Brown (2012), for example, finds that operational risk is a significant factor. To a limited extent, funds can allay these concerns with institutional practices such as managed accounts that segregate customer monies and hold title in the investor's name (Giraud, 2005). Moreover, given the private nature of hedge funds, they may have greater opportunity to invest in technology and internal processes than a publicly traded firm, which may be more concerned with short-term revenues. Given that most hedge fund managers are stakeholders in their funds, their incentive for developing strong internal controls and back-up processes should be motivated as much out of self-interest as fiduciary responsibility to outside investors.

Still, large investors and fund-of-fund managers should and do insist on formal and due-diligence reviews of the operations of candidate hedge funds. These reviews focus on fraud, conflicts of interest, incompetence and the adequacy of processes and systems (Scharfman, 2008). They typically include a detailed pre-investment review, including background checks and interviews of key staff. Scharfman (2008) recommends a scorecard approach to guarantee that all relevant dimensions of the review are covered and comparable across funds. On the other side, a scorecard approach is also modeled by the Managed Funds Association (MFA, 2007) as part of its recommended practice for funds in their interactions with potential investors. Regular (e.g., annual), on-site, post-investment reviews confirm that no material deterioration in operational quality has occurred. Occasional on-site reviews can be supplemented with continuous monitoring of industry news and legal events such as court docket reports and new case filings. MFA (2009), in their best-practice recommendations, focus heavily on operational issues and processes:

- Disclosure to investors
- Valuation governance to mitigate model risk and misreporting
- Risk management (market, credit, liquidity, operational, legal/compliance)
- Operational processes (trading and business)
- Compliance, ethics and conflicts of interest
- Anti-money laundering compliance
- Business continuity and disaster recovery.

Interestingly, both the internal and third-party evaluation and monitoring approaches typical in the hedge-fund industry seem closer to the process-centric COSO approach than the loss-distribution approach of the Basel framework designed specifically for financial institutions.

9.3 Conclusions

Financial risk management is the industry's analytical response to the challenges presented by an imperfect world. The analysis requires information as its key input, which in turn requires enormous amounts of data. The output is an array of concrete decisions, such as whether to make or deny a loan, whether to add capital to the firm, or how to hedge a position. Knight (1921) introduced a simple but fundamental classification of the information challenges we face: between Knightian risks, which can be successfully addressed with statistical tools, and Knightian uncertainties, which cannot. A clear emphasis on statistical tools emerges in our overview of risk management models and practices. This is largely an artifact of a modern financial culture, which is awash in data and heavily informed by a deeply rooted mindset of empirical and statistical models of markets (Mackenzie, 2008). These statistical models of financial risk are valuable and appropriate, but they are not all-encompassing. Risk managers should also give due weight to uncertainties and the tools that can address them. For example, there are statistical tools to apply extreme value theory to rare events, but these may often be better approached using scenario analysis and stress testing. Similarly, subtle organizational hazards can often be tracked in data on operational risk events, but may often be better approached with "softer" tools such as formal processes, risk controls and governance, internal and external transparency, and the alignment of risk-taking incentives. Risk managers should never forget that tools are only as good as the people that use them, and that confining oneself to the output of a model when an open-eyed view of the market would direct a different outcome is not effective risk management. Good risk management requires toolkits to handle both Knightian risk and uncertainty.

In summary, this whirlwind tour of risk management information generates two broad lessons. The first is simply that information matters. While there will always be a large component of randomness in the world – and especially in financial markets – there are also areas of ignorance that can be illuminated simply by applying oneself thoughtfully and diligently to better understand the data. To paraphrase Arnold Palmer, the more you inform yourself, the luckier you will get. Where data are available, statistical tools will likely help, but the absence of high-quality numeric data does not imply the absence of potential hazards. Second, risk management is an enormous and enormously intricate topic. The long list of topics mentioned here is merely the tip of an iceberg. We hope the references provided will serve as useful pointers to more detailed investigations.

Bibliography

Addy, Christopher, 2008, Operational Due Diligence for Hedge Funds, presented at the Public Pension Financial Forum.

Alexander, Carol, 2008, *Market Risk Analysis, Volume II: Practical Financial Econometrics*, Wiley Finance.

Alexander, Carol, 2009, *Market Risk Analysis, Volume IV: Value At Risk Models*, Wiley Finance.

Alexander, Carol and Elizabeth Sheedy, 2008, Developing a stress testing framework based on market risk models, *Journal of Banking and Finance*, **32**, 2220–2236.

Allen, Linda, Gayle DeLong and Anthony Saunders, 2004, Issues in the credit risk modeling of retail markets, *Journal of Banking and Finance*, **28**(4), 727–752.

Altman, Edward, 1968, Financial ratios, discriminant analysis and the prediction of corporate bankruptcy, *Journal of Finance*, **23**(4), 589–609.

Altman, Edward, Andrea Resti and Andrea Sironi, 2004, Default recovery rates in credit risk modelling: a review of the literature and empirical evidence, *Economic Notes*, **33**(2), 183–208.

Altman, Edward, Brooks Brady, Andrea Resti and Andrea Sironi, 2005, The link between default recovery rates: theory, empirical evidence, and implications, *Journal of Business*, **78**(6), 2203–2227.

Amihud, Yakov, Haim Mendelson and Lasse Pedersen, 2005, Liquidity and asset prices, *Foundations and Trends in Finance*, **1**(4), 270–364.

Andersen, Torben, Tim Bollerslev, Francis X. Diebold and Paul Labys, 2000, Great realizations, *Risk*, **18**, 105–108.

Andersen, Torben, Tim Bollerslev, Peter Christoffersen and Francis X. Diebold, 2007, Practical volatility and correlation modeling for financial market risk management, in: M. Carey and R. M. Stulz (eds.), *The Risks of Financial Institutions*, University of Chicago Press, ch. 11, 513–548, `www.nber.org/chapters/c9618.pdf`

Andersen, Torben, Tim Bollerslev, Peter Christoffersen and Francis X. Diebold, 2011, Financial Risk Measurement for Financial Risk Management, PIER Working Paper 11–037, University of Pennsylvania, November, `http://ssrn.com/abstract=1955935`

Ang, Andrew and Joseph Chen, 2002, Asymmetric correlations of equity portfolios, *Journal of Financial Economics*, **63**(3), 443–494.

Ashcraft, Adam and Til Schuermann, 2008, Understanding the securitization of subprime mortgage credit, *Foundations and Trends in Finance*, **2**(3), 191–309.

Avery, Robert, Paul Calem, Glenn Canner and Rafael Bostic, 2003, An overview of consumer data and credit reporting, *Federal Reserve Bulletin*, February, 47–73.

Baesens, Bart and Tony van Gestel, 2009, *Credit Risk Management: Basic Concepts*, Oxford University Press.

Bank for International Settlements (BIS), 2010, Triennial Central Bank Survey Report on Global Foreign Exchange Market Activity in 2010, Technical Report, December, `www.bis.org/publ/rpfxf10t.htm`

Bao, Jack, Jun Pan and Jiang Wang, 2011, The illiquidity of corporate bonds, *Journal of Finance*, **66**(3), 911–946.

Basel Committee on Banking Supervision (BCBS), 2005, Amendment to the Capital Accord to Incorporate Market Risks, updated November, Technical Report, Bank for International Settlements, `www.bis.org/publ/bcbs119.htm`

Basel Committee on Banking Supervision (BCBS), 2006, *International Convergence of Capital Measurement and Capital Standards: A Revised Framework*, Bank for International Settlements, June, `www.bis.org/publ/bcbs128.htm`

Basel Committee on Banking Supervision (BCBS), 2008, Liquidity Risk: Management and Supervisory Challenges, Technical Report, Bank for International Settlements, February, `www.bis.org/publ/bcbs136.pdf`

Basel Committee on Banking Supervision (BCBS), 2010, Basel III: International Framework for Liquidity Risk Measurement, Standards and Monitoring, Technical Report, Bank for International Settlements, December, `www.bis.org/publ/bcbs188.htm`

Basel Committee on Banking Supervision (BCBS), 2011a, Revisions to the Basel II Market Risk Framework, updated as of 31 December 2010, Technical Report, Bank for International Settlements, February, `www.bis.org/publ/bcbs193.htm`

Basel Committee on Banking Supervision (BCBS), 2011b, Messages From the Academic Literature on Risk Measurement for the Trading Book, Working Paper No. 19, Bank for International Settlements, January.

Basel Committee on Banking Supervision (BCBS), 2012a, Monitoring Indicators for Intraday Liquidity Management, Consultative Document, Bank for International Settlements, July.

Basel Committee on Banking Supervision (BCBS), 2012b, Fundamental Review of the Trading Book, Consultative Document, Bank for International Settlements, May, `www.bis.org/publ/bcbs219.htm`

Bernstein, Peter, 1996, *Against the Gods: The Remarkable Story of Risk*, Wiley.

Bervas, Arnaud, 2006, Market liquidity and its incorporation into risk management, *Banque de France Financial Stability Review*, **8**, May, 63–79.

Bianchetti, Marco, 2009, Bootstrapping the illiquidity: multiple yield curves construction for market coherent forward rates estimation, in: F. Mercurio (ed.), *Modeling Interest Rates: Latest Advances for Derivatives Pricing*, Risk Books.

Bielecki, Tomasz and Marek Rutkowski, 2010, *Credit Risk: Modeling, Valuation, and Hedging*, Springer Verlag.

Bisias, Dimitrios, Mark Flood, Andrew Lo and Stavros Valavanis, 2012, A survey of systemic risk analytics, *Annual Review of Financial Economics*, **4**, 255–296. www.annualreviews.org/doi/abs/10.1146/annurev-financial-110311-101754

Board of Governors of the Federal Reserve (FRB), 2008, Interagency Statement – U.S. Implementation of Basel II Advanced Approaches Framework, Policy Statement, July 8, www.federalreserve.gov/boarddocs/srletters/2008/SR0804.htm

Bollerslev, Tim, 2010, Glossary to ARCH (GARCH*), in: T. Bollerslev, J. Russell and M. Watson (eds.), *Volatility and Time Series Econometrics*, Oxford University Press, ch. 8.

Bookstaber, Richard, 2009, *Testimony of Richard Bookstaber Submitted to the U.S. House of Representatives*, Committee on Science and Technology, Subcommittee on Investigations and Oversight, For the Hearing: 'The Risks of Financial Modeling: VaR and the Economic Meltdown', Congressional Testimony, September 10, 2009, http://bookstaber.com/rick/Testimony_of_Richard_Bookstaber_to_House_20090910.pdf

Bootle, Roger, 2011, *The Trouble with Markets: Saving Capitalism from Itself*, Nicholas Brealey Publishing.

Brigo, Damiano and Fabio Mercurio, 2006, *Interest Rate Models – Theory and Practice: With Smile, Inflation and Credit*, Springer Verlag.

Briner, Beat and Gregory Connor, 2008, How much structure is best? A comparison of market model, factor model and unstructured equity covariance matrices, *Journal of Risk*, **10**(4), 3–30.

Brown, Stephen, 2012, Quantitative Measures of Operational Risk: an Application to Funds Management, Working Paper, New York University, May, http://papers.ssrn.com/sol3/papers.cfm?abstract_id=2025112

Brown, Stephen, W. Goetzmann, B. Liang and C. Schwarz, 2008, Estimating Operational Risk for Hedge Funds: The ω-Score, Working Paper, Yale University, May, http://papers.ssrn.com/sol3/papers.cfm?abstract_id=1086341

Brownlees, Christian, Robert Engle and Bryan Kelly, 2011, A practical guide to volatility forecasting through calm and storm, *Journal of Risk*, **14**(2), 3–22.

Bruche, Max and Carlos González-Aguado, 2010, Recovery rates, default probabilities, and the credit cycle, *Journal of Banking and Finance*, **34**(4), 754–764.

Brunnermeier, Markus, 2009, Deciphering the liquidity and credit crunch 2007–2008, *Journal of Economic Perspectives*, **23**(1), 77–100, www.aeaweb.org/articles.php?doi=10.1257/jep.23.1.77

Brunnermeier, Markus and Lasse Pedersen, 2009, Market liquidity and funding liquidity, *Review of Financial Studies*, **22**(6), 2201–2238.

Brunnermeier Markus, Thomas Eisenbach and Yuliy Sannikov, 2012, Macroeconomics with Financial Frictions: A Survey, Working Paper, Princeton University, March, http://scholar.princeton.edu/markus/announcements/macroeconomics-financial-frictions

Campbell, John, 1995, Some lessons from the yield curve, *Journal of Economic Perspectives*, **9**(3), 129–152.

Castrén, Olli, John Fell and Nico Valckx, 2009, Cross-border macro stress-testing: progress and future challenges for the EU, in: M. Quagliariello (ed.), *Stress-testing the Banking System: Methodologies and Applications*, Cambridge University Press, ch. 15.

Center for Research in Securities Prices (CRSP), 2000, CRSP Data Description Guide for the CRSP US Stock Database and CRSP US Indices Database, Technical Manual, University of Chicago, www.crsp.com/documentation/product/stkind/data_descriptions_guide.pdf

CFA Institute, 2012, Global Investment Performance Standards: The GIPS Standards and Interpretations, Internet resource, accessed August 18, 2012, www.gipsstandards.org/standards/index.html

Champ, Norm, 2012, What SEC Registration Means for Hedge Fund Advisers, speech delivered May 11, 2012 to the New York City Bar, www.sec.gov/news/speech/2012/spch051112nc.htm

Chaplin, Geoff, 2010, *Credit Derivatives: Trading, Investing, and Risk Management*, Second edition, Wiley.

Chen, Long, David, Lesmond and Jason, Wei, 2007, Corporate yield spreads and bond liquidity, *Journal of Finance*, **72**(1), 119–149.

Chernih, Andrew, Steven Vanduffel and Luc Henrard, 2006, Asset Correlations: A Literature Review and Analysis of the Impact of Dependent Loss Given Defaults, Working Paper, Katholieke Universiteit Leuven, November.

Chernobai, Anna S., Svetlozar Rachev and Frank J. Fabozzi, 2007, *Operational Risk: A Guide to Basel II Capital Requirements, Models, and Analysis*, Wiley.

Chernobai, Anna, P. Jorion and F. Yu, 2011, The determinants of operational risk in U.S. financial institutions, *Journal of Financial and Quantitative Analysis*, **46**(6), 1683–1725.

Choo, Chun Wei, 2005, *The Knowing Organization: How Organizations Use Information to Construct Meaning, Create Knowledge, and Make Decisions*, Oxford University Press.

Choudhry, Moorad, 2007, *Bank Asset and Liability Management: Strategy, Trading, Analysis*, Wiley Finance.

Cogneau, Philippe and Georges Hübner, 2009, *The 101 Ways to Measure Portfolio Performance*, Technical Report, University of Liege, January.

Coleman, Thomas, 2012, *Quantitative Risk Management: A Practical Guide to Financial Risk*, Wiley Finance.

Committee of Sponsoring Organizations of the Treadway Commission (COSO), 1987, Report of the National Commission on Fraudulent Financial Reporting, Technical Report, October, www.coso.org/Publications/NCFFR.pdf

Committee of Sponsoring Organizations of the Treadway Commission (COSO), 1992, COSO Report: Internal Control – An Integrated Framework, Technical Report, www.coso.org/IC.htm

Committee of Sponsoring Organizations of the Treadway Commission (COSO), 2004, Enterprise Risk Management – Integrated Framework: Executive Summary, Technical Report, September.

Committee of Sponsoring Organizations of the Treadway Commission (COSO), 2009, Internal Control – Integrated Framework: Guidance on Monitoring Internal Control Systems, Technical Report, American Institute of CPAs (AICPA).

Committee on the Global Financial System (CGFS), 2005, Stress Testing at Major Financial Institutions: Survey Results and Practice, Technical Report, January, www.bis.org/publ/cgfs24.pdf

Committee on the Global Financial System (CGFS), 2011, Global liquidity – Concept, Measurement and Policy Implications, CGFS Publications No 45, Bank for International Settlements, November, www.bis.org/publ/cgfs45.htm.

Commodity Futures Trading Commission and Securities and Exchange Commission (CFTC-SEC), 2010, Findings Regarding the Market Events of May 6, 2010: Report of the Staffs of the CFTC and SEC to the Joint Advisory Committee on Emerging Regulatory Issues, Technical Report, September, www.sec.gov/news/studies/2010/marketevents-report.pdf

Commodity Futures Trading Commission and Securities and Exchange Commission (CFTC-SEC), 2011, Reporting by Investment Advisers to Private Funds and Certain Commodity Pool Operators and Commodity Trading Advisors on Form PF, *Federal Register*, **76**(221), November 16, 71128–71239, www.gpo.gov/fdsys/pkg/FR-2011-11-16/pdf/2011-28549.pdf

Cook, Timothy and Robert Laroche, 1993, *Instruments of the Money Market*, Federal Reserve Bank of Richmond, www.richmondfed.org/publications/research/special_reports/instruments_of_the_money_market/index.cfm

Cope, Eric W., G. Mignola, G. Antonini and R. Uggocioni, 2009, Challenges and pitfalls in measuring operational risk from loss data, *Journal of Operational Risk*, **4**(4), 3–27.

Coval, Joshua, Jakub Jurek and Erik Stafford, 2009, The economics of structured finance, *Journal of Economic Perspectives*, **23**(1), 3–25.

Dai, Qiang and Kenneth Singleton, 2003, Term structure dynamics in theory and reality, *Review of Financial Studies*, **16**(3), 631–678.

Daníelsson, Jón, 2002, The emperor has no clothes: Limits to risk modelling, *Journal of Banking and Finance*, **26**(7), 1273–1296.

Daníelsson, Jón, 2011, *Financial Risk Forecasting: The Theory and Practice of Forecasting Market Risk with Implementation in R and Matlab*, Wiley.

Der Kiureghian, Armen and Ove Ditlevsen, 2009, Aleatoric or epistemic? Does it matter? *Structural Safety*, **31**(2), 105–112.

Domanski, Dietrich, Ingo Fender and Patrick McGuire, 2011, Assessing global liquidity, *BIS Quarterly Review*, December, 57–71, www.bis.org/publ/qtrpdf/r_q1112.htm

Dow Jones Credit Suisse, 2012, Core Hedge Fund Indexes, Internet resource, accessed August 18, 2012, www.hedgeindex.com/hedgeindex/en/default.aspx?cy=USD

Duffie, Darrell and Kenneth Singleton, 2003, *Credit Risk: Pricing, Measurement, and Management*, Princeton University Press.

The Economist, 2010, The gods strike back, February 11, www.economist.com/node/15474137/print

The Economist, 2012, The LIBOR scandal: the rotten heart of finance, July 7, www.economist.com/node/21558281/print

Elton, Edwin, Martin Gruber, Stephen Brown and William Goetzmann, 2009, *Modern Portfolio Theory and Investment Analysis*, Wiley.

Embrechts, Paul and J. Nešlehová, Quantitative Models for Operational Risk, www.math.ethz.ch/~baltes/ftp/OpRisk-talk.pdf

Engle, Robert, 2001, GARCH 101: The use of ARCH/GARCH models in applied econometrics, *Journal of Economic Perspectives*, **15**(4), 157–168.

Fabozzi, Frank and Steven Mann, 2011, *The Handbook of Fixed Income Securities*, McGraw Hill.

Federal Reserve Bank of New York (FRB-NY), 2012a, FX Volume Survey, Internet resource, downloaded July 11, 2012, www.newyorkfed.org/FXC/volumesurvey/

Federal Reserve Bank of New York (FRB-NY), 2012b, Intradaily Liquidity Flows, Report of the Payments Risk Committee, March, www.newyorkfed.org/prc/prc_120329.pdf

Financial Products Markup Language (FpML), 2011, *FpML 5.2 Recommendation – February 28, 2011*, Technical Document, February, www.fpml.org/spec/fpml-5-2-6-rec-2/

Fisher, Mark, 2001, Forces that Shape the Yield Curve: Parts I and II, Working Paper 2001–3, Federal Reserve Bank of Atlanta, March, www.frbatlanta.org/pubs/wp/working_paper_2001-3-abstract.cfm

Fisher, Mark, 2004, Modeling the term structure of interest rates: an introduction, *Federal Reserve Bank of Atlanta Economic Review*, third quarter, 41–62.

Fixler, Dennis J. and Teresa L. Weadock, 2006, Gross domestic product: revisions and source data, *Survey of Current Business*, **86**(2), 11–15. www.bea.gov/scb/pdf/2006/02february/0206_gdp.pdf

Fleming, Michael, 2003, Measuring treasury market liquidity, *Federal Reserve Bank of New York Economic Policy Review*, September, 83–108, www.ny.frb.org/research/epr/03v09n3/0309flem.pdf

Flood, Mark, Allan Mendelowitz and William Nichols, 2012, Monitoring financial stability in a complex world, in: V. Lemieux (ed.), *Financial Analysis and Risk Management: Data Governance, Analytics and Life Cycle Management*, Springer Verlag, ch. 2, 15–46.

Foust, Dean and Aaron Pressman, 2008, Credit scores: not-so-magic numbers, *Business Week*, February 7.

Frésard, Laurent, Christophe Pérignon and Anders Wilhelmsson, 2011, The pernicious effects of contaminated data in risk management, *Journal of Banking and Finance*, **35**(10), 2569–2583.

Gapper, John, 2011, *How to Be a Rogue Trader*, Portfolio/Penguin Books.

Gates, Susan Wharton, Vanessa Gail Perry and Peter M. Zorn, 2002, Automated underwriting in mortgage lending: good news for the underserved? *Housing Policy Debate*, **13**(2), 369–391.

Getmansky, Mila, Andrew Lo and Shauna Mei, 2004, Sifting through the wreckage: lessons from recent hedge-fund liquidations, *Journal of Investment Management*, **2**(4), 6–38.

Giesecke, Kay, 2004, Credit risk modeling and valuation: an introduction, in: D. Shimko (ed.), *Credit Risk: Models and Management*, Volume 2, Risk Books, 487–525.

Giesecke, Kay, 2009, An Overview of Credit Derivatives, Working Paper, Stanford University, March.

Giraud, Jean-Renė, 2005, Mitigating Hedge Funds' Operational Risks: Benefits and Limitations of Managed Account Platforms, Technical Report, EDHEC Business School, June.

Goetzmann, William, Jonathan Ingersoll, Matthew Spiegel and Ivo Welch, 2007, Portfolio performance manipulation and manipulation-proof performance measures, *Review of Financial Studies*, **20**(5), 1503–1546.

Gonzalez, Fernando, François Haas, Ronald Johannes, Mattias Persson, Liliana Toledo, Roberto Violi, Carmen Zins and Martin Wieland, 2004, Market dynamics associated with credit ratings: a literature review, *Banque de France Financial Stability Review*, **4**, 53–76.

Goodhart, Charles, 2008, Liquidity risk management, *Banque de France Financial Stability Review – special issue on liquidity*, 11, February, 39–44.

Gordy, Michael and Bradley Howells, 2006, Procyclicality in Basel II: Can we treat the disease without killing the patient? *Journal of Financial Intermediation*, **15**(3), 395–417.

Gorton, Gary, 2010, *Slapped by the Invisible Hand: The Panic of 2007*, Oxford University Press.

Gup, Benton, 2012, Operational risk and large internal frauds at financial institutions, *International Research Journal of Applied Finance*, **3**(7), 954–970.

Hanson, Samuel and Til Schuermann, 2006, Confidence intervals for probabilities of default, *Journal of Money, Credit, and Banking*, **30**(8), 2281–2301.

Harris, Larry, 2002, *Trading and Exchanges: Market Microstructure for Practitioners*, Oxford University Press.

Hasbrouck, Joel, 2007, *Empirical Market Microstructure: The Institutions, Economics, and Econometrics of Securities Trading*, Oxford University Press.

Heber, Gerd, Asger Lunde, Neil Shephard and Kevin Sheppard, 2009, Oxford-Man Institute's realized library, version 0.2, *Oxford-Man Institute*, University of Oxford.

Hedge Fund Operational Due Diligence: What Managers Need to Know, *Hedge IT Blog*, posted by Kaleigh Brousseau on November 11, 2010.

Hedge fund operational risk: Meeting the demand for higher transparency and best practice, 2006, The Bank of New York (David Aldrich) and Amber Partners, *Alternative Investment Management Association Journal*, at www.aima.org/en/knowledge_centre/education/aima-journal/past-articles/index.cfm/jid/EEB25D21-C625-4242-A5CA29556ADA3CBF

Hedge Fund Research (HFR), 2012, *HFR Database*, Internet resource, accessed August 18, 2012, www.hedgefundresearch.com/index.php

Hegde, Shantaram and John McDermott, 2003, The liquidity effects of revisions to the S&P500 index: an empirical analysis, *Journal of Financial Markets*, **6**(3), 413–459.

Hey, Tony, Stewart Tansley and Kristin Tolle, 2009, *The Fourth Paradigm: Data-Intensive Scientific Discovery*, Microsoft Research, http://research.microsoft.com/en-us/collaboration/fourthparadigm/

Hilbert, Martin and Priscila López, 2012, The world's technological capacity to store, communicate, and compute information, Part I: results and scope, *International Journal of Communication*, **6**, 956–979, http://ijoc.org/ojs/index.php/ijoc/article/view/1562

Hilscher, Jens, Robert Jarrow and Donald van Deventer, 2008, Measuring the risk: a modern approach, *RMA Journal*, July–August, 60–65.

Hong, Gwangheon and Arthur Warga, 2000, An empirical study of bond market transactions, *Financial Analysts Journal*, **56**(2), 32–46.

Hull, John, 2006, VAR versus expected shortfall, *Risk Magazine*, **19**, 48–49.

Hull, John, 2012a, *Risk Management and Financial Institutions*, Wiley.

Hull, John, 2012b, *Options, Futures, and Other Derivatives*, Eighth edition, Prentice Hall.

Hunt, Robert, 2006, Development and regulation of consumer credit reporting in the United States, in: G. Bertola, R. Disney and C. Grant (eds.), *The Economics of Consumer Credit*, MIT Press, ch. 9, 301–345.

Institute of International Finance, 2011, Risk and IT Operations: Strengthening Capabilities, Technical Report, June, www.iif.com/regulatory/article+968.php

International Capital Markets Association (ICMA), 2012, Repo Market Surveys, Internet resource, accessed July 11, 2012, www.icmagroup.org/Regulatory-Policy-and-Market-Practice/short-term-markets/Repo-Markets/repo/

International Swap Dealers' Association (ISDA), 2002, 2002 ISDA Master Agreement, Technical Document, www.isda.org/publications/isdamasteragrmnt.aspx

International Swap Dealers' Association (ISDA), 2003, 2003 ISDA Credit Derivatives Definitions, Technical Document, www.isda.org/publications/isdacredit-deri-def-sup-comm.aspx

International Swap Dealers' Association (ISDA), 2012, OTC Derivatives Data Sources, Internet resource, accessed July 28, 2012, www2.isda.org/functional-areas/research/data-sources/

James, Jessica and Nick Webber, 2000, *Interest Rate Modelling*, Wiley.

Jarrow, Robert, 2009, Credit risk models, *Annual Review of Financial Economics*, **1**, September, 37–68.

Jarrow, Robert, 2011, Risk management models: construction, testing, usage, *Journal of Derivatives*, **18**(4), 89–98.

Jarrow, Robert, 2012, Problems with Using CDS to Infer Default Probabilities, Working Paper, Cornell University, January, www.kamakuraco.com/Portals/0/Research/Implied%20Default%20Prob.pdf

Jobst, Andreas, 2010, The credit crisis and operational risk – implications for practitioners and regulators, *Journal of Operational Risk*, **5**(2), http://papers.ssrn.com/sol3/papers.cfm?abstract_id=1491193

Johnson, Simon and James Kwak, 2009, Seduced by a model, *New York Times: Economix*, October 1, http://economix.blogs.nytimes.com/2009/10/01/seduced-by-a-model/

Jorion, Philippe, 2006, *Value at Risk: The New Benchmark for Managing Financial Risk*, Third edition, McGraw-Hill.

Kashyap, Anil and Jeremy Stein, 2004, Cyclical implications of the Basel II capital standards, *Federal Reserve Bank of Chicago Economic Perspectives*, First Quarter, 18–31.

Knight, Frank, 1921, *Risk, Uncertainty and Profit*, Houghton-Mifflin.

Krishnamurthy, Arvind, 2002, The bond/old-bond spread, *Journal of Financial Economics*, **66**(2–3), 463–506.

Kyle, Albert, 1985, Continuous auctions and insider trading, *Econometrica*, **53**(6), 1315–35.

Kyle, Albert and Anna Obizhaeva, 2011, Market Microstructure Invariants: Empirical Evidence from Portfolio Transitions, Working Paper, University of Maryland, December, http://papers.ssrn.com/sol3/papers.cfm?abstract_id=1978943

Lando, David, 2004, *Credit Risk Modeling: Theory and Applications*, Princeton University Press.

Langohr, Herwig and Patricia Langohr, 2009, *The Rating Agencies and Their Credit Ratings: What They Are, How They Work, and Why They are Relevant*, Wiley.

Lewis, Michael, 2011, *The Big Short*, W. W. Norton.

Lipper TASS, 2012, Lipper TASS Web Tool, Internet resource, accessed August 18, 2012, http://tass.lipperweb.com

Livingston, Miles, Jie Wei and Lei Zhou , 2010, Moody's and S&P ratings: are they equivalent? Conservative ratings and split rated bond yields, *Journal of Money, Credit and Banking*, **42**(7), 1267–1293.

Lo, Andrew, 2010, *Hedge Funds: An Analytic Perspective*, Princeton University Press.

Lo, Andrew and A. Craig MacKinlay, 2012, *A Non-Random Walk Down Wall Street*, Princeton University Press.

Mackenzie, Donald, 2008, *An Engine, Not a Camera: How Financial Models Shape Markets*, MIT Press.

Malkiel, Burton, 2012, *A Random Walk Down Wall Street: The Time-Tested Strategy for Successful Investing*, Tenth edition, W. W. Norton.

Managed Funds Association (MFA), 2007, Model Due Diligence Questionnaire for Hedge Fund Investors, Internet resource, accessed August 18, 2012, `www.managedfunds.org/wp-content/uploads/2011/06/Due-Dilligence-Questionnaire.pdf`

Managed Funds Association (MFA), 2009, Sound Practices for Hedge Fund Managers: 2009 Edition, Technical Report, `www.managedfunds.org/wp-content/uploads/2011/06/Final_2009_complete.pdf`

Matz, Leonard, 2011, Monitoring and controlling liquidity risk, in: L. Matz and P. Neu (eds.), *Liquidity Risk Management: A Practitioner's Guide to Global Best Practices*, Wiley Finance, ch. 4, 66–99.

Matz, Leonard and Peter Neu, 2011, Liquidity risk management strategies and tactics, in: L. Matz and P. Neu (eds.), *Liquidity Risk Management: A Practitioner's Guide to Global Best Practices*, Wiley Finance, ch. 5, 100–120.

McAleer, Michael and Marcelo Medeiros, 2008, Realized volatility: a review, *Econometric Reviews*, **27**(1–3), 10–45.

McCauley, Robert, 2001, Benchmark tipping in the money and bond markets, *BIS Quarterly Review*, March, 39–45, `www.bis.org/publ/r_qt0103.htm`

McNeil, Alexander, 1999, Extreme Value Theory for Risk Managers, Technical Report, ETH Zentrum, May, `www.math.ethz.ch/~mcneil/ftp/cad.pdf`

McNeil, Alexander, Rüdiger Frey and Paul Embrechts, 2005, *Quantitative Risk Management: Concepts, Techniques, and Tools*, Princeton University Press.

Medova, Elena A. and Berg-Yuen, P., 2008, Banking capital and operational risks: comparative analysis of regulatory approaches for a bank, *Journal of Financial Transformation*, Capco Institute.

Mehra, Yash, 2006, Inflation uncertainty and the recent low level of the long bond rate, *Federal Reserve Bank of Richmond Economic Quarterly*, **92**(3), 225–253, `www.rich.frb.org/publications/research/economic_quarterly/2006/summer/pdf/mehra.pdf`

Mengle, David, 2007, Credit derivatives: an overview, *Federal Reserve Bank of Atlanta Economic Review*, **92**(4), Fourth Quarter, 1–24.

Moody's, 2009, Rating Symbols and Definitions, Technical Report, June.

Moody's, 2011, Corporate Default and Recovery Rates, 1920–2010, Technical Report, February.

Mora, Nada, 2012, What determines creditor recovery rates? *Federal Reserve Bank of Kansas City Economic Review*, Second Quarter, 1–31.

Mulherin, J. Harold, Jeffry Netter and James Overdahl, 1991, Prices are property: the organization of financial exchanges from a transaction cost perspective, *Journal of Law and Economics*, **34**(2), 591–644.

National Association of Realtors (NAR), 2005, Home Price Analysis for Las Vegas-Paradise, Technical Report, September/October, `www.realtor.org/research.nsf/pages/anti-bubblereports`

Office of the Comptroller of the Currency (OCC), 2001, Rating Credit Risk, Comptroller's Handbook, April.

Peter, Christian, 2006, Estimating loss given default – experiences from banking practice, in: B. Engelmann and Rl. Rauhmeier (eds.), *The Basel II Risk Parameters: Estimation, Validation, and Stress Testing*, Springer Verlag, ch. 8, 143–174.

Poon, Ser-Huang and Clive Granger, 2003, Forecasting volatility in financial markets: a review, *Journal of Economic Literature*, **41**(2), 478–539.

Poon, Ser-Huang and Clive Granger, 2005, Practical issues in forecasting volatility, *Financial Analysts Journal*, **61**(1), 45–56.

Pritsker, Matthew, 2006, The hidden dangers of historical simulation, *Journal of Banking and Finance*, **30**(2), 561–582.

RiskMetrics Group, 2007, CreditMetrics, Technical Document, `www.msci.com/resources/technical_documentation/CMTD1.pdf`

Roll, Richard, 1984, A simple implicit measure of the effective bid ask spread in an efficient market, *Journal of Finance*, **39**, 1127–1139.

Rösch, Daniel and Harald Scheule, 2006, A multi-factor approach for systematic default and recovery risk, in: B. Engelmann and Rl Rauhmeier (eds.), *The Basel II Risk Parameters: Estimation, Validation, and Stress Testing*, Springer Verlag, ch. 6, 105–125.

Rowe, David, 2002, Basel II and procyclicality, *Risk*, **15**(11), `www.dmrra.com/riskmag.php`

Rowe, David, 2003a, The role of correlation, *Risk*, **16**(1), `www.dmrra.com/riskmag.php`

Rowe, David, 2003b, Don't count on buffers, *Risk*, **16**(2), `www.dmrra.com/riskmag.php`

Rowe, David, 2003c, No cure through the cycle, *Risk*, **16**(3), `www.dmrra.com/riskmag.php`

Rowe, David, 2003d, Is 8% for all seasons, *Risk*, **16**(4), `www.dmrra.com/riskmag.php`

Rowe, David, 2009, To VaR or not to VaR? *Insurance Risk and Capital*, Internet resource, March, `www.dmrra.com/otherarticles.php`

Rowe, David, 2012, Not all hedges are created equal, *Risk*, March, 58.

Rubinstein, Mark, 2006, *A History of the Theory of Investments: My Annotated Bibliography*, Wiley.

Saunders, Anthony and Linda Allen, 2010, *Credit Risk Management In and Out of the Financial Crisis: New Approaches to Value at Risk and Other Paradigms*, Wiley Finance.

Saunders, Anthony and Marcia Cornett, 2010, *Financial Institutions Management: A Risk Management Approach*, McGraw-Hill.

Scharfman, Jason A., 2008, *Hedge Fund Operational Due Diligence: Understanding the Risks*, Wiley Finance.

Schmidt, Anatoly, 2011, *Financial Markets and Trading: An Introduction to Market Microstructure and Trading Strategies*, Wiley Finance.

Schönbucher, Philipp, 2003, *Credit Derivatives Pricing Models: Model, Pricing and Implementation*, Wiley.

Schuermann, Til, 2005, What do we know about loss given default? in: E. Altman, A. Resti and A. Sironi (eds.), *Recovery Risk: The Next Challenge in Credit Risk Management*, Risk Books.

Securities and Exchange Commission (SEC), 2011, 17 C.F.R. Parts 275 and 279, Release No. IA-3221; File No. 87-36-10, July 21, 2011, www.sec.gov/rules/final/2011/ia-3221.pdf

Securities and Exchange Commission (SEC), 2012, Credit Rating Agencies – NRSROs, Internet resource, accessed July 28, 2012, www.sec.gov/answers/nrsro.htm

Securities Industry and Financial Markets Association (SIFMA), 2012, *Statistics*, Internet resource, accessed July 11, 2012, www.sifma.org/research/statistics.aspx/

Senior Supervisors Group (SSG), 2010, Observations on Developments in Risk Appetite Frameworks and IT Infrastructure, Technical Report, December, www.newyorkfed.org/newsevents/news/banking/2010/an101223.pdf

Shaw, Helen, 2006, The trouble with COSO, *CFO Magazine*, March 15, www.cfo.com/article.cfm/5598405/c_2984409/?f=archives

Sheppard, Kevin, 2012, Financial Econometric Notes, manuscript, Oxford University, November, accessed December 16, 2012, www.kevinsheppard.com/images/c/c0/Financial_Econometrics_2012-2013.pdf

Shleifer, Andrei and Robert Vishny, 2011, Fire sales in finance and macroeconomics, *Journal of Economic Perspectives*, **25**(1), 29–48.

Standard & Poor's (S&P), 2012a, Standard & Poor's Ratings Definitions, Technical Report, June.

Standard & Poor's (S&P), 2012b, Index Mathematics Methodology, Technical Report, January.

Stephanou, Christou, 2004, Measuring and Managing Operational Risk under Basel II, *Presentation, The World Bank Risk Management Workshop*, Colombia.

Stigum, Marcia and Anthony Crescenzi, 2007, *Stigum's Money Market*, Fourth edition, McGraw Hill.

Stowell, David, 2010, *An Introduction to Investment Banks, Hedge Funds, and Private Equity: The New Paradigm*, Academic Press.

Sufi, Amir, 2009, The real effects of debt certification: evidence from the introduction of bank loan ratings, *Review of Financial Studies*, **22**(4), 1659–1691.

Taksler, Glen, 2008, Credit Default Swap Primer, Fourth edition, Technical Report, Bank of America Securities, May.

Taylor, Stephen, 2005, *Asset Price Dynamics, Volatility and Prediction*, Princeton University Press.

Tirole, Jean, 2011, Illiquidity and all its friends, *Journal of Economic Literature*, **49**(2), 287–325.

Treacy, William and Mark Carey, 2000, Credit risk rating systems at large US banks, *Journal of Banking and Finance*, 167–201.

Tsaig, Yaakov, Amnon Levy and Yashan Wang, 2011, Analyzing the impact of credit migration in a portfolio setting, *Journal of Banking and Finance*, **35**(12), 3145–3157.

Tsay, Ruey S., 2010, *Analysis of Financial Time Series*, Wiley.

UBS, 2008, Shareholder Report on UBS's Write-Downs, Technical Report, April 18, 2008, www.ubs.com/1/ShowMedia/investors/agm?contentId=140333name=080418ShareholderReport.pdf

PART III

REGULATORY DATA

EDITORS' INTRODUCTION TO PART III

Margarita S. Brose and Mark D. Flood

This section focuses on the details of regulatory data, especially in the USA and Europe, slicing the topic several ways: historically, functionally, and geographically. The regulatory system in the USA is famously complex, and so US institutions receive special attention here. Regulators and government offices at all levels are an important feature on the data and information landscape for financial services. Supervisors have specific data and information challenges as they measure adherence to capital standards, implement monetary policy, estimate system-wide risk concentrations, enforce insider trading prohibitions, and perform the sundry other duties of the office. More broadly, the collected data help to define legal claims, monitor risk, enhance market transparency, resolve failing firms, and enforce the laws and rules. Regulators provide, in varying degrees, a service of centralization and cross-comparison of information. In many cases the data are published, perhaps in aggregated form, increasing transparency in the system.

Moreover, many of these information requirements cascade directly or indirectly to the financial firms that must comply with regulatory reporting mandates, thus ramifying their impact. In short, there is a "multiplier effect" to data-driven regulation. The central role of supervisory data is vital for both regulator and regulatee to understand the nature and uses to which data are put. One need look no further than the calculation of risk-weighted assets under the Basel capital rules to see how assumptions about the nature of risk might feed back into the system through risk-based capital requirements to discourage *measured* (by the regulators) risk-taking while implicitly encouraging *unmeasured* risk-taking.

The odd patchwork quilt that describes the US regulatory system is largely the outcome of long series of historical events. In Chapter 10, Alejandro Komai and Gary Richardson survey this history, describing how legislation adapted to historical events – many of them financial crises – by creating, enhancing, and occasionally removing or replacing regulatory institutions. For example, the Great Depression was the impetus for a radical transformation of the regulatory landscape. The current supervisory "framework" is impossible to understand fully without this historical perspective on the origins of the regulatory institutions. Unfortunately, the resulting fragmented regulatory authority is itself a source of financial instability.

In Chapter 11, Mark Flood, Simon Kwan and Irina Leonova provide an overview of one important slice of the regulatory system, banks and banking, with a particular emphasis on the data and information requirements. Most of the key data collections are itemized in detail. The chapter closes with an appendix on the international context for US banking regulation, which is dominated by but certainly not limited to the Basel capital rules.

In Chapter 12, Margarita Brose and Jesse Weintraub take a focused view of how the securities markets, including the stock and commodities markets, are regulated in the USA. They emphasize the legal underpinning of federal securities regulation, which centers on disclosure, and the process and entities that must comply with these rules and regulations. The authors also look at the supervisory regime at the SEC and CFTC, and how risk assessments and data collection have begun to drive supervisory reviews for these agencies.

In Chapter 13, Per Nymand-Andersen, Nicola Antoniou, Oliver Burkart and Jarl Kure present an in-depth view of the important changes in the European framework for both microprudential and macroprudential supervision in the post-financial crisis environment. They describe the details of the new European Supervisory Authorities (ESAs) for banking, securities and markets, and insurance and pensions. This represents an important Europe-wide restructuring of financial regulation. The chapter discusses interactions among the ESAs, and also with the new European System Risk Board. The financial crisis highlighted the inadequacy of data sharing, a shortcoming that the restructuring should address. In view of this, the authors focus on the need to collect and share information systematically across the jurisdictions.

By design, microprudential supervision tends to respect the boundaries of the firm, and the resulting focus of attention is both a strength and a weakness of a purely microprudential regime. In Chapter 14, Laura Kodres discusses the current environment and future needs for systemic data, as macro-supervisory bodies struggle with the challenge of maintaining the safety and soundness of the financial system as a whole. For obvious reasons, the recent crisis has reoriented supervisors much more heavily on the macroprudential perspective. Reviewing the financial crisis and its components, the chapter considers both old and new challenges for supervisors as they grapple with systemic risk.

10

A history of financial regulation in the USA from the beginning until today: 1789 to 2011

Alejandro Komai and Gary Richardson

10.1 Introduction

In the USA today, the system of financial regulation is complex and fragmented. Responsibility to regulate the financial services industry is split between about a dozen federal agencies, hundreds of state agencies, and numerous industry-sponsored self-governing associations. Regulatory jurisdictions often overlap, so that most financial firms report to multiple regulators; but gaps exist in the supervisory structure, so that some firms report to few, and at times, no regulator. The overlapping jumble of standards, laws, and federal, state, and private jurisdictions can confuse even the most sophisticated student of the system. At times, it can be unclear exactly who regulates whom, what rules apply in which instances, and where to turn for a resolution of these questions. This confusion occasionally inhibits innovation in the financial services industry and investments in some sectors of the economy. At other times, this confusion enables firms and investors to fly under the radar and profit from regulatory arbitrage. Whether this confusion promotes economic growth or causes economic instability is an open question.

How this confusion arose can be explained. The history of financial regulation is long but well documented. Responsibility for overseeing the financial services industry evolved in the USA during the last two centuries. Debate about how to regulate financial activity began at the Constitutional Convention in 1787 and continued unabated for two centuries. The political debate dictated the structure of the financial system; scholars have long noted this fact. An example comes from Jacob Viner's address at the American Economic Association's annual meeting in 1936. Viner (1936) argued that America's fragmented financial system,

Handbook of Financial Data and Risk Information, ed. Margarita S. Brose, Mark D. Flood, Dilip Krishna and Bill Nichols. Published by Cambridge University Press. © Cambridge University Press 2014.

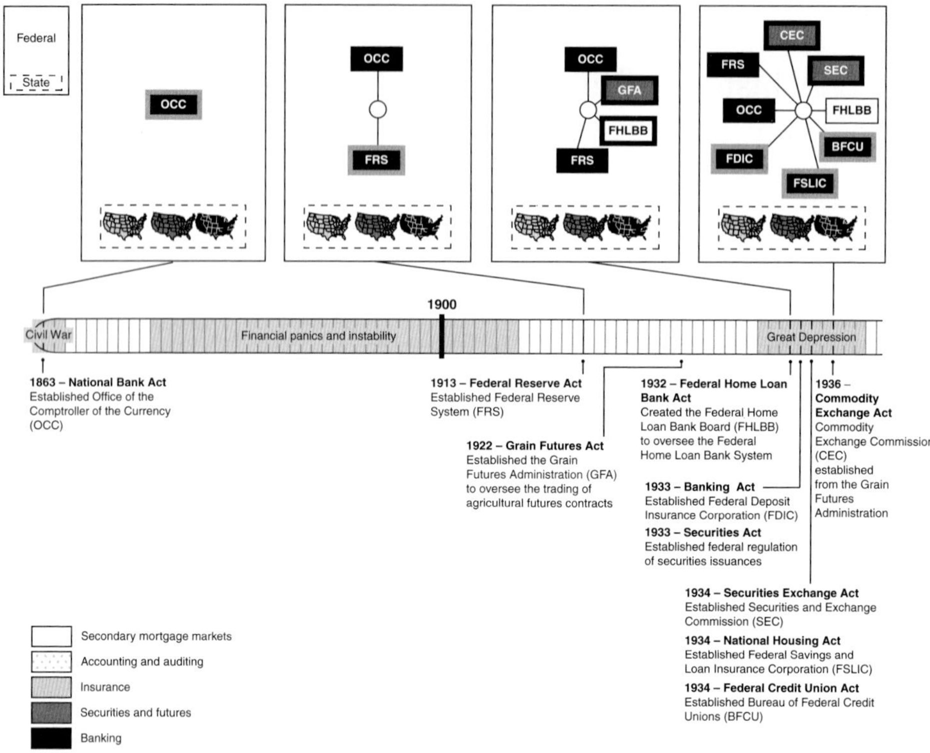

Figure 10.1 Graphical representation of the evolution of financial regulation in the USA, from the Great Depression through to today.

> [...] has deep roots in our history, in our regional diversities, and local loyalties. Its persistence is due to the support it derives from state jealousy of encroachments on state autonomy, from agrarian and small-town jealousy of the metropolitan areas, and from the nation-wide fear of undue concentration of financial power in the great metropolitan centers, and especially fear of Wall Street domination.

This chapter summarizes that history. Section 10.2 briefly describes the foundations of the financial system in the eighteenth and nineteenth centuries. The story begins with the United States Constitution, which establishes the parameters of the debate. Section 10.3 examines the response of the system to financial crisis in the early decades of the twentieth century, focusing on the creation of the Federal Reserve System. Section 10.4 examines the reform of the system in response to the financial crises of the Great Depression of the 1930s. Sections 10.5 and 10.6 discuss the creation of the modern financial system from

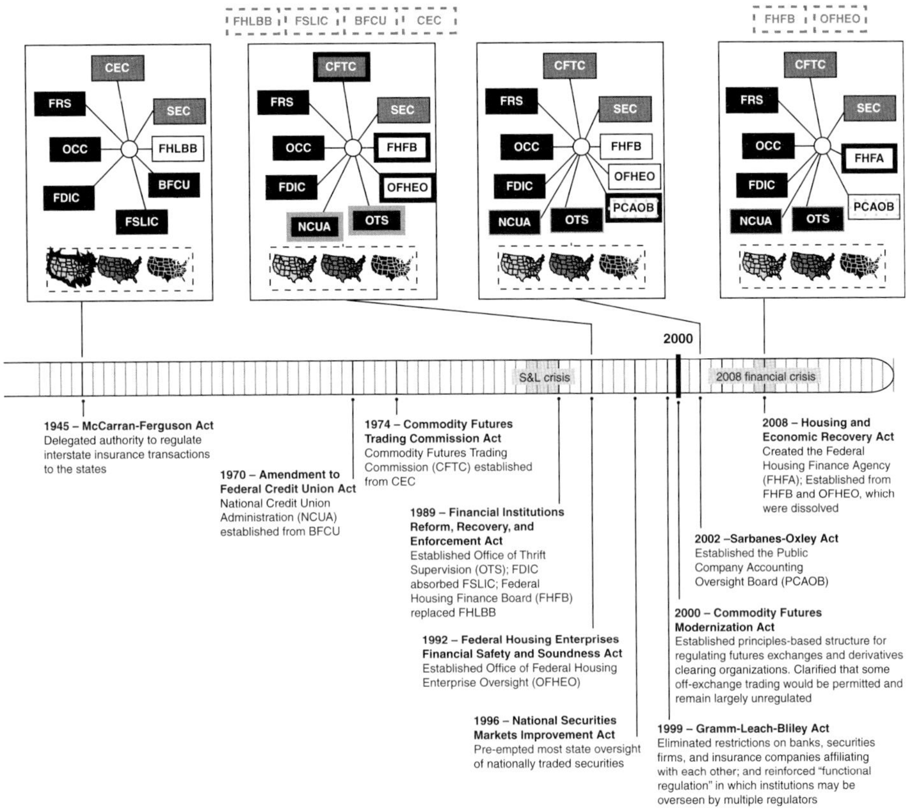

Figure 10.1 (cont.)

the 1940s through the 1990s. Section 10.7 discusses attempts to plug leaks that arose in the modern financial system during the first decade of the twenty-first century.

To illuminate the information in our narrative, three tables appear at the end of our essay. Table 10.1 lists the government agencies that oversee financial markets today (or oversaw financial activity in the past). The columns of the table indicate the legislation that authorized the agency (and major revisions), the purpose of the agency, and (some of) its major data collections. Table 10.2 lists the principal non-governmental organizations that set standards for financial markets. Table 10.3 lists the principal laws that influenced the regulation of financial markets over the last 200 years, with a focus on legislation since the founding of the Federal Reserve System in 1913. Of course, in constructing these tables, choices had to be made, because publishing complete lists would be prohibitive. In this regard, the authors decided to exclude information about hundreds of state agencies and thousands of state laws and to exclude references to the 12 Federal Reserve District banks.[1]

To illuminate the story that we tell, we reproduce as Figure 10.1 a figure previously published in January 2009 by the Government Accountability Office of the United States (Government Accountability Office, 2009). This figure is a useful depiction of the history of the system. The figure begins in 1860s, at the time of the US Civil War. But the complexity of financial regulation in the USA begins before that date. Understanding why requires a discussion of the founding of our nation.

10.2 Constitutional foundations of our financial system

At the Constitutional Convention in 1787, delegates debated how to regulate financial activity. Some delegates advocated the creation of a national currency and a national bank. Others opposed those proposals, arguing that the regulation of financial activity should be left to state governments. Bitter divisions engendered broad compromises. These appear in the portion of the Constitution that delineates powers of the federal legislature. Article 1, Section 8 provides Congress with powers to

- Borrow money on the credit of the USA
- Coin money, regulate the value thereof, and of foreign coin
- Regulate commerce with foreign nations and among the several states
- Establish uniform laws on the subject of bankruptcies throughout the USA.

In 1791, Congress chartered the First Bank of the United States to handle the financial needs of the federal government and the credit and coinage of the nation. In 1811, the charter expired, and by one vote Congress defeated the bill reauthorizing the institution. In 1816, Congress chartered the Second Bank of the United States, whose charter expired in 1836. These charters expired because politicians disagreed about the federal government's role in financial regulation. Politicians from northern industrial states favored federal – centralized – rather than state government. Politicians from southern and western states feared financial conglomerates and favored regulating financial activity through state legislatures. Their opposition prevented the establishment of a central bank, a uniform fiat currency, or uniform nationwide regulations for financial institutions (Dewey, 1902; Hammond, 1991; Myers, 1970; Russell, 1991).

State governments filled these gaps. State courts enforced financial contracts. States chartered corporations to provide financial services, particularly banks, and

[1] The authors would be interested to hear readers' thoughts about our decisions, particularly what information should have been excluded and what information should have been added in its place.

regulated their behavior. The profitability of these charters created problems with political corruption, which were mitigated when states adopted general incorporation and free banking laws, enabling anyone meeting specific criteria to obtain a charter and open a bank. By the mid-1830s, financiers had little difficulty chartering banks. Banks opened in large numbers. Each bank issued its own currency, which traded at an exchange rate that reflected the bank's reputation and risk of default. State governments left bank regulation to market forces. About half of all banks failed. Most failures occurred within a few years of the opening of an institution. Average longevity for financial firms appears to have been about five years. Because of these characteristics, financial historians refer to this era, from the expiration of the Second Bank of the United States to the Civil War, as one of wildcat banking (Dwyer, 1996; Rockoff, 1974).

The Civil War provided an opportunity to reform the financial system, because southern politicians who opposed federal regulation of financial markets withdrew from Congress. In their absence, Congress passed the National Currency Act (ch. 58, 12 Stat. 665; February 25, 1863) establishing a national currency printed by the US Treasury and issued by commercial banks (Friedman and Schwartz, 1971; Hammond, 1993; White, 1982). A bank could issue notes in proportion to the value of the capital that the bank deposited with the Treasury. To discourage the circulation of privately printed currencies, the Act taxed currencies of all other types, effectively forcing them out of circulation. One year later, Congress passed the National Banking Act (ch. 106, 13 Stat. 99; June 3, 1864), which established a system for issuing federal charters to commercial banks and authorized the Office of the Comptroller of Currency to supervise those banks. The Comptroller's annual reports created a systematic national data collection covering a large swath of the banking industry. The act established a pyramid structure of reserves cities which shaped the financial landscape of the USA during the decades that followed.

The first panel of Figure 10.1 depicts the regulatory landscape after the passage of the National Banking and Currency Acts. The Office of the Comptroller of Currency (OCC) appears as the sole federal regulator of financial activity. The governments of the 36 states regulated financial activities within their borders, typically under a Superintendent of Banks or similar agency regulating commercial banks, trust companies, and building and loan corporations. Many states also possessed (or soon established) an agency that regulated insurance companies.

10.3 Responses to financial panics, 1890 to 1930

In the decades between the Civil War and World War I, financial panics occurred frequently, including major panics in 1873, 1893, and 1907 (Jalil, 2011, 2012).

After these panics, legislators (both federal and state) debated reforming financial regulation. The panics of the 1890s contributed to the passage of the Bankruptcy Act of 1898 (July 1, 1898, ch. 541, 30 Stat. 544). The act established federal procedures for court-supervised liquidation of corporations unable to pay creditors, but left the liquidation of commercial banks in the hands of state bank supervisors and the Office of the Comptroller of Currency.

The panic of 1907 inspired further reform. An attempt to corner the market on the stock of the United Copper Company failed, and the banks that had lent money to the speculators sustained substantial losses. Depositor runs on these institutions spread rapidly to associated banks and trust companies, culminating in the collapse of the Knickerbocker Trust Company, New York City's third-largest trust institution. In turn, Knickerbocker's failure provoked a nationwide run on bank deposits (Sprague, 1910). In the absence of a central bank, the famous financier J. P. Morgan interceded by convincing New York's bankers to pledge funds to shore up depositories beset by the cash crunch. One year later, Congress established a commission, chaired by Senator Nelson Aldrich, to investigate the crisis and propose solutions. The commission studied financial systems in numerous nations. Its exhaustive report inspired the creation of the Federal Reserve System. The environment was ripe for this proposal thanks to the findings of the Pujo Commission, which publicly uncovered the connections between influential Wall Street financiers and major industries in the USA (FRASER, 2011).

The Federal Reserve continued the compromise between advocates of local and national regulation. The system consists of 12 district banks, each of which originally acted as the central bank for a region of the nation. The system has a board of directors located in Washington, DC, which coordinated (but did not control) the activities of the system. Federal Reserve district banks possessed authority to conduct monetary policy – including discount lending and open market operations – at its own discretion under rules different from today. Then, Federal Reserve banks could only purchase (or loan money on the security of) what is now known as short-term commercial paper. Federal Reserve banks did not deal in overnight loans among private banks (federal funds) or US government securities, which are the principal monetary policy levers today (Richardson and Troost, 2009; Carlson et al., 2011).

District banks also supervised the commercial banks that joined the system. The Federal Reserve Act required all nationally chartered banks to join the system, putting them under a dual system (Fed and OCC) of regulation and examination. State chartered banks could also join, if they fulfilled all federal regulations – such as levels of required reserves and restrictions on risky investments – which tended to be stricter than those imposed by state statutes (Meltzer, 2003, 2009).

The second panel in Figure 10.1 depicts the regulatory landscape after the creation of the Federal Reserve in 1913. State governments continued to be the principal regulators of financial activity. State courts enforced most financial contracts. State legislators chartered most financial corporations. State regulators supervised roughly two out of three commercial banks and all other financial institutions, including insurance companies, trust companies, mutual savings banks, credit unions, mortgage originators, and building and loan societies (White, 1981, 1983). Some states even dabbled in state-run deposit insurance programs (Chung and Richardson, 2006).

An array of private entities also supervised financial intermediation. Clearing houses operated in nearly one hundred cities. Banks that belonged to clearing houses had to obey their rules regarding reserves, risk, and regular inspections. State banking associations and the American Banking Association imposed codes of conduct on the behavior of their members. Financial conglomerates – some of which began to form bank holding companies – began to purchase shares of stock in large numbers of banks and place directors on the banks' boards. Rating agencies (e.g., Moody's) and business information providers (e.g., Rand McNally) began to collect and disseminate balance sheet information from most banks operating in the USA. Stock exchanges operated in dozens of cities; some, such as New York, possessed several. Exchanges regulated transactions in equity, bond, and futures markets, ensuring that those who bought and sold in those venues fulfilled the terms of their contracts.[2]

The third panel of Figure 10.1 depicts further changes in the regulatory landscape during the 1910s and 1920s. On September 21, 1922, Congress passed the Grain Futures Act (ch. 369, 42 Stat. 998, 7 U.S.C. § 1), which established the Grain Futures Administration (GFA). The GFA supervised trading of commodities futures contracts. The City of Chicago challenged the constitutionality of the Act, which was upheld by the Supreme Court in *Board of Trade of City of Chicago v. Olsen*, 262 US 1 (1923). The Supreme Court had ruled against an earlier version (the Futures Trading Act of 1921) in *Hill v. Wallace*, 259 U.S. 44 (1922). These court cases illustrate the political tension generated by the federal government's increasing attempts to regulate – and at times shape – the financial markets.

The government's broadest intervention at the time may have been in agricultural credit. In 1916, Congress passed the Federal Farm Loan Act. This act established a Federal Farm Loan Board to supervise 12 Federal Intermediate Credit Banks. These banks extended short-term, seasonal loans to farms,

[2] One of the contemporary descriptions of the US financial system appears in Rand McNally Bankers Directory, which was published biannually, in January and July, beginning in the 1890s.

ranches, and companies that processed agricultural products. Funds for these loans came from bonds with similar maturities sold on securities markets in major cities. The Farm Loan Act also established Federal Land Banks and National Farm Loan Associations. These organizations raised funds for farm mortgages by selling mortgage-backed bonds in cities with sizeable securities markets. Capital for the land banks and intermediate credit banks came from the US Treasury and from farmers who were the customers, who were required to purchase stock in the corporations (Spahr, 1932).

The federal government's intervention into farm lending coincided with the rapid expansion of agricultural production in the Great Plains and western states. This expansion led to the creation of new banks, businesses, and even towns that served as conduits for farm commodities. The expansion continued throughout World War I and for several years afterward, until climatic shocks (particularly droughts) reduced crop yields and international competition (particularly from European farms put back into production after the war) lowered crop prices. The ensuing contraction of farm incomes impeded the repayment of agricultural loans, contributing to the failure of thousands of small banks that operated in farm communities (Alston et al., 1994).

While the US Congress debated a wide array of financial reforms during the 1920s, the most important legislation was the McFadden Act of 1927 (Preston, 1927). The Act's provisions fell into three groups. First, it extended the life of the Federal Reserve System. The Act extended indefinitely the original 20-year charters of Federal Reserve Banks. Second, the McFadden Act revised national banking laws to conform with legal rulings, administrative decisions, and current business practices.[3] Third, the McFadden Act expanded national banks' authority to operate branches. Before the Act, national banks could not operate branch facilities except in a few special circumstances. After the Act, national banks could operate branches in the same city as their headquarters to the same extent as state-chartered banks under state law, but could not branch outside their home city or across state lines. While the McFadden Act had broad effects, it is remembered primarily for its impact on branch banking, which persisted for nearly 70 years, until modified by the Riegle–Neal Interstate Banking and Branching Efficiency Act of 1994 (Rajan and Ramcharan, 2011).

[3] Some of these changes appeared merely technical, such as allowing subordinate officers to sign reports to the Comptroller of Currency, rather than only the president or cashier (CEO and CFO) of the bank. Others provided firm legal footing for services increasingly offered by national banks, such as investment departments which marketed bonds and similar securities and trust departments which served as custodians for customers' financial assets.

10.4 Policy responses to the Great Depression

Thousands of banks failed during the contraction beginning in 1929 and continuing until March 1933 (Richardson, 2007; Richardson and Troost, 2009). While the Federal Government made a few attempts to address the financial calamity during the years of 1930–1932, their efforts proved to be too little and too late. In early 1933, the public lost faith in banks in general. Depositors fled from the financial system, forcing 28 states to close all financial institutions and eventually forcing the president to declare a national banking holiday, which shut down all financial institutions for seven business days before gradually resuscitating the financial system. In response to this disaster, the federal government changed the structure of financial regulation (Galbraith, 1954; White, 1990).

On January 22, 1932, Congress passed an act (c. 8, 47 Stat. 5) chartering the Reconstruction Finance Corporation (RFC) and authorized the RFC to extend loans to all financial institutions in the USA, including state-chartered banks lacking links to the Federal Reserve, and to accept as collateral an array of assets, as long as the RFC's leaders deemed the loans to be "amply" secured. The RFC's mandate emphasized loaning funds to solvent but illiquid institutions, whose assets appeared to have sufficient long-term value to pay all obligations, but which in the short run could not be sold at a price high enough to repay current creditors. The RFC also loaned funds to the receivers of banks in liquidation, which enabled receivers to repay depositors as soon as possible, and repay the RFC in the future, when assets could be sold at higher prices. The RFC also loaned funds to Federal Land Banks, which financed farm mortgages, and Federal Intermediate Credit Banks, which financed seasonal agricultural lending. The RFC also advanced funds to railroads, which indirectly aided banks, since numerous banks possessed portfolios of railroad bonds, which declined in value as rail traffic declined during the Depression, and to insurance companies, which also aided banks, since banks often purchased insurance on the values of their bond portfolios.

The Reconstruction Finance Corporation was a quasi-public corporation, staffed by professionals recruited outside of the civil service system, but owned by the federal government, which appointed the corporation's executive officers and board of directors. The RFC's initial capital came from $500 million in stock sold to the US Treasury. The RFC raised an additional $1.5 billion by selling bonds to the Treasury, which the Treasury in turn sold to the public. In the years that followed, the RFC borrowed $51.3 billion from the Treasury and $3.1 billion directly from the public. All of the RFC's obligations were guaranteed by the federal government (Jones, 1951). On July 21, 1932, an amendment authorized the RFC to loan funds to states and localities for self-liquidating public relief projects, such as the construction of utilities and bridges, whose construction costs would be repaid by user charges and tolls. The amendment also authorized the RFC to loan funds to states and localities to provide relief for the unemployed, when those loans could be repaid by future tax receipts.[4]

On February 27, 1932, Congress passed the Banking Act of 1932.[5] The Banking Act of 1932 expanded the Federal Reserve's lending powers, allowing Federal Reserve district banks to loan funds to member banks on the security of a broad range of assets equivalent to the assets accepted by the RFC. Loans secured by collateral previously ineligible for rediscount had to be approved by a minimum of five members of the Federal Reserve Board and had to pay a rate at least 1% above the prevailing discount rate. Federal Reserve districts could also loan funds to individuals, firms, and corporations, under restrictions mentioned above, and with the added stipulation that the borrowers prove that they had applied for but could not obtain credit from commercial banks in their own communities.

Congress passed the Emergency Banking Relief Act on March 9, 1933, in the week following Franklin Roosevelt's inauguration, in the midst of the national banking holiday, to facilitate reopening the nation's banks. Herbert Hoover's subordinates in the Department of Treasury and the Reconstruction Finance Corporation wrote the Act, which sat on Herbert Hoover's desk for many months to be used as a last resort in dire circumstances. Hoover never implemented the plan, although the economic crisis deepened during the months after his electoral defeat. During the interregnum, currency and gold fled from the USA. Bank runs swept the nation. Governors of more than 20 states declared financial holidays. Franklin Roosevelt implemented the disaster plan immediately after his inauguration. The Emergency Banking Relief Act clarified the Federal Government's authority to act during a national financial emergency under legal authority created by the Trading with the Enemy Act which Congress passed during World War I.

The Emergency Banking Act contained five titles. Title I authorized the President to declare an emergency, during which he could control the national finances and foreign exchange of the USA, prohibit the hoarding and export of gold, and dictate which banks would reopen, merge, or remain closed. Title II authorized the Comptroller of the Currency to seize and operate any bank in the USA. The

[4] To accomplish its goals, the RFC established several subsidiary and allied corporations. These include the Metals Reserve Company, the Defense Plant Corporation (DPC), Defense Homes Corporation (DHC), War Damage Corporation (WDC), Rubber Reserve Company (RRC), Electric Home and Farm Authority (EHFA), Lafayette Building Corporation (LBC), and Federal Facilities Corporation (FFC). Many of these agencies played important roles in financing economic expansion during World War II. In 1953, Congress passed an act that disbanded the RFC, transferring most of its functions to the Treasury Department effective June 1954, to wind down its affairs. Treasury completed that task in 1954. Vestiges of the RFC survive in the federal bureaucracy today. Successor agencies include the National Science Foundation, General Services Administration, and the Office of Defense Lending.

[5] Senator Carter Glass and Representative Henry Steagall coauthored the legislation, which was initially called the Glass–Steagall Act, until that label became the universal appellation for the act that Senator Glass and Representative Steagall co-sponsored in the summer of 1933.

Comptroller used this authority to appoint conservators for banks deemed unfit to resume operations but with the potential to recover. Title III authorized national banks to issue preferred stock, to pay dividends not exceeding 6% per year and did not subject holders to double liability (Macey and Miller, 1992; Grossman, 2001). The Reconstruction Finance Corporation could purchase preferred stock, and during the years that followed did so in large quantities. Title IV greatly expanded the range of eligible collateral Federal Reserve Banks could use for Federal Reserve notes. Fed banks also received expanded lending powers, allowing loans to member banks under "exceptional and exigent circumstances" whenever the loan was secured to the satisfaction of the Federal Reserve Bank. Title V of the Act contained three sections. Section 1 allowed Federal Reserve banks to convert debt instruments of the US federal government into currency at par value and to convert any circulating liability of a commercial bank (e.g., check, draft, or banker's acceptance) into cash at 90% of its apparent value. Section 2 authorized Federal Reserve banks to make unsecured loans to member banks at a rate at least 1% above the discount rate. Section 3 authorized Federal Reserve banks to loan funds to any individual or corporation for 90 days if the loan was secured by US government securities. On March 24, 1933, an amendment expanded these powers, enabling Federal Reserve banks to loan funds directly to non-member banks and trust companies for the duration of the existing emergency.

On June 16, 1933, the Banking Act of 1933 (Pub.L. 73–66, 48 Stat. 162) became law (Flood, 1992). Congress had considered progenitors of the legislation during preceding years. The final act merged in conference the two bills introduced by Carter Glass in the Senate and Henry Steagall in the House. Congress approved the Act on June 13 and the President signed it on June 16. The final bill contains most of the provisions proposed by Senator Glass, with the exception of branch banking, which Steagall opposed, and the addition of deposit insurance, which Steagall advocated.

The Glass–Steagall Act contained several provisions which shaped the financial landscape in the USA during the next decades. First, the act established nationwide deposit insurance, replacing the temporary insurance fund established by the Emergency Banking Act. This provision created the Federal Deposit Insurance Corporation (FDIC), under the management of a board of directors appointed by the President. The corporation's capital came from the US Treasury, Federal Reserve District Banks, and banks that joined the insurance system. All banks that belonged to the Federal Reserve had to join the deposit insurance system. Non-member banks could join the system by subscribing to stock in the association. Those that joined had to meet the requirements for Federal Reserve membership by July 1, 1936. This provision, in effect, induced all FDIC-insured banks to join the Federal Reserve System. Carter Glass – the legislative proponent of this provision – had intended this to happen. He hoped to induce all state-chartered banks to join the Federal Reserve, which would have effectively ended the dual banking system (Flood, 1992, pp. 70–71). All insured banks could be charged assessments, if the stock of the corporation proved insufficient to cover required insurance payouts. The FDIC insured all deposits up to $10,000 and of larger deposits, 100% of the first $10,000, 75% of the next $40,000, and 50% of any deposit over $50,000.

Second, the act separated commercial from investment banking. The act required commercial banks to sell their securities affiliates within one year and restricted their bond departments to the purchase and sale of securities on the order of and for the account of customers. Underwriting investment securities was prohibited. Interlocking directorates between commercial banks and securities companies was also forbidden. Firms engaged in selling securities were prohibited from taking deposits one year after the enactment of the law (i.e., after June 16, 1934). The use of bank credit for the purchase of securities and speculation in securities markets was restricted. The Federal Reserve received powers to prevent member banks from extending loans for investment in securities markets.

These provisions were motivated, in part, by Congressional investigations of the causes of the Wall Street Crash of 1929 and the collapse of the commercial banking system from the fall of 1930 through the winter of 1933. These hearings came to be known by the name of the chief counsel, Ferdinand Pecora, hired to pursue the investigations and write the final report. Pecora personally interviewed an array of high profile witnesses, including some of the most influential financiers in the USA. His efforts uncovered abusive practices and conflicts of interest on the part of banks, bank holding companies, and their financial affiliates. The worst abuses included the underwriting of unsound securities to pay off defaulted bank loans, insider trading, and fraudulent manipulation of securities prices. These revelations garnered widespread media attention. The ensuing public outcry galvanized support for banking reforms during the 1930s (Pecora, 1939; Perino, 2010).

Third, the Banking Act of 1933 imposed stricter regulations on financial institutions. Some of these regulations sought to reduce conflicts of interest among officers and directors. For example, the act prohibited officers and directors of member banks from borrowing from their own institutions and required them to report all borrowing from all other organizations. The act also prohibited officers and directors of member banks from associating with corporations that loaned funds on the security of stocks and bonds. Officers and directors of federally insured banks also had to conform to these regulations.

Other regulations sought to alter conditions that engendered bank failures, particularly among small banks. One example is an increase in minimum capital requirements. Another example was the prohibition of payments on demand deposits, which legislators expected would reduce the cost of funds for commercial banks and encourage depositors to place more of their funds in time deposits (i.e., savings accounts and certificates of deposit), providing commercial banks with a stable source of funds that was less subject to panics and runs. Additional examples were the restriction upon the use of bank credit for speculation, authorization of state-wide branch banking, federal supervision of group banking, modification of double liability, and increased authority of bank examiners.

Another restriction was the prohibition of private banking. Private bankers were individuals (or partnerships) that accepted demand deposits. The act required private bankers, after one year, to surrender either their deposit business or their dealing in investment securities. If they elected to conduct a deposit business, the law required them to submit to periodic examination by the Comptroller of the Currency.

Two years later, the Banking Act of 1935 contained two key sections. Title I modified deposit insurance so that the FDIC insured the first $5,000 of all deposits and nothing over that amount. The FDIC collected an annual assessment of 1/12 of 1% of all deposits in insured banks with no provision for collecting "special assessments" to cover periodic losses. Insured state chartered banks with deposits over $1,000,000 were still required to join the Federal Reserve System, but the deadline for doing so was pushed from 1936 back to 1942. Banks with deposits less than $1,000,000 were no longer required to join the Fed. Those that had joined were given the option to depart, but only 50 of the roughly 7,500 banks that joined the system chose to leave it.

While Title I made minor modifications to the FDIC, Title II made major changes to the structure of the Federal Reserve System. These changes centralized control of supply of money and credit in the hands of the Federal Reserve Board of Governors (the "Board"). Title II gave the Board the power to approve governors and vice-governors of the 12 district banks. The Board also received the authority to set discount rates and establish lending policies. The act provided that "subject to such regulations as to maturity and other matters as the Federal Reserve Board may prescribe," a Federal Reserve Bank might discount any commercial, agricultural, or industrial paper for member banks, and might make advances to member banks secured by "any sound asset." The act also permitted the Federal Reserve to purchase securities issued or guaranteed by the US government.

The new Board dominated a new Federal Reserve Open Market Committee (FOMC), consisting of the seven members of the Board, the President of the Federal Reserve Bank of New York (FRB-NY), and the presidents of four other Federal Reserve districts on a rotating basis. Title II gave the FOMC the authority to establish policies pertaining to the purchase of securities in the open market. The decisions of the FOMC became binding on Federal Reserve banks, which in the past needed to participate in programs of open-market purchases and sales recommended by the Federal Reserve Board. Other Depression-era legislation dealt with deposit-taking institutions other than banks, such as savings and loans and credit unions. These organizations differed in the types of deposits that they accepted and the types of assets that they held. Commercial banks accepted deposits payable upon demand and provided customers with the opportunity to circulate those liabilities by writing checks. Commercial banks invested the preponderance of their short-term liabilities in short-term commercial loans, providing credit (often seasonal) to manufacturers, wholesalers, retailers, and farmers. Savings and loans accepted only savings deposits and invested the bulk of these long-term liabilities in long-term investments like home mortgages. Credit Unions did not accept deposits. Instead, members of credit unions (and related entities such as mutual savings banks and building and loan societies) held stock in a non-profit credit cooperative. The cooperative typically treated the shares of stock like savings accounts, allowing members to buy and sell shares just like individuals deposited and withdrew funds from commercial banks. In 1934, Congress passed the National Housing Act, which established the Federal Savings and Loan Insurance Corporation (FSLIC), which insured deposits in savings and loans (S&Ls) and regulated the S&L industry. Congress also passed the Federal Credit Union Act, which established the Bureau of Federal Credit Unions to insure and regulate member-owned credit cooperatives (Wheelock, 2008).

The federal government's intervention into mortgage markets expanded in 1932, when the Congress passed the Federal Home Loan Bank Act. The act created the Federal Home Loan Bank Board to oversee 12 government-backed banks with the authority to purchase mortgages loans issued by originators – primarily building and loan and savings and loan organizations – operating within their jurisdiction (Quigley, 2006). In 1938, federal government intervened in mortgage markets again, when Congress amended the National Housing Act to create the Federal National Mortgage Association (FNMA), colloquially known as Fannie Mae. Fannie Mae operated as a government sponsored entity whose mission was to purchase home loans originated by commercial banks and mortgage brokers, create a liquid second mortgage market, reduce the cost of home ownership, and encourage middle-class home ownership. Fannie Mae principally purchased mortgages insured by the Federal Housing Administration (FHA). For 30 years, Fannie Mae was the preponderant purchaser of mortgages, exercising a virtual monopoly over the secondary mortgage market (Fabozzi and Modigliani, 1992).

Another series of acts regulated stock exchanges and securities markets. In 1933, Congress passed the Securities Act, which established federal regulation of securities issues. In 1934, Congress passed the Securities Exchange Act which established the Securities and Exchange Commission (SEC) to regulate the issuance, purchase, and sale of securities, particularly equities and debt instruments. The act required all public companies to submit periodic financial statements under penalty of perjury. In 1936, Congress passed the Commodity Exchange Act (ch. 545, 49 Stat. 1491, enacted June 15, 1936) which required all commodities futures and options to be traded on organized exchanges; see Stassen (1982) for a legislative history. To regulate those exchanges, the legislation established the Commodity Exchange Commission (CEC). This organization absorbed and assumed the legal authority of the Grain Futures Administration (GFA), which had been established in 1922.

10.5 Constructing the modern financial system, 1940 to 1980

Financial markets operated calmly from the 1940s through the 1980s. Institutions created in the wake of the New Deal held significant sway over the financial world and regulatory practices during this period.

One of the hallmarks of the SEC's oversight of securities markets was the preference for industry self-regulation. While the SEC has authority to create accounting standards for publicly traded companies, it often defers to private accounting standards boards such as the Financial Accounting Standards Board (FASB) and its predecessors (Moehrle et al., 2002). The most prominent example of this tradition is the FASB, which sets the Generally Accepted Accounting Principles (GAAP) for the industry. In 1972, the American Institute of Certified Public Accountants (AICPA) issued a report calling for an end to the Accounting Principles Board (APB) and the creation of a fully independent FASB (Seidler, 1972).[6]

GAAP comprises a five-tiered hierarchy of rulings and opinions:

1. Category A, the highest category
   (a) FASB-issued statements
      - Statements of Financial Accounting Standards (SFAS)
      - Financial Accounting Standards Board Interpretations (FINs)
   (b) APB-issued statements
      - Accounting Principles Board Opinions

[6] Whereas the members of the APB were part-time, unpaid members who worked for firms as accountants, members of the FASB are paid to work full-time devising standards. This set of distinctions is the intended source of independence, though Meyer concludes from his examination of APB Opinions that the APB was not systematically influenced by the connections between members and their "external constituencies" (Meyer, 1974). Seidler points out that members may be influenced on an industry and national level rather than an individual and company level, and that this influence would not be broken when transitioning to the FASB (Seidler, 1972).

   (c) AICPA Committee on Accounting Procedure (CAP) statements (APB's predecessor)
      - Accounting Research Bulletins (ARBs)
2. Category B
   (a) FASB Technical Bulletins (FTB)
      - Created by FASB staff, rather than the actual board
   (b) AICPA Industry Audit and Accounting Guides
      - Created by committees and task forces in AICPA
   (c) AICPA Statements of Positions (SOPs)
      - Created by AICPA Accounting Standards Executive Committee (AcSEC)
3. Category C
   (a) Consensus positions of FASB's Emerging Issues Task Force (EITF)
   (b) AcSEC practice bulletins
4. Category D
   (a) AICPA Accounting Interpretations (AINs)
   (b) FASB Staff Implementation Guides
   (c) "[P]ractices that are widely recognized and prevalent either generally or in the industry" (Moehrle et al., 2002; AICPA, 2000)
5. Category E
   (a) Catchall encompassing all other written sources of accounting authority (Moehrle et al., 2002; AICPA, 2000).

The Norwalk Agreement with the International Accounting Standards Board (IASB) marked a shift in FASB goals. Issued September 18, 2002 (FASB website), the Norwalk Agreement made the alignment of US GAAP with international standards a priority for the FASB.

From the 1940s through the 1980s, practices of the FDIC influenced bank behavior. Banks insured by the FDIC must submit a Report of Condition and Income, also known as a Call Report, each quarter. The FDIC is responsible for maintaining and correcting these data, as well as making them available for the public. This responsibility is set down in the Federal Deposit Insurance Act of 1950. The bank's Call Report must follow Federal Financial Institutions Examination Council (FFIEC) and FASB rules, as enforced by the FDIC. FFIEC was created by the Financial Institutions Regulatory and Interest Rate Control Act of 1978 (FFIEC, 2011) and is tasked with creating uniform principles and standards across the several financial regulatory bodies such as the Fed, the OCC, and the FDIC. These data from "insured national and state nonmember commercial banks and state-chartered savings banks," are the principal source of information on the banking system available to the public and are often used by regulators as a measure of the system (FDIC website).

When a bank fails, the FDIC has a few standard procedures for resolution. Typically, the FDIC will either perform a deposit payoff, a purchase and assumption (P&A), or open bank assistance.[7] There are two types of deposit payoff. In a straight deposit payoff, the FDIC pays insured depositors up to the limit. In an insured deposit transfer, "insured deposits and secured liabilities of a failed bank or thrift are transferred to a healthy institution, and service to insured depositors is uninterrupted" (FDIC, 2012). The main alternative is a purchase and assumption (P&A). A P&A resembles an assisted merger in that a failed financial institution is absorbed by a healthy one (James, 1991). The FDIC auctions a package of assets from the failed institution and replaces them on the balance sheets with "good" assets (Glaessner and Mas, 1995). To determine which method to employ in failure resolution, the FDIC estimates the cost of covering uninsured depositors of the failed institution, plus FDIC administrative costs of executing a payoff and liquidation and sees whether it can find bidders among healthy institutions willing to cover those costs to absorb the failed institution. P&A is typically preferred, mainly because payoff and liquidation leads to a loss of going-concern value (Buck, 1984). The Dodd–Frank Act will change resolution requirements – in particular through the introduction of orderly liquidation authority (OLA) for systemically important financial institutions – but these rules are still being implemented.

In 1974, Congress amended the Commodities Exchange Act and created the Commodity Futures Trading Commission (CFTC). The CFTC succeeded the Commodity Exchange Commission, as seen in the sixth panel of Figure 10.1. The CFTC consists of five commissioners appointed by the President and confirmed by the Senate. The CFTC has the "authority to regulate futures trading in all goods, articles, services, rights, and interests traded for future delivery" (Michigan, 1975). To that end it has power of injunction, giving it authority to pursue its own matters. The CFTC is also granted power to take special action in emergencies to maintain order. The proximate cause of the creation of the CFTC was the leap in prices in 1973 attributed to the action of speculators. In order to ease passage of the CFTC Act, some issues were left for the Commission to decide. Among these issues were the regulation of option trading in previously unregulated futures commodities, time-stamping, and regulation of whether to permit futures commission merchants (FCMs) to dual trade – "trading for their own accounts as well as for the accounts of their customers" (Michigan, 1975).

Other new regulations and regulatory agencies were created in the 1970s. The Employment Retirement Income Security Act (ERISA) was passed to protect retirees. ERISA also created the Pension Benefit Guaranty Corporation (PBGC),

[7] Open bank assistance has not been used since the 1992 implementation of least cost analysis (US FDIC, 2012).

"to encourage the continuation and maintenance of private-sector defined benefit pension plans, provide timely and uninterrupted payment of pension benefits, and keep pension insurance premiums at a minimum" (PBGC, 2012). The PBGC acts to implement data collection and disclosure requirements mandated in ERISA (US DoL, 2008). The PBGC publishes statistics on its programs annually in its *Pension Insurance Data Book* since 1996 (PBGC, 2009).

ERISA required retirement plans to disclose information to participants, provide a system by which participants can file grievances and appeals, and set minimum standards (US DoL, 2011). Some information required by ERISA includes "corporate plan sponsors provide participants with audited annual reports, summaries of plan descriptions, and other disclosures" (Langbert, 1994). ERISA also asserted the right of retirement plan participants to sue their plan providers for delivery of services (DoL website). Additionally, ERISA marked a movement in setting retirement standards from the state to the federal level. ERISA superseded state laws on employee benefits and gave jurisdiction over disputes regarding employee benefit claims to the federal courts (Langbert, 1994). Over time, several amendments to ERISA have been passed, including the Consolidated Omnibus Budget Reconciliation Act of 1974 (COBRA), the Health Insurance Portability and Accountability Act (HIPAA), the Newborns' and Mothers' Health Protection Act, the Mental Health Parity Act, and the Womens' Health and Cancer Rights Act (DoL website).[8] COBRA continues a participant's health coverage for a limited time after loss of a job and HIPAA provides protections for workers who "might otherwise suffer discrimination in health coverage based on factors that relate to an individual's health" (DoL website). HIPAA is an important example of regulatory data standardization. Administrative simplification provisions "required HHS to adopt national standards for electronic health care transactions and code sets, unique health identifiers, and security" (US DHHS, 2012).

The Home Mortgage Disclosure Act (HMDA) of 1975 was initially implemented to help discern whether depository institutions were reinvesting in their local communities. Financial institutions covered by HMDA were required to report aggregates of dollars and locations. The required information did not include racial or other demographic data on loan applications.This model prevailed from HMDA's passage to the 1989 Financial Institutions Reform, Recovery, and Enforcement Act (FIRREA). FIRREA was one of many amendments to HMDA that expanded its

[8] Other pieces of legislation that affected ERISA include the Tax Equity and Fiscal Responsibility Act, the Retirement Equity Act, the Revenue Act of 1978, and the Tax Reform Act of 1986 (Langbert, 1994).

scope to more mortgage lenders. FIRREA's main impact on HMDA was to require lenders to report demographic information such as race, sex, and income, including reports on rejected applications. From FIRREA up through the twenty-first century, HDMA was amended so the data collected could be used to address issues of discrimination. In the early 2000s, HDMA was further amended through changes in regulation rather than through passage of laws. Most significantly, starting in 2004 HMDA required reporting of pricing of loans (Kolar and Jerison, 2005), to support the use of the HMDA loan/application register (LAR) data to assess predatory lending and discriminator pricing.

The Securities Acts Amendments became law on June 4, 1975 (Saunders, 1985). In these Amendments Congress directed the SEC to promote the creation of National Market and National Clearing Systems (Gillis, 1975) encompassing several goals. The National Market System was meant to improve competition, liquidity, efficiency and stability in securities markets (O'Hara, 2004; Werner, 1975). The SEC sought methods to exploit advances in computer technology to combine the several regional exchanges into the Intermarket Trading System (ITS) (Macey and Haddock, 1985; Gillis and Dreher, 1982). To that end, on April 28, 1981 the SEC ordered the ITS and the National Association of Securities Dealers (NASD)'s Computer Assisted Execution System (CAES) be automatically linked (Gillis and Dreher, 1982). Additional SEC-backed projects include the Consolidated Transaction Reporting System to provide real-time transaction reports for NYSE, AMEX, and regional exchanges, and the Composite Quotation System to display quotations and quotation sizes for the Consolidated Transaction System (Gillis and Dreher, 1982). In an effort to end anti-competitive regulations, the SEC ended fixed minimum commission rates by 1975 (Werner, 1975).

Despite its orders from Congress, ten years after the passage of the Securities Acts Amendments, a National Market System had not materialized. Macey and Haddock pointed to off-board trading restrictions as a key way the SEC could act to improve competition, and noted, "[t]he SEC [. . .] was concerned that too much freedom in the marketplace might be detrimental. The Commission was particularly concerned with three phenomena: fragmentation of orders, overreaching, and market surveillance. These considerations are the only policy reasons that the SEC has advanced in defense of its failure to ban off-board trading restrictions" (Macey and Haddock, 1985). By "fragmentation of orders," Macey and Haddock mean the SEC was concerned with consequences of transactions occurring outside organized exchanges. "Overreaching" refers to internalization of order flows, meaning that a broker-dealer may sell stock to a client at higher than market prices drawing from shares held in inventory by the broker-dealer. "Market surveillance" means SEC market monitoring responsibilities. Macey and Haddock (1985) go on to argue that "fragmentation is an unwarranted fear," by applying arbitrage logic, that overreaching is a fallacy, and that the SEC had ample market surveillance facility, concluding that there was no valid reason for the SEC's failure to enact a National Market System.

10.6 Constructing the modern financial system, 1980 to 1995

During the 1980s and 1990s, the structure of financial regulation in the USA changed dramatically, as summarized in the sixth panel of Figure 10.1. Impetus for change came from three directions. First, free-market thinking increasingly prevailed in policy debates. Second, globalization forced financial institutions in the USA to compete in ever more competitive international markets against institutions operating in more permissive regulatory environments. US institutions incessantly lobbied to loosen regulations and level the playing field. Third, during the 1980s, the S&L industry collapsed. The Fed applied the Depression-era authority to set deposit interest rate ceilings (Regulation Q) to passbook deposits at S&Ls for the first time in 1966 (Gilbert, 1986). As nominal interest rates rose with inflation through the 1970s, deposits gradually fled to uncapped money market funds, a trend exacerbated by the Volcker Fed's inflation-fighting, which pushed short-term interest rates above 20% in 1981. This put S&L balance sheets under severe stress, because the asset portfolios of long-term mortgages could not adjust in pace with the higher funding costs and outflows in deposit liabilities. Congress responded in March 1980 with the Depository Institutions Deregulation and Monetary Control Act (DIDMCA), which phased out interest-rate ceilings on deposits and more than doubled deposit insurance coverage. Later that year, the Federal Home Loan Bank announced the first of several reductions in capital requirements for the S&Ls. In 1982, Congress enacted the Garn–St. Germain Depository Institutions Act, which fully eliminated interest rate ceilings and broadly expanded S&Ls' asset powers. These efforts were inadequate to prop up an industry designed to absorb much lower levels of interest-rate risk. Many savings and loans became insolvent. Most teetered on the edge of the abyss. Ultimately the results were disastrous, as bankrupt but fully insured institutions expanded operations into areas of the financial service industry in which they had little or no experience.

FIRREA was passed in response to the S&L crisis of the 1980s, mainly to restore confidence to the public. FIRREA amended the Home Owner's Loan Act of 1933, replacing the Federal Home Loan Bank Board, and creating the Office of Thrift Supervision (OTS), as seen in the sixth panel of Figure 10.1. The OTS was created in the Department of the Treasury. To monitor interest-rate risk, the OTS implemented the "Net Present Value" (NPV) risk model, which required thrifts to submit aggregated data on the terms and conditions of the full portfolio of assets and liabilities. FIRREA also created the Resolution Trust Corporation to resolve failed S&Ls. This corporation was put under the management of the FDIC. While the FDIC took over the FSLIC's operations through the Resolution Trust Corporation, the explicit successor to FSLIC in FIRREA was the Savings Associations Insurance Fund (SAIF) (Providenti, 1991).

Changes in regulations in that period tended to reduce restrictions on the operations of financial institutions, allowing them to enter new lines of business. Interstate and intrastate bank branching was heavily restricted up to the 1970s. Intrastate branching was limited, but deregulation began in the 1970s; interstate branching restrictions began to be lifted starting in 1978. The shift to deregulation was slow and gradual, culminating in the Riegle–Neal Interstate Banking and Branching Efficiency Act of 1994. This act allowed nationwide branching starting June 1997, but it also had opt-in and opt-out provisions. The opt-in provision allowed states to pass legislation to allow branching earlier than the June 1997 deadline. Only two states, Texas and Montana, passed legislation to opt out of the provisions of the act (Dick, 2006).

Since the Banking Act of 1933, banks were prevented from engaging in universal banking, or banking in both commercial and investment industries. The Gramm–Leach–Bliley (GLB) Act sought to allow increased competition by removing barriers between the banking sectors of commercial, investment, and insurance, repealing limitations set by the Banking Act and Bank Holding Company Act of 1956. It did this by allowing financial institutions to form a "financial holding company" which can engage in all three industries. Anticipating the potential problems associated with the merging of personal financial data in these three industries, GLB included several privacy provisions. These provisions included required annual privacy notices for customers and an opt-out provision for customers to disallow financial institutions from sharing personal information with non-affiliates. GLB also required financial institutions "develop policies to promote data security." A right of enforcement was assigned to federal agencies including the Federal Trade Commission, the Board of Governors of the Federal Reserve, the office of the Comptroller of the Currency, and the SEC. Despite these efforts, the privacy provisions were universally considered a failure soon after the passage of GLB (Janger and Schwartz, 2002).

The Federal Deposit Insurance Corporation Improvement Act of 1991 reformed rules for bank regulators and aimed to implement principles of prompt corrective action (PCA) and least-cost resolution (LCR). The principle of PCA was a response to banking and savings and loan troubles of the 1970s and 1980s when regulators delayed taking action. PCA and LCR were designed to realign the incentives of regulators – who may be jockeying for industry jobs – to oversee the industry more conscientiously. To that end, FDICIA mandated more inspections of banks by the FDIC and annual audits of the FDIC by GAO. Subverting this structure, regulators were given discretion, under FDICIA, to set capital/asset thresholds which would trigger additional regulation and restrictions on banks, and immediately set the thresholds low enough for most banks to be considered "adequately capitalized" (Benston and Kaufman, 1997). "The act also addressed such issues as the need for higher capital levels, risk-based deposit insurance, and a strengthening of the regulatory responsibility for early intervention" (Gupta and Misra, 1999). Risk-based deposit insurance was intended to minimize moral hazard distortions on the part of bankers. In this same vein, FDICIA prevented the FDIC from protecting deposits of uninsured depositors – depositors with deposits in excess of $100,000 (Benston and Kaufman, 1997).

FDICIA addressed the doctrine of "too big to fail." Effective in 1995, the FDIC was prohibited from protecting "uninsured depositors or creditors at a failed bank if it would result in an increased loss to the deposit insurance fund," with the exception being the case that the institution is considered "too big to fail" (Benston and Kaufman, 1997). "Articulated by the Comptroller of the Currency after the failure of the Continental Illinois in 1984, the too-big-to-fail policy is based on the premise that the failure of a large institution could have a domino effect, starting bank runs that could bring down the financial system" (Gupta and Misra, 1999). FDICIA actually weakens this policy by requiring, "written approval of two-thirds of the Board of Directors of the FDIC and the Board of Governors of the Federal Reserve System [...] permission from the Secretary of the Treasury, after the secretary has consulted with the President of the United States. The FDIC is also required to recover any losses incurred from protecting uninsured claimants" (Gupta and Misra, 1999). It was considered that "too big to fail" policy would only rarely, if ever, be used.

10.7 Fine tuning the system: 2000 until today

During the last decade, the regulatory system continued to evolve. Some policy-makers intended these changes to loosen restrictions on the behavior of financial institutions. These changes are visible in the seventh and eighth panels of Figure 10.1.

The Commodity Futures Modernization Act of 2000 (December 21, 2000) clarified regulatory jurisdictions between the Commodity Futures Trading Commission (CFTC) and the Securities and Exchange Commission (SEC) over many financial instruments. Title I amends the Commodity Exchange Act, limiting its scope. Title II amends the Securities Act of 1933, Securities Exchange Act of 1934, Commodity Exchange Act, and Shad–Johnson Jurisdictional Accord, "to provide implementing rules necessary for shared oversight by the SEC and CFTC of single stock futures trading" (Kloner, 2001). "Title III provides additional legal certainty for swap agreements by providing guidelines for SEC regulation of equity based swaps" (Kloner, 2001). Title IV further limits the Commodity Exchange Act by clarifying that it does not apply "to certain swap agreements (including credit and equity swaps), hybrid instruments and other products commonly offered by banks" (Kloner, 2001).

The Public Company Accounting Reform and Investor Protection Act of 2002, more commonly referred to as Sarbanes–Oxley (SOX), passed on July 25, 2002 after several "prominent companies [were] involved in financial scandals and bankruptcies: Enron, Worldcom, Xerox, Sunbeam, Waste Management, Adelphia, Tyco, HealthSouth, Global Crossing, and others" (Coates, 2007). SOX was a direct response to these scandals. SOX contains three main components. First, in an attempt to provide market participants with access to identical information and a level playing field, SOX "forbids preferential disclosures to market analysts," although this provision may have the unintended consequence of "less total disclosure" (Easterbrook, 2009). Second, in an attempt to create accountability and monitoring within corporations, SOX requires the CEO and CFO of all publicly traded corporations to sign the balance sheets that they submit to the SEC, opening them up to criminal penalties for perjury should the forms prove fraudulent.[9] In addition, SOX requires publicly traded companies have an independent board of directors. Third, SOX mandated "more monitoring by accountants, in addition to monitoring by independent directors" (Easterbrook, 2009). SOX created the Public Company Accounting Oversight Board (PCAOB) to "enlist auditors to enforce existing laws against theft and fraud by corporate officers" (Coates, 2007). The PCAOB is charged with "registering, setting standards for, inspecting, investigating, and disciplining audit firms for public companies" (Coates, 2007). The PCAOB appears in the seventh panel of Figure 10.1.

The Sarbanes–Oxley Act also gave the SEC the task of reviewing the Financial Accounting Standards Board's (FASB's) process of creating Generally Accepted Accounting Principles (GAAP). Specifically, it was a commonly held belief that

[9] This provision returned banks to circumstances that prevailed in the past. From the 1860s through the 1920s, the president and chief financial officer of a nationally chartered bank had to sign their bank's financial statement and faced civil and criminal liability if the financial statement proved to be inaccurate. The McFadden Act of 1927 allowed the president and chief financial officer to delegate this task to a subordinate who then assumed this legal liability.

Enron had avoided detection for so long by adhering to the letter of the rules set down by GAAP. The SEC was asked to determine how long it would take to move from a rules-based system to a principles-based system, the reasoning being that a principles-based system would have exposed Enron earlier than the rules in place under GAAP (Bratton, 2003). After Enron, the FASB was asked, under SOX, to seek an alignment of US GAAP with international standards (Bratton, 2003). Doubts have been raised as to whether convergence is a feasible goal and whether it will ever happen.

The Dodd–Frank Wall Street Reform and Consumer Protection Act, (Pub.L. 111–203, H.R. 4173) (DFA), is a hodgepodge of several unrelated regulations. Passed in July 2010 in response to the financial crisis at the end of the first decade of the twenty-first century, DFA restructured the regulatory system. A few highlights from this act include an overhaul of the bankruptcy code, a re-regulation of most derivatives previously deregulated, and regulations disallowing bailouts in many cases. The act also led to the creation and destruction of many new government agencies. For example, the act led to the creation of the Financial Stability Oversight Council, the elimination of the Office of Thrift Supervision, the creation of the Bureau of Consumer Financial Protection (CFPB), and the creation of the Federal Insurance Office (Davis and Wardwell, 2010). Especially noteworthy from a financial data perspective is the creation of the Office of Financial Research (OFR) under Title I of the Act, with mandates to monitor systemic risk and to standardize and collect positions and transactions data from market participants. One of the OFR's first major initiatives has been the promotion of a global system of legal entity identifiers (LEIs), to facilitate the management and communication of identifiers for counterparties and other obligors. Many of the final implications of DFA will not be known until regulatory agencies create rules to implement their respective mandates.

"*The Volcker Rule* prohibits proprietary trading and certain fund activities by bank holding companies and their affiliates and imposes enhanced capital and other quantitative limits on such activities by systemically important nonbank financial companies, including systemically important hedge funds" (Davis and Wardwell, 2010, emphasis in the original). This effectively repeals Gramm–Leach–Bliley's deregulation of banking restrictions imposed by the Banking Act of 1933.

As previously mentioned, most derivatives deregulated under the Commodity Futures Modernization Act of 2000 were re-regulated under Dodd–Frank. "Largely following the historical jurisdictional divisions between the CFTC and the SEC, the Act categorizes the derivatives transactions within its scope as either 'swaps', which are subject to primary regulation by the CFTC, 'security-based swaps', which are subject to primary regulation by the SEC, or 'mixed swaps', which are subject to joint regulation by the CFTC and SEC" (Davis and Wardwell, 2010). The requirement for centralized clearing has the side-effect of centralizing information on these transactions, facilitating regulatory data collection.

Nationally recognized statistical rating organizations (NRSROs) have held government-backed significance since the Great Depression. The banking acts of the 1930s required financial institutions to hold high-quality assets, with quality determined by ratings given by commonly used ratings firms. The act did not state the identities of these firms. Over time, this requirement evolved into the notion of NRSRO, and the SEC became the organization that determined which organizations fit into this category. The history of the evolution of this process is opaque, but it seems that the SEC had assumed this role before 1975. An SEC public memo, File No. S7–23–94, from the year 1994, describes the evolution of the commissions oversight authority (SEC File No. S7–23–94, available at `www.sec.gov/rules/concept/34-34616.pdf`).

After Dodd–Frank, Fed investigators are not allowed to use NRSRO ratings at all in their evaluation of the risk of any securities. Additionally, Dodd–Frank "requires each NRSRO Board to oversee: policies and procedures for management of conflicts of interest; policies and procedures for determining ratings and the effectiveness of internal controls with respect to such policies and procedures; and policies and procedures for compensation and promotion" (Davis and Wardwell, 2010). NRSROs are now liable for their ratings. "The Act establishes that the enforcement and penalty provisions of the Exchange Act apply to statements made by credit rating agencies in the same manner and to the same extent as they apply to statements made by registered public accounting firms or securities analysts under the securities laws" (Davis and Wardwell, 2010).

Government agencies are now writing the rules that implement the Dodd–Frank legislation. Whether these rules will prevent future financial crises remains to be seen. The authors of this essay are skeptical. Prior to the Great Depression, the US financial system experienced periodic financial panics. Their cause was, in part, the complex and fragmented regulatory system created by the constitutional structure of the US government. During the Great Depression, policymakers prohibited all practices that they believed contributed to financial instability. That regulatory structure prevented financial panics from occurring for 50 years. In the 1990s, our nation dismantled the last of the Depression-era restrictions but took no actions to solve the systemic problems that caused financial instability in the past, and that appears to be causing financial instability in the present. Unless our nation deals with the root cause of the problem – fragmented regulatory authority – we should expect financial panics as regularly in the future as they were before the Great Depression.

Table 10.1 *Government agencies or abbreviations*

Abbreviation	Name of agency (and successor)	Date founded (and date disbanded)	Authorizing legislation and principal reforms	Function	Data collections
BFCU	Bureau of Federal Credit Unions (National Credit Union Administration)	1934 (1970)	Federal Credit Union Act (1934)	Oversight of Federal Credit Unions	
FRB	Federal Reserve Board of Governors	1935	Banking Act of 1935	Oversight of Federal Reserve System	
CEC	Commodity Exchange Commission	1936	Commodity Exchange Act (1936)	Required commodities futures and options to be traded on organized exchanges to be regulated by the GFA	
CFPB	Bureau of Consumer Financial Protection	2010	Wall Street Reform and Consumer Protection Act (2010)	"Congress established the CFPB to protect consumers by carrying out Federal consumer financial protection laws."[1]	Reports: `www.consumerfinance.gov/reports/` Guidance documents: `www.consumerfinance.gov/guidance/` Regulations: `www.consumerfinance.gov/regulations/`
CFTC	Commodity Futures Trading Commission	1974	Amendments to Commodity Exchange Act (1974) Commodity Futures Modernization Act (2000) Wall Street Reform and Consumer Protection Act (2010)	Regulates commodity futures and options markets[2]	Market reports: `www.cftc.gov/MarketReports/index.htm`

FDIC	Federal Deposit Insurance Corporation	1933	Banking Act of 1933 Banking Act of 1935 Federal Deposit Insurance Act of 1950 Financial Institutions Reform, Recovery, and Enforcement Act (1989) FDIC Improvement Act of 1991	Insures deposits at commercial banks	Quarterly call reports: `www2.fdic.gov/call_tfr_rpts/`
FFIEC	Federal Financial Institutions Examination Council	1979	Financial Institutions Regulatory and Interest Rate Control Act of 1978	Interagency body empowered to set uniform standards for FRB, FDIC, NCUA, OCC, CFPB; uniform standards for financial institutions[3]	Financial institutional information: `www.ffiec.gov/financial.htm`
FHFA	Federal Housing Finance Agency	2008	Housing and Economic Recovery Act of 2008[4]	Oversight of secondary mortgage markets – Fannie Mae, Freddie Mac, and the Federal Home Loan Banks. Combined OFHEO, FHFB, and the GSE mission office of HUD[5]	Office of Policy Analysis and Research (OPAR): `www.fhfa.gov/Default.aspx?Page=5`
FHFB	Federal Housing Finance Board	1989 (2008)	Financial Institutions Reform, Recovery, and Enforcement Act (1989)	Oversight of Federal Home Loan Banks	News Center archive: `www.fhfa.gov/Default.aspx?Page=244`
FHLBB	Federal Home Loan Bank Board	1932 (1989)	Federal Home Loan Bank Act (1932) Financial Institutions Reform, Recovery, and Enforcement Act (1989)	Oversight of Federal Home Loan Banks	

Table 10.1 (*cont.*)

Abbreviation	Name of agency (and successor)	Date founded (and date disbanded)	Authorizing legislation and principal reforms	Function	Data collections
Fed, FRS	Federal Reserve System	1913	Federal Reserve Act (1913) Banking Act of 1932 Emergency Banking Relief Act (1933) Banking Act of 1933 Banking Act of 1935 Financial Institutions Reform, Recovery and Enforcement Act of 1989 Federal Deposit Insurance Corporation Improvement Act of 1991 Gramm–Leach–Bliley Act (1999) Wall Street Reform and Consumer Protection Act (2010)	Determines and executes US monetary policy	Data download program: `www.federalreserve.gov/datadownload/`
FSLIC	Federal Savings and Loan Insurance Corporation (Savings and Insurance Fund)	1934 (1989)	National Housing Act (1934) Financial Institutions Reform, Recovery, and Enforcement Act (1989)	Insurance for savings accounts in S&Ls	
GAO	Government Accounting Office, Government Accountability Office	1921 (Renamed 2004)	Budget and Accounting Act of 1921 (42 Stat. 20) GAO Human Capital Reform Act of 2004, Pub. L. 108–271, 118 Stat. 811 (2004)[6]	Monitor of Federal activity	Reports and testimonies, legal decisions, and other resources: `www.gao.gov/researchers.html`

GFA	Grain Futures Administration	1922(1974)	Grain Futures Act *Board of Trade of City of Chicago v. Olsen*, 262 US 1 (1923)	Supervises the trading of commodities futures contracts	
LCR	Least-cost resolution	1991	FDIC Improvement Act of 1991	A rule for bank failure resolution to align regulator actions with proper monitoring	
NCUA	National Credit Union Administration	1970	Federal Credit Union Act (1934)	Facilitates the availability of credit union services[7]	
OCC	Office of the Comptroller of the Currency	1863	National Currency Act (creation 1863) National Banking Act (1864) Emergency Banking Relief Act (1933) Banking Act of 1933 Gramm-Leach-Bliley (1999)	"Charter, regulate, and supervise all national banks and federal savings associations"[8]	Annual reports: `www.occ.gov/publications/publications-by-type/annual-reports/index-annual-reports.html`
OFHEO	Office of Federal Housing Enterprise Oversight	1992 (2008)	Federal Housing Enterprises Financial Safety and Soundness Act of 1992 Housing and Economic Recovery Act of 2008	Oversight of Fannie Mae and Freddie Mac	
OTS	Office of Thrift Supervision	1989 (2011)	Financial Institutions Reform, Recovery, and Enforcement Act (1989) Wall Street Reform and Consumer Protection Act (2010)	Oversight of S&Ls and savings banks	
PBGC	Pension Benefit Guaranty Corporation	1974	Employment Retirement Income Security Act of 1974	"[T]o encourage the continuation and maintenance of private-sector defined benefit pension plans, provide	*Pension Insurance Data Book* (since 1996)

Table 10.1 (*cont.*)

Abbreviation	Name of agency (and successor)	Date founded (and date disbanded)	Authorizing legislation and principal reforms	Function	Data collections
				timely and uninterrupted payment of pension benefits, and keep pension insurance premiums at a minimum."[9]	
PCA	Prompt corrective action	1991	FDIC Improvement Act of 1991	A rule for bank failure resolution to align regulator actions with proper monitoring	
PCAOB	Public Company Accounting Oversight Board	2002	Public Company Accounting Reform and Investor Protection Act of 2002	"[O]versee the audits of public companies in order to protect investors and the public interest by promoting informative, accurate, and independent audit reports."[10]	Board public reports: `http://pcaobus.org/Inspections/Pages/PublicReports.aspx`
RFC	Reconstruction Finance Corporation	1932	Glass–Steagall Act of 1932 (c. 8, 47 Stat. 5)	Authorized to extend loans to all financial institutions in the US	
SAIF	Savings Associations Insurance Fund	1989	Financial Institutions Reform, Recovery, and Enforcement Act (1989)	Successor to FSLIC after the S&L crisis of the 1980s	

SEC	Securities and Exchange Commission	1934	Securities Exchange Act (1934) Securities Acts Amendments (1975) Gramm–Leach–Bliley (1999) Commodity Futures Modernization Act (2000) Public Company Accounting Reform and Investor Protection Act of 2002 Wall Street Reform and Consumer Protection Act (2010)	Oversight of securities markets and firms	EDGAR

[1] www.consumerfinance.gov/the-bureau/, 2012.
[2] www.cftc.gov/About/MissionResponsibilities/index.htm, 2012.
[3] www.ffiec.gov/, 2012.
[4] www.fhfa.gov/Default.aspx?Page=5
[5] www.fhfa.gov/Default.aspx?Page=4
[6] www.gao.gov/about/history/articles/working-for-good-government/01-introduction.html, 2012.
[7] www.ncua.gov/about/History/Pages/History.aspx, 2012.
[8] www.occ.gov/about/what-we-do/mission/index-about.html, 2012.
[9] www.pbgc.gov/about/who-we-are.html, 2012.
[10] http://pcaobus.org/About/Pages/default.aspx, 2012.

Table 10.2 *Non-government organizations or abbreviations with a role in financial standards and oversight*

Abbreviation	Full name	Key dates	Purpose
AcSEC	Accounting Standards Executive Committee		Relevant Data: AICPA reSOURCE Lexis-Nexis: CCH U.S. Master GAAP Guide[10] Issues opinions on behalf of AICPA[11]
AICPA	American Institute of Certified Public Accountants	1887[12]	Relevant Data: AICPA Online Professional Library[13] "[S]ets ethical standards for the profession and U. S. auditing standards for audits of private companies, non-profit organizations and federal, state and local governments."[14]
AINs	AICPA Accounting Interpretations		
APB	Accounting Principles Board	1958–1973	Sets accounting standards.
ARB	Accounting Research Bulletins		Official rulings by the Accounting Procedures Committee[15]
CAES	Computer Assisted Execution System		
CAP	Committee on Accounting Procedure	1939–1959[16]	Created to settle accounting problems as they arose[17]
EITF	Emerging Issues Task Force		Resolves urgent issues before they become widespread[18]
FASB	Financial Accounting Standards Board	1972	Relevant legislation: Public Company Accounting Reform and Investor Protection Act of 2002. "[. . .] to establish and improve standards of financial accounting and reporting that foster financial reporting by nongovernmental entities that provides decision-useful information to investors and other users of financial reports."[19]

[10] Moehrle et al., 2002. [11] Bragg, 2006, p. 31. [12] www.aicpa.org/About/Pages/About.aspx, 2012.
[13] www.cpa2biz.com/AST/Main/CPA2BIZ_Primary/Accounting/Standards/PRDOVR-PC-WCF-FS/PC-WCF-FS.jsp, 2012.
[14] www.aicpa.org/About/Pages/About.aspx, 2012. [15] Bragg, 2006, p. 34.
[16] Bragg, 2006, pp. 27–28. [17] Bragg, 2006, p. 27. [18] Bragg, 2006, p. 29.
[19] www.fasb.org/facts/index.shtml#mission, 2012.

Table 10.2 (*cont.*)

Abbreviation	Full name	Key dates	Purpose
FINs	FASB Interpretations		"[C]larify or expand upon any accounting pronouncements that have previously been issued, usually addressing very specific topics."[20]
FTB	FASB Technical Bulletins		"[I]ntended to clarify or elaborate upon underlying accounting standards."[21]
GAAP	Generally Accepted Accounting Principles		Relevant legislation: Public Company Accounting Reform and Investor Protection Act of 2002 Accounting standards set by private accounting standards agencies
IASB	International Accounting Standards Board	2001	Sets standards for the International Financial Reporting Standards Foundation[22]
NASD	National Association of Securities Dealers	1938	Relevant legislation: Maloney Act (1938) "Nonprofit organization formed under the joint sponsorship of the investment bankers' conference and the SEC to comply with the Maloney Act, which provides for the regulation of the OTC."[23]
NRSROs	Nationally Recognized Statistical Ratings Organizations		Relevant legislation: Wall Street Reform and Consumer Protection Act (2010) Organizations that give risk ratings to securities
SFAS	Statements of Financial Accounting Standards		"[P]rimary source of GAAP to the extent that they supersede any previous pronouncements [. . .]"[24]
SOPs	AICPA Statements of Positions		A set of standards for a particular industry issued by the AcSEC[25]

[20] Bragg, 2006, p. 37. [21] Bragg, 2006, p. 51.
[22] www.ifrs.org/The+organisation/IASCF+and+IASB.htm, 2012.
[23] www.nasdaq.com/investing/glossary/n/national-association-of-securities-dealers, 2012.
[24] Bragg, 2006, pp 40. [25] Bragg, 2006, p. 31.

Table 10.3 *Regulatory legislation*

Abbreviation	Full name	Key dates	Purpose
COBRA	Consolidated Omnibus Budget Reconciliation Act of 1974	1974	Amendment to ERISA
DFA	Dodd–Frank Wall Street Reform and Consumer Protection Act (2010)	2010	Response to the first financial crisis of the twenty-first century
ERISA	Employment Retirement Income Security Act	1974	Designed to aid and protect retirees. Related legislation: Consolidated Omnibus Budget Reconciliation Act of 1974 Health Insurance Portability and Accountability Act (HIPAA) Newborns' and Mothers' Health Protection Act Mental Health Parity Act Womens' Health and Cancer Rights Act Tax Equity and Fiscal Responsibility Act Retirement Equity Act Revenue Act of 1978 Tax Reform Act of 1986
FDICIA	FDIC Improvement Act of 1991	1991	Reforms rules for bank regulators and implements PCA and LCR
FIRREA	Financial Institutions Reform, Recovery, and Enforcement Act	1989	Response to the S&L crisis of the 1980s
GLB	Gramm–Leach–Bliley Act (Financial Services Modernization Act of 1999)	1999	Removed barriers between commercial banking, investment banking, and insurance sectors
HIPAA	Health Insurance Portability and Accountability Act	1996	Regulatory data standardization[26]
HMDA	Home Mortgage Disclosure Act	1975	Collects data on lending to test for discriminatory practices Related Legislation: FIRREA (1989)
SOX	Sarbanes–Oxley Act (Public Company Accounting Reform and Investor Protection Act of 2002)	2002	Direct response to financial and accounting scandals, including Enron

[26] www.dhhs.gov/ocr/privacy/hipaa/administrative/index.html, 2012.

Table 10.3 (*cont.*)

Abbreviation	Full name	Key dates	Purpose
	Amendments to Commodity Exchange Act (1974)	1974	Created CFTC, which succeeded the GFA and CEC
	Banking Act of 1932	1932	Expanded lending powers of the FRS, aligning their powers with those granted to the RFC
	Banking Act of 1933	1933	Created the FDIC. Split commercial and investment banks. Imposed stricter regulation on financial institutions. Also known as the Glass–Steagall Act.
	Banking Act of 1935	1935	Increased FDIC coverage. Centralized control of the money supply in the FRS.
	Budget and Accounting Act of 1921 (42 Stat. 20)	1921	Created GAO
	Commodity Exchange Act (1936)	1936	Required all commodities futures and options be traded on organized exchanges
	Commodity Futures Modernization Act (2000)	2000	Clarified regulatory jurisdictions between the CFTC and SEC
	Emergency Banking Act (1933)	1933	Granted expansive powers to the President. Authorized OCC to seize and operate any US bank. Authorized national banks to issue preferred stock. Expanded powers of FRS.
	Federal Credit Union Act (1934)	1934	Established BFCU to regulate credit unions
	Federal Deposit Insurance Act of 1950	1950	Required FDIC to collect and maintain call reports from banks
	Federal Home Loan Bank Act (1932)	1932	Created FHLBB to oversee the 12 government-backed home loan banks
	Federal Housing Enterprises Financial Safety and Soundness Act of 1992	1992	Established OFHEO, which oversees Freddie Mac and Fannie Mae
	Federal Reserve Act (1913)	1913	Created FRS. Required all national banks to join the FRS, gave permission for state banks to join.
	Financial Institutions Regulatory and Interest Rate Control Act of 1978	1978	Created FFIEC, which is tasked with creating uniform principles and standards

Table 10.3 (*cont.*)

Abbreviation	Full name	Key dates	Purpose
			across several different government agencies
	GAO Human Capital Reform Act of 2004, Pub. L. 108–271, 118 Stat. 811 (2004)[27]	2004	Enacted reforms on GAO
	Grain Futures Act	1922	Established the GFA to oversee commodities futures trading
	Housing and Economic Recovery Act of 2008[28]	2008	Enacted reforms on OFHEO
	National Banking Act (1864)	1864	Allowed for national charters of banks to be regulated by the OCC
	National Currency Act	1863	Established a national currency printed by the Treasury and issued by commercial banks
	National Housing Act (1934)	1934	Created FSLIC to insure and regulate the S&L industry
	Securities Acts Amendments (1975)	1975	Tasked SEC with creating a National Market and National Clearing System
	Securities Exchange Act (1934)	1934	Established SEC to regulate securities, especially equities and debt instruments

Bibliography

Alston, Lee J., Wayne A. Grove and David C. Wheelock, 1994, Why do banks fail? Evidence from the 1920s, *Explorations in Economic History*, **31**, 409–431.

Anson, Mark J. P., 2001, Performance presentation standards: which rules apply when? *Financial Analysts Journal*, **57**(2), 53–60.

Benston, George J. and George G. Kaufman, 1997, FDICIA after five years, *The Journal of Economic Perspectives*, **11**(3), 139–158.

Bradford, Frederick A., 1935, The Banking Act of 1935, *American Economic Review*, **25**(4), 661–672.

Bragg, Steven M., 2006, *The Ultimate Accountants' Reference: Including GAAP, IRS & SEC Regulations, Leases, and More*, John Wiley & Sons.

Bratton, William W., 2003, Enron, Sarbanes-Oxley and accounting: rules versus principles versus rents, Villanova University School of Law's Law Review Symposium Issue,

[27] www.gao.gov/about/history/articles/working-for-good-government/01-introduction.html, 2012.

[28] www.fhfa.gov/Default.aspx?Page=5

Lessons from Enron, How did Corporate and Securities Law Fail? Volume **48**(4), 1023–1056.

Buck Jr., Willis R., 1984, Bank insolvency and depositor setoff, *The University of Chicago Law Review*, **51**(1), 188–221.

Calomiris, Charles W., 1990, Is deposit insurance necessary? A historical perspective, *The Journal of Economic History*, **50**(2), 283–295.

Calomiris, Charles W., 2000, *U.S. Bank Deregulation in Historical Perspective*, Cambridge University Press.

Carlson, Mark, Kris Mitchener and Gary Richardson, 2011, Arresting banking panics: federal reserve liquidity provision and the forgotten panic of 1929, *Journal of Political Economy*, **119**(5), 889–924.

Chung, Ching-Yi and Gary Richardson, 2006, Deposit insurance altered the composition of bank suspensions during the 1920s: Evidence from the archives of the Board of Governors, *Contributions to Economic Analysis & Policy*, **5**(1), Article 34, 1–42.

Coates IV, John C., 2007, The goals and promise of the Sarbanes-Oxley Act, *The Journal of Economic Perspectives*, **21**(1), 91–116.

Davis, Polk and Wardwell, L. L. P., 2010, Summary of the Dodd–Frank Wall Street Reform and Consumer Protection Act, Enacted into Law on July 21, 2010.

Department of Health and Human Services, HIPAA Administrative Simplification Statute and Rules, downloaded 2012, www.dhhs.gov/ocr/privacy/hipaa/administrative/index.html

Dewey, Davis R., 1902, *Financial History of the United States*, Longmans Green.

Dick, Astrid A., 2006, Nationwide branching and its impact on market structure, quality, and bank performance, *The Journal of Business*, **79**(2), 567–592.

Dwyer, G., 1996, Wildcat banking, banking panics, and free banking in the United States, *Federal Reserve Bank of Atlanta Economic Review*, December.

Easterbrook, Frank H., 2009, The race for the bottom in corporate governance, *Virginia Law Review*, **95**(4), 685–706.

Fabozzi, Frank J. and Franco Modigliani, 1992, *Mortgage and Mortgage-backed Securities Markets*, Harvard Business School Press.

Farm Credit Administration: History of FCA and FCS, www.fca.gov/about/history/historyFCA_FCS.html

FASB, www.fasb.org/news/memorandum.pdf

FDIC: FFIEC Central Data Repository (CDR) Public Data Distribution (PDD), https://cdr.ffiec.gov/public/HelpFileContainers/FAQ.aspx

Federal Deposit Insurance Corporation, *Resolutions Handbook*, downloaded 2012, www.fdic.gov/bank/historical/reshandbook

Federal Financial Institutions Examination Council, About the FFIEC, downloaded 2011, www.ffiec.gov/about.htm

Federal Reserve Archival System for Economic Research, 1912–1913: Money Trust Investigation: Investigation of Financial and Monetary Conditions in the United States Under House Resolutions Nos. 429 and 504, downloaded 2011, http://fraser.stlouisfed.org/publications/montru/

Flood, Mark, 1992, The Great Deposit Insurance Debate, *Federal Reserve Bank of St. Louis Review*, July/August.

Friedman, M. and A. Schwartz, 1971, *A Monetary History of the United States*, Princeton University Press.

Galbraith, J. K., 1954, *The Great Crash: 1929*, Houghton Mifflin.

Gayer, A. D., 1935, The Banking Act of 1935, *Quarterly Journal of Economics*, **50**(1), 97–116.

Gilbert, R. A., 1986, Requiem for regulation Q: What it did and why it passed away, *Federal Reserve Bank of St. Louis Review*, February, 22–37.

Gillis, John G., 1975, Securities law and regulation: securities acts amendments of 1975, *Financial Analysts Journal*, **31**(5), 12–15.

Gillis, John G. and Robert G. Dreher, 1982, Securities law and regulation: national market system, *Financial Analysts Journal*, **38**(5), 13–15.

Glaessner, Thomas and Ignacio Mas, 1995, Incentives and the resolution of bank distress, *The World Bank Research Observer*, **10**(1), 53–73.

Gorton, Gary, 2009, *Slapped in the Face by the Invisible Hand: Banking and the Panic of 2007*, Federal Reserve Bank of Atlanta, 2009.

Government Accountability Office, 2009, Financial Regulation: A Framework for Crafting and Assessing Proposals to Modernize the Outdated U.S. Financial Regulatory System, Washington, DC.

Grossman, R., 2001, Double liability and bank risk taking, *Journal of Money, Credit and Banking*, **33**(2), May.

Gupta, Atul and Lalatendu Misra, 1999, Failure and failure resolution in the US thrift and banking industries, *Financial Management*, **28**(4), 87–105.

Hammond, Bray, 1991, *Banks and Politics in America from the Revolution to the Civil War*, Princeton University Press.

Hammond, Bray, 1993, *Sovereignty and an Empty Purse*, Princeton University Press.

Jalil, A., 2011, *A New History of Banking Panics in the United States, 1825–1929: Construction and Implications*, Reed College, mimeo, http://academic.reed.edu/economics/jalil/ANewHistoryOfBankingPanics.pdf

Jalil, A., 2012, The economic history of financial panics, *Handbook of Modern Economic History* (forthcoming).

James, Christopher, 1991, The losses realized in bank failures, *The Journal of Finance*, **46**(4), 1223–1242.

Janger, Edward J. and Paul M. Schwartz, 2002, The Gramm-Leach-Bliley Act, information privacy, and the limits of default rules, *Minnesota Law Review*, **86**, 1219–1261.

Jones, Jesse H., 1951, *Fifty Billion Dollars: My Thirteen Years with the RFC, 1932–1945*, Macmilan.

Kloner, Dean, 2001, The Commodity Futures Modernization Act of 2000, *Securities Regulation Law Journal*, **29**, 286–297.

Kolar, Joseph M. and Jonathan D. Jerison, 2005, The Home Mortgage Disclosure Act: its history, evolution, and limitations, *Buckley Kolar LLP*, **7** March.

Langbert, Mitchell, 1994, ERISA: law, interests, and consequences, *Journal of Economic Issues*, **28**(1), 129–153.

Macey, Jonathan R. and David D. Haddock, 1985, Shirking at the SEC: the failure of the national market system, *Faculty Scholarship Series*, Paper 1767, http://digitalcommons.law.yale.edu/fss_papers/1767

Macey, J. and G. Miller, 1992, Double liability of bank shareholders: history and implications, *Wake Forest Law Review*, **31**.

Mason, Joseph, 2002, Savings banks, *Dictionary of American Economic History*, Charles Scribner's & Sons.

Mason, Joseph, 1998, American banks during the Great Depression, *Federal Reserve Bank of St. Louis Review*, May/June.

Meltzer, A. H., 2003, *A History of the Federal Reserve*, Volume 1, 1913–1951, University of Chicago Press.

Meltzer, A. H., 2009, *A History of the Federal Reserve*, Volume 2, 1951–1986, University of Chicago Press.

Meyer, Philip E., 1974, The APB's independence and its implications for the FASB, *Journal of Accounting Research*, **12**(1), 188–196.

The Michigan Law Review Association,1975, The role of the Commodity Futures Trading Commission under the Commodity Futures Trading Commission Act of 1974, *Michigan Law Review*, **73**(4), 710–757.

Moehrle, Stephen R., Jennifer A. Reynolds-Moehrle and Wilbur L. Tomlinson, 2002, Is there a gap in your knowledge of GAAP? *Financial Analysts Journal*, **58**(5), 43–47.

Mulherin, Harold, Jeffry Netter and James Overdahl, 1991, Prices are property: the organization of financial exchanges from a transaction cost perspective, *Journal of Law and Economics*, **34**(2), Part 2, Contracts and the Activities of Firms: A Conference Sponsored by the John M. Olin Foundation and the Lynde and Harry Bradley Foundation, 591–644.

Myers, Margaret G., 1970, *A Financial History of the United States*, Columbia University.

Office of the Comptroller of the Currency, www.occ.treas.gov/about/what-we-do/history/history.html

O'Hara, M., 2004, Searching for a new center: U.S. securities markets in transition, *Federal Reserve Bank of Atlanta Economic Review*, Fourth Quarter, 37–52.

Palmer, Walter B., 1916, The Federal Farm Loan Act, in: *Publications of the American Statistical Association*, **15**(115), 292–312.

Pecora, Ferdinand, 1939, *Wall Street Under Oath*, Cresset Press.

Pension Benefit Guaranty Corporation, 2009, Pension Insurance Data Book 2009, downloaded 2012, www.pbgc.gov/documents/2009databook.pdf

Pension Benefit Guaranty Corporation, Who We Are, downloaded 2012, www.pbgc.gov/about/who-we-are.html

Perino, Michael, 2010, *The Hellhound of Wall Street: How Ferdinand Pecora's Investigation of the Great Crash Forever Changed American Finance*, Penguin Press HC.

Preston, Howard H., 1927, The McFadden Banking Act, *American Economic Review*, **17**(2), 201–218.

Preston, Howard H. 1933, The Banking Act of 1933, *American Economic Review*, **23**(4), 585–607.

Providenti Jr., Anthony C., 1991, Playing with FIRREA, Not Getting Burned: Statutory Overview of the Financial Institutions Reform, Recovery and Enforcement Act of 1989, 59 Fordham L. Rev. S323, http://ir.lawnet.fordham.edu/flr/vol59/iss6/12

Putnam, George F., 1916, The Federal Farm Loan Act, in: *American Economic Review*, **6**(4), 770–789.

Quigley, J., 2006, Federal credit and insurance programs: housing, *Federal Reserve Bank of St. Louis Review*, July/August, 1–29.

Rajan, Raghuram G. and Rodney Ramcharan, 2011, Constituencies and Legislation: the Fight over the McFadden Act of 1927, NBER Working Paper No. 17266, August.

Rand McNally and Company, 1898 to 1940, *Rand McNally Bankers Directory*, Rand McNally, various issues,.

Richardson, Gary, 2006, The records of the Federal Reserve Board of Governors in the national archives of the United States, *Financial History Review*, **13**(1), 123–134.

Richardson, Gary, 2007, Bank distress during the Great Depression: The illiquidity-insolvency debate revisited, *Explorations in Economic History*, **44**(4), 586–607.

Richardson, Gary, 2007, The check is in the mail: correspondent clearing and the banking panics of the Great Depression, *Journal of Economic History*, **67**(3), 643.

Richardson, Gary, 2008, Quarterly data on the categories and causes of bank distress during the Great Depression, *Research in Economic History*, **25**, 37–115.

Richardson, Gary and William Troost, 2009, Monetary intervention mitigated banking panics during the Great Depression: quasi-experimental evidence from the federal reserve district border in Mississippi, 1929 to 1933, *Journal of Political Economy*, **117**(6), 1031–1073.

Rockoff, Hugh, 1974, The free banking era: a reexamination, *Journal of Money, Credit and Banking*, **6**(2), 141–167.

Russell, Steve, 1991, The U.S. currency system: a historical perspective, *Federal Reserve Bank of St. Louis Review*, September/October.

Saunders, A., 1985, Securities activities of commercial banks: the problem of conflicts of interest, *Federal Reserve Bank of Philadelphia Business Review*, July/August, 17–27.

Seidler, Lee J., 1972, Accounting for financial analysis: the Wheat Report: good or bad for analysts? *Financial Analysts Journal*, **28**(3), 93–95.

Shulman, Stuart W., 1999, The Origin of the Federal Farm Loan Act: Agenda-Setting in the Progressive Era Print Press, University of Oregon, Department of Political Science, Dissertation.

Spahr, W., 1932, Bank failures in the United States, *American Economic Review*, **22**(1).

Sprague, O. M. W., 1910, *History of Crises under the National Banking System*, GPO, Washington, DC.

Stassen, J., 1982, The Commodity Exchange Act in perspective, *Washington and Lee Law Review*, **39**(3), 825–843.

Thompson, C.W., 1917, The Federal Farm Loan Act, in: *American Economic Review*, Supplement, Papers and Proceedings of the Twenty-ninth Annual Meeting of the American Economic Association, **7**(1), 115–131.

United States Department of Labor: Health Plans & Benefits: Employment Retirement Income Security Act – ERISA, downloaded 2011, `www.dol.gov/dol/topic/health-plans/erisa.htm`

United States 2008, Department of Labor, Reporting and Disclosure Guide for Employee Benefit Plans, Washington, DC, `www.dol.gov/ebsa/pdf/rdguide.pdf`

Van Horn, Patrick and Gary Richardson, 2009, Intensified regulatory scrutiny and bank distress in New York City during the Great Depression, *Journal of Economic History*, **69**(2).

Viner, Jacob, 1936, Recent legislation and the banking situation, *American Economic Review*, **26**(1), 106–119.

Werner, Walter, 1975, Adventure in social control of finance: the national market system for securities, *Columbia Law Review*, **75**(7), 1233–1298.

Wheelock, D., 2008, The federal response to home mortgage distress: lessons from the Great Depression, *Federal Reserve Bank of St. Louis Review*, May/June, 133–148.

White, Eugene N., 1981, State-sponsored insurance of bank deposits in the United States, 1907–1929, *Journal of Economic History*, **41**, 537–557.

White, Eugene N., 1982, the political economy of banking regulation, 1864–1933, *Journal of Economic History*, **42**, 33–40.

White, Eugene N., 1983, *The Regulation and Reform of the American Banking System, 1900–1929*, Princeton University Press.

White, Eugene N., 1990, The stock market boom and crash of 1929 revisited, *Journal of Economic Perspectives*, **4**(2), 67–83.

Williams, John H., 1936, The Banking Act of 1935, *American Economic Review*, **26**(1), 95–105.

Willis, H. Parker, 1935, The Banking Act of 1933 in operation, *Columbia Law Review*, **35**(5), 697–724.

11

Data for microprudential supervision of US banks

Mark D. Flood, Simon Kwan and Irina S. Leonova

This chapter provides an overview of microprudential data collection in one of the most highly regulated portions of the US financial sector, the banking industry.[1] The public policy objectives for banking supervision and regulation include a safe and sound banking system, stability in the financial markets, and fair and equitable treatment of consumers in their financial transactions. The relatively intrusive supervisory framework for depository institutions contrasts with the regulation of securities markets, where the focus is instead on transparency and investor protection.

Microprudential supervision in banking faces several fundamental challenges, including the inherent complexity of the largest institutions stemming from their scale, organizational structure, and their portfolio of financial services and products, the limitations of the standard accounting framework for measuring risk, the incentive to engage in arbitrage activities to lessen the regulatory impacts or supervisory scrutiny, and the possibility that some important hazards to the financial system may not be obvious at the firm level but become apparent at the system level upon viewing the financial firms collectively (Flood et al., 2012, section 2).

The significance of these issues is apparent in the financial crisis of 2007–2009, which erupted from securitization markets designed, ironically, to remove credit-risky loans from bank balance sheets. Banks, of course, were nonetheless badly hurt by the crisis. During the crisis, the nation's largest investment banks either were acquired by banking organizations or received regulatory approvals to become bank holding companies, thus gaining access to the bank safety net while subjecting themselves to banking supervision and regulation. The financial crisis provided a painful but useful lesson to policymakers. It has had significant ramifications both

[1] Chapter 10 of this Handbook details the origins of US banking regulation.

Handbook of Financial Data and Risk Information, ed. Margarita S. Brose, Mark D. Flood, Dilip Krishna and Bill Nichols. Published by Cambridge University Press. © Cambridge University Press 2014.

for the philosophy and for approaches to banking supervision and regulation. For example, there is a general recognition that an over-reliance on microprudential supervision may expose the gaps in supervisory coverage and lead to regulatory arbitrage by shifting activities from regulated to unregulated entities, such as so-called "shadow banking." Moreover, there is renewed emphasis on macroprudential supervision to address the risks and vulnerabilities that emerge from the interaction among financial firms, and more broadly, on financial stability.[2]

The Dodd–Frank Wall Street Reform and Consumer Protection Act (Dodd–Frank Act, or DFA) of 2010 acknowledged the innovations in the financial sector, and took important steps to eliminate gaps in supervision and regulation, for example through the creation of the Financial Stability Oversight Council (FSOC) and the new Office of Financial Research (OFR). The OFR in particular has data collection among its most important mandates. While the Dodd–Frank Act has vast implications for data collection and reporting requirements, the rule-making process is still ongoing at the time of the writing of this chapter (FRB-STL, 2011b, provides an up-to-date DFA implementation tracker). Moreover, it remains to be seen how banking regulators embrace more data-driven supervision to supplement the more traditional, judgment-based supervisory approaches, as well as how they leverage the expertise of accountants, economists, lawyers, and risk managers to supervise the handful of largest systematically important financial firms.

11.1 Banks and other depository institutions

At the turn of the last century, traditional commercial banks accounted for most of the financial activities in the USA (see Kaufman and Mote, 1994). Partly because of this head start, and partly because of its capacity to have outsize effects on the broader economy, banking is one of the most heavily regulated sectors in the financial system.[3] We focus on the data collection implications of banking supervision.

Banking warrants special scrutiny due to its central role in the monetary system, alongside access to a bank safety net provided by the government that includes deposit insurance and the Federal Reserve's discount window.[4] Bank deposits are still the dominant component of the money supply. Occasional interbank clearing crises (e.g., in 1907, 1929, and 2008) that freeze the banks can propagate quickly to

[2] Macroprudential supervision is addressed in Chapter 14 of this Handbook.

[3] For detailed coverage of the legal aspects of regulation, see De Ghenghi et al. (2010), Greenlee (2008), and Malloy (2011).

[4] The literature on the so-called "specialness" of banks originates with Corrigan's (1983) article. Three salient rationales advanced as possibly justifying special treatment are: the potential for taxpayer losses through the safety net; the potential for systemic spillovers via a general collapse of liquidity and credit in the event of a banking crisis; and the desire to protect small depositors from mishaps beyond their control.

disrupt a much broader range of economic activities. Although they are largely macroeconomic in scope, monetary policy and regulation require active microprudential monitoring of funding structures and credit availability at the level of individual banks, with an implicit or explicit assumption that a supervisory framework that protects the individual component institutions will work from the bottom up to protect the system as a whole. In consequence, firm-level regulation, examination, and data collection has dominated the supervisory framework.

Banking in the USA has evolved historically into a *dual banking system*, in which banks and thrifts may be chartered at either the federal or state level. Institutions have some discretion in choice of charter, see Blair and Kushmeider (2006).[5] Basic "demographic" data on specific institutions, including their charter type, are available online at the FDIC's (2012b) Institution Directory. Chartering authority is the purview of the OCC at the national level, and various state authorities for state banks and thrifts. State banks are also subject to federal regulatory jurisdiction, via (optional) membership in the Federal Reserve System and/or federal deposit insurance. To handle the various cases, banks and thrifts operate under a layered system of primary and secondary federal regulators. The OCC is the primary regulator for national banks and thrifts. The Federal Reserve is the primary regulator for state-chartered "member banks." The FDIC has jurisdiction over all insured depositories (banks and thrifts) and is the primary regulator for state non-member banks and thrifts.[6] The primary regulator coordinates the scheduling, scoping and staffing of supervisory activities, including administrative and enforcement actions (see, e.g., FRB, 2012e, section 5000; OCC, 2007).

Shadow banking

Partly in response to regulatory burden, and partly exploiting recent financial and legal innovations, recent years have seen significant growth in the so-called "shadow banking" system, exemplified by loan securitization markets and their various funding sources (Poszar et al., 2012): finance companies, asset-backed commercial paper (ABCP) conduits, structured investment vehicles (SIVs), credit hedge funds, money market mutual funds, securities lenders, limited-purpose finance companies (LPFCs), and the government-sponsored enterprises (GSEs). Shadow banks exist outside the heavily supervised structure of traditional banking, yet provide many of the same

[5] National banks (i.e., those chartered by the OCC) are known officially as "national associations" and carry the abbreviation "N.A." "Thrift institution" or "thrift" is an umbrella term covering savings banks, savings and loan institutions, and building and loan societies, all of which are depository institutions with a traditional focus on mortgage lending. The Dodd–Frank Act closed the Office of Thrift Supervision (OTS), giving the OCC regulatory authority over thrifts, and the Fed authority over thrift holding companies.

[6] The phrase "member bank" means a member of the Federal Reserve system. Banking regulators have other duties as well. The general structure of the bank supervision and regulation is well covered elsewhere. See, for example, Saunders and Cornett (2010), FRB (2005, especially ch. 5), Spong (2000, especially ch. 4), and the references provided in Table 11.1 below.

economic services, such as maturity, credit and liquidity transformation (see Noeth and Sengupta, 2011).

Shadow banking played a central role in the recent financial crisis (see Gorton, 2010). In August 2007, the inability of Countrywide Financial to roll over funding in the repo markets for its mortgage securitization machine was followed within a week by BNP Paribas's inability to value billions of dollars worth of mortgage collateral in several of its investment funds, leading it to suspend redemptions. A mark-to-market valuation of fund shares was impossible, because the market for the underlying assets did not exist: buyers had fled (FCIC, 2011, ch. 13).

Shadow banking is emblematic of the challenges faced by traditional banking regulation. Supervisors have a relatively powerful set of examination and enforcement tools for institutions within regulatory scope, but shadow banks operate beyond the pale of this scrutiny. Ironically, shadow banking can emerge from the shadows at the worst moment. Citigroup, for example, executed liquidity puts to pull billions of dollars of the SIVs it had created back onto its balance sheet in December 2007 (FCIC, ch. 14); this action has provoked a class-action lawsuit.

At the direction of the G20, the Financial Stability Board (FSB) initiated an investigation of the shadow banking system (FSB, 2011a). The FSB issued its recommendations in October 2011 (FSB, 2011b), allocating workstreams related to banks, money market funds, securitization markets, securities lending and repos, and other shadow banking entities. The recommendations establish seven high-level principles for monitoring the shadow banking system.

1. *Scope*: A system-wide oversight framework for the shadow banking system.
2. *Process*: Monitoring for risks on a regular and continuous basis.
3. *Data/Information*: Authorities should have powers to collect all necessary data and information, as well as the ability to define the regulatory perimeter for reporting.
4. *Innovation/Mutation*: Monitoring should be flexible and adaptable to capture innovations and mutations that could lead to emerging risks.
5. *Regulatory arbitrage*: Authorities should be mindful of the incentives to expand shadow banking created by changes in regulations.
6. *Jurisdiction-specific features*: Authorities should abide by the rules of both their local jurisdiction as well as their international connections.
7. *Information exchange*: Authorities should exchange information both within and across jurisdictions on a regular basis to be able to assess the risks of the shadow banking system.

Unsurprisingly, a salient focus of these recommendations is on improved data, information, and monitoring. The FSB is recommending the elimination of the shadows.

Individual depository institutions (i.e., banks and thrifts) make up the core of the banking system, but other institution types play an important role, most notably bank holding companies (BHCs).[7] A BHC is defined under the Bank Holding Company Act simply as a company that has control over a bank.[8] A BHC may be a legal wrapper around a single bank – a structure known as a unitary BHC – or it may include scores of other subsidiaries, including both banks and non-banks. For example, Citicorp – the BHC controlling Citibank N.A. – has over 2,000 subsidiaries, even after a major internal "legal vehicle simplification" project completed in 2006 (Herring and Carmassi, 2010; Dugan, 2010, Appendix E).

The Federal Reserve has supervisory authority over BHCs, which it implements through its Regulation Y (see FRB, 2011a). As part of this supervision, the Federal Reserve collects quarterly BHC financial statements through the Y-9 filings, which are publicly available in different formats through the Federal Financial Institutions Examination Council (FFIEC, 2011a) and the Chicago Fed (FRB-CHI, 2011).[9] A parallel system exists for savings and loan holding companies (SLHCs), oversight for which has passed from the now defunct (since 1989) Federal Home Loan Bank Board (FHLBB) to the now defunct (since the DFA) Office of Thrift Supervision (OTS) to the Federal Reserve.

The Gramm–Leach–Bliley Act of 1999 (GLBA, see Chapter 12 of this Handbook) upended BHC supervision (see Vir Bhatia, 2011). In addition to repealing the Depression-era Glass–Steagall separation of commercial and investment banking, the GLBA created the financial holding company (FHC), which is allowed to maintain a full range of financial subsidiaries, including commercial banks, merchant banks, broker-dealers, and insurance underwriters. FHCs must meet additional eligibility requirements: they must be well capitalized, well managed and have satisfactory (or better) ratings under the Community Reinvestment Act. Significantly, the GLBA required the Fed to defer fully to the functional regulators (the Securities and Exchange Commission (SEC), Commodity Futures Trade Commission (CFTC), and state regulators) in the supervision of non-bank subsidiaries. The net effect was a compartmentalized system in which no one regulator had a complete picture of the activities of the consolidated holding company. The DFA rectifies this shortcoming by empowering the Fed with supervisory authority over all BHC subsidiaries, including the so-called significant non-bank financial companies (SNBFCs).

[7] There are other institution types in the banking ecosystem that we do not cover here due to lack of space. These range from the relatively familiar (e.g., investment banks, mortgage banks, mutual savings banks, and credit unions) to the relatively obscure: industrial loan corporations, Edge Act corporations, industrial banks, limited-purpose trust companies, agreement corporations, bankers' banks, and agricultural credit corporations.

[8] The definition of "control" in the Act focuses on percentage of voting shares and influence over the selection of directors, see FDIC (2012f).

[9] The FFIEC was created in 1978 to coordinate supervision across the various regulators of depository institutions, including the Federal Reserve, OCC, FDIC and National Credit Union Administration (NCUA). One of FFIEC's main tasks is to coordinate the collection and publication (where appropriate) of supervisory data.

The Dodd–Frank Act also created a special new class of systemically important financial institutions (SIFIs), subject to additional scrutiny and regulation. Section 165 of the DFA defines two kinds of SIFIs, large BHCs and designated non-bank SIFIs, and requires the Federal Reserve to set stricter prudential standards for them. "Large" BHCs are those with more than $50 billion in consolidated assets. Non-bank SIFIs are to be designated individually by the Financial Stability Oversight Council (FSOC). In designating non-bank SIFIs, the FSOC has organized the factors listed in the Act into six categories. Three (leverage, liquidity risk, and existing regulatory scrutiny) evaluate a company's susceptibility to financial distress; the others (size, interconnectedness, and sustainability) evaluate the impact that a company's financial distress could have on the industry and the broader economy (see FSOC, 2012). Whether large BHCs or designated non-banks, all SIFIs are subject to the so-called "living will" requirements of the DFA, under which they must maintain plans for their own rapid and orderly resolution in the event of material distress or failure. The "rapid and orderly" rule implicitly requires a higher degree of data standardization and more robust reporting infrastructure than has been available heretofore. The Federal Reserve Board and the FDIC's new Office of Complex Financial Institutions (OCFI) will manage the living will process (FRB-FDIC, 2011).

The FSOC is a college of microprudential regulators created by the DFA to coordinate supervision and eliminate gaps in supervision system-wide.[10] A central mandate for the FSOC and the new Office of Financial Research (OFR) is to facilitate the sharing and collection of information across the member agencies. While many of the efforts of the FSOC and the OFR are devoted to macroprudential concerns, many also have microprudential implications, such as the data standardization work that will help to support living wills and the initiative, coordinated by the Financial Stability Board (FSB), to establish a global legal entity identifier (LEI) that will provide unique identification of market participants for the first time (FSB, 2012a).[11]

11.2 The supervisory process and on-site examination

The traditional mode of supervisory information collection is on-site examination. The supervisory agencies maintain large examination staffs, whose main role is to

[10] The FSOC is a who's who of microprudential regulators, chaired by the Secretary of the Treasury and including as voting members the heads of: Federal Reserve, the OCC, SEC, CFPB, FDIC, CFTC, FHFA, NCUA and an independent insurance expert appointed by the President. Non-voting members are the heads of the new OFR and the new Federal Insurance Office, and representatives (one each) selected by state insurance commissioners, state banking supervisors, and state securities supervisors.

[11] Parallel developments in microprudential supervision in Europe after the financial crisis are addressed in Chapter 13.

ensure the safety and soundness of the regulated entities. On-site supervision can create tension with data-driven off-site analysis. Flood et al. (2012) point out that data-driven regulation favors "hard" information over "soft," and that vital nuance is often lost in the process.[12] On the other hand, standardized off-site data collection offers the opportunity for systematic comparisons and statistical analysis of large numbers of institutions, frequently revealing important trends and patterns. For example, starting with its first Annual Report in 1863, the Comptroller of the Currency has collected and published basic institution-level financial details; see FRB-STL (2012). Ideally, off-site analysis and on-site supervision should complement each other, with data-driven analysis helping to target on-site supervision and examinations to identify emerging activities that may warrant more systematic analysis.

Frequency of on-site examination varies considerably across firms. Most banks have relatively slowly evolving portfolios of loans and deposits, and are typically subject to full-scope on-site examinations annually (e.g., FDIC, 2012g; FRB, 2012d). Large firms – so-called large, complex banking organizations (LCBOs) and large banking organizations (LBOs) – are additionally subject to continuous on-site supervision (FRB, 2012e; OCC, 2007, 2010). Supervisors may also initiate occasional targeted examinations to address specific concerns. Targeted examinations may occur in the context of a "horizontal review," in which the regulator simultaneously performs the same examination procedures across a group of institutions.

Targeted examinations focus on a specific vulnerability or supervisory concern, which could arise from macroeconomic or financial developments. Horizontal targeted examinations are also key to macroprudential supervision. For example, while an exposure at one institution may not raise a supervisory concern, the same exposure at many institutions could imply a systemic concern that warrants policy makers' attention. Furthermore, a horizontal review could uncover interconnectedness among institutions that pose fragility to the financial system.

Banking examinations – especially the annual full-scope variety – follow a certain prescribed protocol. While the details may vary across regulators, across firms (for example, de novo banks and large banks get special treatment), and across examination types, the Fed's (FRB, 2012d, section 1000.1) six-step process for large-bank examinations is illustrative; see Table 11.1.

A key preliminary action in this process is the appointment of a "central point of contact" – the Fed's equivalent of an Examiner in Charge (EIC) – to coordinate the

[12] Chapter 5 of this Handbook offers a perspective from industry on data-driven approaches. Kuritzkes and Schuermann (2010) develop a simple framework for what is knowable and unknowable about bank risks, with a focus on data-driven statistical analysis.

Table 11.1 *Banking examinations, outline of the on-site examination process*

Step	Description	Output
1	Understanding the institution	Institutional overview
2	Assessing the institution's risk	Risk matrix and risk assessment
3	Planning and scheduling supervisory activities	Supervisory plan and examination program
4	Defining examination activities	Scope memorandum and First Day letter
5	Performing examination procedures	Functional examination modules
6	Reporting the findings	Examination report(s)

on-site examination, which will likely extend over multiple offices and jurisdictions for large banks. Steps 1–3 delineate the extensive preparatory work that culminate in the scope memorandum and entry letter in Step 4. The First Day letter is the formal introduction of the examination to the institution, and it itemizes a range of documents and reports the bank must prepare for the exam team in preparation for the review. These reports become part of the overall exam documentation, often referred to as the exam "work papers," which also includes correspondence with the bank, records of meetings, and the examination report itself. Exam work papers are typically held as confidential supervisory information (CSI).

On-site examinations are a key input into a bank's CAMELS rating, a high-level summary score that facilitates comparison across institutions and is in an input into numerous other regulatory processes, such as the Prompt Corrective Action (PCA) framework described below.[13] While CAMELS ratings can incorporate significant subjective judgments, they represent an early attempt to measure bank quality on a numerical scale that is directly comparable across institutions. There is a broader trend toward "data-driven" supervision and regulation, which include off-site monitoring and early-warning tools and the emphasis in the Basel capital accords on formal models of credit and market risk.[14] The Federal Reserve's 2010 Annual Report highlights stress testing as another example (FRB, 2011h, p. 75). Supervisory agencies have very broad authority in their role as

[13] CAMELS is an acronym for a standardized ratings system (formally, the Uniform Financial Institutions Ratings System, UFIRS) that assesses banks on each of six dimensions: capital, assets, management, earnings, liquidity and sensitivity to market risk. The CAMELS program (originally CAMEL) dates to 1979, and is coordinated by the FFIEC.

[14] On early-warning models, see Gramlich et al. (2010). For BHCs, the Federal Reserve maintains the Supervision and Regulation Statistical Assessment of Bank Risk early-warning model (SR-SABR), which outputs a two-component surveillance rating to each subsidiary bank in a BHC. The rating consists of the current CAMELS rating and a separate letter grade that rates the bank's strength or weakness relative to other institutions with the same CAMELS rating. The SR-SABR model replaces the predecessor early-warning System to Estimate Examination Ratings (SEER); see FRB (2012e, section 4080). King et al. (2006) provide a catalog of off-site surveillance and early-warning models, including SEER, the FDIC's Statistical CAMELS Off-site Rating (SCOR), and the OCC's Canary model.

safety-and-soundness examiners to demand reports and documents from the institutions they regulate. However, the Gramm–Leach–Bliley Act of 1999 explicitly divided supervisory authority by functional area, giving jurisdiction over insurance matters to state insurance supervisors, over securities to the SEC, and over commodities to the CFTC (OCC, 2007, pp. 20–22). As noted above, this explicit codification of functional regulation excludes "functional regulated affiliates" of a bank from the examination scope of the bank's primary supervisory agency, effectively Balkanizing the flow of information to supervisors. The DFA corrected this separation of duties.

11.2.1 Supervisory data collections

The centerpiece of off-site supervisory monitoring is the quarterly financial statements reported systematically by both banks and BHCs. The main data collection for banks is the *Report of Condition and Income*, commonly known as the "Call Report." The quarterly Call Report provides a snapshot of the reporting institution at the end of each calendar quarter, including a comprehensive set of financial statements such as the balance sheet (Schedule RC) and income statement (Schedule RI), as well as other information relevant to prudential supervision including, for example, past due and non-accrual loans (Schedule RC-N), derivatives and off-balance-sheet items (Schedule RC-L), and charge-offs and recoveries (Schedule RI-B). Significant changes to the Call Reports were introduced in 2012, including the expansion of a number of schedules to collect new items (FFIEC, 2012d) and a proposal to add two entirely new schedules (FFIEC, 2011d).

All national banks, state member banks, and insured state non-member banks (and savings associations) are required to complete the Call Reports. Call Reports have been collected since 1976, and coordinated by the FFIEC since its inception in 1978. Over the years, the Call Reports have expanded quite significantly to gather increasing details of the financial condition of the reporting institution. The data are publicly available at the FFIEC's (2012b) online Central Data Repository (CDR).[15] Note that the Call Report covers only banks (not BHCs or thrifts); for large BHCs with many non-bank subsidiaries, the Call Report therefore provides a limited view of the overall financial activities. Between its inception in 1989 and its elimination in 2010 by the Dodd–Frank Act, the OTS maintained a parallel data collection, the quarterly Thrift Financial Report (TFR) for federally chartered thrift institutions.

[15] Call Report data are also available in bulk from the Chicago Fed in SAS XPORT format; see FRB-CHI (2012). The Federal Reserve Board maintains a data dictionary, the Micro Data Reference Manual (MDRM), that covers most of the institution-level data collections; see FRB (2012m).

The Federal Reserve integrates the Call Report data with a range of other supervisory collections in its National Information Center (NIC), portions of which are made public by the FFIEC (2011a). Separately, the FDIC (2011b) has done the difficult work of mapping TFR data definitions to the Call Report to provide more comprehensive coverage of insured depositories. This combined database, known as the Research Information System (RIS), is available from the FDIC upon request. RIS also includes information on institutional structure and merger histories for all banks and thrifts. To support bank examiners, the FFIEC (2008) maintains an augmented version of the Call Reports, known as the Uniform Bank Performance Report (UBPR). The UBPR combines the Call Report with peer group analyses and percentile ranking to provide context for comparisons. In 2005, the FFIEC and its member agencies initiated a modernization project to migrate the Call Report and UBPR from traditional paper forms to XML-based electronic submissions.[16] Reporting now occurs via XML messages in an extensible business reporting language (XBRL) format, which supports standardized structured metadata known as "XBRL taxonomies." The taxonomies include data definitions and messaging formats, which allow for validation of the data before transmission to the FFIEC (the validation burden is on the reporting institution). The fact that most banks (especially smaller institutions) rely on a relatively small number of large, third-party data-services vendors to handle their regulatory reporting facilitates the automated publication and propagation of XBRL taxonomy revisions, vastly simplifying the management of Call Report revisions.

A parallel system of reporting applies to BHCs, which are supervised by the Federal Reserve. A BHC may contain other BHCs, creating a "tiered" BHC, and the ownership hierarchy is significant. Form FR Y-9C collects the consolidated financial statements for a (possibly tiered) BHC, and is essentially a BHC-level Call Report. For tiered BHCs, the Y-9C is required only for the top-level BHC in the group, the so-called "high holder." FR Y-9LP collects financial statements for each of the individual subsidiary BHCs in large (consolidated assets exceeding $500M), tiered BHCs (see FRB, 2007).[17] The Fed also creates a BHC Performance Report (BHCPR) for peer group comparison, analogous to the UBPR for individual banks. Like the Call Reports, Y-9s are available for individual institutions online at the FFIEC website (FFIEC, 2011a), and in bulk from the Chicago Fed (FRB-CHI, 2011).

Identification of institutions in regulatory collections presents special challenges, in large part due to the diversity of identifiers. In the absence of the LEI's unified

[16] Form FFIEC 041 applies to domestic-only institutions. Banks with foreign offices must file FFIEC 031; see FFIEC (2011b, 2012c). The modernization was unveiled in 2010.

[17] There is a distinct Form FR Y-9SP that applies to small tiered BHCs (consolidated assets below $500m). There is also a specialized form FR Y-9ES for BHCs that are employee stock ownership plans.

Table 11.2 *Important US bank regulatory data collections and reports*

The table presents a non-comprehensive list of the major collections and publications of financial data related to banking and depository institutions. The best sources for detailed and timely listings of information collections remain the individual agency Internet sites. All information in the table comes from public sources, identified in the final column.

Agency	Name	Frequency	Description	Reference
Federal Financial Institutions Examination Council (FFIEC)	Bank Call and Thrift Financial Reports	Quarterly	Every national bank, state member bank, and insured non-member bank is required by its primary federal regulator to file a Consolidated Reports of Condition and Income (Call Report) as of the close of business on the last day of each calendar quarter (the report date). Historically, thrifts have filed a parallel Thrift Financial Report (TFR). The specific reporting requirements depend upon the size of the bank and whether it has any foreign offices. Information reported includes earnings, balance sheet, asset quality, liquidity and capital.	OCC (2011c), FFIEC (2012c), FFIEC (2012b)
	Uniform Bank Performance Report (UBPR)	Quarterly	The Uniform Bank Performance Report (UBPR) is a peer comparison report for banks and thrifts, based on the Call Report. The UBPR includes dollar values, ratios, peer group averages and percentile rankings for a bank and a variety of benchmark groups.	FFIEC (2008), FFIEC (2012b)
	Bank Holding Company Performance Report (BHCPR)	Quarterly	The Bank Holding Company Performance Report (BHCPR) is a computer-generated report of current and historical financial information produced quarterly for three groups of bank holding companies: top-tier	FRB (2011g)

		bank holding companies with consolidated assets of $500 million or more; top-tier bank holding companies required to file the FR Y-9C and FR Y-9LP; and top-tier bank holding companies that are not subject to the Board's risk-based capital guidelines but elect voluntarily to file the FR Y-9C and FR Y-9LP report forms.	
Country Exposure Report (Forms 009 and 009a)	Quarterly	The Country Exposure Report (FFIEC 009) is filed quarterly as of the last business day and provides information on international claims of US banks, savings associations, and bank holding companies. The information is used to monitor country exposure of banks to assess risk and possible impact on US banks of adverse developments in particular countries. The Country Exposure Information Report (FFIEC 009a) is a supplement to the FFIEC 009, providing publicly available information on material foreign country exposures of US banks, thrifts and BHCs filing the FFIEC 009. Reporting institutions must also furnish a list of countries in which they have material lending exposures.	FFIEC (2011c)
Central Data Repository (CDR)	Quarterly	The Central Data Repository provides current and historical financial and structural information for most FDIC-insured institutions on the FFIEC website. Publicly available information includes Call Report and UBPR data. The earliest data provided are from March 31, 2001.	FFIEC (2006)
Community Reinvestment Act (CRA)	Annual	Under the Community Reinvestment Act (CRA, 12 U.S.C. 2901), all institutions regulated by the OCC, FDIC and Federal Reserve that meet	FFIEC (2010a), FFIEC (2012d), FDIC (2012a)

Table 11.2 (*cont.*)

Agency	Name	Frequency	Description	Reference
			the asset size threshold are subject to CRA data collection and reporting requirements, which are updated annually. The CRA website houses disclosure reports from filing institutions and provides compilation reports on aggregate data from such institutions.	
FFIEC/CFPB	Home Mortgage Disclosure Act (HMDA)	Annual	The Home Mortgage Disclosure Act (HMDA) was enacted by Congress in 1975 and was implemented by the Federal Reserve Board's Regulation C. On July 21, 2011, rule-writing authority was transferred to the Consumer Financial Protection Bureau (CFPB). HMDA requires covered depository and non-depository institutions to collect and disclose information about applications for, originations of, and purchases of home purchase loans, home improvement loans, and refinancings. Reporting requirements depend on an institution's asset size, location, and whether it is active in residential mortgage lending. The FFIEC website supports search for disclosure reports by institutions as well as aggregate data for all reporting institutions.	FFIEC (2010b)
Federal Deposit Insurance Corporation (FDIC)	Institution Directory	Weekly	The FDIC Institution Directory (ID) provides the latest comprehensive financial and demographic data for every FDIC-insured institution, including the most recent quarterly financial statements, with performance and condition ratios.	FDIC (2012b)

Summary of Deposits (SOD)	Annual	The Summary of Deposits (SOD) report is the annual survey of branch office deposits as of June 30 for all FDIC-insured institutions, including insured US branches of foreign banks. All institutions with branch offices are required to file; institutions with only a main office are exempt. Users can access these data by: single institution, institutions within a geographical area, or aggregated within a geographical area. SOD features include custom market share reports and downloads.	FDIC (2010a), FDIC (2012c)
Research Information System (RIS)	Quarterly	The Research Information System (RIS) is a data warehouse developed and maintained by the FDIC's Division of Insurance and Research (DIR). RIS information is the basis for DIR publications, which include the Institution Directory, Quarterly Banking Profile, Summary of Deposits, and the Statistics on Banking. RIS covers all FDIC-insured institutions including commercial banks, thrifts, and insured branches of foreign banks. RIS information is quarterly and begins with the December 31, 1972 Call Report. RIS contains almost 2,800 variables for each FDIC-insured institution as well as over 7,500 pre-calculated industry aggregates for user convenience.	FDIC (2011b)
Historical Statistics on Banking (HSOB)	Annual	The Historical Statistics on Banking (HSOB) provides annual statistical information on the banking industry beginning in 1934. The HSOB contains aggregated data from individual financial reports filed by FDIC-insured commercial banks and thrifts to	FDIC (2012e)

Table 11.2 (*cont.*)

Agency	Name	Frequency	Description	Reference
			support the analysis of long-term trends and benchmarks in the industry.	
Office of the Comptroller of the Currency (OCC)	Shared National Credits Program (SNC)	Transactional	Established in 1975, the Shared National Credits (SNC) Program was designed to provide an efficient and consistent review and classification of SNC, which includes any loan and/or formal loan commitment, and any asset such as real estate, stocks, notes, bonds, and debentures taken as debts previously contracted, extended to borrowers by a federally supervised institution, its subsidiaries, and affiliates that aggregates to $20 million or more and is shared by three or more unaffiliated supervised institutions. Many of these large loan commitments are also shared with foreign banking organizations and non-banks, including securitization pools, hedge funds, insurance companies, and pension funds. The program provides uniform treatment and increases efficiencies in SNC risk analysis and classification.	FRB / FDIC / OCC (2011), OCC (1998b)
	Cost of Funds Report	Monthly	National and state average and national median cost of funds for OTS-regulated, SAIF-insured thrifts required to file. The average cost of funds is the sum of the period interest paid and accrued on deposits (excluding escrow deposits), FHLB advances, and other	OCC (2011d)

			borrowings, divided by the average period balance in those accounts. The median cost of funds is the midpoint value of individual institutions' cost of funds ratios, adjusted for variation in length of month.	
	Survey of Credit Underwriting Practices	Annual	An annual survey to identify trends in lending standards and credit risk for the most common types of commercial and retail credit offered by national banks. The survey generally includes examiner assessments of credit underwriting standards at large national banks with assets of $3 billion or more. OCC examiners assigned to each bank assess overall credit trends for 13 commercial (e.g., residential construction, asset-based lending, small business, etc.) and 7 retail (e.g., credit cards, home equity, etc.) credit products. Underwriting standards refers to the terms and conditions under which banks extend or renew credit, such as financial and collateral requirements, repayment programs, maturities, pricing, and covenants.	OCC (2011e)
Board of Governors of the Federal Reserve (FRB)	Beige Book	FOMC schedule	First published in 1983, the Beige Book or, "Summary of Commentary on Current Economic Conditions by Federal Reserve District," is produced by local Reserve Bank staff and released to the public approximately two weeks prior to each regularly scheduled FOMC meeting. Each Reserve Bank collects anecdotal information on current economic conditions through reports from Bank and Branch directors and interviews with key business contacts, economists, market experts,	FRB (2012g)

Table 11.2 (*cont.*)

Agency	Name	Frequency	Description	Reference
			and other sources. The Beige Book presents summaries by District and a national summary of the information.	
	Blue Book	FOMC schedule	First published in 1965, the Blue Book or, "Monetary Policy Alternatives," provides background and context on monetary policy alternatives for the FOMC. Prepared by Board staff, it is distributed to FOMC participants the week before an FOMC meeting.	FRB (2012g)
	Green Book	FOMC schedule	The Green Book or, "Current Economic and Financial Conditions," is an in-depth analysis of the US and international economy, produced by Board staff and distributed to FOMC attendees the week before the meeting. The Green Book comprises a summary/ outlook and recent developments, plus a supplement.	FRB (2012g)
	FOMC Minutes	FOMC schedule	A summary of significant policy issues addressed by FOMC meeting participants. The minutes record all decisions taken by the Committee and explain the reasoning behind these decisions. From 1967–1992, the minutes were split into two: the Record of Policy Actions and the Minutes of Actions. From February 1993 until December 2004, the minutes were published approximately three days after the Committee's subsequent meeting. Since December 2004, the minutes have been published three weeks after the date	FRB (2012g)

		of the policy decision, to reduce the publication lag. The minutes are subsequently published in the Board's *Annual Report*.	
Senior Loan Officer Opinion Survey on Bank Lending Practices (SLOOS)	Quarterly	The survey covers approximately 60 large domestic banks and 24 US branches and agencies of foreign banks. The Federal Reserve generally conducts the survey quarterly, timing it so that results are available for the January/February, April/May, August, and October/November meetings of the Federal Open Market Committee. The Federal Reserve occasionally conducts one or two additional surveys during the year. Questions cover changes in the standards and terms of the banks' lending and the state of business and household demand for loans. The survey often includes questions on one or two other topics of current interest.	FRB (2011c)
Senior Credit Officer Opinion Survey on Dealer Financing Terms (SCOOS)	Quarterly	The survey, patterned on the SLOOS, canvasses dealers in the securities financing and over-the counter (OTC) derivatives markets for qualitative information on credit terms and conditions in securities financing and OTC derivatives markets, which are important conduits for leverage in the financial system. The survey panel for the SCOOS currently includes 20 dealers, accounting for almost all of the dealer activity in dollar-denominated securities financing and OTC derivatives markets. The survey is directed to senior credit officers responsible for maintaining a consolidated perspective on the management of credit risks.	FRB (2012j)

Table 11.2 (*cont.*)

Agency	Name	Frequency	Description	Reference
	Survey of Consumer Finances (SCF)	Triennial	The Survey of Consumer Finances (SCF) is normally a triennial cross-sectional survey of US families, but over the 1983–1989 and 2007–2009 periods, the survey collected panel data. The survey data include information on families' balance sheets, pensions, income, and demographic characteristics. Information is also included from related surveys of pension providers and the earlier such surveys conducted by the Federal Reserve Board.	FRB (2011c)
	Survey of Small Business Finances	Occasional	The Survey of Small Business Finances (SSBF) periodically collects information on small businesses (fewer than 500 employees) in the USA. Owner characteristics, firm size, use of financial services, and the income and balance sheets of the firm are just some examples of the types of information collected.	FRB (2011c)
	FR 2900 Report of Transaction Accounts, Other Deposits and Vault Cash	Weekly	The FR 2900 collects information on transaction accounts, time and savings deposits, vault cash, and other reservable obligations from depository institutions. These are the primary source for calculation of required reserves and for construction of the monetary and reserves aggregates. The panel consists of all depository institutions with total reservable liabilities exceeding a de minimis exemption threshold.	FRB (2009b)

Commercial Paper (CP) release	Daily, weekly, monthly	CP issuance rates and volumes are updated daily. Weekly data on CP outstanding are available as of the close of business each Wednesday and monthly as of the last business day of the month. Data are typically posted with a one-day lag. Data are supplied by the Depository Trust and Clearing Corporation (DTCC), a national clearinghouse for the settlement of securities trades and a custodian for securities. In addition, the Board publishes one-, two-, and three-month rates on AA non-financial and AA financial CP in its weekly H.15 release.	FRB (2012h), FRB (2012i)
G.19 Consumer Credit	Monthly (fifth business day of the month)	The G.19 release includes estimates of consumer credit outstanding in the USA by credit type (revolving and non-revolving) and by lender type (commercial banks, finance companies, credit unions, federal government, savings institutions, non-financial business, and pools of securitized assets). In addition, the release contains data on terms of credit at commercial banks and finance companies.	FRB (2012h), FRB (2012i)
H.3 Aggregate Reserves and Monetary Base	Weekly	The H.3 release provides data on aggregate reserves of depository institutions, including required reserves, total reserves, excess reserves, non-borrowed reserves, and borrowings by depository institutions from the Federal Reserve's discount window. The release also provides data on the monetary base, which includes currency and reserves.	FRB (2012h), FRB (2012i)

Table 11.2 (*cont.*)

Agency	Name	Frequency	Description	Reference
	H.6 Money Stock Measures	Weekly	The H.6 release provides measures of the monetary aggregates (M1 and M2) and their components. M1, the more narrowly defined measure, is included in M2 and consists of the most liquid forms of money, namely currency and checkable deposits. The non-M1 components of M2 are primarily household holdings of savings deposits, small time deposits, and retail money market mutual funds. Monthly data are available back to January 1959; for most series, weekly data are available back to January 1975.	FRB (2012h), FRB (2012i)
	H.8 Assets and Liabilities of Commercial Banks in the USA	Weekly	The H.8 release provides an estimated weekly aggregate balance sheet for all commercial banks in the USA. The release also includes separate balance sheet aggregations for several bank groups: domestically chartered commercial banks; large domestically chartered commercial banks; small domestically chartered commercial banks; and foreign-related institutions in the USA. Foreign-related institutions include US branches and agencies of foreign banks as well as Edge Act and agreement corporations. Published weekly, the release is typically available to the public by 4:15 pm each Friday.	FRB (2012h), FRB (2012i)

H.10 Foreign Exchange Rates	Weekly	The H.10 weekly release contains daily rates of exchange of major currencies against the US dollar. The data are noon buying rates in New York for cable transfers payable in the listed currencies. The rates have been certified by the Federal Reserve Bank of New York for customs purposes as required by section 522 of the amended Tariff Act of 1930.	FRB (2012h), FRB (2012i)
H.15 Selected Interest Rates	Weekly	The H.15 release contains daily interest rates for selected US Treasury and private money market and capital market instruments. It is published weekly.	FRB (2012h), FRB (2012i)
Z.1 Flow of Funds Accounts	Quarterly	Report issued by the Federal Reserve with information on availability of funds for the latest quarter, credit market statistics by sector, distribution of gross domestic product, distribution of national income, and savings and investment, among others. The Z.1 is released the second week of the quarter for the preceding quarter.	FRB (2012h), FRB (2012i)
FR Y-14A Annual Capital Assessments and Stress Testing Report	Annual	The FR Y-14A report collects large bank holding companies' (BHCs) quantitative projections of balance sheet, income, losses, and capital across a range of macroeconomic scenarios, along with qualitative information on methodologies used to develop internal projections of capital across scenarios. The report includes schedules for: Summary, Macro Scenario, Counterparty Credit Risk (CCR), Basel III/Dodd–Frank, and Regulatory Capital Instruments.	FRB (2012k)

Table 11.2 (*cont.*)

Agency	Name	Frequency	Description	Reference
	FR Y-14M Monthly Capital Assessments and Stress Testing Report	Monthly	The FR Y-14M report comprises two loan-level collections, one account- and portfolio-level collection, and one detailed address-matching collection to supplement the two loan-level collections. The report includes schedules for: Domestic First Lien Closed-End 1–4 Family Residential Mortgage, Domestic Home Equity Residential Mortgage, Domestic Credit Card, and Address Matching.	FRB (2012l)
	FR Y-14Q Quarterly Capital Assessments and Stress Testing Report	Quarterly	The FRY-14Q report collects detailed data on bank holding companies' various asset classes and categories of pre-provision net revenue (PPNR) on a quarterly basis, which is used to support supervisory stress testing models and for continuous monitoring efforts. The FR Y-14Q report includes data schedules for Securities Risk, Retail Risk, PPNR, Wholesale Risk, Trading, Private Equity, and Other Fair Value Assets (Trading), Basel III and Dodd–Frank (Basel III), and Regulatory Capital Instruments.	FRB (2012a)
	National Information Center (NIC)	Various	The National Information Center (NIC) provides comprehensive information on banks and other institutions for which the Federal Reserve has a supervisory, regulatory, or research interest including both domestic and foreign banking organizations operating in the USA. The NIC public website, available through the FFIEC, allows one to search for financial and descriptive data on bank holding companies, both current and historical.	FFIEC (2011a)

Federal Reserve Bank of Chicago (FRB-CHI)	Call Reports	Quarterly	The Reports of Condition and Income (Call Reports) are a complete set of quarterly financial statements, required of all national banks, state member banks, and insured state non-member banks (and savings associations). They are available for bulk download from the Chicago Fed.	FRB-CHI (2012)
	FR Y-9 BHC Financial Statements	Quarterly	BHC financial statements are reported on the Y-9 forms. Form FR Y-9C collects the consolidated financial statements for a (possibly tiered) BHC. For tiered BHCs, the Y-9C is required only for high holder BHC. FR Y-9LP collects financial statements for the subsidiary BHCs in large (consolidated assets exceeding $500M) tiered BHCs. The Y-9s are available for bulk download from the Chicago Fed.	FRB-CHI (2011)
Federal Reserve Bank of Philadelphia (FRB-PHL)	Real-Time Data	Quarterly	The real-time dataset comprises snapshots of time series of important macroeconomic variables, recording what the data looked like at earlier points in time, thus allowing assessments of policy and forecasting performance to be based on contemporaneous (rather than subsequently revised) information. Series include most of the basic National Income and Product Accounts (NIPA).	Croushore and Stark (2000), FRB-PHL (2012)
	Risk Assessment Data Analysis and Research (RADAR)	Various	Data warehouse and computational platform established in partnership with the Federal Reserve Bank of Kansas City. RADAR manages confidential loan-level information (scrubbed of personally identifiable information) on tens of millions of mortgage loans and consumer credits.	FRB-PHL (2010)

Table 11.2 (*cont.*)

Agency	Name	Frequency	Description	Reference
Federal Reserve Bank of St. Louis (FRB-STL)	Federal Reserve Economic Data (FRED)	Various	The Federal Reserve Bank of St. Louis makes available online a wide range of economic time series for a variety of categories, including banking and finance, the national income and product accounts (NIPA), labor markets, business activity, prices and inflation, and regional and international statistics. Economists and researchers can use tag identified data to filter for information on individual series.	FRB-STL (2011a)
	Archival Federal Reserve Economic Data (ALFRED)	Various	The ArchivaL Federal Reserve Economic Data website is a resource for researchers and others who wish to access historical economic data. The site stores vintage versions of economic data that were available on specific dates in history. Data are updated as more accurate estimates become available. Current data available include Green Book projections, population, employment and labor data, pricing data, and production and business activity data.	FRB-STL (2011b)
	Federal Reserve Archival System for Economic Research (FRASER)	Various	FRASER provides a wide range of scanned historical documents, including Annual Reports of the Board of Governors, minutes of FOMC meetings, the Annual Statistical Digest, Banking and Monetary Statistics, original documents from the formation of the First and Second Banks of the United States, and historical Budgets of the United States.	FRB-STL (2012)

Federal Housing Finance Administration (FHFA)	Housing Price Index (HPI)	Monthly, Quarterly	The HPI measures the movement of single-family house prices by measuring average price changes in repeat sales or refinancings on the same properties. The HPI is an indicator of house price trends at various geographical levels, and includes house price figures for the nine Census Bureau divisions, for the 50 states and the District of Columbia, and for Metropolitan Statistical Areas (MSAs) and Divisions.	FHFA (2012a), Calhoun (1996), OFHEO (2007)
	Housing Mission and Goals (HMG) Public Use Database (PUDB)	Annual	The Public Use Databases (PUDB) store data on mortgages acquired by Fannie Mae and Freddie Mac and make available information on the flow of mortgage credit and capital. The PUDB single-family dataset includes detailed information on the income, race, and gender of the borrower as well as the census tract location, loan-to-value ratio, age of mortgage note, and affordability of the mortgage. The PUDB multifamily property-level dataset includes information on property size, unpaid principal balance, and type of seller/servicer from which the Enterprise acquired the mortgage. The multifamily unit-class file also includes information on the number and affordability of the units. Both the single-family and multifamily data include indicators of whether the purchases are from "underserved" census tracts.	FHFA (2012b)
	Monthly Interest Rate Survey (MIRS)	Monthly / Quarterly	The MIRS survey provides monthly information on interest rates, loan terms, and house prices by property type (all, new, previously occupied), by loan type (fixed- or adjustable-rate), and by lender type (savings associations,	FHFA (2011b)

Table 11.2 (*cont.*)

Agency	Name	Frequency	Description	Reference
			mortgage companies, commercial banks, and savings banks), as well as information on 15-year and 30-year fixed-rate loans. The survey also provides quarterly information on conventional loans by major metropolitan area and by FHLBank district. It excludes FHA-insured and VA-guaranteed loans, multifamily loans, mobile home loans, and loans created by refinancing another mortgage.	
	National Mortgage Database (NMDB)	Monthly	A new effort, being conducted jointly by the FHFA and CFPB, to assemble a detailed loan-level sample of mortgage characteristics from a variety of sources. This database is not yet operational as of 2012.	FHFA (2012c)
Financial Crimes Enforcement Network (FinCEN)	Bank Secrecy Act Reporting	Various	The Bank Secrecy Act (BSA) mandates specific reporting by financial institutions on certain types of activities to guard against money laundering and financial crime. Reports of suspicious activities, transactions over $10,000, and certain currency transactions are required. Subject institutions include: depository institutions (e.g., banks, credit unions and thrifts); brokers or dealers in securities and/or futures; money services businesses (MSBs) (e.g., money transmitters; issuers of money orders and travelers' checks; check cashers; etc.); casinos and card clubs; insurance companies; mutual funds; and	FinCEN (2012)

			persons engaged in transactions exceeding \$10,000. FinCEN makes these data accessible to law enforcement, counter-terrorism agencies, financial regulators, and the intelligence community.	
Municipal Securities Rulemaking Board	Electronic Municipal Market Access (EMMA)	Various	A centralized source for municipal disclosures, market transparency data and educational materials about the municipal securities market, targeted for the retail investor. EMMA houses municipal disclosures, including offering documents for most new offerings of municipal bonds, notes, 529 college savings plans and other municipal securities issued since 1990. EMMA also provides advance refunding documents and market transparency data. The latter includes real-time prices and yields at which bonds and notes are bought and sold, for most trades occurring on or after January 31, 2005.	MSRB (2012)

identification system, US banking regulators have long maintained several identification schemes in parallel. The Federal Reserve issues RSSD IDs to all institutions in its "Research, Statistics, Supervision and Regulation, and Discount and Credit Database" (RSSD); the FDIC issues insurance Certificate Numbers (often called simply, "CERTs") to insured depositories; the OCC issues Charter Numbers for national banks while retaining the legacy OTS-issued Docket Numbers for thrifts. The CFTC (2012a, 2012b) has already begun requiring participants in swap transactions to obtain "pre-LEIs" known as the CFTC Interim Compliant Identifier (CICI); see also CPSS-IOSCO (2012) and FSB (2012a). Separately, the National Association of Insurance Commissioners (NAIC) recently announced their plans to require the LEI for insurance industry regulatory reporting starting in 2013; see NAIC (2012). For the most part, identifiers resolve to the bank or BHC level; an exception is the FDIC's (2012c) Summary of Deposits, which identifies individual bank branches by Office Number. Commonly used non-supervisory identification schemes include: the American Bankers Association's (ABA) Routing Transit Number (RTN); the Society for Worldwide Interbank Financial Telecommunication (SWIFT)'s Bank Identifier Codes (BICs); and issuer identification numbers (IINs) used in credit-card processing.[18]

In addition to the regular data collections, Call Reports and Y-9s, banking regulators maintain a range of more specialized off-site monitoring tools.[19] The Federal Reserve, for example, maintains a "Watch List" (FRB, 2012d, section 1020) of state member banks, based on a combination of examination ratings, Call Report data, and other information. Similarly, the FDIC maintains a "Problem Bank List" based on examiner CAMELS ratings. The interagency Shared National Credit (SNC) program collects the details on large loan commitments syndicated across three or more lenders (OCC, 1998b; FRB-NY, 2012a). Because they incorporate confidential supervisory information, the SNC database, the Fed's Watch List, and the FDIC's Problem Bank List are not publicly available, although the FDIC announces quarterly the number of banks on its list.

The Home Mortgage Disclosure Act (HMDA) of 1975 and Community Reinvestment Act (CRA) of 1977 both initiated significant data collections of retail banking activities. CRA is designed to encourage firms to reinvest in their local markets, especially in low- and moderate-income communities. CRA data are

[18] There are additional identification schemes for securities issued by corporations (e.g., BHCs), including International Securities Identification Numbers (ISIN) issued by National Numbering Agencies, such as the CUSIP Service Bureau in North America which issues Committee on Uniform Security Identification Procedures (CUSIP) numbers; and the Center for Research in Securities Price's (CRSP) permanent identifiers, PERMNO and PERMCO, for traded securities and companies respectively; see CRSP (2012). The Federal Reserve Bank of New York maintains a mapping between RSSD IDs and PERMCOs; see FRB-NY (2012c).

[19] Table 11.2 summarizes a number of other significant data collections maintained by banking supervisors. These are supplemented by occasional special reports on particular topics (for example, FRB-NY, 2012b).

reported annually, and include details of community development and small business/farm loans (see FDIC, 2012a; FFIEC, 2010a). HMDA data help identify possible discriminatory mortgage lending patterns (see FFIEC, 2012b). Data collected on the Loan Application Register (HMDA-LAR) include type and amount of the loan, type and location of the property, approve/deny status, and ethnicity, sex and income of the applicant(s).[20] The prominent role of mortgage finance in the 2007–2009 financial crisis has prompted significant new initiatives involving mortgage data collection, including the National Mortgage Database (NMDB) initiative of FHFA and CFPB (see FHFA, 2012c) and the Risk Assessment, Data Analysis, and Research (RADAR) project of the Philadelphia and Kansas City Federal Reserve Banks (FRB-PHL, 2010; Cordell et al., 2011). In addition, the DFA transferred HMDA authority from the Federal Reserve to the new Consumer Financial Protection Bureau (CFPB).

11.2.2 Regulation and monitoring of bank risk

Since the 1980s, as financial activity has migrated increasingly into specialized trading markets between institutions – especially derivatives and securitization markets – risk has grown in significance as a focal point for supervision. The OCC's Bank Supervision Handbook (2007, p. 23) defines risk as, "the potential that events, expected or unanticipated, may have an adverse impact on a bank's earnings or capital." The attribution to anonymous "events" implicitly externalizes the hazards confronting the institution. At the same time, risk is intrinsically forward looking and contingent on those (difficult-to-predict) events. Risk exposures are difficult to convey in traditional accounting reports, which necessarily reduce the distribution of possible outcomes for the value of an asset or liability to a single number – typically either historical cost or fair value – for presentation on the balance sheet. Kuritzkes and Schuermann (2010) highlight some of the fundamental challenges here. This difficulty has led supervisors to focus on capital, which is simultaneously regarded as a buffer for unexpected losses and as the most risk--sensitive element on the balance sheet. It also leads to a range of non-accounting

[20] Supervisors also collect housing data through the authority of the Federal Housing Finance Agency (FHFA), which regulates Fannie Mae, Freddie Mac, and the Federal Home Loan Banks (FHL Banks). Fannie and Freddie failed in the summer of 2008, and were placed under conservatorship of the newly created FHFA. The two government-sponsored enterprises (GSEs) maintain large amounts of loan-level detail through their own whole-loan and loan-guarantee portfolios, a repository available to supervisors even before the GSEs entered conservatorship. The FHL Banks administer the Affordable Housing Program (AHP), the Community Investment Cash Advance (CICA) program, and the Community Investment Program (CIP), reporting project details and funding commitments annually to the FHFA. FHFA also publishes in aggregate form a monthly mortgage interest-rate survey (MIRS) and quarterly repeat-sales housing-price index (HPI, Calhoun, 1996). FHFA and CFPB recently announced a joint initiative to assemble a detailed National Mortgage Database (NMDB) combining loan-level information from a variety of sources; see FHFA (2012c).

information sources, including banks' own internal risk models, third-party risk evaluations, and supervisory stress tests.

The key objectives of microprudential supervision related to risk are bank solvency and liquidity. Solvency (or "economic solvency") refers to the value of the institution's assets in excess of the value of its deposits and other debts. This difference appears as net worth in the capital account of the financial statements.[21] The size of this capital buffer relative to the riskiness of assets and liabilities is the key ingredient in the solvency standard imposed by regulators. An underlying insight is that a certain loss experience is to be expected in a portfolio, and this expected loss should be covered by the interest rates that the bank charges. Capital provides a backstop to cover "unexpected losses" when the bad news exceeds projections (Burns, 2004). Economic insolvency occurs when the capital buffer is exhausted. Liquidity addresses the related issue, sometimes called "legal solvency," which is the requirement that the bank meet its obligations not only in expectation (i.e., economic solvency), but in actuality as those obligations come due.

Bank regulatory capital standards are coordinated internationally at the Basel Committee for Banking Supervision (BCBS), which after their adoption are then formally implemented as regulations by the appropriate US agencies. The Basel guidance on regulatory capital requirements has evolved as a series of formal accords, dubbed Basel I (BCBS, 1988), Basel II (BCBS, 2006), and now Basel III (BCBS, 2011b).[22] Since Basel II, the framework for bank capital standards has been organized into three pillars:

- Pillar I: Minimum capital requirements
- Pillar II: Supervisory review
- Pillar III: Disclosure and market discipline.

The first two pillars are the foci of supervision and regulation.

The centerpiece of Pillar I is a collection of minimum allowable capital ratios of eligible capital (the numerator) to risk-weighted assets (RWA, the denominator). These minimums differ in the definition of eligible capital used:[23]

[21] There is a devil in the details, of course, in reducing intricate, multifaceted contractual relationships to simple numerical values for accounting purposes. For example, accounting rules require mark-to-market valuations for many traded securities, but these values can become volatile in distressed market conditions, with the potential to compound an already risky situation; see Laux and Leuz (2010). Similarly, the formal definition of bank "capital" involves a number of details; see Berlin (2011). See Jackson et al. (2001) and Kuritzkes et al. (2003) for a discussion of the risk considerations.

[22] Jackson (2001) describes the intent of the Basel II capital accord early in the negotiations, while BCBS (2006) presents the final negotiated agreement. OCC-FRB-FDIC-OTS (2007) represents the corresponding US implementation. BCBS (2011a, 2011b) is the new, post-crisis Basel III accord.

[23] These are the new Basel III definitions. Basel II, which was less stringent, also had a Tier 3 capital category, which has been dropped in the latest accord.

- Tier 1 Capital (Basel III minimum ratio 6%)
  - Common equity (Basel III minimum ratio 4.5%) – most junior common shares (including surplus), plus retained earnings
  - Additional Tier 1 Capital – other perpetual, subordinated claims
- Tier 2 Capital – other subordinated claims with a minimum original maturity of at least five years
- Total Capital – Tier 1 + Tier 2 (Basel III minimum ratio 8.0%).

The capital categories apply to thousands of institutions in various jurisdictions. The precise definitions of the various capital categories are therefore more intricate than described here (see BCBS, 2011b, pp. 12–19). In addition, Basel III layers on a conservation capital buffer of 2.5% of RWA and a countercyclical capital buffer between 0 and 2.5%, which varies according to local macroeconomic conditions. Finally, Basel III phases in a leverage ratio over the next several years; this will use Tier 1 Capital in the numerator and non-risk-weighted assets in the denominator.

Calculation of RWA (the denominator) is more complicated, and involves adjusting the book values of individual portfolios and positions for their risk levels. These are typically implemented by calculating portfolio-level (or position-level) capital charges, which are summed to get the overall bank-level required capital. The capital charges break down further into specific charges for market risk, credit risk and operational risk. For each of these risk types, both a standardized approach (i.e., simplified) and an advanced approach (described here) are available.[24]

For market risk, Basel prescribes the use of value at risk (VaR), specifying a 10-day exposure horizon and a 99% confidence interval. However, banks have considerable discretion in the type of VaR model (e.g., variance-covariance, historical simulation, or Monte Carlo simulation) and whether and how to incorporate returns correlations in the analysis. Sample data for estimation or calibration of the VaR model must extend back at least a year, although longer estimation periods are commonplace.

Credit risk is typically the most important hazard. The Basel charges focus on two key components, the *probability of failure* (a binary outcome) and *losses given default* (a continuous random variable). The rules treat these two forces as statistically independent, multiplying them together to get a joint probability, which is then scaled by the size of the exposure:

[24] The standardized approach is only available to smaller banks. The advanced approaches for market, credit and operational risk introduce additional nomenclature. For market risk, the advanced framework is known as the internal models approach (IMA). For credit risk, it is the advanced internal risk-based approach (A-IRB); there is also a foundation internal risk-based (F-IRB) approach in the Basel guidance, but F-IRB was never adopted in the USA. For operational risk, it is the advanced measurement approach (AMA). Operational risk also offers a highly simplified basic indicator approach as a third option. Risk-adjustment rules for market, credit and operational risk are little changed in Basel III, and the Basel II guidance (BCBS, 2006) remains the best source for detailed procedures.

$$\mathbf{E}(Loss) = \mathbf{P}(Default) \times \mathbf{E}(Loss\ Given\ Default) \times Exposure\ at\ Default.$$

This is frequently abbreviated as: EL = PD × LGD × EAD. Data requirements play a crucial role in credit analysis. The assumption of statistical independence is often a concession to the lack of sufficient data to estimate a joint density reliably. Defaults are (hopefully) rare events, made rarer still when considered for individual instrument types. Where instead the data exist to calculate correlations between PDs and LGDs, this information is frequently used.[25]

Basel II defines operational risk as "risk of loss resulting from inadequate or failed internal processes, people and systems or from external events" (BCBS, 2006, p. 144). In practice, this breaks out into a 56-cell matrix describing seven Basel-prescribed loss event types (e.g., internal fraud) across eight business lines (e.g., retail banking). Banks must record historical loss events over a minimum five-year window to estimate a combined loss distribution for each cell of the matrix, using data from both internal and external sources. These basic data should be augmented with operational loss scenarios generated by expert opinion, as well as information on business environment and internal control factors that affect the risk profile. There are two fundamental challenges to this data-driven approach to operational risk. First, many of the loss-event types are internal to the operations of the bank, but are treated initially at least as exogenous random events. Second, there is an obvious incentive for firms and employees to under-report large losses both internally and externally, leading to downward bias in the incidence of tail events.

In light of the systemic dynamic in the recent crisis, the latest Basel guidance adds significant new features, in addition to the conservation and countercyclical capital buffers mentioned above. In particular, there are new rules for bank liquidity, including a liquidity coverage ratio (LCR) for monitoring high-quality liquid assets and a net stable funding ratio (NSFR) to assure a ready source of liabilities (BCBS, 2011a). Basel III also adds a new risk charge for counterparty risk exposures.

Still under consideration for future recommendation are provisions for "bail-in debt" – also known as contingent capital, or "cocos" – which would convert automatically to equity capital when the bank nears insolvency, or the pre-specified threshold that triggers conversion. The contingent nature of bail-in debt shares an important feature of prompt corrective action (PCA) mandate enacted in the Federal Deposit Insurance Improvement Act of 1991. PCA targets an implicit lack of will among supervisors by mandating increasingly stringent enforcement actions as a bank's capital nears the insolvency point. The chief practical difficulty with PCA lies in measuring capital, which is a small fraction of a volatile portfolio, data on which are at least partly within the discretion of bank managers. Bail-in debt could

[25] Once again, there are numerous practical considerations in implementing a measurement framework for loss exposures. See Engelmann and Rauhmeier (2006) and the articles therein for details and references.

mitigate this difficulty by becoming actionable even after the bank has failed and any supervisory reticence is moot.

Modern risk analysis, by its nature, relies heavily on computational resources, implying mission-critical roles for input data, analytical models, and derived risk reports, which need to be both accurate and timely. Because of these challenges, the calculation of Basel capital charges is typically (under the advanced approaches) delegated to the bank's data and risk infrastructure rather than having firms report risk inputs to the supervisors. Model risk – the danger of poor decision-making due to mis-calibration or misuse of analytical models for underwriting, mark-to-market valuations, or risk reporting – is similarly a focus area for on-site examinations (FRB, 2012d, section 4027). A side-effect of this dependence on analytics is the challenge it presents to outside supervisors, who do not have computational resources or data access on the same scale as the firms themselves. Another supervisory response has been an increasing reliance on the computational resources of third-party ratings agencies or the regulated entities themselves. The Dodd–Frank Act (section 939A) requires US supervisory agencies to expunge references to credit ratings from their regulations, depriving the ratings of much of their legal impetus.[26]

Both the increase in external risk exposures and the expansion of risk-centric supervision have organizational implications. Governance issues include the internal audit function, compensation structure, and risk management, especially the relationships between senior management, the Board of Directors, and the Chief Risk Officer.[27] These issues are an important focus of on-site examinations (e.g., FRB, 2012d, section 5000).

11.2.3 Stress tests

An important lesson from the 2007–2009 financial crisis is the limitation of static capital regulation using backward-looking accounting data. This was succinctly illustrated by a number of large banking organizations meeting the "well capitalized" regulatory standard during the financial crisis when their market-based capital ratios were dangerously low. In fact, both Washington Mutual and Wachovia were officially well capitalized at the time of their demise. The uncertainty about the solvency of the large financial institutions undermined confidence and deepened the economic malaise. It was not until the release of the results of the Supervisory Capital Assessment Program (SCAP) of the 19 largest financial firms that these

[26] On the other hand, the Basel guidance on risk-based capital regulation (BCBS, 2011b) continues to see a role for external credit assessment institutions (ECAIs). The Basel standardized approach has relied significantly on third-party ratings, while the internal risk-based approaches place the burden more heavily on the banks themselves.

[27] Chapter 3 of this Handbook discusses risk governance in greater detail. Chapter 1 describes the emergence of the Chief Risk Officer (CRO).

Table 11.3 *FR Y-14 data collections for stress testing*

The CCAR stress testing process requires large BHCs to report a great deal of granular information about their risk exposures. Data collections occur on annual, quarterly and monthly frequencies. The table below identifies the broad groups of data collected on each schedule; consult the actual forms and instructions for full details.

Schedule	Description
	FR Y-14A – Annual
Detailed data on BHCs' quantitative projections of balance sheet assets and liabilities, income, losses, and capital across a range of macroeconomic scenarios and qualitative information on methodologies used to develop internal projections of capital across scenarios. Required for large BHCs with more than USD 50B in consolidated assets. Macroeconomic scenarios include the supervisory baseline scenario, supervisory adverse scenario, supervisory severely adverse, BHC baseline scenario, and BHC stress scenario, and any additional scenarios generated by the firm or supplied by the Federal Reserve (FRB, 2012k).	
Summary	Separate projection summaries for each scenario
Contact information	Directory of compliance contacts at the reporting institution
Scenario	Details of scenario variable definitions, as well as scenario projections over nine future quarters
Regulatory capital instruments	Detailed breakdown of the capital instruments, plus 9-quarter projections for the baseline scenario
Basel III and Dodd–Frank	Detailed 9-quarter projection of Basel III capital, RWA and leverage for the baseline scenario
Operational risk historical capital	Historical 4-quarter operational risk capital for the Basel III advanced filers
Counterparty credit risk/CVA	Details of current and expected counterparty credit exposures
	FR Y-14Q – Quarterly
Granular data on BHCs' various asset classes and pre-provision net revenue (PPNR) for the reporting period. BHCs are required to complete a range of schedules for securities risk, retail risk, PPNR, wholesale risk, trading, private equity, and other fair value assets, Basel III/Dodd–Frank, regulatory capital instruments, and operational risk (FRB, 2012l).	
Contact information	Directory of compliance contacts at the reporting institution
Basel III/Dodd–Frank	Detailed 9-quarter projection of Basel III capital, RWA and leverage for the baseline scenario
Regulatory capital instruments	Detailed breakdown of the capital instruments, plus 9-quarter projections for the baseline scenario
Fair value option/held for sale (FVO/HFS)	Current unpaid principal and carrying values for residential and non-residential retail loans
Mortgage servicing rights (MSR)	Current valuation of MSRs and their sensitivities to a range of univariate stresses
Preposition net revenue (PPNR)	Worksheets for calculating current-quarter PPNR
Operational risk	History of operational losses and pending legal events by Basel categories

Auto loans – US	Details on auto loans originated in the USA, bucketed by borrower and loan attributes
Auto loans – international	Details on auto loans originated abroad, bucketed by borrower and loan attributes
Credit card – international	Details on foreign credit card loans, bucketed by borrower and loan attributes
Home equity – international	Details on foreign home equity loans, bucketed by borrower and loan attributes
Mortgage – international	Details on foreign mortgage loans, bucketed by borrower and loan attributes
Other consumer – US	Details on US non-auto, non-student consumer loans, bucketed by borrower and loan attributes
Other consumer – international	Details on foreign non-auto, non-student consumer loans, bucketed by borrower and loan attributes
Small/medium enterprise (SME) – US	Details on US non-credit-card SME loans, bucketed by borrower and loan attributes
SME – international	Details on foreign non-credit-card SME loans, bucketed by borrower and loan attributes
Student loans	Details on student loans, bucketed by borrower and loan attributes
Securities	Cost, market and face values for 19 categories of securities held in portfolio
Supplementary	Carrying values, charge-offs and impairments for a variety of accounts
Trading, private equity and fair value assets	Greeks and DV01 sensitivities for trading book exposures, including equities and FX, by country
Corporate loans	Detailed loan-level or credit-facility-level attributes of corporate loan exposures
Commercial real estate (CRE)	Detailed loan-level or credit-facility-level attributes of CRE loan exposures
FR Y-14M – Monthly	
A monthly collection of account- and portfolio-level data on domestic first-lien closed-end 1–4 family residential mortgages, domestic home equity residential mortgages, domestic credit cards, and address matching (FRB, 2012a).	
Contact information	Directory of compliance contacts at the reporting institution
Retail credit card	Details on retail credit-card exposures, bucketed by account or portfolio segment
Retail first mortgage	Loan-level and portfolio-level details on retail first-lien residential loans
Retail home equity	Loan-level and portfolio-level details on retail home equity residential loans and lines of credit
Retail address matching	Property location information for loans/lines in the first mortgage and home equity schedules

uncertainties were finally contained (FRB, 2009a). The success of the SCAP rooted on the forward-looking nature of the exercise, the credibility of the stress scenarios, and the transparency of the results.

Since the SCAP in 2009, banking regulators conducted the Comprehensive Capital Analysis and Review (CCAR) for the same set of large financial firms in 2010 and 2011 (FRB, 2011i, 2012f). The DFA requires the Federal Reserve to conduct an annual supervisory stress test, on a consolidated basis, of each bank holding company with consolidated assets of at least $50 billion as well as each non-bank financial company that the Financial Stability Oversight Council has designated for supervision by the Board; collectively, these are the "covered companies" (see FRB-FDIC-OCC, 2012). In addition, the DFA requires the Federal Reserve to issue regulations for covered companies to conduct annual and semi-annual stress tests and also requires other financial companies with total consolidated assets of more than $10 billion and for which the Federal Reserve is the primary regulatory agency to conduct stress tests on an annual basis (company-run stress tests). The Act requires the Federal Reserve Board to provide for at least three different sets of conditions, including baseline, adverse, and severely adverse conditions for both supervisory and company-run stress tests.

Stress tests have fundamentally changed banking regulators' approach to prudential supervision, and effectively trumped Basel Capital standards as the binding regulatory requirements. The data requirements to support the stress testing process are unprecedented in scale and granularity; see Cannata and Krüger (2009) for a general overview. As part of the CCAR process, the Federal Reserve now collects detailed quarterly data via Form FRY-14Q on bank holding companies' various asset classes and categories of pre-provision net revenue (PPNR). These quarterly reports were recently augmented with annual reports (FRY-14A) on balance sheet and income projections under a range of stress scenarios, and monthly reports (FRY-14M) on loan-level retail exposures; see FRB (2012a, 2012k, 2012l). The Y-14 reports for stress testing purposes take the granularity of data collection to an entirely different level; see Table 11.3.[28]

11.3 Summary and conclusions

Information collection to support microprudential supervision of banking in the USA is dominated by a broad portfolio of on-site examination methods and off-site data collections. Leading up to and highlighted by the financial crisis of 2007–2009,

[28] The closest parallel is the bucketed portfolio terms and conditions that the OTS maintained to support its Net Portfolio Value (NPV) model for the interest-rate risk exposures of thrift institutions. The OCC discontinued the NPV model at the end of 2011. The various Y-14 collections are still in flux; the most recent updates occurred in November 2012 (FRB, 2012n), and are still a draft for public comment as of this writing.

the fragmented approach to regulation produced a very uneven data collection across financial sectors. In turn, this created data gaps and a compartmentalized view of industry sectors. Policymakers were caught off-guard when the crisis erupted.

The Dodd–Frank Act addressed some of these data inadequacies. It empowered the Federal Reserve to serve as the super regulator for the financial system. It created the FSOC to designate non-bank SIFIs to be subject to supervision and regulation. It expanded the reach of the SEC and the CFTC to regulate and supervise hedge funds and swaps dealers. It created the Office of Financial Research which is charged with data collection as one of its mandates. It also mandated both supervisory stress tests and company-run stress tests at regular frequencies. These wholesale changes to data collection in the financial services industry by the Dodd–Frank Act are important and far reaching, but the implementation of the new rules has only just begun.

As the new era of financial data collection takes shape, two trends are apparent. First is the renewed emphasis on macroprudential supervision for financial stability. Second is the expansion of data-driven supervision, including the identification and elimination of data gaps. The key inputs to macroprudential supervision nonetheless remain at the micro level: connecting the dots begins with identifying the dots. Collection of granular financial data from a large number of institutions at increasing frequencies will require a data infrastructure that is both secure and robust. More importantly, policymakers must have the processing power and expertise to turn the collected data into useful information for making sound decisions to foster financial stability.

11.4 Appendix: Institutions for international coordination of regulation

Domestic supervision exists within a broader context, and many aspects of regulation are coordinated internationally to create a level playing field for international competition in financial services, and to present internationally active financial institutions with a relatively consistent regulatory environment across the jurisdictions in which they do business. This appendix outlines the key bodies for international coordination of financial regulation.

11.4.1 Group of Twenty (G20)

The G20 or the Group of Twenty, is the forum of international cooperation on the most important aspects of the international economic and financial agenda.[29] It comprises representatives of 19 country members and the European Union, including both advanced and emerging economies. As of 2012, the G20 included:

[29] Descriptions in the section are based on material at `www.g20.org/`.

Argentina, Australia, Brazil, Canada, China, France, Germany, India, Indonesia, Italy, Japan, Mexico, Republic of Korea, Russia, Saudi Arabia, South Africa, Turkey, UK, USA and the European Union.The G20 countries together represent about 90% of global GDP, 80% of global trade and two-thirds of the world's population.

The stated objective of the G20 is policy coordination between its members to achieve global economic stability and sustainable growth, to promote financial regulations that reduce risks and prevent future financial crises, and to create a new international financial architecture. The G20 began in December 1999 when the finance ministers and central bank governors of advanced and emerging countries met in Germany for an informal dialog on key issues for global economic stability. Since then the meetings have been held annually. Since 2008, these annual meetings have been called G20 Leaders' Summits, with the most recent taking place in Los Cabos, Mexico in June, 2012.

The work of the G20 is split into two tracks: the financial track and the "Sherpas' track." The Sherpas' track focuses on political, non-financial issues, such as employment, agriculture, energy, the fight against corruption, and development, among others. The finance track addresses issues such as economic stabilization and structural reforms, increasing international coordination for crisis prevention, correction of fiscal and financial imbalances, global liquidity, and strengthening the international financial system. Representatives of the International Monetary Fund, World Bank, Organization for Economic Cooperation and Development, and the Financial Stability Board also attend meetings of the finance track.

The major areas of the finance track change according to current concerns. As of 2012, the structure is as follows:

I. Framework for Strong, Sustainable and Balanced Growth Working Group
II. Financial Regulation
III. Financial Inclusion, Financial Education and Consumer Protection
IV. International Financial Architecture Working Group
V. Energy and Commodities Markets Working Group
VI. Commodities Markets Subgroup
    1. Energy and Growth Subgroup
    2. Disaster Risk Management
VII. Climate Finance Study Group.

11.4.2 Financial Stability Board

The Financial Stability Board (FSB) was established in 2009 as the successor to the Financial Stability Forum to coordinate the work of national financial authorities and international standard setting bodies and to help implement effective regulatory, supervisory and other financial sector policies.[30] Its Secretariat is hosted in Basel, Switzerland by the Bank for International Settlements. Its members come from 24 countries (Argentina, Australia, Brazil, Canada, China, France, Germany, Honk Kong SAR, India, Indonesia, Italy, Japan, Mexico, the Netherlands, Republic of Korea, Russia, Saudi Arabia, Singapore, South Africa, Spain, Switzerland, Turkey, UK, and USA).

Membership includes:

- National authorities responsible for financial stability, usually central banks, ministries of finance, securities regulators or equivalent
- International financial institutions:
  - Bank for International Settlements
  - European Central Bank
  - European Commission
  - International Monetary Fund
  - Organization for Economic Co-operation and Development
  - World Bank
- Sector-specific international organizations:
  - Basel Committee on Banking Supervision
  - Committee on the Global Financial System
  - Committee on Payment and Settlement Systems
  - International Association of Insurance Supervisors
  - International Accounting Standards Board, and
  - International Organization of Securities Commissions.

The FSB promotes and helps coordinate the alignment of the activities of the standard setting bodies to address any overlaps or gaps and clarify demarcations in light of changes in national and regional regulatory structures relating to prudential and systemic risk, market integrity and investor and consumer protection, infrastructure, as well as accounting and auditing.

Among the recent publications released by the FSB (available on the FSB website) are the report on securities lending and repos, global SIFI (G-SIFI) framework, shadow banking, accounting convergence, residential mortgage underwriting, legal entity identifier, peer review, macroprudential policy tools, common data templates for globally systemically important banks (GSIBs), over-the-counter derivatives, compensation, resolution of G-SIFIs, and credit rating agencies ratings, among others. The FSB also collaborates with other international organizations, for example its work with the International Monetary Fund (IMF) on the Early Warning Exercise.

[30] Descriptions in the section are based on material at `www.financialstabilityboard.org/`.

The key standards that are periodically reviewed and updated by the FSB can be divided in three areas: macroeconomic policy and data transparency (monetary and financial policy transparency, fiscal policy transparency, data dissemination), financial regulation and supervision (banking, securities, and insurance supervision), and institutional and market infrastructure (crisis resolution and deposit insurance, insolvency, corporate governance, accounting and auditing, payment, clearing and settlement, market integrity).

One example of the FSB's work is the global LEI initiative. On June 8, 2012, the FSB published a report (FSB, 2012a), which set out global LEI system High Level Principles and 35 recommendations for the development of a unique identification system for parties to financial transactions. The G20 endorsed the recommendations at the Los Cabos Summit and asked the FSB to take forward the work to implement the system. The broad aim of the FSB report is to put in place a strong global governance framework to protect the public interest, while promoting an open, flexible, and adaptable operational model for the global LEI system. The FSB recommended that the global system should support a high degree of federation and extensive reliance on local implementation under agreed and commonly applied global standards. The global LEI system as endorsed by the G20 is a three-tier system with Regulatory Oversight Committee (ROC), Central Operating Unit and Local Operating Units. On November 5, 2012, the G20 Ministers and Central Bank Governors endorsed the Charter of the Regulatory Oversight Committee for the Global Legal Entity Identifier system, thus initiating the process for the ROC to be formed. ROC membership is open to all public sector authorities from across the globe that assent to the Charter (FSB, 2012b).

Separately, the FSB has collaborated with the IMF on the "Data Gaps" initiative. The IMF/FSB report to the G20 (FSB-IMF, 2011b) highlighted 20 recommendations for areas of improvement in financial data. The FSB has a mandate to develop a common data template for global systemically important banks in order to address key information gaps identified during the crisis and to provide the authorities with a strong framework for assessing potential systemic risks (recommendations 8 and 9). In October 2011 the FSB published a consultation paper (FSB, 2011c), proposing a template to address four main questions.

- What were the main gaps in information on the systemic effects of interlinkages and common exposures of global systemically important banks that were apparent in the build-up to and during the crisis?
- Recognizing that the nature and cause of future financial instability are unlikely to be the same as in the past, are there other areas where there is a compelling case to collect additional data?
- How might better data collection improve financial stability policy making?

- How can the incremental costs of the proposals be minimized by building on current data collection frameworks and other existing statistical initiatives?

Global regulators are currently implementing the first stage of the Data Gaps project.

11.4.3 Basel Committee on Banking Supervision

The Basel Committee on Banking Supervision was established as the Committee on Banking Regulations and Supervision at the end of 1974.[31] The member countries are represented by their central banks and authorities with formal responsibility for the prudential supervision of banking business where this is not the central bank. Currently the member countries are Argentina, Australia, Belgium, Brazil, Canada, China, France, Germany, Honk Kong SAR, India, Indonesia, Italy, Japan, Korea, Luxembourg, Mexico, the Netherlands, Russia, Saudi Arabia, Singapore, South Africa, Spain, Sweden, Switzerland, Turkey, the UK and the USA. BCBS is organized into four subcommittees: the standards implementation group, the policy development group, the accounting task force, and the Basel consultative group. BCBS reports to a joint committee of central bank Governors and (non-central) heads of supervision from its member countries.

The Basel Committee does not possess any formal supranational supervisory authority. Its recommendations do not have legal force but rather the BCBS formulates broad supervisory standards and guidelines and recommends statements of best practice for national authorities to implement. The topic that has been the focus of the Committee recently is capital adequacy: Basel Capital Accord.

The Basel Committee also engages in policy research. For example, BCBS (2012) presents the principles and supervisory expectations for a large bank's risk management data, including especially data that allow the bank to manage its risks. The paper recommends that bank risk data and reports should enable management to monitor risks relative to the bank's risk appetite, and that the data should be forward looking to provide early warnings of any potential breaches of risk limits.

11.4.4 International Organization of Securities Commissions

The International Organization of Securities Commissions (IOSCO) was created in 1983.[32] It is generally recognized as the international standard setter for securities markets. Its secretariat is located in Madrid, Spain. IOSCO develops and promotes

[31] Descriptions in the section are based on material at `http://www.bis.org/bcbs`. See Goodhart (2011) for a history of the Basel Committee.

[32] Descriptions in the section are based on material at `www.iosco.org`.

standards of securities regulation in order to maintain efficient and sound markets. It draws on its international membership to establish standards for effective surveillance of international securities markets and provides mutual assistance to promote integrity of markets by a rigorous application of the standards and effective enforcement against offenses.

In June 2010, IOSCO published a report on objectives and principles of securities regulations. The document sets out 38 Principles that have objectives of protecting investors, ensuring that markets are fair, efficient and transparent, and reducing systemic risk. In this regard, the members of IOSCO agreed to cooperate in developing, implementing and promoting adherence to internationally recognized and consistent standards of regulation, oversight and enforcement to protect investors, maintain fair, efficient and transparent markets, and seek to address systemic risk; enhance investor protection and promotion of investor confidence in the integrity of securities markets, through strengthened information exchange and cooperation in enforcement against misconduct and in supervision of markets and market intermediaries; and exchange information at both global and regional levels on their respective experiences to assist the development of markets, strengthen market infrastructure and implement appropriate regulation.

IOSCO also engages in financial data research. For example, CPSS-IOSCO (2012) specifies OTC derivatives recordkeeping and reporting requirements, as well as types of recommended data formats. The report also discusses issues relating to authorities' and reporting entities' access to data, and disseminating selected OTC derivatives data to the public while taking into account any confidentiality constraints. Further, this report addresses data aggregation mechanisms and tools needed to enable authorities to aggregate data in a manner that fulfils their regulatory mandates, including methods, rationale and possible tools to support and implement data aggregation such as legal entity identifiers.

11.4.5 International Association of Insurance Supervisors

The International Association of Insurance Supervisors (IAIS) was established in 1994.[33] It represents insurance regulators and supervisors of some 190 jurisdictions in nearly 140 countries and has also more than 120 insurance professionals, insurers, reinsures and trade associations as observers. The Association's mission is to promote effective and globally consistent regulation and supervision of the insurance industry to develop and maintain fair, safe and stable insurance markets for the benefit and protection of policyholders and to contribute to global financial stability. The IAIS work is divided among three main committees – the technical committee, the

[33] Descriptions in this section are based on material at www.iaisweb.org/.

implementation committee and the budget committee. Among the objectives of the Association are the development of principles, standards and guidance for the supervision of insurance markets, development of methodologies for the assessment of the observance of its principles and standards, and facilitation of assessment processes.

11.4.6 International Monetary Fund

The mandate of the International Monetary Fund (IMF) is the surveillance of its members' macroeconomic and financial policies, as well as of the international monetary system.[34] The IMF develops and monitors international standards relevant for its mandate. Among IMF projects are the development of international standards for data dissemination and transparency practices in fiscal, monetary and financial policies, assistance in development of international standards for banking, insurance and securities supervision. The Bank also conducts assessment of compliance with all the core international financial sector standards through its Financial Sector Assessment Program (FSAP) and Report on the Observance of Standards and Codes (ROSC) initiatives.

Examples of IMF's international collaboration include its work with the FSB on the early warning exercise and Data Gaps project. The Early Warning Exercise is a joint IMF-FSB program conducted every six months and reported to Ministers and Governors at spring and annual meetings of the IMF. The IMF leads the work on macroeconomic and macrofinancial vulnerabilities, while the FSB is responsible for vulnerabilities and regulatory challenges in the financial sector. Together the organizations produce a presentation that addresses a common understanding of systemic tail risk between IMF and FSB staff.

11.4.7 Bank for International Settlements

The Bank for International Settlements was established on May 17, 1930.[35] It is the world's oldest international financial organization. The organization's head office is located in Basel, Switzerland and there are two representative offices in Hong Kong SAR and Mexico City. The mission of the BIS is to serve central banks in their pursuit of monetary and financial stability, to foster international cooperation in those areas and to act as a bank for central banks. Among the applications of this mission BIS promotes discussion and facilitates collaboration among central banks, supports dialog with other authorities that are responsible for promoting financial stability, conducts research on policy issues confronting central banks and financial

[34] Descriptions in this section are based on material at `www.imf.org/`.
[35] Descriptions in this section are based on material at `www.bis.org/`.

supervisory authorities, acts as a prime counterparty for central banks in their financial transactions, and serves as an agent or trustee in connection with international financial operations.

11.4.8 Committee on Payment and Settlement Systems

Formally established in 1990, the Committee on Payment and Settlement Systems has predecessors going as far back as 1980.[36] The Committee is composed of the Reserve Bank of Australia, National Bank of Belgium, Central Bank of Brazil, Bank of Canada, The People's Bank of China, European Central Bank, Bank of France, Deutsche Bundesbank, Hong Kong Monetary Authority, Reserve Bank of India, Bank of Italy, Bank of Japan, Bank of Korea, Bank of Mexico, Netherlands Bank, Monetary Authority of Singapore, Sveriges Riksbank, Swiss National Bank, Central Bank of the Russian Federation, Saudi Arabian Monetary Agency, South African Reserve Bank, Central Bank of the Republic of Turkey, Bank of England, Board of Governors of the Federal Reserve System, and the Federal Reserve Bank of New York. The CPSS provides a forum for cooperation among its member central banks on issues related to payment, clearing and settlement systems. It monitors and analyzes developments in such systems as well as cross-border and multicurrency arrangements and it formulates broad oversight standards in these areas. As suggested from the name of the Committee, CPSS does work in areas of payment systems, large-value payment systems, retail payment instruments and systems, securities clearing and settlement systems, foreign exchange settlements, clearing and settlement arrangements for derivatives transactions and general infrastructure issues.

11.4.9 Joint Forum

The Joint Forum deals with issues common to the banking, securities and insurance sectors, including the regulation of financial conglomerates.[37] It was established in 1996 under the auspies of the BCBS, IOSCO and IAIS. The Forum comprises an equal number of senior bank, insurance and securities supervisors representing each supervisory constituency. There are three main subgroups in the Forum: risk assessment and capital, conglomerate supervisions, and customer suitability.

[36] Descriptions in this section are based on material at `www.bis.org/cpss.htm`.

[37] Descriptions in this section are based on material at `www.bis.org/bcbs/jointforum.htm`.

11.5 Appendix: Alphabet soup of microprudential supervision in the USA

A wide range of government and quasi-governmental institutions are involved in the supervision of financial activities in the USA. Most of them are involved at some level in the collection and dissemination of data information. Although this list is long, it is not comprehensive. For example, it excludes the various financial exchanges designated as self-regulatory organizations by the SEC.

Name	Description
Board of Governors of the Federal Reserve	Also known as the Federal Reserve Board. See Federal Reserve System, below.
Commodity Futures Trading Commission (CFTC)	Created in 1974 via amendments to the Commodity Exchange Act, the CFTC succeeds the Commodity Exchange Commission. The CFTC is the main regulator for futures and options markets in the USA.
Conference of State Bank Supervisors (CSBS)	A national organization to coordinate and advocate for the state banking system. Originally formed in 1902 as the National Association of Supervisors of State Banks, it was renamed as the CSBS in 1971.
Consumer Financial Protection Bureau (CFPB)	The official name is Bureau of Consumer Financial Protection. The CFPB was created by Title X of the Dodd–Frank Act in 2010 to supervise and enforce Federal consumer financial protection laws and regulations. It is an independent bureau within the Federal Reserve System.
Employee Benefits Security Administration (EBSA)	The EBSA administers fiduciary, reporting and disclosure of pension benefits for the US Department of Labor. The mandate for EBSA was created by the Employee Retirement Income Security Act of 1974 (ERISA). Prior to 2003, EBSA was known as the Pension and Welfare Benefits Administration (PWBA), see US-DoL (2011).
Farm Credit Administration (FCA)	Originally created by executive order in 1933, the FCA is an independent agency regulating banks and related entities of the Farm Credit System (FCS), including Farmer Mac. The FCA is governed by a three-person Presidentially appointed Board.
Farm Credit System Insurance Corporation (FCSIC)	Established by the Agricultural Credit Act of 1987, the FCSIC is an independent government-controlled corporation with the primary purpose of insuring the financial obligations of Farm Credit System banks. The FCSIC is governed by a three-person Board composed of the members

Name	Description
	of the FCA Board. The FCSIC manages the Farm Credit Insurance Fund.
Federal Agricultural Mortgage Corporation (Farmer Mac)	Established by the Agricultural Credit Act of 1987, Farmer Mac is a government-sponsored enterprise (GSE) dedicated to providing a secondary market for agricultural real estate mortgage loans, rural housing mortgage loans, and rural utility cooperative loans. It is part of the Farm Credit System regulated by the FCA.
Federal Deposit Insurance Corporation (FDIC)	Created by the Banking Act of 1933, the FDIC insures deposits in banks and thrifts and has primary supervisory authority for state non-member institutions. Under the Dodd–Frank Act of 2010, the FDIC has new responsibilities for managing the orderly resolution of failing institutions.
Federal Financial Institutions Examination Council (FFIEC)	Created by the Financial Institutions Regulatory and Interest Rate Control Act of 1978 (FIRA), the FFIEC is a formal interagency body prescribing uniform principles, standards, and report forms to support the examination of financial institutions. Current membership includes the Federal Reserve Board, the FDIC, the NCUA, the OCC, and the CFPB.
Federal Home Loan Bank Board (FHLBB)	Now defunct, the FHLBB was originally established as an independent agency by the Federal Home Loan Bank Act in 1932. The FHLBB also comprised the system of 12 Federal Home Loan Banks (FHLBanks) and the Federal Savings and Loan Insurance Corporation (FSLIC). The FHLBB was replaced under the Financial Institutions Reform, Recovery, and Enforcement Act of 1989 (FIRREA) by the FHFB and OTS, with the FSLIC passing to the FDIC and supervision of the FHLBanks passing to the FHFB.
Federal Home Loan Mortgage Corporation (FHLMC or Freddie Mac)	Created by the Emergency Home Finance Act of 1970, Freddie Mac guarantees and securitizes conforming home mortgages, to support the expansion of mortgage finance in the USA. In the summer of 2008, both Fannie Mae and Freddie Mac failed, and were placed under conservatorship of the new FHFA.
Federal Housing Administration (FHA)	The FHA was created by Congress in 1934 and incorporated into the Department of Housing and Urban Development in 1965. FHA provides mortgage insurance for qualifying residential

Name	Description
	home buyers, to encourage home ownership and support the mortgage market.
Federal Housing Finance Agency (FHFA)	Created in the summer of 2008 by the Housing and Economic Recovery Act (HERA) after the simultaneous failure of the Federal National Mortgage Association (FNMA, or Fannie Mae) and the Federal Home Loan Mortgage Corporation (FHLMC or Freddie Mac), FHFA assumed the duties of the former Federal Housing Finance Board (FHFB) in overseeing the 12 Federal Home Loan Banks (FHLBanks) as well as the duties of the former Office of Federal Housing Enterprise Oversight (OFHEO) in overseeing Fannie Mae and Freddie Mac.
Federal Housing Finance Board (FHFB)	Created in 1989 to oversee the Federal Home Loan Banks, replacing the Federal Home Loan Bank Board, and was merged into the FHFA by the Housing and Economic Recovery Act (HERA) in 2008. See the entry for Federal Housing Finance Agency.
Federal Insurance Office (FIO)	Created by the Dodd–Frank Act in 2010, the FIO monitors the insurance industry, especially issues that could contribute to a systemic crisis in the insurance industry. The FIO also coordinates Federal policy on certain international insurance matters, including representing the USA in the International Association of Insurance Supervisors. The FIO is not a regulator; insurance regulation in the USA continues to occur at the state level.
Federal National Mortgage Association (FNMA or Fannie Mae)	Created in 1938 via amendments to the National Housing Act, Fannie Mae guarantees and securitizes conforming home mortgages, to support the expansion of mortgage finance in the USA. In the summer of 2008, both Fannie Mae and Freddie Mac failed, and were placed under conservatorship of the new FHFA.
Federal Reserve System	Created by the Federal Reserve Act in 1913, it consists of 12 regional Federal Reserve Banks, each with its own geographical district, and a central Board of Governors in Washington, DC. The system's core responsibility is for US monetary policy and operation of the payments system, with additional responsibilities for financial system stability and supervision of Federal Reserve member commercial banks. The system runs the Federal Open Market Committee (FOMC), which meets regularly to

Name	Description
	implement monetary policy. The Federal Reserve is the primary regulator for bank holding companies as well as state-chartered member banks, and issues a range of banking regulations. The Board of Governors is a federal agency. Each of the 12 regional Federal Reserve Banks is a private institution, owned by the member banks in its district. Each Federal Reserve Bank handles check clearing, currency management, and examination of of state-chartered member banks within its district.
Financial Accounting Standards Board (FASB)	Non-governmental body designated by the SEC for setting accounting standards for public companies in the USA. The FASB was created in 1973 to replace the Accounting Principles Board (APB) of the American Institute of Certified Public Accountants (AICPA).
Financial Crimes Enforcement Network (FinCEN)	Created by Treasury Order Number 105–08 in April 1990, FinCEN was elevated to a bureau of the US Treasury by the USA Patriot Act of 2001. It administers the Bank Secrecy Act and supports other agencies involved in the detection and deterrence of financial crime.
Financial Industry Regulatory Authority (FINRA)	A non-governmental self-regulatory organization (SRO) recognized by the SEC in 2007. FINRA is the product of an SEC-approved merger of two predecessor enforcement organizations, of the National Association of Securities Dealers (NASD) and the New York Stock Exchange. FINRA enforces regulations for brokerage firms across the country.
Financial Stability Oversight Council (FSOC)	Created by the Dodd–Frank Act in 2010 to coordinate information sharing and regulation across the various microprudential regulators. The FSOC has ten voting members: Secretary of the Treasury, Chairman of the Board of Governors of the Federal Reserve System, Comptroller of the Currency, Director of the CFPB, Chairman of the SEC, Chairperson of the FDIC, Chairperson of the CFTC, Director of the FHFA, Chairman of the NCUA Board, and an independent member with insurance expertise appointed by the President. The FSOC also has five non-voting members: Director of the OFR, Director of the Federal Insurance Office, a state insurance commissioner representative, a state banking supervisor representative, and a state securities commissioner representative.

Name	Description
Government National Mortgage Association (GNMA or Ginnie Mae)	Created by the Housing and Urban Development Act of 1968, Ginnie Mae is a government-owned corporation within the Department of Housing and Urban Development (HUD). Ginnie Mae issues government-guaranteed MBS on underlying pools of insured loans issued by other government agencies, such as the FHA.
Department of Housing and Urban Development (HUD)	Created as a Cabinet-level Department by the Department of Housing and Urban Development Act of 1965, as a successor to a range of earlier federal housing programs. HUD develops and implements federal housing policy through dozens of programs and initiatives, including Ginnie Mae, FHA, and enforcement of fair housing and equal opportunity laws.
Municipal Securities Rulemaking Board (MSRB)	Created by the 1975 Securities Act Amendments to write rules affecting participants in the markets for state and municipal debt securities. MSRB does not have enforcement authority; its rules are enforced primarily by the SEC and FINRA, but also by the FDIC, Fed, and OCC.
National Association of Insurance Commissioners (NAIC)	Created in 1871 by state insurance regulators to coordinate rules for multistate insurers. NAIC helps state insurance regulators establish standards and best practices, conduct peer review, and coordinate their regulatory oversight. It is governed by the chief insurance regulators from the 50 states, the District of Columbia and five US territories.
National Credit Union Administration (NCUA)	Created in 1970, replacing the older Bureau of Federal Credit Unions established by the 1934 Federal Credit Union Act. The NCUA was reorganized as an independent federal agency in 1978. The NCUA charters and supervises federal credit unions, and manages the National Credit Union Share Insurance Fund (NCUSIF). The NCUA is governed by a three-member Board appointed by the President.
National Association of State Credit Union Supervisors (NASCUS)	Created by state credit union regulators in 1965, NASCUS promotes the safety and soundness of state-chartered credit unions and is a liaison of the state agencies to Congress and federal agencies, including the NCUA. NASCUS represents all 47 state-level credit union regulators (Delaware, South Dakota and Wyoming do not have state-chartered credit

Name	Description
	unions). NASCUS is governed by an elected nine-person Board of Directors.
National Futures Association (NFA)	Created in 1982, National Futures Association (NFA) is the non-governmental self-regulatory organization (SRO) designated for the US futures industry. The NFA regulates firms and individuals conducting futures trading business with the public. (This is separate from the self-regulation by futures exchanges themselves of their member firms, traders and floor brokers.)
North American Securities Administrators Association (NASAA)	A voluntary association, formed in 1919, of state, provincial, and territorial securities regulators in the USA, Canada and Mexico. NASAA members license securities firms and investigate and enforce state-level securities laws.
Office of Federal Housing Enterprise Oversight (OFHEO)	Created in 1992 to oversee Fannie Mae and Freddie Mac, and merged into the FHFA by the Housing and Economic Recovery Act (HERA) in 2008. See the entry for Federal Housing Finance Agency.
Office of Financial Research (OFR)	Created by the Dodd–Frank Act in 2010, the OFR is an independent federal agency within the Department of the Treasury. The OFR incorporates both a Data Center and a Research and Analysis Center. The OFR is mandated to monitor financial exposures across the system, as well as to support the Financial Stability Oversight Council in its duties. The OFR is headed by a Director appointed by the President.
Office of the Comptroller of the Currency (OCC)	Created by the National Banking Act of 1863, the OCC charters and supervises national banks and thrift institutions. The Dodd–Frank Act of 2010 closed the Office of Thrift Supervision, transferring responsibility for federally chartered thrifts to the OCC.
Office of Thrift Supervision (OTS)	Created in 1989 as an Office within the US Treasury to oversee federally chartered thrifts in the wake of the 1980s savings and loan crisis, assuming many of the duties of the predecessor Federal Home Loan Bank Board. The OTS was run by a Presidentially appointed director. The OTS was merged into the OCC by the Dodd–Frank Act in 2010. See the entry for Office of the Comptroller of the Currency.
Pension Benefits Guarantee Corporation (PBGC)	A federal agency created by the Employee Retirement Income Security Act of 1974 (ERISA) to protect pension benefits in private-sector defined benefit plans. Financed by

Name	Description
	premiums assessed on pension plans, the PBGC assumes the obligations of covered plans that terminate prematurely.
Securities and Exchange Commission (SEC)	Created by the Securities Exchange Act of 1934, the SEC is led by a five-person commission. The SEC focuses on: (a) disclosure for traded securities via offering documents, annual reports, and proxy materials; (b) ensuring fair and orderly markets, working with self-regulatory organizations (SROs) in individual markets; and (c) investor protection and the promotion of capital formation via oversight of the investment management and mutual fund industry.
Securities Investor Protection Corporation (SIPC)	Created by the Securities Investor Protection Act of 1970 in the wake of the Paperwork Crisis of 1968–1970, the SIPC insures investors against loss of funds revealed by the failure of brokerage firms. SIPC does not provide general insurance against investment fraud.
Department of the Treasury	Established by an Act of Congress in 1789, the Treasury is a Cabinet-level Department of the federal government, led by the Presidentially appointed Secretary of the Treasury. The scope of duties is broad, and includes: managing federal finances; collecting taxes; paying all bills of the USA; currency and coinage; managing the public debt; supervising national banks and thrifts; advising on economic policy; enforcing federal finance and tax laws; and prosecuting tax evaders, counterfeiters, and forgers.

Acknowledgements

The authors are very grateful for helpful comments from Margarita Brose, Jesse Weintraub, and Larry Cordell. Any remaining errors pertain to the authors alone.

The views expressed are those of the individual authors and do not necessarily reflect official positions of the Financial Stability Board or the Federal Reserve Bank of San Francisco.

Bibliography

Basel Committee on Banking Supervision (BCBS), 1988, International Convergence of Capital Measurement and Capital Standards, Technical Report, Bank for International Settlements, July, www.bis.org/publ/bcbs04a.pdf

Basel Committee on Banking Supervision (BCBS), 1996, Amendment to the Capital Accord to Incorporate Market Risks, Technical Report, Bank for International Settlements, January, www.bis.org/publ/bcbs24.htm

Basel Committee on Banking Supervision (BCBS), 2006, Basel II: International Convergence of Capital Measurement and Capital Standards: A Revised Framework – Comprehensive Version, Technical Report, Bank for International Settlements, June, www.bis.org/publ/bcbs128.htm

Basel Committee on Banking Supervision (BCBS), 2009a, Results from the 2008 Loss Data Collection Exercise, Technical Report, Bank for International Settlements, July, www.bis.org/publ/bcbs160.htm

Basel Committee on Banking Supervision (BCBS), 2009b, Report on Special Purpose Entities, Joint Forum Report, September, www.bis.org/publ/joint23.pdf

Basel Committee on Banking Supervision (BCBS), 2009c, Principles for Sound Stress Testing Practices and Supervision, Technical Report, May, www.bis.org/publ/bcbs155.htm

Basel Committee on Banking Supervision (BCBS), 2011a, Basel III: International Framework for Liquidity Risk Measurement, Standards and Monitoring, Technical Report, Bank for International Settlements, June, www.bis.org/publ/bcbs188.htm

Basel Committee on Banking Supervision (BCBS), 2011b, Basel III: A Global Regulatory Framework for more Resilient Banks and Banking Systems, Technical Report, Bank for International Settlements, June, www.bis.org/publ/bcbs189.htm

Basel Committee on Banking Supervision (BCBS), 2011c, Global Systemically Important Banks: Assessment Methodology and the Additional Loss Absorbency Requirement, Technical Report, Bank for International Settlements, July, www.bis.org/publ/bcbs201.htm

Basel Committee on Banking Supervision (BCBS), 2012, Principles for Effective Risk Data Aggregation and Risk Reporting, Consultative Document, Bank for International Settlements, June, www.bis.org/publ/bcbs222.pdf

Bank for International Settlements (BIS), 2008, Guidelines to the International Consolidated Banking Statistics, Technical Report, Monetary and Economic Department, December, www.bis.org/statistics/consbankstatsguide.pdf

Bank for International Settlements (BIS), 2009, Guide to the International Financial Statistics, Technical Report, July, www.bis.org/statistics/intfinstatsguide.pdf

Bank for International Settlements (BIS), 2010, Triennial Central Bank Survey: Report on Global Foreign Exchange Market Activity in 2010, December, www.bis.org/publ/rpfxf10t.pdf

Bank for International Settlements (BIS), 2011a, About BIS Statistics, Internet resource, accessed July 14, 2011, www.bis.org/statistics/index.htm

Bank for International Settlements (BIS), 2011b, Guidelines to the International Locational Banking Statistics, Technical Report, Monetary and Economic Department, December, www.bis.org/statistics/locbankstatsguide.pdf

Berlin, Mitchell, 2011, Can We Explain Banks' Capital Structures? *Business Review*, Federal Reserve Bank of Philadelphia, Second Quarter, 1–11, www.phil.frb.org/

research-and-data/publications/business-review/2011/q2/brq211_can-we-explain-banks-capital-structures.pdf

Bhatia, Ashok Vir, 2011, Consolidated Regulation and Supervision in the United States, IMF Working Paper, WP/11/23, International Monetary Fund, January, www.imf.org/external/pubs/ft/wp/2011/wp1123.pdf

Blair, Christine E. and Rose M. Kushmeider, 2006, Challenges to the dual banking system: the funding of bank supervision, *FDIC Banking Review*, **18**(1), 1–20, www.fdic.gov/bank/analytical/banking/2006mar/article1/article1.pdf

Board of Governors of the Federal Reserve (FRB), 2005, *Federal Reserve System: Purposes and Functions*, Board of Governors of the Federal Reserve, www.federalreserve.gov/pf/pf.htm

Board of Governors of the Federal Reserve (FRB), 2007, Instructions for Preparation of Consolidated Financial Statements for Bank Holding Companies, Reporting Form FR Y–9C, Technical Report, March, www.federalreserve.gov/reportforms/forms/FR_Y-9C20120331_i.pdf

Board of Governors of the Federal Reserve (FRB), 2009a, The Supervisory Capital Assessment Program: Overview of Results, Technical Report, May, www.federalreserve.gov/newsevents/press/bcreg/bcreg20090507a1.pdf

Board of Governors of the Federal Reserve (FRB), 2009b, Instructions for the Preparation of the Report of Transaction Accounts, Other Deposits, and Vault Cash Reporting Form FR 2900, Technical Report, July, www.federalreserve.gov/reportforms/forms/FR_2900cb20090930_i.pdf

Board of Governors of the Federal Reserve (FRB), 2011a, *Banking Information and Regulation: All Regulations*, Internet resource, accessed July 12, 2011, www.federalreserve.gov/bankinforeg/reglisting.htm

Board of Governors of the Federal Reserve (FRB), 2011b, Statistics and Historical Data, Internet resource, accessed July 14, 2011, www.federalreserve.gov/econresdata/releases/statisticsdata.htm

Board of Governors of the Federal Reserve (FRB), 2011c, Surveys and Reports, Internet resource, accessed July 14, 2011, www.federalreserve.gov/econresdata/releases/surveysreports.htm

Board of Governors of the Federal Reserve (FRB), 2011d, Board of Governors of the Federal Reserve System: Supervision Manuals, Internet resource, accessed July 14, 2011, www.federalreserve.gov/boarddocs/supmanual/

Board of Governors of the Federal Reserve (FRB), 2011e, Basel II Capital Accord, Basel I Initiatives, and Other Basel-Related Matters, Internet resource, accessed 4 September 2011. http://www.federalreserve.gov/generalinfo/basel2/

Board of Governors of the Federal Reserve (FRB), 2011f, Reporting Forms, Internet resource, accessed September 4, 2011, www.federalreserve.gov/reportforms/default.cfm

Board of Governors of the Federal Reserve (FRB), 2011g, A User's Guide for the Bank Holding Company Performance Report, Technical Report, March, www.federalreserve.gov/boarddocs/supmanual/bhcpr/UsersGuide11/

Board of Governors of the Federal Reserve (FRB), 2011h, *Annual Report 2010*, June, www.federalreserve.gov/publications/annual-report/files/2010-annual-report.pdf

Board of Governors of the Federal Reserve (FRB), 2011i, Comprehensive Capital Analysis and Review: Objectives and Overview, Technical Report, March, www.federalreserve.gov/newsevents/press/bcreg/bcreg20110318a1.pdf

Board of Governors of the Federal Reserve (FRB), 2012a, General Instructions for the Reporting of the Capital Assessments and Stress Testing information collection (FR Y-14Q/M), Technical Report, accessed December 22, 2012, www.federalreserve.gov/reportforms/forms/FR_Y-14Q20120930_f.zip

Board of Governors of the Federal Reserve (FRB), 2012b, Survey of Consumer Finances, Internet resource, accessed May 11, 2012, www.federalreserve.gov/econresdata/scf/scfindex.htm

Board of Governors of the Federal Reserve (FRB), 2012c, Survey of Small Business Finances, Internet resource, accessed May 11, 2012, www.federalreserve.gov/pubs/oss/oss3/nssbftoc.htm

Board of Governors of the Federal Reserve (FRB), 2012d, Commercial Bank Examination Manual, Technical Report, Division of Banking Supervision and Regulation, April, www.federalreserve.gov/boarddocs/supmanual/cbem/cbem.pdf

Board of Governors of the Federal Reserve (FRB), 2012e, Bank Holding Company Supervision Manual, Technical Report, Division of Banking Supervision and Regulation, January, www.federalreserve.gov/boarddocs/supmanual/bhc/bhc.pdf

Board of Governors of the Federal Reserve (FRB), 2012f, Comprehensive Capital Analysis and Review 2012: Methodology and Results for Stress Scenario Projections, Technical Report, March, www.federalreserve.gov/newsevents/press/bcreg/bcreg20120313a1.pdf

Board of Governors of the Federal Reserve (FRB), 2012g, Federal Open Market Committee: Transcripts and other historical materials, Internet resource, accessed June 27, 2012, www.federalreserve.gov/monetarypolicy/fomc_historical.htm

Board of Governors of the Federal Reserve (FRB), 2012h, Data Releases, Internet resource, accessed June 30, 2012, www.federalreserve.gov/econresdata/statisticsdata.htm

Board of Governors of the Federal Reserve (FRB), 2012i, Data Download Program, Internet resource, accessed June 30, 2012, www.federalreserve.gov/datadownload/

Board of Governors of the Federal Reserve (FRB), 2012j, Senior Credit Officer Opinion Survey on Dealer Financing Terms, (SCOOS), Technical Report, June, www.federalreserve.gov/econresdata/files/SCOOS_201206.pdf

Board of Governors of the Federal Reserve (FRB), 2012k, General Instructions for the Preparation of the Annual Capital Assessments and Stress Testing Information Collection (FR Y–14A), Technical Report, accessed December 11, 2012, www.federalreserve.gov/reportforms/forms/FR_Y-14A20120930_f.zip

Board of Governors of the Federal Reserve (FRB), 2012l, General Instructions for the Preparation of the Annual Capital Assessments and Stress Testing Information

Collection (FR Y–14Q/M), Technical Report, accessed December 11, 2012, www.federalreserve.gov/reportforms/forms/FR_Y-14M20120930_f.zip

Board of Governors of the Federal Reserve (FRB), 2012m, Micro Data Reference Manual (MDRM), Internet resource, accessed December 11, 2012, www.federalreserve.gov/reportforms/mdrm/

Board of Governors of the Federal Reserve (FRB), 2012n, General Instructions for the Reporting of the Capital Assessments and Stress Testing information collection (FR Y-14A): November 26, 2012 – DRAFT, Technical Report, accessed December 22, 2012, www.federalreserve.gov/reportforms/formsreview/FRY14A_20121220_i.pdf

Board of Governors of the Federal Reserve System (FRB) and Federal Deposit Insurance Corporation (FDIC), 2011, Resolution Plans Required, *Federal Register*, **76**(211), 67323–67340, www.fdic.gov/regulations/laws/federal/2011/11FINALNov1.pdf

Board of Governors of the Federal Reserve System (FRB), Federal Deposit Insurance Corporation (FDIC), Office of the Comptroller of the Currency (OCC), 2011, Shared National Credits Program: 2011 Review, Technical Report, August, www.occ.treas.gov/topics/credit/commercial-credit/shared-national-credits-reports.html

Board of Governors of the Federal Reserve System (FRB), Federal Deposit Insurance Corporation (FDIC), Office of the Comptroller of the Currency (OCC), 2012, Supervisory Guidance on Stress Testing for Banking Organizations With More Than $10 Billion in Total Consolidated Assets, Final Regulatory Guidance, *Federal Register*, **77**(96), 29458–29472, www.gpo.gov/fdsys/pkg/FR-2012-05-17/pdf/2012-11989.pdf

Burns, Robert L., 2004, Economic capital and the assessment of capital adequacy, *FDIC Supervisory Insights*, **1**(2), www.fdic.gov/regulations/examinations/supervisory/insights/siwin04/siwin04.pdf

Calhoun, Charles, 1996, OFHEO House Price Indexes: HPI Technical Description, Technical Report, Office of Federal Housing Enterprise Oversight, March, www.fhfa.gov/webfiles/896/hpi_tech.pdf

Cannata, Francesco and Ulrich Krüger, 2009, Data needs for stress testing, in: M. Quagliariello (ed.), *Stress-testing the Banking System: Methodologies and Applications*, Cambridge University Press, ch. 6.

CCH Editorial Staff, 2010, *Dodd-Frank Wall Street Reform and Consumer Protection Act: Law, Explanation and Analysis*, CCH, http://onlinestore.cch.com/default.asp?ProductID=7342

Center for Research in Securities Prices (CRSP), 2012, Data Descriptions Guide: CRSP US Stock & US Index Databases, Technical Report, March, accessed December 11, 2012, www.crsp.com/documentation/product/stkind/

Committee on Payments and Settlements Systems and International Organization of Securities Commissions (CPSS-IOSCO), 2012, Report on OTC Derivatives Data Reporting and Aggregation Requirements, Technical Report, Bank for International Settlements, January, www.bis.org/publ/cpss100.pdf

Commodity Futures Trading Commission (CFTC), 2012a, Determination of the Availability of a Legal Entity Identifier Meeting the Requirements of the Regulations of the

Commodity Futures Trading Commission and Designation of the Provider of Legal Entity Identifiers to Be Used in the Recordkeeping and Swap Data Reporting Pursuant to Those Regulations, December, www.cftc.gov/LawRegulation/DoddFrankAct/Dodd-FrankFinalRules/ssLINK/federalregister072412c

Commodity Futures Trading Commission (CFTC), 2012b, *CFTC Interim Compliant Identifier Utility*, Internet resource, accessed December 14, 2012, www.ciciutility.org/index.jsp

Consumer Financial Protection Bureau (CFPB), 2011, Supervision and Examination Manual, Technical Report, October, www.consumerfinance.gov/wp-content/themes/cfpb_theme/images/supervision_examination_manual_11211.pdf

Consumer Financial Protection Bureau (CFPB), 2012a, Examination Procedures: Mortgage Origination, Technical Report, January, www.consumerfinance.gov/wp-content/uploads/2012/01/Mortgage-Origination-Examination-Procedures.pdf

Consumer Financial Protection Bureau (CFPB), 2012b, Examination Procedures: Short-term, Small-dollar Lending, Technical Report, January, www.consumerfinance.gov/wp-content/uploads/2012/01/Short-Term-Small-Dollar-Lending-Examination-Manual.pdf

Cordell, L., Y. Huang and M. Williams, 2011, Collateral damage: sizing and assessing the subprime CDO crisis," Working Paper number 11–30, Federal Reserve Bank of Philadelphia, Philadelphia, PA, www.philadelphiafed.org/research-and-data/publications/working-papers/2011/wp11-30.pdf

Corrigan, Gerald, 1983, Are Banks Special? *Annual Report 1982*, Federal Reserve Bank of Minneapolis, January, www.minneapolisfed.org/publications_papers/pub_display.cfm?id=684

Croushore, Dean and Tom Stark, 2000, A funny thing happened on the way to the data bank: a real-time data set for macroeconomists, *Federal Reserve Bank of Philadelphia Business Review*, 15–27.

De Ghenghi, Luigi, R. Agrawal Sahni and C. Fong, 2010, United States, in: J. Putnis (ed.), *The Banking Regulation Review*, First edition, Law Business Research Ltd., London, ch. 35, 434–489, http://tinyurl.com/68h2d4n

Dugan, John, 2010, Statement of John C. Dugan, Comptroller of the Currency Before the Financial Crisis Inquiry Commission, April 8, 2010, written testimony, www.occ.gov/news-issuances/news-releases/2010/nr-occ-2010-39.html

Employee Benefits Security Administration (EBSA), 2008, Reporting and Disclosure Guide for Employee Benefit Plans, Technical Report, US Department of Labor, October, www.dol.gov/ebsa/pdf/rdguide.pdf

Engelmann, Bernd and Robert Rauhmeier (eds.), 2006, *The Basel II Risk Parameters: Estimation, Validation, and Stress Testing*, Springer.

Federal Deposit Insurance Corporation (FDIC), 1997a, Banking legislation and regulation, in: *History of the Eighties – Lessons for the Future: Volume 1: An Examination of the Banking Crises of the 1980s and Early 1990s*, FDIC Division of Research and Insurance, ch. 2, 87–136, www.fdic.gov/bank/historical/history/87_136.pdf

Federal Deposit Insurance Corporation (FDIC), 1997b, Banking examination and enforcement, in: *History of the Eighties – Lessons for the Future: Volume 1: An Examination of the Banking Crises of the 1980s and Early 1990s*, FDIC Division of Research and

Insurance, ch. 12, 421–476, www.fdic.gov/bank/historical/history/421_476.pdf

Federal Deposit Insurance Corporation (FDIC), 1997c, Off-site surveillance systems, in: *History of the Eighties – Lessons for the Future: Volume 1: An Examination of the Banking Crises of the 1980s and Early 1990s*, FDIC Division of Research and Insurance, ch. 13, 477–520, www.fdic.gov/bank/historical/history/477_520.pdf

Federal Deposit Insurance Corporation (FDIC), 2010a, Highlights from the 2010 summary of deposits, *FDIC Quarterly*, **4**(4), www.fdic.gov/bank/analytical/quarterly/2010_vol44.html

Federal Deposit Insurance Corporation (FDIC), 2011a, Bank Data Guide, Internet resource, accessed July 14, 2011, www.fdic.gov/bank/statistical/guide/index.html

Federal Deposit Insurance Corporation (FDIC), 2011b, Research Information System (RIS), Internet resource, accessed July 14, 2011, www2.fdic.gov/dict/RisHome.asp

Federal Deposit Insurance Corporation (FDIC), 2011c, Risk Management Manual of Examination Policies Data Guide, Internet resource, accessed July 14, 2011, www.fdic.gov/regulations/safety/manual/

Federal Deposit Insurance Corporation (FDIC), 2012a, "CRA Statute and Tools," Internet resource, accessed 22 April 2012. http://www.fdic.gov/regulations/community/community/index.html

Federal Deposit Insurance Corporation (FDIC), 2012b, Institution Directory, Internet resource, accessed April 23, 2012, www2.fdic.gov/idasp/index.asp

Federal Deposit Insurance Corporation (FDIC), 2012c, Summary of Deposits, Internet resource, accessed April 23, 2012, www2.fdic.gov/sod/

Federal Deposit Insurance Corporation (FDIC), 2012d, Bank Data Guide, Internet resource, accessed April 23, 2012, www.fdic.gov/bank/statistical/guide/index.html

Federal Deposit Insurance Corporation (FDIC), 2012e, Historical Statistics on Banking, Internet resource, accessed April 23, 2012, www2.fdic.gov/hsob/index.asp

Federal Deposit Insurance Corporation (FDIC), 2012f, Bank Holding Company Act of 1956, 12 USC 1841, Internet resource, accessed June 8, 2012, www.fdic.gov/regulations/laws/rules/6000-100.html

Federal Deposit Insurance Corporation (FDIC), 2012g, Basic Examination Concepts and Guidelines, Technical Report, March, www.fdic.gov/regulations/safety/manual/section1-1.pdf

Federal Financial Institutions Examinations Council (FFIEC), 2006, Improved Business Process Through XBRL: A Use Case for Business Reporting, Technical Report, www.xbrl.org/us/us/FFIEC%20White%20Paper%2002Feb2006.pdf

Federal Financial Institutions Examinations Council (FFIEC), 2008, A User's Guide for the Uniform Bank Performance Report, Technical Report, December, www.ffiec.gov/ubprguide.htm

Federal Financial Institutions Examinations Council (FFIEC), 2010a, A Guide to CRA Data Collection and Reporting, Technical Report, www.ffiec.gov/cra/pdf/2010_CRA_Guide.pdf

Federal Financial Institutions Examinations Council (FFIEC), 2010b, A Guide to HMDA Reporting: Getting It Right! Technical Report, www.ffiec.gov/hmda/pdf/2010guide.pdf

Federal Financial Institutions Examinations Council (FFIEC), 2011a, The National Information Center (NIC), Internet resource, accessed July 14, 2011, www.ffiec.gov/nicpubweb/nicweb/nichome.aspx

Federal Financial Institutions Examinations Council (FFIEC), 2011b, Reporting Forms, Internet resource, accessed July 14, 2011, www.ffiec.gov/ffiec_report_forms.htm

Federal Financial Institutions Examinations Council (FFIEC), 2011c, Instructions for the Preparation of the Country Exposure Report (FFIEC 009), Technical Report, March, www.ffiec.gov/PDF/FFIEC_forms/FFIEC009_201103_i.pdf

Federal Financial Institutions Examinations Council (FFIEC), 2011d, Bank Reports: Proposed Revisions to Consolidated Reports of Condition and Income (Call Report) for 2012, supervisory correspondence: FIL-72–2011, December 7, 2011, www.ffiec.gov/pdf/FFIEC_forms/FFIEC031_FFIEC041_FIL72_201112.pdf

Federal Financial Institutions Examinations Council (FFIEC), 2012a, FFIEC IT Examination Handbook Infobase, Internet resource, accessed February 10, 2012, http://ithandbook.ffiec.gov/

Federal Financial Institutions Examinations Council (FFIEC), 2012b, Central Data Repository: Public Data Distribution, Internet resource, accessed April 22, 2012, https://cdr.ffiec.gov/public/Default.aspx

Federal Financial Institutions Examinations Council (FFIEC), 2012c, Instructions for Preparation of Consolidated Reports of Condition and Income (FFIEC 031 and 041), Technical Report, March, www.fdic.gov/regulations/resources/call/crinst/callinst2012_mar.html

Federal Financial Institutions Examinations Council (FFIEC), 2012d, Depository Institution Regulatory Reports: Revisions to the Consolidated Reports of Condition and Income for March and June 2012, supervisory correspondence: FIL-10–2012, March 2, 2012, www.ffiec.gov/pdf/FFIEC_forms/FFIEC031_FFIEC041_FIL10_201203.pdf

Federal Housing Finance Agency, 2011a, Low-Income Housing and Community Development Activities of the Federal Home Loan Bank System, 2010, Technical Report, December, www.fhfa.gov/webfiles/22844/2010_Final_2010_Low-Income_Housing_and_Community_Development_Activities_of_the_FHLBank_System.pdf

Federal Housing Finance Agency, 2011b, Monthly Survey of Rates and Terms on Conventional 1-Family Nonfarm Mortgage Loans (MIRS), OMB Number 2590–0004, Supporting Statement, memorandum, March, www.reginfo.gov/public/do/PRAViewDocument?ref_nbr=201008-2590-002

Federal Housing Finance Agency, 2012a, Downloadable HPI data, Internet resource, accessed April 27, 2012, www.fhfa.gov/Default.aspx?Page=87

Federal Housing Finance Agency, 2012b, HMG Public Use Databases, Internet resource, accessed April 27, 2012, www.fhfa.gov/Default.aspx?Page=137

Federal Housing Finance Agency, 2012c, FHFA and CFPB Partner on Development of National Mortgage Database, news release, November 1, 2012, www.fhfa.gov/webfiles/24621/NMDFHFACFPB110112F.pdf

Federal Housing Finance Board, 2005, Report of the Horizontal Review of the Affordable Housing Programs of the Federal Home Loan Banks, Technical Report, March, www.

fhfa.gov/webfiles/2981/AHP_HR_Report_Master3-15-05_--_FINAL%5B1%5D.pdf

Federal Housing Finance Board, 2006, Affordable Housing Program/Community Investment Cash Advance (AHP/CICA) Data Reporting Instructions, Technical Report, www.fhfa.gov/webfiles/13095/AHP_Data_Reporting_Instructions.pdf

Federal Reserve Bank of Chicago (FRB-CHI), 2011, Bank Holding Company Data, Internet resource, accessed July 14, 2011, www.chicagofed.org/webpages/banking/financial_institution_reports/bhc_data.cfm

Federal Reserve Bank of Chicago (FRB-CHI), 2012, Commercial Bank Data, Internet resource, accessed June 15, 2012, www.chicagofed.org/webpages/banking/financial_institution_reports/commercial_bank_data.cfm

Federal Reserve Bank of New York (FRB-NY), 2012a, Shared National Credit Program: Guidelines For Preparing Shared National Credit Data, Technical Report, April, www.newyorkfed.org/banking/reportingforms/guidelines.pdf

Federal Reserve Bank of New York (FRB-NY), 2012b, Intradaily Liquidity Flows, Report of the Payments Risk Committee, March, www.newyorkfed.org/prc/prc_120329.pdf

Federal Reserve Bank of New York (FRB-NY), 20112c, Banking Research Datasets; CRSP-FRB LINK, Internet resource, accessed December 20, 2012.

Federal Reserve Bank of Philadelphia (FRB-PHL), 2010, Keeping credit markets on the RADAR screen, *Navigating Change: Federal Reserve Bank of Philadelphia Annual Report 2010*, 26–29, www.philadelphiafed.org/publications/annual-report/2010/2010-annual-report.pdf

Federal Reserve Bank of Philadelphia (FRB-PHL), 2012, Real-Time Data Set for Macroeconomists, Internet resource, accessed June 30, 2012, www.phil.frb.org/research-and-data/real-time-center/real-time-data/

Federal Reserve Bank of St. Louis (FRB-STL), 2011a, Federal Reserve Economic Data, Internet resource, accessed July 14, 2011, http://research.stlouisfed.org/fred2/

Federal Reserve Bank of St. Louis (FRB-STL), 2011b, Dodd-Frank Act Regulatory Reform Rules: A Roadmap for Tracking the Rulemaking Process from Start to Finish, Internet resource, accessed October 29, 2011, www.stlouisfed.org/regreformrules/

Federal Reserve Bank of St. Louis (FRB-STL), 2012, Federal Reserve Archival System for Economic Research (FRASER), Internet resource, accessed April 23, 2012, http://fraser.stlouisfed.org/

Financial Crimes Enforcement Network (FinCEN), 2012, Bank Secrecy Act Forms and Filing Requirements, Internet resource, accessed April 23, 2012, www.fincen.gov/forms/bsa_forms/

Financial Crisis Inquiry Commission (FCIC), 2011, The Financial Crisis Inquiry Report: Final Report of the National Commission on the Causes of the Financial and Economic Crisis in the United States, January, www.gpo.gov/fdsys/pkg/GPO-FCIC/pdf/GPO-FCIC.pdf

Financial Stability Board (FSB), 2011a, Shadow Banking: Scoping the Issues, Technical Report, April, www.financialstabilityboard.org/publications/r_110412a.pdf

Financial Stability Board (FSB), 2011b, Shadow Banking: Strengthening Oversight and Regulation, Technical Report, October, www.financialstabilityboard.org/publications/r_111027a.pdf

Financial Stability Board (FSB), 2011c, Understanding Financial Linkages: A Common Data Template for Global Systemically Important Banks, Consultation Paper, October, www.financialstabilityboard.org/publications/r_120328j.pdf

Financial Stability Board (FSB), 2012a, A Global Legal Entity Identifier for Financial Markets, Report to G20 Leaders, June, www.financialstabilityboard.org/publications/r_120608.pdf

Financial Stability Board (FSB), 2012b, Charter of the Regulatory Oversight Committee for the Global Legal Entity Identifier System, Technical Document, November, www.financialstabilityboard.org/publications/r_121105c.pdf

Financial Stability Board and International Monetary Fund (FSB-IMF), 2011a, The Financial Crisis and Information Gaps: Report to the G-20 Finance Ministers and Central Bank Governors, Technical Report, October, www.financialstabilityboard.org/publications/r_091107e.pdf

Financial Stability Board and International Monetary Fund (FSB-IMF), 2011b, The Financial Crisis and Information Gaps: Implementation Progress Report, Technical Report, June, www.imf.org/external/np/g20/pdf/063011.pdf

Financial Stability Oversight Council (FSOC), 2011, Authority to Require Supervision and Regulation of Certain Nonbank Financial Companies: Second notice of proposed rule-making and proposed interpretive guidance, *Federal Register*, **76**(201), 64264–64282, www.gpo.gov/fdsys/pkg/FR-2011-10-18/pdf/2011-26783.pdf

Financial Stability Oversight Council (FSOC), 2012, Authority to Require Supervision and Regulation of Certain Nonbank Financial Companies, Final rule and interpretive guidance, *Federal Register*, **77**(70), 21637–21662, www.gpo.gov/fdsys/pkg/FR-2012-04-11/pdf/2012-8627.pdf

Flood, Mark, Allan Mendelowitz and William Nichols, 2012, Monitoring financial stability in a complex world, in: V. Lemieux (ed.), *Financial Analysis and Risk Management: Data Governance, Analytics and Life Cycle Management*, Springer Verlag, ch. 2, 15–45, www.springer.com/business+%26+management/finance/book/978-3-642-32231-0

Goodhart, Charles, 2011, *The Basel Committee on Banking Supervision: A History of the Early Years, 1974–1997*, Cambridge University Press.

Gorton, Gary, 2010, *Slapped by the Invisible Hand: The Panic of 2007*, Oxford University Press.

Gramlich, Dieter, Gavin Miller, Mikhail Oet and Stephen Ong, 2010, Early warning systems for systemic banking risk: critical review and modeling implications, *Banks and Bank Systems*, **5**(2), 199–211.

Greenlee, Mark, 2008, Historical review of 'umbrella supervision' by the Board of Governors of the Federal Reserve, *Review of Banking and Financial Law*, **27**(2), 407–460, www.bu.edu/law/central/jd/organizations/journals/banking/archives/documents/volume27/Issue2Greenlee.pdf

Hazen, Thomas, 2005, *The Law of Securities Regulation*, Fifth edition, West Publishing.

Herring, Richard and Jacopo Carmassi, 2010, The corporate structure of international financial conglomerates: complexity and its implications for safety & soundness, in: A. Berger, P. Molyneux and J. Wilson (eds.), *The Oxford Handbook of Banking*, Oxford University Press, ch. 8, 195–229.

Jackson, Patricia, 2001, Bank capital standards: the new Basel Accord, *Bank of England Quarterly Bulletin*, **41**(1), 55–63. www.bankofengland.co.uk/publications/quarterlybulletin/qb0101.pdf

Jackson, Patricia, William Perraudin and Victoria Saporta, 2001, Regulatory and 'economic' solvency standards for internationally active banks, *Journal of Banking and Finance*, **26**(5), 953–976, www.sciencedirect.com/science/issue/5967-2002-999739994-305435

Kaufman, George and Larry Mote, 1994, Is banking a declining industry? A historical perspective, *Economic Perspectives*, Federal Reserve Bank of Chicago, May, 2–21, http://ideas.repec.org/a/fip/fedhep/y1994imayp2-21nv.18no.3.html

King, Thomas, Daniel Nuxoll and Timothy Yeager, 2006, Are the causes of bank distress changing? Can researchers keep up? *Federal Reserve Bank of St. Louis Review*, January-February, 57–80, http://research.stlouisfed.org/publications/review/06/01/KingNuxollYeager.pdf

Kirilenko, Andrei, 2011, Presentation slides, *Conference on Systemic Risk and Data Issues*, hosted by the University of Maryland, New York University, Carnegie-Mellon University and University of California at Berkeley, Washington DC, 5–6 October 2011, www.rhsmith.umd.edu/cfp/events/2011/confSystemicRisk1011/recap/docs/KirilenkoSlides.pdf

Kuritzkes, Andrew and Til Schuermann, 2010, What we know, don't know, and can't know about bank risk: a view from the trenches, in: *The Known, the Unknown, and the Unknowable in Financial Risk Management: Measurement and Theory Advancing Practice*, Princeton University Press, ch. 6, 103–144.

Kuritzkes, Andrew, Til Schuermann and Scott Weiner, 2003, Risk measurement, risk management, and capital adequacy in financial conglomerates, in: R. Herring and R. Litan (eds.), *Brookings-Wharton Papers on Financial Services: 2003*, Brookings Institution Press.

Laux, Christian and Christian Leuz, 2010, Did fair-value accounting contribute to the financial crisis? *Journal of Economic Perspectives*, **24**(1), 93–118, www.aeaweb.org/atypon.php?return_to=/doi/pdfplus/10.1257/jep.24.1.93

Malloy, Michael, 2011, *Principles of Bank Regulation*, Third edition, West Publishing.

Municipal Securities Rulemaking Board, 2012, Electronic Municipal Market Access (EMMA), Internet resource, accessed August 22, 2012, http://emma.msrb.org/

National Association of Insurance Commissioners (NAIC), 2012, NAIC Advances Transparency to Track Risk Globally, press release, September 11, 2012, www.naic.org/Releases/2012_docs/naic_legal_entity_identifier.htm

Noeth, Bryan J. and Rajdeep Sengupta, 2011, Is shadow banking really banking? *Regional Economist, Federal Reserve Bank of St. Louis*, October, 8–13, www.stlouisfed.org/publications/re/articles/?id=2165

Office of Federal Housing Enterprise Oversight, 2007, Highlights – Part 2: Monthly Housing Price Indexes, Quarterly HPI Report (Q2 2007), news release, August 30, www.fhfa.gov/webfiles/1175/2q07hpi.pdf

Office of Information and Regulatory Affairs (OIRA), 2012, Information Collection Review, Internet site, accessed June 22, 2012, www.reginfo.gov/public/do/PRAMain

Office of the Comptroller of the Currency (OCC), 1998a, Sampling Methodologies: Comptroller's Handbook, Technical Report, August, www.occ.gov/publications/publications-by-type/comptrollers-handbook/sampmeth.pdf

Office of the Comptroller of the Currency (OCC), 1998b, SNC Program Description and Guidelines, memorandum, May, www.occ.treas.gov/news-issuances/bulletins/1998/bulletin-1998-21.html

Office of the Comptroller of the Currency (OCC), 2007, Bank Supervision Process: Comptroller's Handbook, Technical Report, September, www.occ.gov/publications/publications-by-type/comptrollers-handbook/_pdf/banksupervisionprocess.pdf

Office of the Comptroller of the Currency (OCC), 2010, Large Bank Supervision: Comptroller's Handbook, Technical Report, January, www.occ.gov/publications/publications-by-type/comptrollers-handbook/lbs.pdf

Office of the Comptroller of the Currency (OCC), 2011a, Comptroller's Handbook Booklets, Internet resource, accessed July 14, 2011, www.occ.treas.gov/publications/publications-by-type/comptrollers-handbook/index-comptrollers-handbook.html

Office of the Comptroller of the Currency (OCC), 2011b, Concentrations of Credit: Comptroller's Handbook, Technical Report, December, www.occ.gov/publications/publications-by-type/comptrollers-handbook/Concentration-HB-Final.pdf

Office of the Comptroller of the Currency (OCC), 2011c, Thrift Financial Report Instruction Manual, Technical Report, December, www.occ.gov/publications/index-publications.html

Office of the Comptroller of the Currency (OCC), 2011d, Cost of Funds Report, Announcement Date: December 15, 2011, Technical Report, www.occ.treas.gov/news-issuances/news-releases/2011/nr-occ-2011-147a_v2.pdf

Office of the Comptroller of the Currency (OCC), and the Office of Thrift Supervision (OTS), 2011, OCC and OTS Mortgage Metrics Report: Disclosure of National Bank and Federal Thrift Mortgage Loan Data, Fourth Quarter 2010, Technical Report, March, www.occ.gov/publications/publications-by-type/other-publications-reports/mortgage-metrics-2010/mortgage-metrics-q4-2010.pdf

Office of the Comptroller of the Currency (OCC), the Board of Governors of the Federal Reserve System (FRB), the Federal Deposit Insurance Corporation (FDIC), and the Office of Thrift Supervision (OTS), 2007, Risk-based capital standards: advanced capital adequacy framework – Basel II," *Federal Register*, **72**(235), 69288–69445, www.federalregister.gov/articles/2007/12/07/07-5729/risk-based-capital-standards-advanced-capital-adequacy-frameworkbasel-ii

Parker, David C., 2011, *Closing a Failed Bank: Resolution Practices and Procedures*, International Monetary Fund.

Pozsar, Zoltan, Tobias Adrian, Adam Ashcraft and Hayley Boesky, 2012, Shadow banking, *Staff Report* No. 458, Federal Reserve Bank of New York, February, www.newyorkfed.org/research/staff_reports/sr458.pdf

Saunders, Anthony and Marcia Cornett, 2010, *Financial Institutions Management: A Risk Management Approach*, McGraw-Hill.

Seldin, Sian, 2011, Federal Reserve Board Statistical Release Publications History, Technical Report, January, www.federalreserve.gov/econresdata/Publications_History_2011.pdf

Senior Supervisors Group (SSG), 2008, Leading-Practice Disclosures for Selected Exposures, Technical Report, April, www.occ.gov/publications/publications-by-type/other-publications-reports/pub-other-lending-practice-disc-sel-exposures-2008.pdf

Senior Supervisors Group (SSG), 2009a, Risk Management Lessons from the Global Banking Crisis of 2008, Technical Report, October.

Senior Supervisors Group (SSG), 2009b, Self-Assessment Template: A Supplement to: Risk Management Lessons from the Global Banking Crisis of 2008, Technical Report, October.

Senior Supervisors Group (SSG), 2010, Observations on Developments in Risk Appetite Frameworks and IT Infrastructure, Technical Report, December.

Spong, Kenneth, 2000, *Banking Regulation: Its Purposes, Implementation, and Effects*, Fifth edition, Federal Reserve Bank of Kansas City. www.kansascityfed.org/publications/banking/bankingregulation/

Sunstein, Cass, 2010, Memorandum for the Heads of Executive Departments and Agencies, and Independent Regulatory Agencies: Information Collection under the Paperwork Reduction Act, memorandum, April 7, www.whitehouse.gov/sites/default/files/omb/assets/inforeg/PRAPrimer_04072010.pdf

US Department of Labor, 2011, History of EBSA and ERISA, Internet resource, accessed September 3, 2011, www.dol.gov/ebsa/aboutebsa/history.html

12

Microprudential supervisory data in the USA

Securities and derivatives

Margarita S. Brose and Jesse T. Weintraub

12.1 Introduction

This chapter provides an overview of microprudential data collection and supervision in the securities industry. The origins and development of securities regulation can be found elsewhere in this Handbook.[1] Securities regulation was enacted in the USA to protect individual investors from unfair practices and potential abuses in the public markets, with the foundation of that protection being transparency in the markets, through mandatory disclosures.[2] The concept is two pronged: one, that the act of disclosure itself makes the one disclosing more honest, and two, that by having relevant information available, an investor can make a more informed decision before making an investment. The model developed by the USA has been followed in many other countries.[3]

The disclosure requirements ensure that there is a wealth of data available to investors and to the regulators themselves, who are responsible for enforcing the laws and maintaining fair and orderly markets. Crises and breakdowns in the markets have led to new legislation to ensure the continued transparency that is the hallmark of the capital market system; for example, in 2000, after it was revealed that companies were providing analysts and other insiders with information that investment banks were then sharing selectively with their clients, Regulation FD was enacted to require issuers to simultaneously disclose material information to analysts, large investors and the public.[4]

[1] See Chapter 10. [2] See Paredes (2011). See also Hu (2012).
[3] *See* Karmel (2008): "Since the SEC has served as the gold standard of securities regulation, it is not surprising that as the EU has striven to improve and integrate European capital markets, it has looked to US securities regulation as a model."
[4] www.sec.gov/rules/final/33-7881.htm

Handbook of Financial Data and Risk Information, ed. Margarita S. Brose, Mark D. Flood, Dilip Krishna and Bill Nichols. Published by Cambridge University Press. © Cambridge University Press 2014.

Similarly, the financial crisis also led to new legislation, the Dodd–Frank Wall Street Reform and Consumer Protection Act (DFA), which was passed and signed into law in July 2010. There are many provisions of the Act for which the Securities and Exchange Commission (SEC) and Commodity Futures Trade Commission (CFTC) are writing rules and regulations, including those with respect to swap dealers and major swap participants, swaps clearing, swaps data repositories registration standards, hedge fund registration and disclosures, asset-backed securities, and credit rating agencies. The new regulation addresses the gaps identified in existing rules and statutes over these types of securities and groups, for which transparency was deemed to be required to improve the markets. Implementation of provisions of the Dodd–Frank Act is ongoing, and while specific relevant areas will be discussed here, the reader is encouraged to consult more timely information on recent rule-making. (The SEC and CFTC websites both provide up-to-date information regarding their activities with respect to the DFA.)[5]

Given the philosophy behind the US capital markets of providing full transparency for investors, the collection and distribution of data by the SEC, whether financial, governance, or compensation related, is expected to continue. What the financial crisis of 2007 highlighted once again for regulators was that financial innovation will continue to push the markets and challenge regulators to identify new activity for which existing regulations need to be enforced or new rules need to be written. For example, a recent study looks at Special Purpose Entities (SPEs), which were used by First Executive Corporation (a life insurance company owned by a close personal acquaintance of Michael Milken) as a vehicle for packaging and selling junk bonds as safe investments, then by Enron to take losses off its balance sheet, and subsequently by investment bankers to package and sell asset-backed securities during the housing boom (Bratton, 2012). The authors of the study contend that the SPE legal vehicle was abused and regulators should take another look at this construct. Indeed, Michael Milken (who created the junk bond issuers market and sold junk bonds to a circle of savings and loans and insurance companies, which led to the need for companies like First Executive to look to create the SPE), Enron and Goldman Sachs were all investigated for potential violations of the securities laws, and enforcement actions were brought by the SEC in those cases.[6]

Recognizing that financial innovation drives today's markets, the SEC's division of Risk, Strategy and Innovation today supports the agency in analyzing trends based on existing data in the markets and provides support to other divisions. The

[5] `www.sec.gov/spotlight/dodd-frank.shtml`; `www.cftc.gov/LawRegulation/DoddFrankAct/index.htm`

[6] Bratton and Levitin (2012) review the transactions which were at the center of the three enforcement proceedings and suggest that an accounting view of SPEs should be adopted over the contractual legal view.

CFTC's Technology Advisory Committee looks at technology changes in the markets, for example, high frequency trading, and makes recommendations to the Commission based on their work. The self-regulatory agencies, of which the Financial Industry Regulatory Authority (FINRA) is the largest, are also involved in surveillance and risk analysis activities. While disclosure is what makes our markets transparent, regulators must look to adapt to the changing modes of information dispersal to ensure that that goal is achieved (see Hu, 2012). In an orderly market environment, regulation, supervision and financial innovation must coexist and thrive.

12.2 Registration, reporting and disclosure for securities and derivatives

Securities regulation was born out of the Great Stock Market Crash of 1929. Public investors in securities in the markets leading up to the Crash had little information available to assist them in understanding the nature of the securities being sold or their financial underpinning. Congress enacted a new set of laws to protect the individual investor and restore confidence in the public markets. The Securities and Exchange Commission (SEC) was created to oversee the Securities Act of 1933 (Securities Act) and the Securities Exchange Act of 1934 (Exchange Act) and was empowered to develop, oversee and enforce compliance with the new rules and regulations. The Exchange Act also gave the SEC authority to regulate and supervise exchanges and broker-dealers; the Investment Act of 1940 and the subsequent Investment Advisors Act of 1940 provided additional regulatory and rule-making authority over investment funds and asset managers, respectively.[7]

The development of federal rules and regulations concerning the commodities markets came about differently than the securities markets, but was similarly enacted in response to market events. The creation of the Commodity Futures Trade Commission in 1974 was largely in reaction to the speculative (and allegedly manipulative) markets in grains and soybeans in 1973, which caused lawmakers to revisit the legislative regime for commodities. While trading in agricultural futures contracts had been occurring on exchanges for nearly 100 years (including on the Chicago Mercantile Exchange, the New York Cotton Exchange, and the New York Mercantile Exchange), and federal rules regarding those markets had been administered by various administrations and agencies under the Department of Agriculture, the creation of the CFTC was an acknowledgment that the futures market had

[7] Other legislative developments in securities regulation have included the Private Securities Litigation Reform Act (1995), the National Securities Markets Improvement Act (1996), and the Sarbanes–Oxley Act (2002).

grown beyond agricultural commodities into many different types of derivative products.[8] Following the creation of the CFTC, the market for derivatives expanded from foreign currency futures to Eurodollar options, from Equity Index Swaps to synthetic collateralized debt obligations.

The financial markets regulated by the SEC and CFTC and the exchanges have led capital formation over the last 30 years, and the sums being invested in investment funds and in the stock market have increased exponentially over that time. Confidence in the functioning of the markets led in part to the passage of the Graham–Leach–Bliley Act in 1999, which finalized the repeal of the Glass–Steagall Act and allowed for the combining of banking and securities businesses in bank holding companies. The financial crisis of 2007 brought attention back to the regulators and the roles that they played or did not play in preventing the crisis.

For historical reasons (see Chapter 10 of this Handbook), the regulation of securities and financial derivatives in the USA has been largely divided between the SEC and the CFTC. While the Dodd–Frank Act has increased their joint activities in important areas, including "securities-based swaps" such as credit default swaps, the two agencies' regulatory turf remains largely distinct, with the SEC having oversight over securities and the CFTC oversight of financial derivatives. The scope of their duties is large, and we do not provide a comprehensive overview here. Instead we focus on the related issues of registration, reporting, disclosure and examinations. The Securities Act, Exchange Act and Commodities Futures Act provide the basis for registration and reporting requirements; the supervisory activities of the agencies are predicated on their responsibility for enforcing the laws and regulations under those acts. Table 12.1 summarizes the main securities-related regulatory information collections by the SEC, CFTC and other agencies (see also SEC, 2012a). Other data collection efforts conducted at the agencies are discussed in following sections.

12.2.1 Securities disclosure requirements

Registration, reporting, and disclosure requirements for securities issuance and transactions are mandated by three crucial statutes and implemented by the rules and regulations of the SEC. The Securities Act and the Exchange Act set all requirements for securities issuance, ownership and sale of securities. Subsequent regulation has improved transparency in smaller markets (so-called "penny stocks"

[8] The Grain Futures Act (GFA) of 1922 (enacted after the Future Trading Act of 1921 was found unconstitutional) provided for the regulation of grain and the designation of contract markets. The Commodity Exchange Act of 1936 replaced the Grain Futures Act, and widened the definition of traded products from grains to all types of commodities. The Commodity Exchange Commission was also created under the Act, and was empowered to establish speculative position limits, which would be applied first to grains in 1938. See `www.cftc.gov`

Table 12.1 *Important US securities markets regulatory data collections*[1]

Name	Agency	Description
Interactive Data to Improve Financial Reporting	Securities and Exchange Commission (SEC)	Adopted by the SEC in 2009 to enhance the usefulness of financial statement information. The rule requires domestic public companies as well as foreign private issuers that prepare their financial statements in accordance with US GAAP, IFRS, or IASB standards to electronically report, to the Commission and on their corporate websites, interactive data using the extensible Business Reporting Language (XBRL). This data format makes financial information easier for investors to analyze, and increases the speed and accuracy of financial disclosure.
Interactive Data for Mutual Fund Risk/ Return Summary	SEC	Adopted by the SEC in 2009 to enhance the usefulness of mutual fund risk/return summary information. The rule requires mutual funds to provide the risk/ return summary section of their prospectuses to the Commission and on their websites using XBRL format. This data format makes the risk/return summary easier for investors to analyze, and increases the speed and accuracy of mutual fund disclosures. The data are provided in connection with registration statements and prospectuses.
Form N-MFP	SEC	Form N-MFP requires money market funds to provide certain information to the Commission on a monthly basis. Form N-MFP must include, for each security held on the last business day of the prior month: the issuer; title of the issue, coupon, and yield; CUSIP identifier; category of security; the credit rating agency designated by the fund and the ratings given each security; maturity date; special features of the investment; principal; amortized cost value; percentage of the fund

Table 12.1 (*cont.*)

Name	Agency	Description
		invested in the security; liquidity information. The form must be filed with the Commission within five business days following the end of each month.
Form S-1	SEC	Registration statement required to be filed with the SEC for the issuance of securities. Information required to be disclosed includes financial information, company description, officers and directors, management's discussion and analysis, quantitative and qualitative disclosures about market risks, executive compensation and corporate governance.
Form 10-K	SEC	Publicly traded companies must file annually a report that provides a comprehensive overview of the company's business and financial condition and includes audited financial statements.
Form 10-Q	SEC	Form 10-Q includes unaudited financial statements and provides a continuing view of the company's financial position during the year. The report must be filed for each of the first three fiscal quarters of the company's fiscal year.
Form 8-K	SEC	In addition to filing annual reports on Form 10-K and quarterly reports on Form 10-Q, public companies must report material corporate events on a more current basis. Form 8-K is the "current report" companies must file with the SEC to announce major events that would be material to a shareholder considering an investment in that security. The types of events that trigger a public company's obligation to file a current report, include business and operations events (i.e., bankruptcy or entry into certain material agreements), financial information (acquisition or disposition of assets,

Table 12.1 (*cont.*)

Name	Agency	Description
		off-balance-sheet arrangements, etc), securities issues (i.e., modification of shareholder rights), accounting events, corporate governance changes, issuance of financial statements, and other material events.
Forms 3 and 4	SEC	Forms 3 and 4 must be filed by corporate officers and directors and other beneficial owners to disclose a greater than 10% ownership in the company (i.e., corporate insiders). Form 3 is the initial form required at the same time as the registration of the securities or no later than 10 days after a qualified person acquires the securities. Changes of ownership must be reported on Form 4 within 2 days of the change. Disclosure must also be made on a company's website, in addition to electronic filing with the SEC.
Form 13F	SEC	Institutional investment managers who exercise investment discretion over \$100 million or more in Section 13 (f) securities must report their holdings on Form 13F (Information Required of Institutional Investment Managers Form) with the SEC. In general, an institutional investment manager is: (1) an entity that invests in, or buys and sells, securities for its own account; or (2) a person or an entity that exercises investment discretion over the account of any other person or entity. Institutional investment managers can include investment advisers, banks, insurance companies, broker-dealers, pension funds, and corporations. Section 13(f) securities generally include equity securities that trade on an exchange or are quoted on the Nasdaq National Market, some equity options and warrants, shares of closed-end investment companies, and some convertible debt securities.

Table 12.1 (*cont.*)

Name	Agency	Description
		Form 13F requires disclosure of the names of institutional investment managers, the names of the securities they manage and the class of securities, the CUSIP number, the number of shares owned, and the total market value of each security on a quarterly basis.
Schedule 13D	SEC	When a person or group of persons acquires beneficial ownership of more than 5% of a class of a company's equity securities registered under Section 12 of the Securities Exchange Act of 1934, they must file a Schedule 13D with the SEC. The term "beneficial owner" is defined under SEC rules. It includes any person who directly or indirectly shares voting power or investment power (the power to sell the security).
		The form must be submitted to the SEC within 10 days after the acquisition and provide other material information. The schedule is provided to the company that issued the securities and each exchange where the security is traded. Any material changes in the facts contained in the schedule require a prompt amendment. The schedule is often filed in connection with a tender offer.
Form N-SAR	SEC	The N-SAR is a semi-annual report filed by investment management companies and unit investment trusts, and provides a wide range of portfolio management information, including fees, turnover and loads. The N-SAR is actually a cluster of related forms, the most important of which are: (1) N-SAR-A: semi-annual report (first six months of the fiscal year) for management companies; (2) N-SAR-B: annual report for management companies;

Table 12.1 (*cont.*)

Name	Agency	Description
		(3) N-SAR-U: annual report for unit investment trusts.
Form PF	SEC and Commodity Futures Trade Commission (CFTC)	Adopted jointly by the SEC and CFTC in October 2011, Form PF requires registered investment advisors and commodity pool operators that advise "private funds" (i.e., hedge funds) and have at least $150 million under management to provide substantial information about each fund they advise. Though data reporting requirements vary based on whether the advisor is a "large advisor" or not, information that must be reported includes: assets under management, leverage and performance information for each fund, derivatives, borrowings, concentration of equity holders, investments in private funds (funds of funds) and parallel investment accounts, beneficial ownership, investment strategies, and a host of related information. Large advisors have heightened reporting requirements. Most private fund advisors must file Form PF annually, but large advisors may have to file quarterly.
FOCUS Report, Form X-17A-5	SEC	Pursuant to Rule 15c3–1 under the Exchange Act, broker-dealers must at least annually file a Financial and Operational Combined Uniform Single (FOCUS) report of their compliance with the rule's net capital requirements. Originally enacted in 1975, the rule was amended in 2004 to provide broker-dealers with an alternative model for computing net capital requirements for market and derivatives related credit risk. Use of the alternative method requires a broker-dealer's ultimate holding company to consent to firm-wide supervision by the SEC. Reports are filed with the

Table 12.1 (*cont.*)

Name	Agency	Description
		SEC or the designated regulatory authority.
Form BD	SEC	Broker-dealer registration statement disclosing financial position, intended activities, officers and directors, and other relevant information, filed with the SEC, SRO or other jurisdictions, as appropriate.
Index Investment Data	CFTC	Swap dealers and index traders that receive a "special call" under CFTC Rule 18.05 must file monthly reports with the CFTC's Division of Market Oversight within 5 business days after the end of the month. A "special call" is a request for details of a market position in commodity futures, and includes both notional values and number of futures contracts for the entire non-netted book.
Commitments of Traders (COT)	CFTC	The Commitments of Traders (COT) report provides a breakdown of each Tuesday's open interest for markets in which 20 or more traders hold positions equal to or above the reporting levels established by the CFTC. The report is typically published each Friday. The report is available in both short and long formats. The short report shows both reportable and non-reportable open interest positions. Additional data are provided about the reportable positions, where data are broken down into categories including commercial and non-commercial holdings, as well as changes from the previous report. For each of the positions taken, the COT report provides information on the size of each position as well as whether it is long or short. The long report includes all information published in the short report, but further groups the data by

Table 12.1 (*cont.*)

Name	Agency	Description
		crop year and shows the positions of the largest traders.
COT-Supplemental	CFTC	The COT Supplemental report contains all information disclosed in the short format and adds the positions of Index Traders. It is published for futures and options combined in selected agricultural markets.
Bank Participation Reports (BPR)	CFTC	The CFTC provides monthly reports of aggregate large-trader positions of banks trading certain commodity futures to US banking authorities and the Bank for International Settlements. The BPR is based on the reporting information that the CFTC uses to create the COT report and to monitor the futures and options markets. If any reportable trader is hedged by the futures or options markets, it must report its business activities on Schedule 1 of CFTC Form 40. The BPR includes data for every market where five or more banks hold reportable positions, and shows each aggregate gross long and short position.
Cotton On-Call Report	CFTC	The Cotton On-Call Report shows the number of physical cotton futures contracts in the market as well as the quantity of the underlying call cotton.
Financial Data for FCMs	CFTC	Futures commission merchants (FCMs) must file monthly financial reports with the CFTC's Division of Swap Dealer and Intermediary Oversight (DSIO). The financial reports must contain (1) the level of available adjusted net and excess capital and (2) the amount of funds that need to be segregated for customers trading US futures and foreign futures.
Net Position Changes	CFTC	The Large Trader Net Position Changes and the Trading Account Net Position Changes data reflect trading that changes or creates an

Table 12.1 (*cont.*)

Name	Agency	Description
		end-of-day position, as contrasted with trading that does not change a trader's end-of-day net position, such as spread or day trading. Data for the Large Trader Net Position Changes is drawn from the CFTC's Large-Trader Reporting System (LTRS). The weekly-reported information identifies (1) the daily-average net position changes for large traders in 27 physical and eight financial futures markets from January 2009 through May 2011and (2) amounts for net position changes. Data for the Trading Account Net Position Changes are drawn from transaction data provided to the CFTC by the exchanges. The data show the daily-average net position change at the trading account level for 20 physical and eight financial futures markets from April 2010 through May 2011. It is reported on a weekly basis.
Swap Data Repositories (SDRs)	CFTC	Swap data repositories ("SDRs") are new entities created by the Dodd–Frank Wall Street Reform and Consumer Protection Act (the "Act") to provide a centralized swap record-keeping facility. All swaps, whether cleared or uncleared, are required to be reported to registered SDRs under the Act. SDRs must register with the CFTC and provide real-time public reporting of data. Data collected include transaction closing, volume, and pricing data. An SDR must provide regulators with direct access to the information; validate the information and keep accurate records; and ensure its safekeeping.
Large Trader Reporting System (LTRS)	CFTC	The LTRS forms the backbone of the CFTC's COT reports. Through the LTRS, the CFTC reviews an individual trader's activities and

Table 12.1 (*cont.*)

Name	Agency	Description
		potential to move the market, and makes necessary adjustments to speculative position limits. Under the LTRS, reporting firms (clearing members, FCMs, and foreign brokers) are required to file daily reports showing positions at or above specific reporting thresholds. The CFTC will raise or lower the reporting levels in specific markets to strike an appropriate balance between data collection and burdensome reporting. Forms 102 and 40 allow the CFTC to identify the name and address of the account, who controls trading, contact persons, whether the account is an omnibus or individual account, whether the reported account is related to another account, the business of the owner of the account, and whether the account is being used for hedging in a certain cash market or markets.

[1] A full listing of all SEC forms can be found at `www.sec.gov/about/forms/secforms.htm`

traded over-the-counter) and enhanced various provisions, including providing the Commission with extended monitoring authority over the holding companies of broker-dealers.[9] Certain provisions of the Dodd–Frank Act also modify, augment, and supplement the requirements in the Securities Act and the Exchange Act.

While the above statutes create a comprehensive regulatory regime, this section will discuss only the requirements related to the disclosure of information to the relevant regulator. This section will not discuss theoretical foundations of securities regulation, exemptions from registration, liability issues, state law, takeovers, proxy issues, jurisdictional considerations, or other topics covered under the relevant securities laws.[10]

[9] See The Securities Enforcement Remedies and Penny Stock Reform Act of 1990 and The Market Reform Act of 1990.

[10] For a comprehensive review of securities law, see Hazen (2005) or Soderquist and Gabaldon (2007).

Securities registration

The Securities Act requires all exchange-traded securities (that is, securities to be traded on a regulated stock exchange) to be registered with the SEC. Registration is the first step in the process of offering securities to the public, and thus generally triggers periodic reporting requirements – under the Exchange Act – in which pertinent information is disclosed to the regulator. Once a company registers, it is called a "reporting company." However, not all issuers of securities are subject to the same reporting requirements. There are four different categories of issuer, each with different standards: non-reporting issuers, unseasoned issuers, seasoned issuers, and well-known seasoned issuers (WKSIs). As a general rule, the more information is available publicly about a potential issuer, the lower the bar for registration compliance.

Non-reporting issuers are not required to file periodic reports with the SEC, and are typically first-time issuers. Unseasoned issuers are required to file reports under the Exchange Act if one of the following is true: (1) their securities are traded on a national exchange, (2) they have more than $10 million in assets and have at least 500 shareholders, (3) they have registered securities within the last year, or (4) they have registered securities owned by at least 300 shareholders. Unseasoned issuers are not permitted to use Form S-3 to register offerings of securities (see below).

Seasoned issuers are so called because they have been providing periodic disclosures under the Exchange Act for at least one year. Seasoned issuers either (1) have more than $75 million in common equity held by non-affiliates, or (2) are offering investment grade debt. Seasoned issuers are permitted to use Form S-3 to register securities offerings.

WKSIs are a subcategory of seasoned issuer, but must meet additional requirements. A WKSI must have either (1) $700 million in common equity held by non-affiliates or (2) issued $1 billion in non-convertible securities over the last three years and be offering such securities in the proposed offering.

The crucial differences between classes of issuers lie in the permissions and limitations on communication about the security during the registration process, with non-reporting issuers able to share the least amount of information, while WKSIs are permitted more leeway in discussions of the offering. The classifications are relevant in determining the forms available and information to be disclosed during the reporting process itself.[11]

Issuers use the registration statement to register securities offerings. The registration statement is primarily composed of a prospectus, and the prospectus is initially drafted for this purpose; in fact, it contains most of the disclosure

[11] We do not address communication issues occurring during the registration period.

required in the registration statement. It should be noted, however, that at the time of registration certain information normally appearing on a complete prospectus is not known, such as the identity of the underwriting syndicate or the price at which the security will be offered. Thus, a registration statement is typically prepared using a preliminary prospectus that omits some pertinent information. This is permitted by the SEC as long as that information is added when it becomes known.

The construction of a registration statement is governed by regulation S-K, which details which information must be disclosed in the registration statement, and what is not required.[12] Required disclosures include those regarding the business of the issuer, information about the securities offered, the issuer's financial statements, certain large shareholders, management, and merger-and-acquisition activities, amongst various exhibits. Additionally, a section called Management's Discussion and Analysis (MD & A) asks the company to provide both short-term and long-term analysis of prospects in an evaluation of the issuer. Regulation S-K provides several different registration forms for issuers' use, depending on specific situations.

Form S-1 is the basic registration form and asks for the highest level of disclosure, as it is typically used by non-reporting and unseasoned issuers – companies with a limited number of shareholders or first-time issuers. In recognition of the extensive disclosures required on this "long-form" registration, some issuers are permitted to "integrate" their S-1 filing with previously filed reports, meaning the prior reports are incorporated by reference into the current filing so as to lessen the reporting burden.

Form S-3 is used by issuers that meet the requirements for the seasoned issuer class, typically large public companies.[13] This form allows these issuers to incorporate by reference any company information contained in previous filings that has not materially changed, and thus allows registrants to disclose only transaction-specific, securities-related facts required by Regulation S-K. To use the form, the issuer must have timely filed all Exchange Act reports for the past 12 months.

All of the registration forms filed with the SEC are reviewed for accuracy and completeness by its Division of Corporation Finance (CorpFin). When a company is making an initial public offering (IPO), preliminary information under review by CorpFin may be made available to investors in what is known as a "red herring" – a preliminary prospectus for the securities. The prospectus is not "official" until

[12] Regulation S-B is an analog of regulation S-K and applies to small issuers only. Forms SB-1 and SB-2 are simplified registration forms that apply only to small business issuers, and will not be addressed in this chapter.

[13] Form S-2 was discontinued in 2005.

CorpFin has finished its review and declared the offering effective, at which point the deal is priced and offered to the public.

While the process of filing registration statements since the 1930s has only been altered by advances of electronic communication, the means of communication to investors of that information has found many different avenues. For example, the explosion of initial public offerings during the dot-com boom of the late 1990s created a large demand for information about those IPOs that investment banks worked to satisfy. Research departments at the banks courted clients with information gleaned from discussions with potential issuers not available to everyone else, that is, from companies with whom they were also trying to gain business as an underwriter of their securities. Regulation FD was the outcome of that period, as the SEC looked to put all investors back on an equal footing; information that could be considered material by investors could not be selectively released by issuers.[14]

Periodic reporting and other reporting company disclosures

After registration, the issuer is now a reporting company and must meet periodic filing requirements with respect to its business and its exchange-traded securities.[15] The SEC has wide latitude in determining the nature of documents to be filed. The most common filings are annual reports (Form 10-K), quarterly reports (Form 10-Q), and reports on materially important events (Form 8-K).

A note on materiality and plain English considerations

At this point it is appropriate to discuss the concept of materiality, as it provides the standard in deciding which information should be disclosed not only in using Form 8-K, but for disclosure considerations on the whole. Generally speaking, the rule is that any information that is material must be disclosed.

Material information is information regarding "those matters to which there is a substantial likelihood that a reasonable investor would attach importance in determining whether to purchase the security."[16] Essentially, if it would be important to an objective investor in determining whether to invest, it should be disclosed.

This is not to say that a policy of blanket disclosure is mandated, advised or realistic. Issuers would be wise to take into consideration compliance with the letter and spirit of the law without over-disclosing information or exposing themselves to liability. A well-trained attorney should be able to provide a sense of where that balance lies with each particular set of facts and with each form.

14 `www.sec.gov/rules/final/33-7881.htm`

15 Certain directly traded securities, called over-the-counter securities, may be subject to registration and reporting requirements under the Exchange Act.

16 See *TSC Industries v. Northway*, 426 U.S. 438, 449 96 S.Ct. 2126, 48 L.Ed.2d 757.

A well-drafted disclosure document must also be in "plain English." While this requirement technically applies only to the cover page, summary, and risk factors sections of a registration statement, it is prudent to draft simply and clearly in all sections. This means use of short sentences, everyday language, active voice, visual presentation of information, and no multiple negatives.

It should be noted that the use of plain English in disclosures should not come at the price of specificity and should in no way lead to blanket statements. On the contrary, simplicity of language should be accompanied by an abundance of declamatory language as a guard against liability.

Form 10-K contains much of the same information as the S-1, and much like the S-1, it references Regulation S-K's reporting requirements. A 10-K must be filed within 60 days of the end of the fiscal year. The Sarbanes–Oxley Act of 2002 modified the relevant portions of the Exchange Act to enhance the disclosure regarding off-balance-sheet transactions, contingent obligations, or any transaction that may affect the company in the future and is not otherwise reported. Form 10-Q is nearly identical in content to the 10-K, but with a shortened, quarterly time frame.

Form 8-K is a flexible filing that can incorporate any material information required by a triggering event or volunteered for disclosure. The Sarbanes–Oxley Act increased the importance of the 8-K by requiring companies to report in "real time" any material change in financial condition. Specific instances when an 8-K is required may include when a company is trading in its own securities, and when it must update or correct previously filed information.

In part because of the large volume of registration documents and periodic reports, the SEC has been at the forefront of automating the collection of regulatory data. Because of its mandated focus on transparency and disclosure, the SEC has similarly acted to automate the publication of this information as well. A centerpiece of this program is the Electronic Data Gathering, Analysis and Retrieval (EDGAR) system, first piloted in 1984 and fully established by 1996; see SEC (2012b, 2012c). EDGAR accepts, stores, and publishes individual electronic files – based on paper disclosure documents – in HTML and ASCII formats. More recently, in 2004, the SEC began exploring reporting in the form of tagged data using the XBRL standard (SEC, 2012d). This led to the creation of the Office of Interactive Disclosure in 2007 (SEC, 2012c) and ultimately to a formal rule requiring the use of XBRL submissions to the SEC (SEC, 2009a). The program is still relatively new, but "interactive" reports are beginning to appear in the "next-generation" EDGAR system online. EDGAR data are also available for bulk download from the SEC via FTP.

Disclosures by other registrants

Other entities subject to the Exchange Act, Investment Advisors Act of 1940 and Investment Company Act of 1940, including broker-dealers, stock exchanges, transfer agents, asset managers and asset management firms, must also register with the Securities and Exchange Commission and comply with its reporting requirements. Broker-Dealers must file Form BD with the Commission, which provides information regarding the officers and directors of the broker-dealer and its business and funding, and also must provide annual updates on the SEC's FOCUS report (see discussion below). Investment Advisors and investment companies similarly are required to file certain information when registering with the Commission. Investment companies wishing to offer shares for public sale must file a Form N1-A, disclosing investment objectives of the fund, fees, risks and performance data, and information about management of the fund, in addition to financial data. Investment Advisors wishing to register with the Commission must file Form ADV, disclosing specific information about the advisor and the business it conducts. All of these forms are available to the investing public on the SEC's website.

Pursuant to Exchange Act Rule 17a-5, broker-dealers who are member firms of FINRA are required to file with the SRO reports concerning their financial and operational status using SEC Form X-17a-5, Financial and Operational Combined Uniform Single (FOCUS) Report. Exchange Act Rule 17a-5 generally requires members that clear transactions or carry customer accounts to file a FOCUS Report Part II. Pursuant to Rule 15c3–1 under the Exchange Act, broker-dealers must also report their compliance (to the SEC or the appropriate self-regulatory organization) with the rule's net capital requirements at least annually. Broker-dealers in their FOCUS filings must demonstrate that they have enough capital to conduct business as a broker, as well as prove that they are appropriately segregating customer funds. The FOCUS report is filed quarterly and annually by registered broker-dealers and includes information on customer accounts, clearing method, exchange memberships, number of employees, number of public customer transactions, and affiliations. Part I of the report calculates and reports the net capital held by the broker-dealer; Part II reports on the financial condition of the firm; and Part III provides the annual independent audit report for the broker-dealer. All of this information is required as part of the SEC's mission to protect investors, that is, by ensuring that the broker-dealers transacting business and holding investor funds are adequately capitalized and appropriately segregating those investments.

In 2011, the SEC approved Rule 4524, proposed by FINRA, which would require member firms to provide supplementary information in their FOCUS Report filing.[17] The rule also requires certain designated firms to provide a supplemental

[17] The rule was effective as of February 28, 2012.

statement of income (SSOI), which is a detailed categorization of the revenue and expense line items that are on the Statement of Income (Loss) page of the FOCUS Report Part II. The additional information was sought by FINRA so that it could better understand a firm's revenue sources and expenses on an ongoing basis. The lesson of the financial crisis for regulators was that certain types of information not currently available for review would be helpful in assessing risk, especially profit and loss information, which could help analyze the actual cash position of the institution.

Under the provisions of the Dodd–Frank Act, hedge funds must also now register with the SEC (and certain funds with the CFTC) on Form PF, supplying similar information, including assets under management, leverage and performance information for each fund, derivatives, borrowings, concentration of equity holders, investments in private funds (funds of funds) and parallel investment accounts, beneficial ownership, investment strategies, and a host of related information.[18] While hedge fund registration had been debated for several years in light of the increasing number of funds started over the last decade as a protection for investors (and indeed had been made voluntary with the SEC), it was not until after the financial crisis that disclosure by these funds was mandated by federal legislation. The Financial Stability Oversight Council, responsible for monitoring emerging financial risks, directed the SEC and CFTC to compel registration of certain types of private fund advisors. Transparency of the investments and leverage of these firms was determined to be a critical component in managing system-wide risk in the capital markets.[19]

12.2.2 Derivative products

Registration, reporting, and disclosure requirements for commodity derivative products are mandated by the Commodity Exchange Act, the Securities Exchange Act, and Securities Act, all as amended by the Dodd–Frank Act. Much, but not all, of derivatives activity falls within the jurisdiction of the CFTC. Some supervision and monitoring of derivatives is the purview of the SEC, and certain matters such as securities-based swaps are the subject of joint rulemaking.

Much like the preceding section on securities disclosure, this section covers only the disclosure of regulatory information as it relates to derivative products. It will not discuss theoretical underpinnings of derivatives or commodities regulation, hedging and speculating, market disruptions, liability, or other pertinent subjects in any great detail, beyond what is necessary for an understanding of reporting,

[18] `www.sec.gov/rules/final/2011/ia-3308.pdf`. See also Chapter 9 discussing Form PF and hedge funds.
[19] `www.sec.gov/rules/final/2011/ia-3308.pdf`

disclosure, and transparency requirements for market participants.[20] As regulators are at this time in the process of implementing the sweeping changes mandated by the Dodd–Frank Act, this section will provide a broad overview only.

Finally, a note on terminology. Much of the jargon used in the derivatives regulatory sphere is new, and many established terms have changed meaning and connotation in the wake of the 2008 financial crisis and regulatory response. For instance, the term "swap" previously designated a specific type of derivative transaction, where participants exchanged cash flows (typically variable or fixed). The term has recently come to encompass more than just this specific type of transaction, often – incorrectly, in the view of some – referring to any type of derivative transaction. At this point there is no clear-cut definition, though the cautious would treat any transaction even distantly related to a swap as exactly that, in addition to seeking legal counsel on the matter.

Reporting firms

Banks that handle customer assets as part of the derivatives trading process are likely to fall into one of three categories: futures commission merchant (FCM), clearing member, or foreign broker.[21] Each functions as an intermediary in the trading process and faces reporting requirements with the goal of enhancing the transparency and accuracy of market surveillance. The reporting requirements are collectively referred to as large trader reporting, and the firms to which these requirements apply are called reporting firms.

A financial entity must register as an FCM if it accepts swap customer money, securities, property, or guarantees transaction or extends credit. A futures commission merchant is much like a broker-dealer in the securities context. A customer has an account at an FCM, which employs traders to trade on the customer's behalf. Given that FCMs handle large orders and enormous notional amounts, they must provide large trader reports to the CFTC on a daily basis, quantifying the contracts that are used for non-hedging purposes only.[22] The Dodd–Frank Act mandates the creation of Swap Data Repositories (SDRs) to collect this information and report it to the relevant agency.

The CFTC requires this because it uses large trader reports to monitor large positions in the given market and compare those positions to open interests (contracts that have not yet been "delivered," paid out, netted out, or otherwise closed).[23]

[20] As a preliminary note, a derivative product is a contract which derives its value from an underlying concept, often a commodity. The market and pricing of the "underlier" itself – often misleadingly referred to as the "cash market" – is thus crucial to determining the value of a derivative contract, and can have massive impact on the derivatives market. While some financial institutions participate in the cash market, this section will address the more widespread use of derivatives and the regulatory requirements thereof.

[21] Foreign brokers will not be addressed in this section, as they are the subject of significant ongoing rulemaking.

[22] See CFTC Reg 1503.

[23] Position refers to a quantity of individual instruments held by an entity.

This enables the CFTC to set position limits and monitor compliance with these limits, in addition to aiding with market congestion. Any large trader that exceeds a position limit must file a monthly report with the CFTC or SEC, depending on the kind of swap; these reports determine whether an entity is engaged in bona fide hedging (versus speculating). All trading reports are aggregated and published in the weekly Commitments of Traders (COT) reports.

Further, FCMs are required to file monthly unaudited financial reports with their self-regulatory organization (SRO), typically the National Futures Association (NFA), and monthly audited financial reports with the SEC and/or CFTC. The FCM must also maintain detailed records of all derivative transactions, "cash" transactions, and positions, which must be made available to the CFTC and/or SEC upon request.

FCMs often need to trade swaps and other derivative products on an exchange, where the exchange in turn clears each trade with a clearinghouse. The clearinghouse is a financial institution that guarantees each transaction by interposing itself between the counterparties. Banks with large derivatives trading businesses are likely to be owners or members of a clearinghouse, as the clearinghouse is required to be well capitalized to ensure margining and daily settlement of trades. "Clearing members" are participants in a clearinghouse. Exchanges report to the CFTC confidential information on the aggregate positions and trading activity of their clearing members. This includes open positions, purchases and sales, swaps, and futures delivery notices, and is reported electronically on a daily basis.[24]

As mentioned above, the large trader reports are used to set position limits and ensure that markets and their participants are functioning properly and within the law. One of the great advantages to having a trading process with multiple participants is that each entity's account of a trade may be compared to another's for accuracy. To that end, the CFTC will periodically cross-check traders' reports against counterparties', clearing members' against exchanges', and FCMs' against other broker-dealers' reports. Any discrepancies may spur an investigation by the commission or an exchange.

Commodity pool operators and commodity trading advisors

A commodity pool is an entity to which a number of other entities contribute funds to form a pool for trading commodity derivatives. The Dodd–Frank Act expanded the concept of commodity pool operator (CPO) to include any person or entity that "solicits, accepts, or receives from others, funds, securities, or property, either directly. . .or otherwise, for the purpose of trading in commodity interests, including

[24] This process is subject to a final joint rulemaking by the SEC and CFTC under the Dodd–Frank Act. See also CFTC-SEC (2012).

any. . . ‘swap’” (Dodd–Frank Act, 721(5)). A CPO must register with the CFTC and file disclosure documents detailing its pools, but each individual pool that it operates need not file its own disclosures. The disclosure documents must be filed upon formation of the pool and when it is seeking new members, and must include a Risk Disclosure Statement, identification of principals and managers, types of commodities, investment program, use of proceeds, the FCM that will execute trades, anticipated counterparties, fees and expenses, conflicts of interest, and other considerations. The CPO must also distribute periodic account statements to pool members themselves regarding the status and health of the pool, namely the net gain or loss, fees and expenses, and net asset value. Finally, a CPO must prepare and distribute to members audited financial statements, and must track all participants, transactions, and operations of the pool.

A commodity trading advisor (CTA) is any person or entity that advises others regarding the trading of a “swap.”[25] A CTA will typically advise a CPO in its trading activities, and thus must register and file a disclosure document. Rather than filing with the CFTC, however, a CTA must register and file with the NFA. Much like a CPO, a CTA must file disclosure information for each pool advised; the documents contain much the same information as CPO filing documents, but emphasize fee structure given the advisory role of the CTA. As described in the last part of Section 12.2.1 above, under the rules promulgated by the CFTC under the Dodd–Frank Act, CPOs and CTAs now have to file certain sections of Form PF with the regulator, as a means of providing additional financial and investment information. The disclosures were mandated and the rule promulgated as part of the risk management activities of the Financial Stability Oversight Council.

Customer-based disclosure

Finally, banks may have disclosure mandates that are activated when they deal with certain specific customers or counterparties. Banks engaging in derivative transactions with any counterparty that is *not* a swap dealer or major swap participant (MSP)[26] must disclose to that counterparty the risks, incentives, conflicts of interest, and the daily market value of the transaction as it changes.

Special disclosure duties also exist for “special entities,” to include pension funds, endowments, retirement plans, and government entities. If dealing with one of the above as a swaps dealer or MSP, a bank must disclose the capacity in which it is acting prior to the closing of the transaction and on whose behalf it is interested.

[25] As noted above, the term “swap” as used in the applicable section of the Dodd–Frank Act is most likely intended as a stand-in for the word “derivative” or “derivative product”. See Dodd–Frank Act 721(7).

[26] A major swap participant is a non-dealer that maintains a significant position in swaps.

Further, if a customer of any type chooses not to require segregation of its margin, dealers and MSPs must produce a quarterly report to the customer indicating compliance with margin segregation requirements.

12.2.3 Supervision and enforcement

While the securities laws are focused on providing transparency to investors through the disclosure by companies of relevant financial and other information that would be material to an investor making investment decisions, the agencies are also responsible for maintaining fair and orderly markets, and to that end engage in supervisory activities to ensure compliance with the applicable laws and to assess risks in the activities of their registrants. Given the number of registrants and the limited staffing at the agencies, however, the number of examinations conducted is limited. The SEC reported in its 2011 Annual Report, for example, that it conducted examinations of 13% of its registered investment companies, and (with the SROs) 42% of its registered broker-dealers. Similarly, the percentage of public companies whose disclosures were reviewed by the SEC staff was only 48%.[27] Unlike the banking supervisory regime, the securities regulators rely to a great deal on the onus of disclosure as a means to encourage compliance with the securities laws. But the SEC and CFTC are also using the lessons of the financial crisis to look across many types of data and leverage experienced staff to develop risk assessments as a means to focus their enforcement and surveillance activities.

Securities and exchange commission examinations

The disclosures and reporting mandated by the Securities Act and Exchange Act are designed to protect investors by making information about publicly held companies available to all investors. The enforcement of the rules and regulations is delegated to the SEC, and through the Commission to other self-regulated organizations (SROs), including FINRA (the US Financial Industry Regulatory Authority; formerly the National Association of Securities Dealers).

The SEC has six divisions that carry out the mission of the agency:

- the Division of Corporation Finance reviews and ensures the accurate and fair disclosure of information to be filed in registration statements and other related financial filings;
- the Division of Investment Management oversees the registration and disclosures and compliance by asset managers and investment funds;

[27] See SEC (2011b).

- the Division of Trading and Markets regulates broker-dealers, SROs, and transfer agents and sets standards for fair and orderly markets;
- the Division of Enforcement works with the other divisions to investigate possible violations of the securities laws and bring civil cases in the federal courts and administrative proceedings, as appropriate;
- the Office of Compliance, Inspections and Examinations conducts regular examinations of the various entities regulated by the SEC; and
- the Division of Risk, Strategy and Innovation (created in 2009), assists the other divisions by providing financial economics and data analytics to the policy-making, enforcement and examination activities of the agency.

The Division of Risk, Strategy and Innovation collects and analyzes data to support the ongoing efforts at the SEC. One of its first visible efforts was conducting research and producing a report on the May 6, 2010 "Flash Crash"; it has also been involved in providing research to support ongoing enforcement and research efforts at the Commission.[28]

The Office of Compliance, Inspections and Examinations (OCIE) conducts a National Examination Program (NEP), the purview of which extends to all entities regulated by the SEC, including broker-dealers, investment advisors, and SROs.[29] The Dodd–Frank Act further expanded the reach of the OCIE to conduct examinations of swaps dealers and security-based data repositories, and nationally recognized statistical rating organizations (NSROs), among others. Specifically, Dodd–Frank mandated the creation of an Office of Credit Ratings (OCR) for the direct supervision of the NSROs.[30]

The OCIE conducts examinations from its home office in Washington, DC and from the 11 regional SEC offices across the USA. Today, the office takes a risk-based approach to choosing its areas of focus and registrants for review. The Division of Risk, Strategy and Innovation provides expertise and advice in this risk review. The OCIE has also established an Office of Risk Assessment and Surveillance (ORAS) that assists in evaluating risks across the markets and registrants to identify possible areas of focus for the NEP. The risk assessments by ORAS play a major role in determining the areas and registrants for the examination program.

OCIE examinations are designed to evaluate a registrant's (e.g., a broker-dealer or investment advisor) compliance with the federal securities laws and regulations and protection of client assets. The examiners review the entity's books and records,

[28] See CFTC-SEC (2010).

[29] FINRA conducts examinations of its members in the same manner as the SEC. See the final part of Section 12.2.3.

[30] The newly created OCR monitors the activities and conducts examinations of registered Nationally Recognized Statistical Ratings Organizations to assess and promote compliance with statutory and Commission requirements. See `www.sec.gov/about/offices/ocr.shtml`

Table 12.2 *Process flow for a typical examination conducted by the Office of Compliance, Inspections and Examinations (SEC, 2012e)*

Step	Description	Product(s)
1	Scheduling	Agreed upon timing of exam on-site visit with registrant; visit may also occur unannounced
2	Entrance Interviews	Examiners talk to management to gain understanding of business, compliance culture, tone at the top, and identify possible risk areas of focus
3	Document Requests	Books and records of the firm are requested, identified by the examiners as pertinent to the review
4	Questions	In reviewing books and records, examiners may ask follow-up questions to better understand a process or the business
5	Exit Interviews/Exit Conference Calls	Opportunity for the exam team to meet with the Compliance Officer or General Counsel to discuss identified weaknesses or failures and plan for remediation
6	Results	Final results of the exam are provided to the registrant and may include: • No findings/No further action • Identified weaknesses/control deficiencies – notice to registrant of areas to improve • Referral to the Division of Enforcement – where deficiencies are determined to put investor funds at risk

interview management and firm employees, and analyze its operations. Given the direction provided by ORAS, the exam will generally include in its scope risks presented by the registrant. The examiners may also seek to identify risks or gain a better understanding of the entity's internal control environment. Typically examinations are noticed to the entity so management may prepare for the visit, anywhere from a few days to a few weeks in advance, but occasionally the exam is unannounced. The program follows a process of notice, review and analysis of the target entity, with results communicated to the Chief Compliance Officer or General Counsel at the conclusion of the visit. The examination steps are outlined in Table 12.2 (SEC, 2012e).

Commodity futures trade commission supervision

The CFTC has a similar organization to the SEC in terms of rule-making, surveillance and enforcement. The Division of Clearing and Risk oversees derivative clearing organizations and others, including futures commission merchants, monitors the clearing of futures and derivatives, and monitors compliance with the agency rules and regulations. The Division of Market Oversight oversees trade execution facilities and data repositories in order to ensure fair and orderly markets, and evaluates new products and new rule filings. The DMO also conducts examinations of exchanges to ensure compliance with the core principles of the Commission. The CFTC's market surveillance program monitors trade activity to prevent manipulation, enforce speculative position limits, and to ensure compliance with reporting requirements. This is done by monitoring the activities of large traders, price relationships, and supply and demand in the market. The Division of Enforcement investigates possible violations of the Commodity Exchange Act and associated regulations, and brings cases as appropriate against both individuals and CFTC registrants. The Division of Swap Dealer and Intermediary Oversight oversees the registration and compliance of intermediaries and futures industry self-regulatory organizations (SROs). Under Dodd–Frank, DSIO also will be responsible for developing and monitoring compliance with regulations addressing registration, business conduct standards, capital adequacy, and margin requirements for swap dealers and major swap participants.

Surveillance at the agency is carried out to identify possible manipulation in the market and take appropriate preventative actions so that these situations do not occur. The agency monitors large traders, supply and demand factors and other key relationships each trading day to identify possible manipulative and other fraudulent activity. The CFTC works with the exchanges to review activity on the markets and allows the exchanges to conduct their own investigations first before the agency commences its own investigation.

Other regulatory supervision

As noted earlier, designated SROs also play a key role in the regulation of the securities markets, through responsibilities delegated to them by the SEC. Exchanges and organizations such as FINRA have surveillance and enforcement divisions that carry out these responsibilities, ensuring compliance with the securities laws and rules promulgated by the Exchanges. FINRA, for example, maintains information on its Central Repository Database (CRD) on all registered brokers (as filed on Form U), information publicly available for review, including reports of any enforcement actions or fines assessed against individual brokers. Brokers must also file form BD to register as an entity with the SEC, SRO or other jurisdictions, information which includes the directors and officers of the entity, including financial information and a description of the business activity to be undertaken. Broker-dealers are required periodically (at least annually) to provide reports on their liquidity, under the net capital rules of the Exchange Act. The FOCUS (Financial and Operational Consolidated Uniform Single) report contains financial information on the broker-dealer, including the calculated net capital being held, in addition to reports on customer accounts and other pertinent information. The

FOCUS report is reviewed by the examining authority (SRO or SEC) for compliance, and to identify any risks associated with a broker-dealer's net positions and ability to meet its obligations.

12.3 Summary

Data collection by the SEC and CFTC is part of the disclosure regime that is the foundation of our securities laws. Whether it is the business and financial information provided by a new issuer in its registration statement, or a Form 3 filing detailing the holdings of an officer of a publicly held company, or whether it is a proxy statement with executive compensation data or a net capital report filed by a broker-dealer, data are made available to investors to enable them to make informed judgments about their investments. As part of their mission, regulatory agencies conduct examinations and surveil the markets to ensure fair and orderly markets. While risk management was not traditionally thought to be the mandate of the agencies, the need to identify emerging trends based on market activity, and focus on potential risks in certain areas or entities, are approaches that the agencies now use to fulfill their mandates. While the philosophy of disclosure continues to play an important role in our capital markets, regulators must continue their efforts to keep pace with the changing landscape as market participants innovate securities products, create new ways to deliver information, and look for ways to evade the securities laws. The agencies' microprudential activities will continue to contribute to the macroprudential supervision ongoing by other groups, including the European Financial Stability Board and the US Financial Stability Oversight Council. Data collection and risk management are now inextricably linked in our regulatory environment.

Bibliography

Bratton, William W. and Adam J. Levitin, 2012, A Transactional Genealogy of Scandal: From Michael Milken to Enron to Goldman Sachs, August 13, http://ssrn.com/abstract=2126778

CCH Editorial Staff, 2010, *Dodd-Frank Wall Street Reform and Consumer Protection Act: Law, Explanation and Analysis*, CCH, http://onlinestore.cch.com/default.asp?ProductID=7342

Commodity Futures Trading Commission (CFTC), 2012a, Commitments of Traders, Internet resource, accessed May 7, 2012, www.cftc.gov/MarketReports/CommitmentsofTraders/index.htm

Commodity Futures Trading Commission (CFTC), 2012b, Index Investment Data, Internet resource, accessed May 7, 2012, www.cftc.gov/MarketReports/IndexInvestmentData/index.htm

Commodity Futures Trading Commission (CFTC), 2012c, Technology Advisory Committee, Internet resource, accessed June 22, 2012, www.cftc.gov/About/CFTCCommittees/TechnologyAdvisory/index.htm

Commodity Futures Trading Commission and Securities and Exchange Commission (CFTC-SEC), 2010, Findings Regarding the Market Events of May 6, 2010: Report of the Staffs of the CFTC and SEC to the Joint Advisory Committee on Emerging Regulatory Issues, September 30, www.sec.gov/news/studies/2010/marketevents-report.pdf

Commodity Futures Trading Commission and Securities and Exchange Commission (CFTC-SEC), 2011a, Joint Study on the Feasibility of Mandating Algorithmic Descriptions for Derivatives: A Study by the Staff of the Securities and Exchange Commission and the Commodity Futures Trading Commission as Required by Section 719(B) of the Dodd-Frank Wall Street Reform and Consumer Protection Act, April 7, 2011, Technical Report, www.cftc.gov/LawRegulation/DoddFrankAct/ReportsandStudies/StudyAlgorithmicDescriptions/index.htm

Commodity Futures Trading Commission and Securities and Exchange Commission (CFTC-SEC), 2011b, Reporting by investment advisers to private funds and certain commodity pool operators and commodity trading advisors on Form PF, Final rule, *Federal Register*, **76**(221), 71128–71239, www.gpo.gov/fdsys/pkg/FR-2011-11-16/pdf/2011-28549.pdf

Commodity Futures Trading Commission and Securities and Exchange Commission (CFTC-SEC), 2012, Joint Report on International Swap Regulation, Technical Report, January, www.sec.gov/news/studies/2012/sec-cftc-intlswapreg.pdf

Financial Crisis Inquiry Commission (FCIC), 2011, The Financial Crisis Inquiry Report: Final Report of the National Commission on the Causes of the Financial and Economic Crisis in the United States, January, www.gpoaccess.gov/fcic/fcic.pdf

Financial Stability Board (FSB) and International Monetary Fund (IMF), 2011a, The Financial Crisis and Information Gaps: Report to the G-20 Finance Ministers and Central Bank Governors, Technical Report, October, www.financialstabilityboard.org/publications/r_091107e.pdf

Financial Stability Board (FSB) and International Monetary Fund (IMF), 2011b, The Financial Crisis and Information Gaps: Implementation Progress Report, Technical Report, June, www.imf.org/external/np/g20/pdf/063011.pdf

Financial Stability Oversight Council (FSOC), 2011, Authority to require supervision and regulation of certain nonbank financial companies: Second notice of proposed rulemaking and proposed interpretive guidance, *Federal Register*, **76**(201), 64264–64282, www.gpo.gov/fdsys/pkg/FR-2011-10-18/pdf/2011-26783.pdf

Financial Stability Oversight Council (FSOC), 2012, Authority to require supervision and regulation of certain nonbank financial companies, Final rule and interpretive guidance, *Federal Register*, **77**(70), 21637–21662, www.gpo.gov/fdsys/pkg/FR-2012-04-11/pdf/2012-8627.pdf

Grant Thornton, 2009, SEC adopts final rule for XBRL reporting, *New Developments Summary*, 2009–10, March, www.grantthornton.com/staticfiles/GTCom/files/services/Audit%20and%20assurance%20services/Assurancepublications/New%20Development%20Summaries/NDS%202009/NDS_2009-10.pdf

Hazen, Thomas, 2005, *The Law of Securities Regulation*, Fifth edition, West Publishing.

Hu, Henry T. C., 2012, Too complex to depict? Innovation, 'pure information' and the SEC disclosure paradigm, *Texas Law Review*, **90**(7).

Karmel, Roberta S., 2008, The EU challenge to the SEC, *31 Fordham International Law Journal*, **1692**, 1711.

Kirilenko, Andrei, 2011, Presentation slides, *Conference on Systemic Risk and Data Issues* hosted by the University of Maryland, New York University, Carnegie-Mellon University and University of California at Berkeley, Washington DC, 5–6 October 2011, www.rhsmith.umd.edu/cfp/events/2011/confSystemicRisk1011/recap/docs/KirilenkoSlides.pdf

Laux, Christian and Christian Leuz, 2010, Did fair-value accounting contribute to the financial crisis? *Journal of Economic Perspectives*, **24**(1), 93–118. www.aeaweb.org/atypon.php?return_to=/doi/pdfplus/10.1257/jep.24.1.93

Loss, Louis, 1988, *Fundamentals of Securities Regulation 33*, Second edition.

Office of Information and Regulatory Affairs (OIRA), 2012, Information Collection Review, Internet site, accessed June 22, 2012, www.reginfo.gov/public/do/PRAMain

OTC Derivatives Regulators' Forum (ODRF), 2009, Framework for Information Sharing and Cooperation Among OTC Derivatives Regulators, Technical Report, September, www.otcdrf.org/documents/framework_sept2009.pdf

OTC Derivatives Regulators' Forum (ODRF), 2010, Scope and Relationship with International Bodies, Technical Report, March, www.otcdrf.org/documents/scope_relationships_mar2010.pdf

Paredes, Troy, 2011, Speech by SEC Comissioner: Twelfth Annual A. A. Sommer, Jr. Lecture on Corporate, Securities and Financial Law, October 27, www.sec.gov/news/speech/2011/spch102711tap.htm

Saunders, Anthony and Marcia Cornett, 2010, *Financial Institutions Management: A Risk Management Approach*, McGraw-Hill.

Securities and Exchange Commission, 2009a, Interactive data to improve financial reporting; Final rule, *Federal Register*, **74**(26), 6776–6821, www.sec.gov/rules/final/2009/33-9002fr.pdf

Securities and Exchange Commission, 2009b, Interactive data for mutual fund risk/return summary; Final rule, *Federal Register*, **74**(32), 7748–7776, www.sec.gov/rules/final/2009/33-9006fr.pdf

Securities and Exchange Commission, 2011a, Security ratings; Final rule, *Federal Register*, **76**(149), 46603–46621, www.gpo.gov/fdsys/pkg/FR-2011-08-03/pdf/2011-19421.pdf

Securities and Exchange Commission, 2011b, Performance Measures Results Summary, United States Securities and Exchange Commission FY 2011 Performance and Accountability Report, www.sec.gov/about/secpar/secpar2011.pdf#2011review

Securities and Exchange Commission, 2012a, Securities and Exchange Commission Forms List, Internet resource, accessed April 30, 2012, www.sec.gov/about/forms/secforms.htm

Securities and Exchange Commission, 2012b, EDGAR Filer Manual – Volume II: EDGAR Filing, Version 19, Technical Report, March, www.sec.gov/info/edgar/edmanuals.htm

Securities and Exchange Commission, 2012c, Office of Interactive Disclosure: History, Internet resource, accessed June 22, 2012, www.sec.gov/spotlight/xbrl/oid-history.shtml

Securities and Exchange Commission, 2012d, XBRL.sec.gov, Internet resource, accessed June 22, 2012, http://xbrl.sec.gov/

Securities and Exchange Commission, 2012e, Examinations by the Securities and Exchange Commission's Office of Compliance, Inspections and Examinations, www.sec.gov/about/offices/ocie/ocieoverview.pdf

Securities and Exchange Commission, 2012f, Office of Credit Ratings, www.sec.gov/about/offices/ocr.shtml

Soderquist, Larry and Theresa Gabaldon, 2007, *Securities Law*, Third edition, Foundation Press.

Sommer Jr., A. A., 1978, The U.S. Securities and Exchange Commission Disclosure Study, 1 J. Comp. Corp. L. & Sec. Reg. 145, 147.

13

Financial data and risk information needed for the European system of financial supervision

Per Nymand-Andersen, Nicola Antoniou, Oliver Burkart and Jarl Kure

13.1 Introduction to the European System of Financial Supervision

One of the European responses to the financial crisis which started in 2007, was the creation of a new architecture for financial supervision in Europe.[1] The new European supervisory structure came into force on January 1, 2011, less than two years after the publication of the "de Larosière report"[2] which recommended the new structure in response to the financial crisis. With the goal of maintaining financial stability in Europe, the European System of Financial Supervision (ESFS) is built on a two pillar supervisory approach. One pillar refers to microprudential supervision and consists of three European Supervisory Authorities[3] (ESAs) and the European Central Bank (ECB). The other, a macroprudential pillar, is the European Systemic Risk Board[4] (ESRB) and the ECB. At the microprudential level, the European Banking Authority (EBA) is created for the banking sector, the European Securities and Markets Authority (ESMA) for securities markets, and the European Insurance and Occupational Pensions Authority (EIOPA) for the insurance and occupational pensions sector. Furthermore, a Joint Committee of the ESAs is established for coordinating the microprudential activities across Europe.

[1] The term Europe in this article refers to the European Economic Area (EEA), which includes 28 European Union Member States and Norway, Iceland and Liechtenstein.

[2] The "de Larosière" report was The High-Level Group of Financial Supervision in the EU, February 25, 2009.

[3] Regulation (EU) No 1093/2010 of the European Parliament and of the Council of 24 November 2010 establishing a European Supervisory Authority (European Banking Authority). Regulation (EU) No 1094/2010 of the European Parliament and of the Council of 24 November 2010 establishing a European Supervisory Authority (European Insurance and Occupational Pensions Authority). Regulation (EU) No 1095/2010 of the European Parliament and of the Council of 24 November 2010 establishing a European Supervisory Authority (European Securities and Markets Authority).

[4] Regulation (EU) No 1092/2010 of the European Parliament and of the Council of 24/11/2010 on European Union macroprudential oversight of the financial system and establishing a European Systemic Risk Board.

Handbook of Financial Data and Risk Information, ed. Margarita S. Brose, Mark D. Flood, Dilip Krishna and Bill Nichols. Published by Cambridge University Press. © Cambridge University Press 2014.

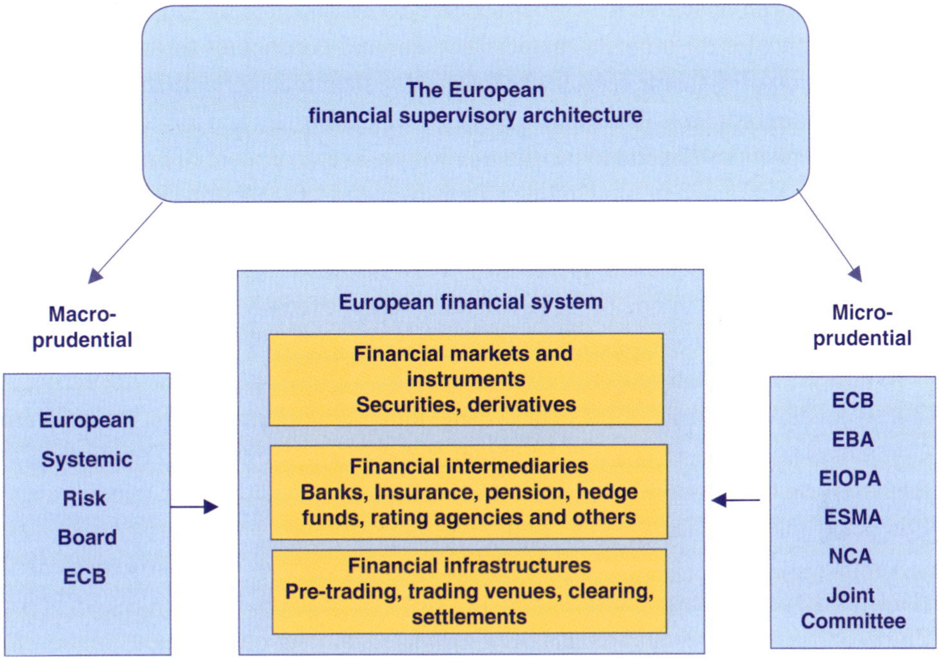

Figure 13.1 The European system of financial supervision.

The three ESAs and the ECB coordinate supervision of, and regulations for, European markets and do not replace existing national competent authorities (NCA), which continue to be involved in the day-to-day supervision. The new European microprudential authorities aim to improve the quality and consistency of supervision, reinforce the supervision of cross-border groups, strengthen crisis prevention across the European Union (EU) and create a coherent set of European rules, standards[5] and associated guidance and recommendations which can be adopted as EU law and applied consistently throughout Europe. Where necessary, the ESAs can investigate and – via arbitration between national authorities – resolve disputes by making a decision that is binding on national authorities to ensure compliance with EU law ("mediation"). Furthermore, in a crisis management situation, the ESAs will provide EU-wide coordination and can, in specific circumstances, issue binding decisions on national supervisors, institutions and certain financial activities. As part of their mandate, the ESAs collect information from national supervisors to be used for analyzing markets, for coordinating EU-wide stress tests and for the macroprudential analysis needs of the ESRB.

[5] These standards will harmonize many technical areas where rules currently diverge between Member States and will contribute to the development of a Single Rulebook for the financial sector.

The European Systemic Risk Board (ESRB) is responsible for the macroprudential oversight of the European financial system and contributes to the prevention and mitigation of systemic risks to the financial system together with the ESAs. The ESRB monitors systemic developments in the financial system and can issue warnings or recommendations directly to the EU as a whole, to one or more Member States, to the ESAs and/or to one or more of the national supervisory authorities. Systemic risk can originate, for instance, from large macroeconomic shocks, contagion effects or growing financial imbalances. Financial imbalances could include the development of asset price booms in one or more Member States, excessive concentration in the financial sector or excessive reliance on risky types of financial instruments.

Whilst the ESAs and the ESRB are independent institutions, there is full accountability to the European Parliament, where they will appear before the relevant Committee known as ECON (The Economic and Monetary Affairs Committee), at their request for formal hearings. Full accountability towards the Council of the European Union and European Commission also exists.

Given its expertise on macroprudential and statistical issues, the European Central Bank (ECB) contributes significantly to the effective macroprudential oversight of the European financial system and has been mandated, among other responsibilities, to collect, process and provide statistical information and give analytical support to the ESRB.[6]

A further step towards a European banking union was reached by the European Council on December 14, 2012 with the proposed creation of the Single Supervisory Mechanism (SSM) for the euro area. Under the SSM, the European Central Bank (ECB) is responsible for the supervision of all banks within the euro area[7] and for the consistent application of the Single Rulebook, as developed by the European Banking Authority (EBA), to this new framework. The ECB carries out its tasks within the SSM with the assistance of the national competent authorities (NCAs). Other non-participating Member States may also join the SSM. The establishment of the SSM is one of the preconditions for direct recapitalization by the European Stability Mechanism (ESM)[8].

As part of the SSM, the ECB shall ensure the coherent and consistent application of the Single Rulebook in the euro area. The ECB will directly supervise banks holding assets in excess of EUR 30 billion or which constitute at least 20% of their home country's GDP, or banks which have requested or received direct public financial assistance from the European Financial Stability Facility (EFSF) or the

[6] Council Regulation (EU) No 1096/2010 of 17 November, 2010 conferring specific tasks upon the European Central Bank concerning the functioning of the European Systemic Risk Board.

[7] There are approximately 6,000 banks within the euro area: `http://www.ecb.europa.eu/stats/money/mfi/general/html/elegass.en.html`

[8] The European Stability Mechanism (ESM), which was launched in October 2012, is the permanent financial assistance fund in the euro area.

ESM. The ECB will monitor the NCAs' supervision of the remaining banks and may decide to supervise directly one or more of these banks to ensure the consistent application of high supervisory standards. The ECB will send instructions to the NCAs, which are obliged to notify the ECB of supervisory decisions of material consequence. For cross-border banks active both within and outside Member States participating in the SSM, existing home/host supervisor coordination procedures will continue to exist as they do today. To the extent that the ECB has taken over direct supervisory tasks, it will carry out the functions of the home and host authorities for all participating Member States. The SSM governance structure of the ECB will consist of a separate Supervisory Board supported by a steering committee, the ECB Governing Council and a mediation panel to solve disagreements that may arise between NCAs and the Governing Council. A clear separation between the ECB's monetary and supervisory tasks will be fully ensured.

The ECB will continue to be involved in the macroprudential supervision of the European financial system as part of the SSM. An important feature of the SSM Regulation is that the ECB may, if it deems necessary, also apply macroprudential measures. These measures include higher requirements for capital buffers as well as more stringent measures to address systemic or macroprudential risks at bank level, subject to the Capital Requirements Directive and the Capital Requirements Regulation. The toolbox that is envisaged is rather broad. The set of available instruments includes: (i) counter-cyclical capital buffers; (ii) a systemic risk buffer; (iii) a capital surcharge for systemically important financial institutions; (iv) Pillar 2 measures applied to groups of institutions; (v) capital requirements; (vi) sector risk weights; (vii) large exposure limits; (viii) leverage ratio and liquidity requirements; and (ix) public disclosure requirements.

An integrated banking union will also include proposals on bank restructuring, a common deposit guarantee scheme and a single bank resolution mechanism, underpinned by a Single Rulebook to deal effectively and efficiently with cross-border bank resolutions, thus avoiding the need for public financial support to rescue banks. This is an important step as banking supervision and resolution need to be aligned and implemented at the same central level by applying a substantive set of rules for bank recovery and resolution in order to mitigate uncertainty and therefore prevent bank runs and contagion to other parts of the euro area.

The ECB, as the euro area bank supervisor, will signal when a bank is in severe financial difficulties and needs to be resolved. A Single Resolution Board will prepare a resolution plan for the respective bank, including a recommendation as to whether or not the European Resolution Fund should be involved. For legal reasons, the Commission will decide if and when a bank should be placed in resolution and would set out a framework for the use of resolution tools and the

Fund. National resolution authorities will be in charge of the execution of the resolution plan under the supervision of the Single Resolution Board. The Resolution Fund would be set up by the Single Resolution Board to ensure the availability of medium-term funding support. The Fund will be funded by the banking sector, replacing national resolution funds as part of the draft Bank Recovery and Resolution Directive. It is expected that the single bank resolution mechanism will enter into force in January 2015.

The interplay between the macroprudential and microprudential perspective allows for better supervision of individual firms, financial infrastructures and markets across Europe as well as maintaining an overview of the interconnectedness and systemic risks within the European financial system and its possible impact on the real economy. This new European surveillance and supervision structure will be the "control tower" and "radar screen" to provide early warnings, suggest corrective measures and timely actions.

The European supervisory architecture has triggered a substantial review of the types of statistics, data and risk information required from the financial system and analysis of the best means to share and compare statistics from the financial system across various jurisdictions based on the network of national and European authorities. This has also fed into similar global initiatives to strengthen supervision of the financial system as led by the Group of Twenty (G20)[9] The key statistical challenge lies in the ability to coordinate and synchronize the various statistical initiatives among the network of European and national authorities and to create a common set of statistical concepts which can be used by all for the multi-analytical prudential supervision assessments required in Europe and beyond.

Several other European initiatives are ongoing to improve stability and governance measures in Europe, such as the discussion on reporting requirements and certain regulation of derivatives, credit default swaps, short selling and credit rating agencies, the outcome of which contributes to the necessary collection of financial statistics and risk information in Europe. Section 13.11 provides an overview of such ongoing European initiatives.

In the subsequent sections, we present an overview of the financial statistics and risk data requirements for both macroprudential and microprudential supervision, followed by a discussion of the main statistical challenges. This is followed by a description of the corresponding supervisory construction in the USA, and initiatives taken at the G20 level, before concluding.

[9] G20 was established in 1999, in the wake of the 1997 Asian Financial Crisis, to bring together major advanced and emerging economies to stabilize the global financial system, `www.G20.org`.

13.2 Statistics and data needs for European macroprudential supervision

The financial crisis revealed that macrofinancial factors, such as the interconnectedness of markets, institutions and financial globalization, play an important role in determining the size, nature and propagation of systemic risk. Therefore, it is essential that the effective framework for macroprudential supervision ensures a systematic, all-encompassing and integrated analysis of systemic risks as well as the formulation of appropriate policies to address such risks. The macroprudential approach to supervision focuses on the financial system as a whole and involves the monitoring, assessment and mitigation of systemic risk. It is important to recognize that systemic risk is partly endogenous – as it depends on the collective behavior of financial intermediaries, markets, infrastructures and their interconnectedness – and partly exogenous – as it can originate from outside the financial system itself and depends on the interaction between the financial system and the real economy. The choice of the appropriate framework and tool-sets to address emerging systemic risks continues to be developed as part of identifying the most efficient and effective instruments for the implementation of macroprudential policies.

Financial system stability requires that the principal components of the system – including financial intermediaries, markets and financial infrastructures – are jointly capable of absorbing adverse disturbances.

Monitoring the smooth functioning of the financial system requires a significant web of statistics and intelligence covering all areas of the financial system and covering the interaction between financial intermediaries (FI), markets and instruments (MI) and/or financial infrastructures (I). Systemic risk can however also originate from other external financial systems and spill-over to and affect the

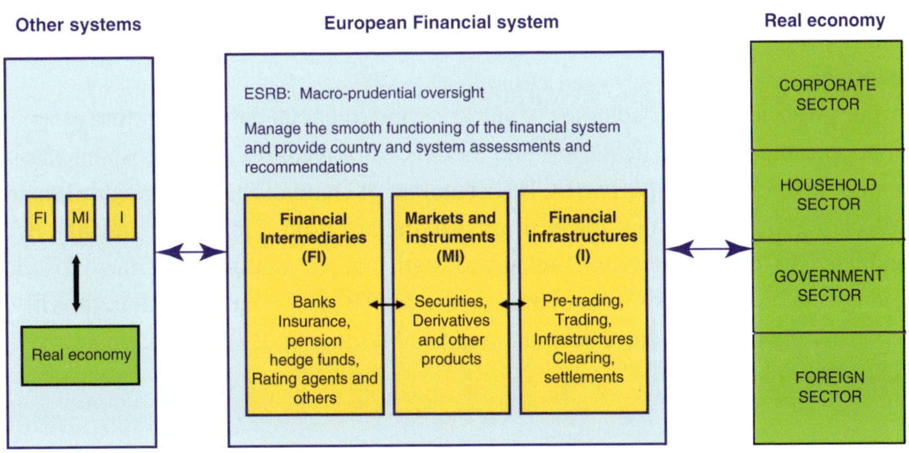

Figure 13.2 The macroprudential oversight challenge.

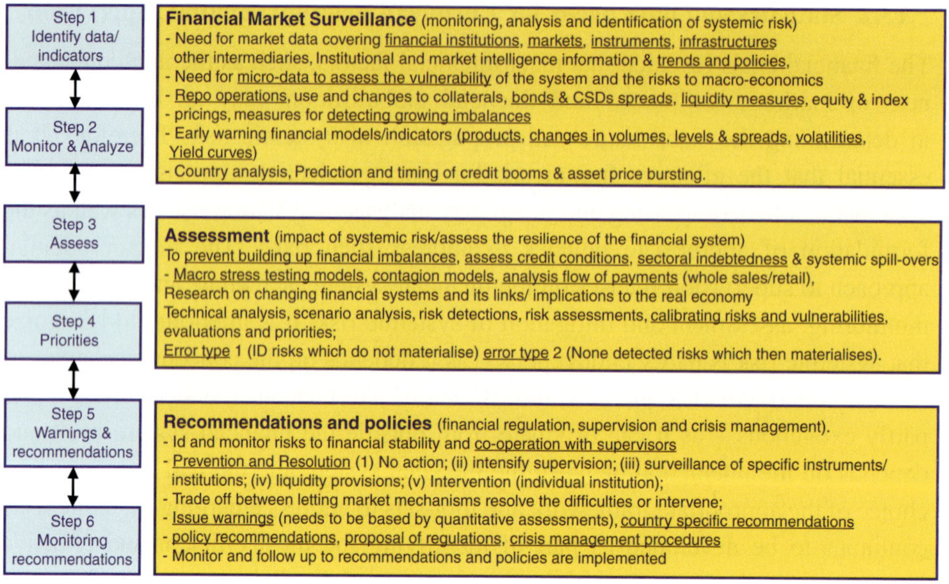

Figure 13.3 The process of managing systemic risks.

performance and interaction of the principal components of other financial systems. Systemic risks can be unexpected, or the likelihood of their occurrence can build up over time in the absence of appropriate timely political responses. The financial crisis has revealed that the interconnectedness of markets and institutions plays an important role in determining the size and nature of systemic risk.

The process of managing systemic risks involves six interactive steps (see Figure 13.3), where the data and risk information relate mainly to the first two steps of the process, namely, (i) to supply the necessary statistics, surveillance information and derived indicators needed to monitor, and (ii) to identify systemic risks in the financial system and links to other financial systems and the real economy.

The provision of a broad range of statistics and financial data is therefore essential for supporting the analytical tools and models employed in the analysis and assessment of financial stability. The ECB Directorate of General Statistics predominantly collects data and statistics for euro area monetary policy purposes and has been mandated to collect new and enhanced statistics to support fully the EU-wide macroprudential analysis framework of the ESRB; a particular challenge will be to expand the coverage from the euro area (17 Member States) to all members of the EU (28 Member States).[10]

[10] This will require additional reporting efforts by the non-euro area member states to comply with the established statistical and methodological framework.

Table 13.1 *Overview of ESRB's financial statistics and risk information requirements.*

ESRB
Four main statistical indicator groups

Macroeconomic and financial statistics	**Market risks on credit, assets and financial conditions**	**Risk information and performance of financial intermediaries**	**Risk information on market infrastructures**
• Macroeconomic indicators; GDP, unemployment; • Debt of governments, households, corporate and forecasts; • Economic and sentiment indicators and asset prices; • Yield curves, slopes and spreads; • Current account balances, Foreign exchange and reserve holdings.	• Intelligence on financial markets; • Volume and price information of financial instruments and derivatives; • Market risks indicators; • Money markets and spreads; • Assets and property prices; • Changes in credit conditions.	• Information on liquidity and funding conditions; • Counterparty interconnectedness; • Profitability and CDS spreads; • Outlook and Risks; • Solvency indicators; • Capital and leverage ratios.	• Interlinkages and systemic risks indicators; • Clearing activities; • Settlement activities; • Market structures.

There are four main statistical indicator groups identified for the purposes of macroprudential analysis (see Table 13.1): (i) macroeconomics and financial statistics; (ii) market risks on credit, assets and financial conditions; (iii) risk information and performance of financial intermediaries; and (iv) risk information on market infrastructures.

These four main statistical indicator groups identified are further elaborated below.

1. *A wide range of macroeconomic and financial statistics is needed to obtain an overview of the financial system and the links to the (real) economy.*

   Historic and current statistics on macroeconomic developments and indicators are required to measure the health of, changes to, and the interaction between the European financial system and the real economy. This requires a monitoring system based on statistics covering among others, actual and forecast real GDPs, unemployment rates, an array of economic and sentiment indicators/indices, financial conditions, changes to the debt level of the household and corporate sectors, and an overview of public finances (stocks, flows and current account balances and associated forecasts).

   It also calls for the calculation of yield curves of various issuer and credit classes of government, banking and corporate sectors and corresponding spreads; exchange rate developments and reserve holdings.

2. *Market intelligence information and financial market statistics are needed for monitoring financial market developments and associated market risks within the European financial system.*

   This relates to the collection and presentation of financial market statistics and indicators covering price and volume information of financial instruments, associated derivatives, various commodities prices, market indices and volatility measures. Of particular interest is the ability to obtain detailed and reliable loan statistics and statistics on securities issues and holdings, which is indispensable to any systemic risk assessment.[11] The financial market statistics leads into a monitoring scheme to easily identify price developments of assets, such as real estate price developments.
3. *Statistics on large and complex financial intermediaries and other actors are needed as they can pose risks to the financial system that can spread to the real economy.*

   Specific and detailed statistics are needed to measure and assess the performance, strength and interconnectedness of (large and complex) financial institutions, as well as distributional information on common exposures. These statistics will be mainly derived from balance sheet statistics, profit, losses and solvency indicators and other similar granular statistics, including statistics collected and supplied by the three European microsupervisory Authorities (ESAs) (see below).
4. *The trading venues, clearing and settlement systems are the lifelines and backbone of the financial system.*

   Statistics and indicators generated from and on financial infrastructures are necessary for systemic risk monitoring and the assessment of the well-functioning of the financial markets. This relates mainly to statistics on value and transactions of financial instruments and derivatives (included standardized OTC derivatives), traded, cleared and settled via post-trading entities.

Following the SSM Regulation, the ECB may apply macroprudential measures to address systemic or macroprudential risks at bank level. This will require the provision of and access to a broad range of statistics and financial micro-level data to determine potential threshold levels for counter-cyclical capital buffers, sector risk weights including large exposure limits for individual banks and/or groups of banks.

These all-encompassing and integrated financial data and risk identification will provide the necessary foundation for the evolving approach to coordinated macroprudential and micro-level analysis and will pose major statistical challenges for the new institutional set-up (see Section 13.6).

[11] Speech by J. C. Trichet; ECB Statistics Conference, October 19, 2010, www.ecb.int/press/key/date/2010/html/sp101019.en.html

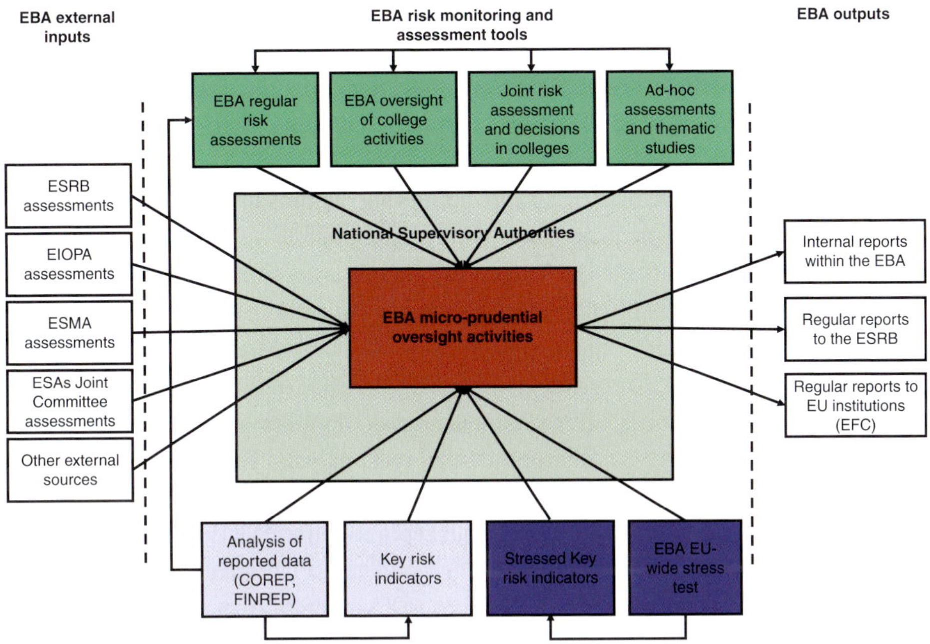

Figure 13.4 Microprudential risk analysis in the European banking sector.

13.3 Statistics and data needs for microprudential supervision of European banks (EBA)

The European Banking Authority (EBA) has established, for the first time, a common framework for supervisory reporting across the EU with the release of its standards on supervisory reporting in 2013. The common framework, with consistent definitions and formats, will reduce reporting costs for cross-border banks, and ensure comparable data across the EU, facilitating effective EU-wide risk assessments. The EBA devotes significant resources to the establishment of a framework for gathering and analyzing microprudential financial information, covering both regular and ad hoc data collections. In fulfilling its tasks of responding to potential risks and contributing to restoring and maintaining a stable and reliable European banking system, the EBA has the tasks of analyzing cross-border microprudential banking sector risks and vulnerabilities, and monitoring and assessing EU-wide banking market developments.

Following the new SSM framework, the EBA will continue to develop the Single Rulebook applicable to all 28 Member States. In order to foster consistency and efficiency of supervisory practices across the European Union, it will develop a single supervisory handbook. The EBA will also ensure that regular stress tests are carried out to assess the resilience of European banks. Though the rules on the functioning of the EBA will be adapted to the new SSM framework, its role will be reinforced. The SSM will benefit from the establishment and implementation of the

EBA's Single Rulebook for European supervision. The EBA will monitor the harmonization of EU supervisory rules and the implementation of the additional rules contained in the Capital Requirements Directive, the Bank Recovery and Resolution Directive and the Deposit Guarantee Directive. The EBA should also be able to carry out its tasks in relation to the ECB. In order to ensure that existing mechanisms for the settlement of disagreements and actions in emergency situations remain effective, a specific procedure will be established.

With the above in mind, the EBA receives data primarily from national supervisory authorities (whose Heads of Banking Supervision form the EBA's governing body, the Board of Supervisors),[12] and also from external sources, such as risk assessments carried out by the other European bodies (ESRB, EIOPA, ESMA, and ECB) and public data sources. The EBA has a number of risk monitoring and assessment tools in place internally to carry out microprudential risk analysis. The outputs of the EBA risk analysis and assessment range from internal reports (within the EBA and to national supervisory authorities), to external reports to the ESRB and the Economic and Finance Committee (EFC) of the EU Council[13] and EU Parliament. For instance, the EBA submits an annual "report on risks and vulnerabilities in the EU banking sector" to the EFC, semi-annual updates to the ESRB General Board meetings. The EBA delivers annual risk and vulnerabilities reports to the EU parliament. In addition to the regular banking risk assessments, the EBA conducts thematic risk analyses on specific themes of supervisory concerns based on the collection and analysis of quantitative and qualitative information.[14]

The EBA collects supervisory data on a centralized and consolidated basis through cooperation with the national supervisory authorities who report bank-specific information to the EBA within their respective jurisdictions. The regular reporting is complemented with ad hoc data collections focusing on specific risks and enables the EBA to carry out thematic reviews of risks and forward-looking stress tests. One example of thematic work is the EBA's review of consistency of risk-weighted assets in banks' portfolios.

As day-to-day banking supervision remains a responsibility of national supervisory authorities, it is necessary to establish a smooth interaction between *banks*, *national supervisory authorities* and the *EBA*. For instance, the EBA conducts the EU-wide stress tests in close cooperation with national supervisory authorities, who are essentially providing data to the EBA and act as intermediaries in contact with

[12] The EBA's Board of Supervisors can be found at `www.eba.europa.eu/Aboutus/Organisation/Members.aspx`

[13] Reports are provided to the EFC to provide microprudential input to the overall assessment of systemic risk in the Europe financial system.

[14] For example, the March 2011 risk assessment report included annexes with two such thematic studies on exposures to the real estate sector and on loan restructuring.

Table 13.2 *Overview of EBA's financial statistics and risk information requirements*

EBA main statistical indicator groups for banks

Supervisory reporting	Risk assessments	Stress testing
Regular reporting on: • Own funds requirements; • Financial information; • Large exposures; • Liquidity; • Leverage ratio; • Asset encumbrance. Key risk indicators on: • Solvency • Credit risk and asset quality; • Liquidity and balance sheet structure; • Profitability; • Other indicators.	• Banks' financial and non-financial data; • Bank vulnerabilities; • Thematic risk assessments; • Individual and joint risk assessments on market, operational, liquidity and credit risk as well as capital requirements for various risk categories; • Capital requirements and capital add-ons for groups and their local entities.	• Applying stress test scenarios and assumptions; • Historical risk and loss data on general and parameters level; • Data on risk exposures.

banks. This approach ensures a uniform European approach and consistent implementation throughout all EEA countries.

In addition to harmonized supervisory reporting, use of common questionnaires and assessment criteria allows the EBA to (i) conduct a horizontal analysis of major banking groups, (ii) flag institution-specific supervisory concerns, (iii) verify whether identified vulnerabilities pose a risk of spill-over to other banking institutions or to the EU financial system as a whole (part of monitoring systemic risk), and (iv) determine institutions' and sector general combined relative ranking of various risks.[15]

Any trends, potential risks and vulnerabilities stemming from the microprudential level must be identified at an early stage, in order to safeguard the financial stability of the banking sector. To achieve this, the EBA has three main areas of use and, at the same time, sources of statistical data for the purpose of its microprudential analysis of the banking sector: regular supervisory reporting, risk assessments and stress testing (see Table 13.2).

These three main areas of EBA bank oversight and sources of statistical data are further elaborated below.[16]

15 The framework for prudential reporting data collection is governed by the following guidelines, all of which will form part of future technical standards on supervisory reporting which will be applicable across Europe: (i) the COREP (commonly agreed EBA reporting frameworks on capital) guidelines, which focus on information on the structure of own funds and capital requirements, as well as information on exposures; (ii) guidelines regarding the reporting of large exposures, which lay down reporting requirements regarding exposures that exceed certain limits and specify additional information to be reported on the composition of the group(s) of connected clients; and (iii) FINREP (financial information) guidelines, which concentrate on the main items of income statements and balance sheets, as well as on information on impairments and past due items, structure of funding and volume of off-balance-sheet commitments.

16 In addition, the EBA has regulatory tasks which require regular supervisory reporting (e.g., calibration on liquidity and leverage thresholds, monitoring small and medium size enterprises (SMEs) and conducting impact assessments).

1. *The EBA has developed a common set of quantitative and qualitative indicators to identify and measure microprudential systemic risk of the banking sector, which is also a part of its statutory responsibility.*[17]

   A set of Key Risk Indicators (KRIs) covers a representative sample of the largest European cross-border banks. These KRIs provide valuable quantitative information of prominent risk areas, essential for the EBA in undertaking its microprudential analysis. The KRIs also meet the ESRB's requirements for aggregated data.

   The KRIs comprise indicators on the solvency, credit risk and asset quality, earnings risk and balance sheet structure of banking institutions. Reporting and calculation of the KRIs are currently based on the guidelines on COREP and FINREP reporting (see footnote 15) to ensure consistency across the EU. Supervisory reporting of capital adequacy, financial information, large exposures, liquidity, leverage ratios and asset encumbrance will be truly harmonized across all EU banks in 2014 when the technical standard on supervisory reporting developed by the EBA enters into force. This standard will specify uniform data elements, frequencies and remittance dates as well as IT solutions to be applied by credit institutions and investment firms in Europe.
2. *The EBA has been performing microprudential bottom-up risk assessments in the European banking sector since November 2008.*[18]

   In addition to addressing institution-specific risks in an aggregate fashion, these assessments consider how global risks and vulnerabilities identified from a macroprudential perspective affect individual banking institutions and vice versa, and how this affects the financial system both nationally and at the EU level. The analysis is conducted in a bottom-up fashion relying on various sources of information, including EBA's own intelligence, market intelligences, analytical work conducted by the experts of national authorities supporting the EBA work, and, most importantly, risk assessment questionnaires addressed to line supervisors of major European cross-border banking groups.[19] To undertake this exercise, a detailed questionnaire is developed by the EBA, including any important themes identified by the EBA's partners, such as the ESRB. The questionnaire is circulated to the consolidating supervisors of the largest European banking groups who are requested to provide up-to-date information (through discussion with the host supervisors in the supervisory colleges, as well as with the bank itself). The information received is mainly of qualitative nature (so-called "soft information"), but risk and vulnerabilities assessments

[17] Article 22 (2) of the EBA Regulation (see footnote 66).

[18] Prior to January 1, 2011, these were undertaken by the EBA's predecessor organization – the Committee of European Banking Supervisors (CEBS).

[19] In 2012 the sample of banks covered by the risk assessment questionnaire included the 40 largest cross-border banks.

increasingly use quantitative statistical data, including KRIs and additional statistics collected through the ad hoc data collections and thematic surveys for the most prominent themes not covered by the regular data collection.

Additionally, there are specific data requirements for European banking groups from their respective colleges of supervisors[20] (colleges). The annual joint risk assessment is the key tool used by these colleges to analyze risks faced by a banking group, as it combines data from the annual individual risk assessments undertaken by the home and host[21] supervisory authorities of a group and assess the adequacy of the group's consolidated and entities' capital to their risk profile.

The colleges' joint risk assessments cover both quantitative and qualitative data on risk and control factors on major risk categories, including market, operational, liquidity and credit risk; risks related to transfer through securitization activities; concentration risk; interest rate risk from non-trading activities; or any additional material risks specific to an institution. Such data feed into an assessment of each entity's financial position, business and risk strategy, risk tolerance and appetite and corporate and internal governance, and the joint decisions on capital adequacy for the group and its respective entities.[22,23]

3. *The most data intensive supervisory tool used by the EBA to analyze risks is the EU-wide stress tests*[24] *which aim to assess the resilience of banks to adverse market developments using hypothetical scenarios and shocks.*

   Examples of data collected for stress testing include granular information on credit and market risk exposures and risk parameters with geographical breakdown, an institution's capital position and exposures and their evolution under stress scenarios for a certain time horizon.

The EBA initiates and coordinates these stress tests in cooperation with the ESRB, the ECB and the EU Commission. So far the EBA EU-wide stress test have been conducted in a constrained bottom-up manner, meaning that banks apply common scenarios and assumptions to their internal risk parameters using internal models, which better reflect institutions' specific situations. The results of the banks' calculations are cross-checked against various benchmarks based on historical risk data, including data on historical losses (e.g., loss and default rates) and results of top-down stress tests conducted by the supervisors.

[20] Colleges of supervisors are permanent, although flexible, structures for cooperation and coordination among national authorities responsible for and involved in the supervision of the different components of cross-border banking groups.

[21] "Home supervisory authority" refers to the consolidating national supervisory authority of a banking group and "host supervisory authority" refers to the national supervisory authority responsible for the supervision of the local entities of a banking group.

[22] Article 129(3) of the Capital Requirements Directive (CRD Directive 2006/48/EC).

[23] This article has also been elaborated in EBA guidelines on the "Joint Assessment of the elements covered by the supervisory review and evaluation process (SREP) and the joint decision regarding the capital adequacy of cross border groups", also known as EBA GL39.

[24] Articles 21 and 32 of the EBA Regulation (see Footnote 67).

Table 13.3 *Preliminary overview of the ECB's financial statistics and risk information requirements*[25]

ECB main statistical indicator groups for banks		
Reporting of banks	**Risk assessments**	**Stress testing**
Standard reporting from COREP and FINREP, covering: • Capital structures and requirements; • Consolidated financial reporting information; • Exposures; • Liquidity; • Leverage ratio; • Balance sheet statistics; • National reporting based on national accounting principle (GAAP); • Non-harmonized micro data as required at national level (Pillar 2 data).	• Financial markets data and instruments (issuance and holding profiles); • Granular credit and loans; • Financial micro-level data; • Sector risk weights values; • Exposures; • Ad hoc surveys and monitoring; • Other risk assessment indicators.	• Applying stress test scenarios and assumptions; • Historical risk and loss data; • Risk exposures.

The results of the stress tests can lead to further analysis of identified vulnerabilities and specific supervisory actions, including setting up appropriate backstop mechanisms[26] for possible recapitalization, if and where needed. The collection of statistical data as a part of the EU-wide stress tests (with limited public release of some data[27]) allows the EBA to build granular data series of essential data on key risk parameters, which can be cross-used in other EBA analytical work, but also used in backtesting of the stress test results and conducting own top-down scenario or sensitivity analyses.

In view of the above, the EBA is constantly working to improve its data collection and analysis mechanisms in order to improve microprudential risk analysis and as part of the overall assessment of systemic risk in Europe by contributing to the macroprudential assessment cycle of the ESRB.

As part of the SSM, the European Central Bank is responsible for the supervision of all banks within the euro area with the assistance of the NCAs. It therefore needs to create a comprehensive set of datasets and statistics involving, inter alia, the NCA and EBA.

The intention of the proposed SSM regulation is for the ECB, as the European supervisor, to be fully operational in its new task from mid-2014.

[25] Under development and subject to change
[26] Backstop mechanism refers to a method that would provide emergency funding in the case of identified funding failures.
[27] The results of the July 2011 EBA stress test can be found at `http://eba.europa.eu/EU-wide-stress-testing/2011/2011-EU-wide-stress-test-results.aspx`

Table 13.4 *Overview of EIOPA's financial statistics and risk information requirements.*

EIOPA
Three main statistical indicator groups for (re-)insurance and pension

(Re-) Insurance	Pension (defined benefit or contribution system)	Stress testing
• Structure and Market developments (company structure, equity price, CDS, defaults and ratings); • Premiums, returns and solvency; • Balance sheet structure, profit and losses; • Trends analysis; • Thematic risks (e.g. banking exposures).	• Contributions and benefits; • Balance sheet structure, profit and losses; • Concentration ratio; • Trends analysis	• Applying agreed stress testing scenarios with common assumptions on shocks; • Risks categories include market risks (interest rates, equities, and real estate), credit risk, life and non-liferisks.

13.4 Statistics and data needs for microprudential supervision for European insurance undertakings and occupational pensions (EIOPA)

The European Insurance and Occupational Pensions Authority (EIOPA) assesses the financial stability in the European insurance and pension fund sector. This is mainly done by collecting data from the industry via national supervisors both on a recurrent basis and via ad hoc surveys. Similar to the EBA, EIOPA conducts a European-wide stress test exercise[28] to supplement the aforementioned ongoing assessment. Further, EIOPA regularly conducts qualitative risk assessments of the insurance and pension sector. Twice a year, EIOPA issues a financial stability report[29] on the outcome of its assessments.

Data will be collected via EIOPA's participation in European supervisory colleges and within the task of crisis management. The Financial Stability data collection will be based on supervisory data on the insurance and reinsurance sector at Member States levels.

There are three main statistical indicator groups identified by EIOPA for the purposes of microprudential analysis of the insurance and pension sector (Table 13.4).

The three main statistical indicator groups identified are further elaborated below.

1. *Market developments and company structure data as they relate to equity prices, interest, CDS spreads, default rates, and rating outlooks of individual European (re-) insurance companies.*

[28] Details of EIOPA European-wide stress test are available at `https://eiopa.europa.eu/activities/insurance/insurance-stress-test/index.html`

[29] The reports are prepared and discussed within EIOPA's Financial Stability Committee and published on EIOPA's website, `https://eiopa.europa.eu/publications/financial-stability/index.html`.

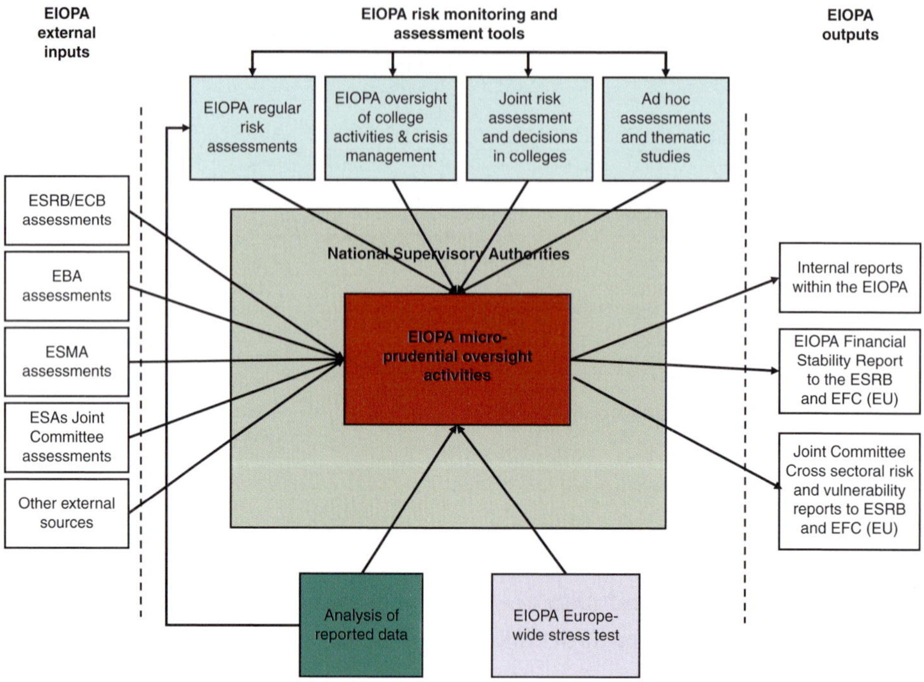

Figure 13.5 Microprudential analysis in the European insurance and pension fund sector.

Supervisory data are collected on return on equity; return on assets; solvency position; cases of insolvency; asset composition; lapses and surrenders; guaranteed products; insurance regulation and supervision; insurance penetration; and risk assessments by individual supervisors.

Specific data are also collected on the reinsurance business, including market trends, specific developments in premium income, combined ratio, developments of catastrophes, company specific information based on quantitative data and qualitative reports by supervisors on key insurance trends[30] regarding premium growth/claims development, lapses/surrenders (i.e., early withdrawals), financial strength and performance, solvency rates, capital injections (the number of undertakings and the total amount of capital injections), changes in asset allocation, merger and acquisition activity, concentration of the insurance market and other possible changes in the market or in the insurance industry.

Information on the internationalization of the insurance sector is collected relating to the most important foreign countries where the insurance industry is conducting its business and on the motives and challenges faced for insurance entities/groups in conducting business abroad and of any particular trends in cross-border activities.

[30] See Section 13.12 for a detailed description.

Since the financial crisis, information is collected on the types of measures taken by the national supervisors. This is mainly descriptive information regarding procyclical or countercyclical effects, combined with descriptions of any new measures to be taken and general risks/challenges in the (re-) insurance sector.

Further data are collected on the volume of reinsurance premiums written and the combined ratio of the seven largest reinsurance companies active in Europe. This is supplemented by market reports and supervisory assessments of the most imminent risks over the last 12 months and expected risks for the next 12 months; statistics and data required for ad hoc theme studies are collected, where applicable.

Due to expected regulatory changes, in particular EU directive on Solvency II with envisaged harmonized reporting formats, data collections to EIOPA are expected to change dramatically, becoming much more granular. Solvency II is envisaged to commence from 2014.

2. *Data requirements of the pensions funds mainly relate to quantitative information and qualitative reports by national supervisors*[31] *on market trends, changes and developments in specific Member States, penetration ratios, contributions received, asset composition, return on assets and cover ratios.*

   Information is collected on the measures taken following the financial crisis and on the corresponding risk assessments carried out by national supervisors and on the risks/challenges of the pension sector and national reforms, which may have an impact on occupational and supplementary pensions. Major trends in the membership rate, the numbers of and composition of active funds are collected, including indications of major risk factors. A description of likely future changes in regulations and supervisory practices is provided on the behavior of pension funds as part of mitigating these risk factors.

   Supervisory assessments of the most imminent risks over the last 12 months and expected for the next 12 months are conducted. Statistics and data required for ad hoc theme studies are also collected.

3. *EIOPA conducts a European-wide stress test to receive information on the current vulnerability of the EU insurance sector to adverse developments.*

   The data collected are capital (own funds) and capital requirements (Solvency Capital Requirements (SCR) and Minimum Capital Requirements (MCR)) as well as data from risk categories, such as market risks (interest rate risk from both assets and liabilities, equity risk and property risk), credit risk (spread risk of corporate bonds and non-EU/EEA government bonds and for structured products), life insurance (mortality risk or longevity risk), and non-life risks (catastrophe or claims reserves deficiency).

[31] See Section 13.13 for a detailed data description.

The results of these stress tests lead to further analyses and policy actions where applicable.

13.5 Statistics and data needs for microprudential supervision of European securities and markets (ESMA)

The European Securities and Markets Authority (ESMA) contributes to safeguarding the stability of the European Union's financial system by ensuring the integrity, transparency, efficiency and orderly functioning of securities markets, as well as enhancing investor protection. In particular, ESMA fosters supervisory convergence amongst securities regulators and across financial sectors by working closely with the other European Supervisory Authorities.

ESMA's work on securities legislation contributes to the development of a single rule book in Europe to ensure consistent treatment of investors and investor protection across Europe and a level playing field of competition for the providers of financial services. Where requested in European law, ESMA undertakes the supervision of certain entities with pan-European reach such as credit ratings agencies and, in the future, possibly also trade repositories.

Finally, ESMA contributes to the financial stability of the EU through monitoring, assessing and measuring systemic risk, as well as through its contribution to the work of the European Systemic Risk Board (ESRB). ESMA is also responsible for coordinating actions of securities regulators and supervisors and, when a crisis situation arises, adopting emergency measures within the scope fixed by the ESMA Regulation.

ESMA will have new competencies such as the ability to:

- propose technical standards that are legally binding in EU Member States once adopted by the European Commission[32];
- issue guidelines and recommendations, whereby national authorities – and in some cases financial market participants – will have to justify compliance publicly within two months or explain the reasons for non-compliance;
- launch a fast track procedure to ensure consistent application of EU law in case a EU Member State fails to apply properly a national provision of EU legislation. ESMA can issue recommendations to a national authority and, in case of persistent non-compliance, it will be able to issue decisions directly to financial market participants, where EU law applies to them directly;

[32] The ability to propose standards is restricted to areas set out in the Omnibus directives for existing legislation (the Omnibus I Directive includes the Markets in Financial Instruments Directive (MiFID), the Prospectus Directive, the Transparency Directive and the Market Abuse Directive) and might be enlarged as new legislative proposals are issued (see Section 13.11).

- issue binding legal decisions to competent authorities as part of resolving disagreements between national authorities and, under certain conditions, in situations of emergency;
- protect consumers, including the ability to temporarily prohibit certain financial activities that threaten financial stability or the orderly functioning of financial markets;
- participate in the activities of Colleges of Supervisors, including on-site visits carried out jointly by two or more competent authorities;
- monitor and assess market developments and contribute to criteria for identifying and measuring systemic risk in consultation with the ESRB, and to cooperate closely with the ESRB and other ESAs and to provide the statistical data and risk information, as necessary;
- supervise credit rating agencies (CRAs) and trade repositories[33] operating in Europe;
- enter into non-EU relationships (administrative arrangements with supervisory authorities, international organisations and the administrations of foreign countries).

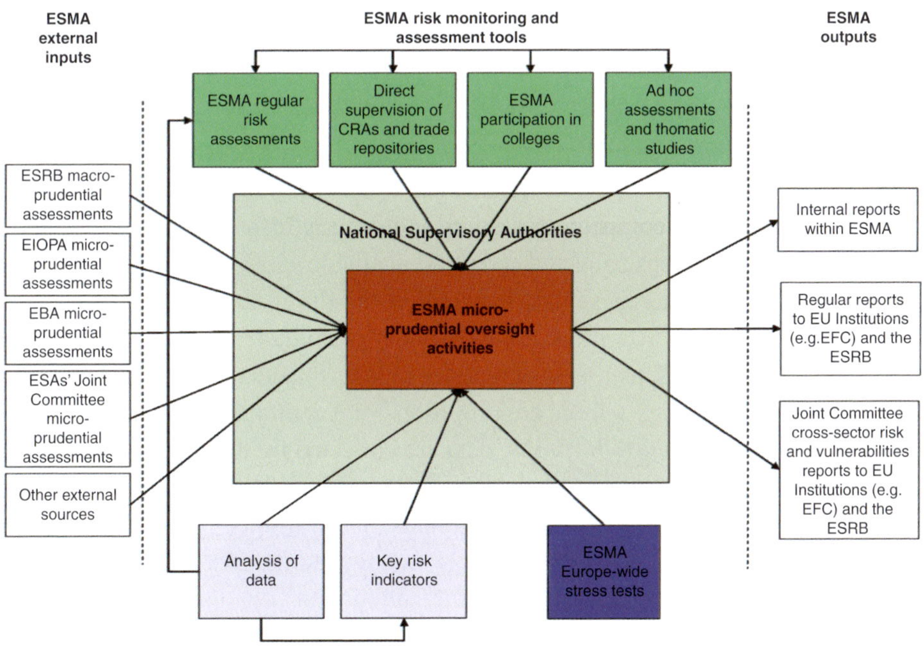

Figure 13.6 Microprudential analysis in the European securities and markets authority. The reference to stress tests is preliminary.

[33] The decision about supervision of trade repositories by ESMA is still pending.

Table 13.5 *Overview of ESMA's financial statistics and risk information requirements.*

ESMA
Four main statistical indicator groups for securities products, markets and actors

Admission to trade and securities products	Market surveillance	Financial markets	Financial actors
• Information on issuance, issuer and maintenance for admission to trade.	• Transactions data of financial products on regulated markets; • Market transparency calculations; • Short selling; • Credit rating agencies; • OTC derivatives; • Data on consumer trends.	• Trading facilities; • Clearing and settlement.	• Alternative investment fund managers; • Central counterparties; • Trade repositories; • Competent authorities; • Rating agents.

In order to identify, monitor and analyse trends, risks and vulnerabilities in securities products and markets, ESMA is active in the following four areas (as detailed in Table 13.5): (i) admission to trade and securities products, (ii) market surveillance, (iii) financial markets and (iv) financial actors.

These four main statistical indicator groups identified are further elaborated below.

1. *ESMA collects and publishes, for each Member State, data on prospectuses approved and prospectuses passported (sent and received*).[34]

   All regulatory information on issuers and issuances for admission to trading on regulated markets is collected and made available at the national level. Efforts are underway to make these data available at the European level.

   Concerning regulated information by issuers admitted to trading on regulated markets (e.g., price sensitive information, regular financial reports and notifications of major holdings), there are proposals for a central access point.
2. *For market surveillance purposes*, data are collected by the respective national authorities covering trade-by-trade data of financial products traded on European regulated markets[35] (such as equities, bonds, and warrants,[36] as well as derivatives whose underlying is traded on a regulated market). MiFID[37] and its Implementing Regulation impose the obligation on national competent

[34] Within the EU there is a passporting regime, which implies that the approval of a prospectus in one Member State allows the issuer to commercial the product in all 28 Member States.

[35] As distinguished from trades on Multilateral Trading Facilities, systematic internalizers, or OTC.

[36] A warrant is a security that entitles the holder to buy/sell the underlying stock of the issuing company at a fixed exercise price until the expiry date. This makes it similar to an option. In contrast to exchange-listed options, warrants are not standardized.

[37] The Markets in Financial Instruments Directive (MiFID), see Section 13.11 for further details.

authorities to collect data on all shares admitted to trading on EU Regulated Markets. The related information, including annual calculations, is published on ESMA's website: name, unique identification number (ISIN[38]), average daily turnover, average daily transactions, free float, average value of transactions, standard market size, etc. That information is targeted for market participants that need the information for recognizing liquid shares (which trigger obligations for systematic internalizers[39] according to Article 27 of MiFID) and for determining the block sizes for exemptions ("waivers") from pre-trade transparency requirements and for delayed post-trade publication.

Under the current proposal for the European Market Infrastructure Regulation (EMIR), which concerns OTC derivatives, ESMA is likely to be given direct supervisory responsibilities on trade repositories (TR) and will be entitled to access or otherwise require information held at TRs, notably on OTC derivatives trades and counterparties for all OTC derivatives (i.e., both eligible and non-eligible for central counterparty (CCP) clearing).[40]

3. *ESMA publishes lists of regulated markets, Multilateral Trading Facilities (MTFs), systematic internalizers and central counterparties on its website.*

   The databases of regulated markets, MTFs and central counterparties allow the identification of the counterparty from each transaction by the means of codes. Each competent authority has to ensure the maintenance and publication of a list of all systematic internalizers, in respect of shares admitted to trading on a regulated market, which it has authorized as investment firms, and review the list at least annually. The list published by ESMA represents the consolidation of national lists communicated by national competent authorities.

   Furthermore, statistics and data will be collected on clearing and settlement activities through trade repositories.[41]

4. *Concerning financial institutions*, ESMA has created a central repository (CEREP) for information on credit rating agencies, which is a reporting system that produces publicly available activity and performance statistics. It contains one-year statistics on ratings activity (for the beginning and the end of the

[38] International Securities Identification Number.

[39] A "systematic internalizer" is an investment firm which, on an organized, frequent and systematic basis, deals on own account by executing client orders outside a regulated market or an multilateral trading facility.

[40] There exists currently a proposal by the European Commission on the Regulation on short selling and certain aspects of credit default swaps (CDS) covering quarterly reporting of net short positions relating to the issued share capital of companies and to the issues sovereign debt. It is envisaged to collect the following data and statistics on net short positions at a lower threshold (0.2%) and for each incremental threshold crossed (0.1%) for supervisory purposes, and at a higher threshold (0.5%) and for each incremental threshold crossed (0.1%) for market disclosure. For sovereign debt securities, the Regulation envisages the collection of net short positions without market disclosure. In addition, when the prohibition of entering into an uncovered sovereign CDS position on a sovereign issuer is temporary lifted by a Competent Authority, a notification to that authority is required above a certain threshold. ESMA would receive summarized data on net short positions in shares and in sovereign debt, as well as, when relevant, on uncovered CDS position from Competent Authorities.

[41] Alternative Investment Funds Managers Directive (AIFMD) and European Market Infrastructure Regulation (EMIR), see Section 13.11 for details.

reporting period, number of ratings, upgrades and downgrades, defaults, withdrawals, etc.), one-year default rates, and one-year rating transition matrices. Short-term and long-term ratings on issuers and issues for at least ten years will be available for three categories of ratings: (a) corporate, (b) structured products (issue basis, except for Special Investment Vehicles; long-term ratings, except for Asset Backed Commercial Paper; six reporting classes: Asset Backed Securities, Residential Mortgage Backed Securities, Commercial Mortgage Backed Securities, Credit Default Obligations, Asset Backed Commercial Paper, and Other) and (c) sovereigns and public finance.[42] The frequency of the data collection will be higher if used for supervisory purposes. Depending on the effective use and interest in CEREP, a number of enhancements could be envisaged in future releases of the system, including, for example, a more granular breakdown of the sector of activity of the issuer or the inclusion of additional information.

The Alternative Investment Funds Managers Directive (AIFMD) requests ESMA to put in place an electronic central public register identifying each alternative investment fund manager (AIFM) authorized under the Directive, a list of the alternative investment funds (AIF) managed and/or marketed in the European Union by such AIFMs and the competent authority for each such AIFM. There are a number of reporting obligations on the principal markets and instruments in which an AIFM trades, etc. At this stage, it is not possible to give any detail on subcategories of data (granularity, historical series, etc).

ESMA collects and publishes information for the identification of Central Counterparties (CCPs) in Europe, so that counterparties can be identified in transaction reporting, and on the corresponding Competent Authorities (CAs) in Europe. Additional data will be collected (a) on alternative investment funds managers and (b) from OTC trade repositories on the basis of new legislation, which is currently being established.[43] In particular, ESMA will release a central public register identifying all alternative investment fund managers authorized under the Directive, and a list of the alternative investment funds managed and/or marketed in the European Union. At the time of writing, details on subcategories of data requirements (granularity, historical series, etc.) are under development.

Furthermore, new data requirements are being developed for OTC trades, short selling, clearing and settlement activities, alternative investment fund managers, and trade repositories.

At a general level, ESMA is obliged to publish and regularly update this information on its website.

[42] The collection of data started as of June 2011.

[43] Directive 2011/61/EU of 8 June 2011 on Alternative Investment Fund Managers; The European Market Infrastructure Regulation (EMIR) has not yet been completely finalized.

13.6 Main data challenges

The statistical needs to fulfil the European financial supervisory architecture are without historical precedence and will create a wealthy source of financial statistics and risk information. However, significant challenges need to be overcome to achieve such standardized, rich, and efficient centralized databases of financial information

A key prerequisite and common feature for all European authorities is the ability to develop ways to collect and share relevant statistics systematically and to determine the granularity of data needed within the network of European and national authorities. This is particularly necessary for cross-border and cross-functional responsibilities.[44] The crisis has clearly demonstrated that barriers to sharing and exchanging statistics within Europe must not impede improvements in the analysis and assessment of systemic risk. Therefore the European Supervisory Authorities (ESAs) and the ECB with their respective networks of national supervisors and central banks must converge towards common statistical frameworks and data standards, which can be used for micro-prudential and macroprudential supervision in Europe.

The process of managing systemic risks starts with obtaining high quality statistics that are comparable across national jurisdictions as presented in Figure 13.7.

There is a clear link between providing high quality statistics and making good quality decisions. Good and factual statistics[45] cannot be taken for granted[46] and to collect such data requires the backing from the statistical community, national authorities and policy makers to achieve the goal of enhancing the statistical framework required for safeguarding borderless European prudential supervision.

Going forward it remains to be seen how the decision makers of the European and national financial supervisory authorities will be able to take their responsibility beyond territorial or sectoral mandate and converge towards the required

[44] In particular there is a need not to overburden market actors with supplementary reporting requirements and administrative costs. The concept of common data standards and applying the principle that data should only be collected once by national authorities and shared among authorities becomes of paramount importance as new reporting requirements emerge.

[45] See for instance, ECB statistics quality framework, ECB, April 2008 (`www.ecb.europa.eu/stats/html/sqf.en.html`) and other similar international statistical quality frameworks (Fundamental Principles of Official Statistics of the United Nations Statistical Commission `http://unstats.un.org/unsd/methods/statorg/FP-English.htm`; the Principles Governing International Statistical Activities `http://unstats.un.org/unsd/methods/statorg/Principles_stat_activities/principles_stat_activities.htm`; The European Statistics Code of Practice `http://epp.eurostat.ec.europa.eu/portal/page/portal/eurostat/home`; and the IMF Data Quality Assessment Framework `www.dsbb.imf.org/Pages/DQRS/DQAF.aspx`).

[46] Schubert and Nymand-Andersen (2013). The role of statistics in generating trust within societies – necessary preconditions.

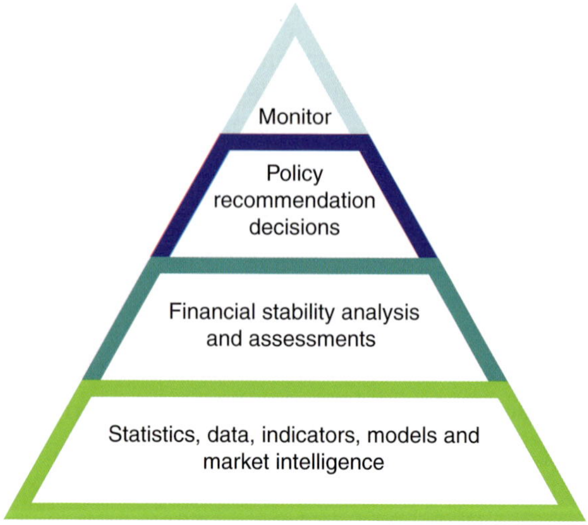

Figure 13.7 The fundamental building block of systemic risk measurement is harmonized, comparable statistics using common data standards.

harmonized data standards on individual institutions, issuers, financial instruments and risk data information.

Five building blocks of standardization are necessary for collecting and sharing relevant European statistics and risk data among European and national authorities:

1. a common conceptual understanding of financial instruments and their specific data characteristics;
2. a unique identification of issuers such as financial institutions;
3. the identification of instruments and their structures;
4. a standardized data model and data structure;
5. a mechanism to transmit and exchange statistics and data.

13.6.1 A common conceptual understanding of financial markets and specific characteristics

It is vital that the national and European authorities have a common conceptual understanding of financial markets, financial instruments, entities and the corresponding underlying data characteristics. This is particularly important when financial markets structures and dynamics differ across national territories affecting (the individual and aggregated reporting of) balance sheet compositions and accounting practices. Communicating in the "same language" when referring to financial markets, entities, instruments and data is a pre-condition for the successful exchange of data and statistics. It is therefore essential that definitions are harmonized[47] across jurisdictions. This will significantly simplify the reconciliation of reporting frameworks across national Member States and contribute to the convergence to one common reporting standard. So far, one key lesson learned from the financial crisis is that financial intermediaries and authorities had considerable difficulties in comparing volumes and levels of statistics and data from different data sources.

13.6.2 Identification of institutions

Identifiers of financial institutions and other issues have been a key topic for the financial industry and public authorities for decades. Several entity identification systems have been developed and applied within the financial market by standards organizations,[48] national and international authorities[49] and by specific industry organizations and by multiple competing commercial firms.

The challenge is for the public authorities and the financial industry and standards organizations to converge by consensus towards a common centralized scheme, thereby avoiding legal obligations. Current classifications and identifiers schemes are not unique and compatible among each other, with different update frequencies and governance structures and the lack of ability to capture institutional relationships and connections among entities globally. A global initiative to create one legal entity identifiers (LEI) scheme for entities operating in the financial sector was launched by the Financial Stability Board in 2011.[50]

13.6.3 Identification of instruments

As with the identification of entities, standards organizations, public authorities and the financial industry have discussed and developed identification schemes for financial instruments over decades. For granular reporting of issuances, transactions and holdings of assets and financial instruments, a common identifier is a necessity. The ISIN system[51] is the most comprehensive and widespread identification network within the financial industry today, covering more than 110 countries.[52]

[47] The difficulties in reaching consensus at the international level have significant implications for the value chain of producing and exchanging data and statistics among multiple institutions as complex conversion tables among various local concepts need to be maintained among authorities, and applied within IT systems, including market participants and data vendors.

[48] Such as work of ISO/TC68.

[49] Company registration schemes such as EU Community VAT systems, European Business Registers (EBR) or sector specific registers such as Monetary Financial Institutions and investment funds and financial vehicle corporations in the EU.

[50] `www.financialstabilityboard.org/list/fsb_publications/tid_156/index.htm`

[51] An International Securities Identification Number (ISIN) uniquely identifies a security (ISO6166), ISO/TC68.

[52] A full list of the national numbering agencies that are members of the Association of National Numbering Agencies can be viewed at `www.anna-web.com`.

Common for all standards is the necessity to be available "free of charge" and to be released and maintained by competition-neutral organizations with no commercial interest. Furthermore, the coverage of the standardization scheme must include other financial instruments, derivatives, loans and credit claims.

13.6.4 A standardized logical data model and data structure

As mentioned in Section 13.6.1, financial markets and instruments can be complex and mapping financial characteristics can be very complicated, as the concepts are used inconsistently throughout the financial industry and across markets, and also because there are hundreds of characteristics (data attributes)[53] associated with financial instruments. Harmonized definitions are necessary and justified, for example in describing accounting, risk information and transactions data and a relational data model and structure should be developed and applied. The lack of a standardized data model, and therefore precision, makes it almost impossible to automate processes and makes it very difficult to compare and analyze data across financial centers and to set rules at an attribute level. Data models also facilitate work on creating links between different reporting frameworks to find synergies for reducing the reporting burden. A logical data model and structure is essential for building prudent databases. An agreed and predefined data structure makes it possible to exchange and manage large amounts of statistics and data between entities.

13.6.5 Transmission and exchange of statistics and data

The ability to manage large volumes of financial market information in a highly automated way from multiple sources is necessary for European and national authorities. Transmission mechanisms and IT infrastructures vary significantly and all stakeholders are required to juggle multiple systems with their associated costs, lack of productivity and loss of timeliness and efficiencies. Also in this field, significant challenges remain in increasing the compatibility of communication facilities to ensure a safe, smooth, and timely communication between the involved parties.

13.7 Other data challenges

There are other vital data challenges. While the ECB statistics collection for the euro area is well developed and comparable across the euro area Member States, a key challenge is to ensure a complete EU-wide coverage, which requires additional

[53] Such as types of pricing, labeling of various corporate actions and interest rate calculation methods.

reporting efforts by the corresponding national authorities to implement and to comply with the established conceptual and methodological framework.

Other data challenges include legal issues related to confidentiality and legal barriers, especially concerning the provision and sharing of micro-level data among the ESAs, the ESRB and the ECB. The sharing of micro-level data is vital because the same statistics can be used for various analytical purposes, thereby avoiding duplication of statistical data collection and the associated administrative burden and costs on the corresponding authorities and the financial intermediaries, which may otherwise be forced to report similar information multiple times. The crisis has clearly demonstrated that data confidentiality must not stand in the way of improvements in systemic risk analysis and assessment.

13.8 Challenges of sharing data

At the core of the financial crisis, analytical and informational gaps limited the ability of authorities to identify the build-up of risks within the financial system, globally. Regulators, supervisors and central banks must now continue to develop ways to collect, share and analyze data systematically from the European financial system. The statistical challenge is to establish a comprehensive and relevant information source shared by, and useful for, the ESRB and the three ESAs and the ECB. Macroeconomic statistics are already available either via the ECB and Eurostat[54] or via other statistical institutes. However, the fundamental challenge relates to sharing micro-level statistics and risk information and the ability to agree on the use and sharing of statistics related to financial markets, instruments and intermediaries. This is particularly important in relation to cross-border financial positions, flows and exposures (for instance of holding of securities or other financial instruments).

One step in this direction of information sharing could be to implement a shared European Union data source starting with a limited set of core basic data for the identification of (i) institutions[55] and their ownership structure (who owns whom) and (ii) financial instruments and their structures (which financial instruments are in the market and who is guaranteeing these), which corresponds to the second and third fundamental building blocks for standardization as presented in Figure 13.8. One lesson learned from the financial crisis is the need to overcome complicated, manual (therefore error-prone) and time-consuming mapping needed to understand and obtain an overview of, for example, the European banks' (the banking sector's) exposure to a handful of large counterparties.

[54] Eurostat is the statistical office of the European Community.
[55] Ideally this should cover all institutions irrespective of whether they issue financial instruments or not.

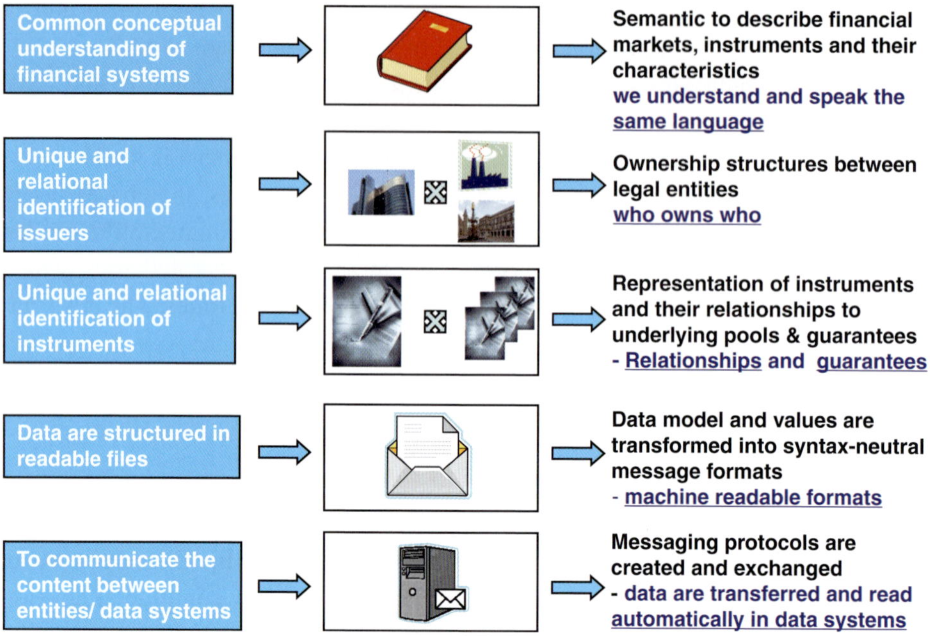

Figure 13.8 Five fundamental building blocks for sharing statistics among European and national authorities.

A few core common identifiers will allow for the swift coordination and exchange of information between the national supervisory authorities and within the new European System of Financial Supervision including the ECB – while ensuring that they can independently merge and link their own reference data and risk information, where applicable. These common identifiers will also allow for swift interaction and linkages between, on the one hand, financial institutions and financial instruments and, on the other hand, the actual activities on regulated markets. This shared data source could pave the way for timely analysis and assessment of risk exposures both at the individual financial intermediary level and at financial sector level, as part of the sharing of vital European statistics and risk information among the new European supervisory landscape. The general principle is illustrated in Figure 13.9. This European data source will protect the individual level data and statistics according to any applicable confidentiality regime, including providing sub-aggregation data marts for groups of regulators at micro- and macro-level. Here again, the limitations are more of a political nature than of database capacities and structures.

In fact, a good starting point for establishing such a European data source, and considering the involved responsible supervisory authorities,[56] is to obtain this

[56] ESMA and the respective national supervisory authority.

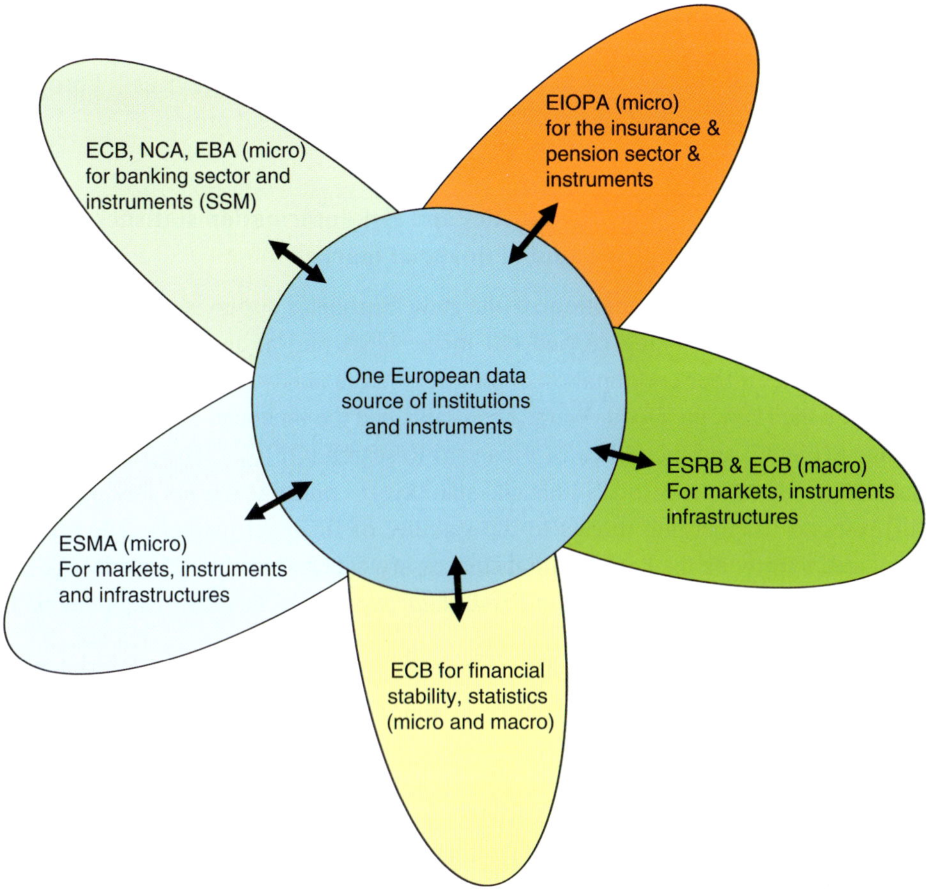

Figure 13.9 One shared data source for European Financial prudential supervision: who owns whom and who issues and guarantees what. ECB = European Central Bank; NCA = National Competent Authority; EBA = European Banking Authority; ESRB = European Systemic Risk Board; ESMA = European Securities and Markets Authorities; EIOPA = European Insurance and Occupational Pensions Authorities; SSM = Single Supervisory Mechanisms.

information systematically for new issues as part of the Prospectus Directive[57] and have this information maintained by the issuer and submitted to the supervisory authorities as required within the Transparency Directive.[58] What is needed is a mandatory requirement by the supervisory authorities for issuers to send this information in a standardized and electronic format for machine processing. This would provide a basis for quickly obtaining a large known-to-all pool of European securities[59] (in terms of total amount outstanding).

[57] Prospectus Directive (Directive 2010/73/EU). [58] Transparency Directive (Directive 2004/109/EC).

[59] As it progresses, the coverage could be extended. The focus of included already issued securities is time and resource intensive.

Other market and public stakeholders could benefit from this shared facility for similar and other important purposes, including the financial institutions themselves, investors or statistical offices, national public services, European Commission Services, and other interested organizations and groups.

13.9 Fitting financial statistics and risk information statistics in a global financial market

In parallel with the construction of the new European supervisory architecture, similar activities are being carried out in the USA and at an international level, which require further coordination among regulators and overseers.

Within the USA, the Dodd–Frank Act established the Financial Stability Oversight Council (FSOC) and the Office of Financial Research (OFR).[60] The FSOC is mandated to (i) identify risks to US financial stability, (ii) promote market discipline, and (iii) respond to emerging threats to the stability of financial markets. The OFR is mandated to provide the FSOC with the necessary data and analyses for this task and *will face very similar statistics and risk data challenges to those encountered in Europe.*

For instance, the OFR includes a Data Center which facilitates information sharing and coordination among US Federal and State agencies to obtain an overarching picture of the US financial system. The OFR Data Center will need to (i) collect and (ii) combine macro-level and micro-level statistics and risk data from multiple US agencies, commercial data vendors, financial intermediaries, and furthermore (iii) publish and maintain a US financial company and financial instrument reference database, based on formats and standards for reporting financial transaction (flows) and position (stocks) of financial instruments traded in the USA. *The mandate for the OFR is very similar to and as complex a statistical challenge as the one faced in Europe.*

At the international level, the Group of Twenty (G20) has been mandated to discuss measures to, among others, promote the financial stability of the world. The scope of financial regulation has been broadened and prudential regulation and supervision are being strengthened together with closer political cooperation and global governance with the International Monetary Fund and the World Bank.

The main priority actions, as agreed at the Seoul Summit,[61] aim to reduce excessive global imbalances and to maintain current account imbalances at sustainable levels. For this purpose, a set of comparable statistical indicators has been agreed on measuring (i) public debt and fiscal deficits, (ii) private savings rate and private

[60] The Dodd–Frank Wall Street Reform and Consumer Protection Act (Pub.L. 111–203, H.R. 4173).
[61] The G20 Seoul summit, November 11–12, 2010, www.G20.org.

debt, and (iii) the external imbalance composed of the trade balance and net investment income flows and transfers, taking due consideration of exchange rate, fiscal, monetary and other policies. Methodological guidance and timetables are being established to assess these indicators as part of the action plan. The G20 has also included a work program on the International Monetary System (IMS) in order to improve and ensure systemic stability, promote orderly adjustment of financial markets and to avoid disruptive fluctuations in capital flows, disorderly movements and misalignment in exchange rates. The G20 is focusing on measures relating to excessive commodity price volatility, in particular to oil, gas and coal, which require comparable data for establishing the causes and effects of price volatilities.

Furthermore, reforms of the financial sector are underway as part of implementing the Basel III rules on new standards for banks to ensure commitment for more effective oversight and supervision, including regular stress testing of banks building on the Basel Committee's principles. The Financial Stability Board (FSB)[62] is providing recommendations on over-the-counter (OTC) derivatives and on reducing the reliance on ratings provided by credit rating agencies. The FSB is also focusing on providing recommendations and an operational framework for monitoring what is called systemically important financial institutions (SIFI) requiring intensive supervisory oversight and effective resolution mechanism cross border and cross economic areas.

Based on reports from the Bank for International Settlements (BIS), the IMF, the FSB and the World Bank, with input from national authorities, further work is underway to extend the macroprudential framework so that it covers financial stability issues in emerging market and developing economies. Simultaneously, there are new recommendations (from the FSB) on regulation and oversight of the shadow banking system to efficiently address the risks, including arbitrage, associated with shadow banking and its interactions with the regulated banking system.

13.9.1 Looking forward

These new frameworks, initiatives and cooperations at a global level will continue to challenge the current structures and practices of collecting statistics and data at national and international level, not only in obtaining comparable and timely data and statistics from different financial structures and practices in various global regions using enhanced regulations and oversight agreements, but also in covering new types of statistics from mainly non-regulated areas such as the shadow banking

[62] The Financial Stability Board (FSB) was established to address vulnerabilities and to develop and implement strong regulatory, supervisory and other policies in the interest of financial stability. The BIS is hosting the FSB secretariat in Basel, Switzerland, `www.financialstabilityboard.org`

system. It is extremely difficult for regulators to provide high quality information on entities outside their regulatory perimeter, therefore the first challenge is the ability to collect basic statistics from these non-regulated entities and to provide meaningful and comparable statistics across borders. Without fundamental comparable statistics, regulators and associated authorities will have difficulties in identifying and assessing the risks associated with these entities and their activities at a regional and global level, thus delaying corrective measures (where necessary).

One of the instruments in improving quality and consistency of statistical data at the European level is the introduction of the set of uniformly applied technical standards, which ESAs will be developing as a part of their mandate (Single Rulebook). Notably, in relation to statistical data, harmonization through the directly applicable EU-wide technical standards will increase consistency of data reported by financial institutions to their supervisors, consequently leading to the creation of the consistent datasets which would improve the quality of risk assessment and analysis performed by the ECB and national supervisors, and ultimately ESAs. The Single Rulebook will also lead to an improved quality and consistency of financial and prudential information publicly disclosed by institutions, as part of their annual disclosures, thus ensuring improvements in the effectiveness of so-called market discipline.

Furthermore, in November 2011, the ESRB Steering Committee approved the creation of a Contact Group on Data Issues dealing with issues related to information exchange between the ESRB (and the ECB in its supporting role of the ESRB) and EBA, EIOPA and ESMA. The Contact Group serves as a network to facilitate further discussion and cooperation in the field of information requirements for the ESRB's macroprudential oversight, and will in particular work on medium-term data requirements (2013 and beyond) and contribute to ensuring that information is made available to the ESRB and the ESAs/ECB with adequate timeliness, consistency and harmonization.

13.9.2 Cooperation across borders

Against this background, data standardization needs to be given high priority by supervisory agents, policy makers and statisticians as a lack of it will constitute the most fundamental barrier for understanding, validating and exchanging core financial statistics and data from global financial intermediaries operating on regional and international markets. In the absence of global coordination, separate European and US solutions will be suboptimal. The new US and European authorities need to take a broad view and work closely together at international level to achieve convergence of data standards and make these mandatory for the collection of statistics and risk information from national, regional and international intermediaries, markets and infrastructures operating within the US and the European financial markets. This will facilitate prudent supervisory oversight and the creation of an effective resolution mechanism for systemically important financial institutions which operate across economic areas.

Therefore, leading agencies within various financial jurisdictions world-wide must introduce coherent and comparable standards at all levels – starting with the fundamental building block of statistical data and risk information. Initial work has started in this direction by the Inter-Agency Group[63] as part of implementing the 20 recommendations in the report *The Financial Crisis and Information Gaps* endorsed by the Group of Twenty (G20) Finance Ministers and Central Bank Governors.

13.10 Conclusion

Europe's Economic and Monetary Union (EMU) is a unique construction with no other historical example. On January 1, 1999, the single monetary policy for the euro area as a whole was implemented by the ECB and the participating national central banks, whereas economic policy was largely left at the national level.

The lessons learnt from the impact of the financial crisis have demonstrated the need to safeguard both the microprudential and macroprudential supervision of the European financial system and, at the same time, to move forward with further coordination of economic policies at the European level to ensure that sufficient coordinated rules and actions with an appropriate response time for mitigating risks are in place and to guarantee that sustainable economic policies are in line with a prudent Economic and Monetary Union.

The crisis has demonstrated that at least two areas of the hitherto EMU need strengthening.

1. First, the dichotomy between, on the one hand, a common European monetary policy (one currency and one system of central banks) and, on the other hand, a system of banking regulation, supervision, safety nets and resolutions organised along national jurisdictions, is too complex a coordination of divergence structures and rules for reacting consistently, prudently and in a timely manner to financial challenges in a stress and crisis mode across national jurisdictions in Europe.

   The crises have shown that despite a regulatory and supervisory framework at national level, the system has failed to identify cross-border interdependencies, the

[63] The Inter-Agency Group on Economic and Financial Statistics (IAG), comprises the Bank for International Settlements (BIS), the European Central Bank (ECB), Eurostat, the International Monetary Fund (IMF, Chair), the Organization for Economic Cooperation and Development (OECD), the United Nations (UN), and the World Bank (WB). The IAG was established in 2008 to coordinate statistical issues and data gaps highlighted by the global crisis and to strengthen data collection.
See the Principal Global Indicators website at www.principalglobalindicators.org/default.aspx

magnitude of spill-overs and contagions in a borderless European financial market and inadequate crisis management mechanisms for banks active in several national jurisdictions.
2. Second, high fiscal deficits and debt levels have challenged the confidence of some sovereign Member States' ability to honour their debt obligations and their ability to support domestic banks in need. Hence, national fiscal instabilities, including unsustainably high debt levels, have been amplified by public funding to banks in need, resulting in a continuous decline in trust by financial markets in the ability of certain Member States to honour their obligations, with the consequence of higher funding costs for both domestic banks and related sovereigns.

Both of these two areas are being reflected upon and are being proposed, inter alia, as part of the new European landscape, although this chapter focuses primarily on the first area and the associated new data and statistical challenges, whilst acknowledging that significant progress is already being made within the second area as part of the integrated budgetary framework for strengthening the Stability and Growth Pact, encompassing both a preventive and corrective leg for prudent economic and fiscal policies.[64]

A European financial supervisory architecture is in the making, establishing new European authorities for safeguarding the financial system. Europe is on the right path towards a banking union building upon a macroprudential and microprudential pillar. The microprudential pillar consists of the European Central bank (ECB) and the National Competent Authorities (NCA) and three new European supervisory authorities; one for banking (EBA), another for securities and markets (ESMA) and a third for Insurance and pension funds (EIOPA). The macroprudential pillar consists of the ECB and the new European Systemic Risk Board (ESRB), which is hosted at the ECB. The transfer of supervisory powers to supranational level is necessary as part of the proposed Single Supervisory Mechanism (SSM) for supporting the Economic and Monetary Union (EMU).

The banking union will consist of the following main building blocks:

1. a Single Supervisory Mechanism at the ECB and the implementation of a Single Rulebook for European supervision;
2. a Single Resolution Mechanism and the possibility of direct capitalisation of banks by European funds and the European Stability Mechanism;
3. a European deposit protection and guarantee scheme.

[64] Including the so-called "six pack," "two pack," "fiscal compact" and the Treaty on Stability, Coordination and Governance in the Economic and Monetary Union.

This new structure of European surveillance and supervision will be the European "control towers" and "radar screens" to detect risks, provide corrective measures and timely actions for safeguarding the well-functioning of the European financial system.

As part of implementing the new and ongoing European financial supervisory architecture, there is a clear call for a substantial review of the type of statistics, data and risk information required. New statistical frameworks are being established in order to provide the statistics and risk information able to support an overarching analysis of risks to the European financial system.

Looking ahead, the key statistical challenges are manifold with no time for complacency. The statistics and risk data needs of the new European financial supervisory architecture will constitute a rich source of macro-level and micro-level financial information. Systematically collecting and sharing relevant statistics and analysis within the network of European and national authorities will allow, in particular:

1. the collection of a complete set of statistics and risk information covering micro-level statistics and data from all 30 members of the European Economic Area (which includes the 28 European Union Member States[65]);
2. the synchronization of the various statistical initiatives among the new European supervisory authorities;
3. the coordination and collection of statistics and risk information across various jurisdictions between the ECB, NCB, the three new ESAs, ESRB and new European supervisory authorities and the associated respective network of national competent authorities;
4. the sharing of statistics and risk information among the European System whilst ensuring confidentiality of statistics and risk information. The crisis has clearly demonstrated that barriers to sharing and exchanging statistics within Europe must not stand in the way of improvements to systemic risk analysis and assessment by policy-makers, which also ensures efficient allocation of tasks between microprudential and macroprudential supervisors;
5. the gathering of good quality and comparable statistics across national jurisdictions. This enables convergence in the collection and sharing of micro-level and micro-aggregation level statistics and risk information among European and national supervisory authorities;
6. the development of micro-aggregation methods which can be applied to micro-level statistics to create useful (non-confidential) sector and other sub-aggregation

[65] ESRB scope relates to the concept of EU 28.

level statistics, by minimizing the loss of information content while preserving the characteristic of the statistical distributions;

7. the convergence above territorial or sectoral barriers among the members of the new financial supervisory architecture in order to share micro-level and macro-level statistics.

Considering the number of Member States within the EU and EEA, the network of European and national authorities may reach close to 40–50 authorities. Such a large network calls for new thinking and approaches with regard to contributing to and extracting relevant information in a timely manner from a single database for the respective purposes of the multi-fold authorities.

There is a strong need to centralize and create one European supervisory micro-level data source of institutions and instruments and associated data, including micro-aggregated data sets which are accessible and tailored to the respective participating authorities, whether national or European. Confidentially is preserved at the micro level in the same way as it is today, whereby one national authority alone will have access to its own micro-level data and to its own tailored micro-aggregated data set.

The implementation phase focuses on working towards fulfilling the short-term needs in order to overcome immediate statistical data gaps, while taking a more structured medium-term to long-term view of standardization, collection, handling and exchanging statistics and risk information. This will contribute to building up the necessary trust and cooperation among the new network of European supervisory authorities and the corresponding network of national supervisors in EEA countries operating in different jurisdictions, financial markets structures, supervisory practices and cultures.

Furthermore, these developments must be viewed in light of the ongoing and supplementary regulatory initiatives to improve stability and governance measures in Europe which will contribute to supplement the rich pool of statistics and risk data requirements.[66]

Furthermore, the European initiatives should not be seen in isolation, as similar activities are being carried out in the USA and at an international level. This requires further coordination among statisticians, regulators and overseers who are faced with similar statistical and data challenges. Within the USA, the Financial Stability Oversight Council (FSOC) and the Office of Financial Research (OFR) have been established by the Dodd–Frank Act as part of identifying risks and mitigating these risks in the US financial system. The US set-up will likewise be faced with similar statistics and risk data challenges as in the European landscape – through their network of various Governmental, State and local agencies. At the international level, the G20 is

[66] See Section 13.11 for a listing of these on-going initiatives.

discussing measures to promote global financial stability and extend the macroprudential framework and financial stability issues in emerging market and developing economies as well as how to efficiently address the risks and arbitrage between financial marketplaces and less regulated but important actors. Both the US and European authorities need to take a broad view and work closely together at the international level in order to converge towards high data standards and make these a mandatory part of collecting statistics and risk information from national, regional and international financial intermediaries. This is a necessity for proper supervisory oversight and effective resolution mechanisms for systemically important financial institutions which operate across economic areas. Leading financial jurisdictions world-wide must introduce compatible and coherent standards that apply at all levels – starting with the fundamental building block of statistical data needs and risk information.

The collection of high quality and standardized statistics beyond sector and national boundaries, in combination with the prudent interplay between a macroprudential and microprudential perspective will allow the members of the new European supervisory architecture to proactively supervise individual financial institutions, financial markets and infrastructures across Europe, while obtaining an overview of the risks to the European financial system and its possible impact on the real economy. This all begins with the harmonization of the conceptual frameworks and the collection of high quality and standardized statistics beyond sector and national boundaries. The road for a banking union has started and will facilitate further economic integration to complete the European Economic and Monetary Union.

The need for quality statistics and data cannot be taken for granted as part of contributing to sound policy decisions and safeguarding our societies.

13.11 Appendix: Overview of on-going initiatives to improve stability and governance measures in Europe requiring financial data and risk information

- The stability of financial institutions is ensured via new rules on capital requirements for banks, investment firms and insurance companies. For banks and investment firms, the **Capital Requirements Directive** (CRD) has been strengthened to reinforce capital rules for the trading book and for complex derivatives. This directive also covers new rules governing **remuneration and bonuses** and reduces incentives for short-term risk taking for enhanced risk management purposes. The CRD implements the **Basel III agreement** to increase the levels of capital which banks and investment firms must hold to cover their risk-weighted assets.
- For insurance corporations the **Solvency II Directive** focuses on more detailed, stringent and risk-sensitive capital rules.

- New legislative proposals on **Corporate governance in financial institutions** are being drafted to improve risk control and oversight as well as enhance supervisory review of risk governance in financial institutions.
- New **Regulation on credit rating agencies** is established to introduce strict authorization requirements and supervision of credit agencies. Additional measures are considered to modify the business model of credit agencies for separating the payments for rating from the allocation of rating activities, the investors and issuers' over-reliance on ratings and to ensure sufficient competition in the rating sector.
- There is a new **Directive on managers of alternative investment funds, including hedge funds (AIFM Directive)** for collecting comprehensive data from market players, such as hedge funds, for supervisory purposes.
- **Accounting standards are being improved** to obtain more accurate information of financial institutions and enhanced disclosure requirements of financial instruments risks have been published by the **International Accounting Standards Board** (IASB), including efforts to obtain disclosure requirements of off-balance-sheet exposures.
- Measures are under way for regulating **auditing agencies**, for enhancing early health warnings of auditing companies.
- The **Markets in Financial Instruments Directive (MiFID)** will be extended to cover new types of trading platform and financial products and enhanced transparency requirements.
- The **Market Abuse Directive (MAD)** will be enhanced for more effective prevention, detection and sanctioning of market abuses of EU Regulations and Directives.
- There ia a new **Regulation for Over-The-Counter (OTC) standardized derivatives markets** for registration of derivatives and to reduce risks.
- There is a new **Regulation on short selling and Credit Default Swaps (CDS)** with the objective increasing transparency of short selling of shares, sovereign debt and related CDSs.
- There is a new **Securities Law Directive (SLD)**.
- **There is legislation on Central Securities Depositories** securing post-trading activities.
- **Considerations** are underway to improve the integrity, oversight, functioning and transparency of commodity trading and associated derivatives.
- **Consumer protection** directives such as **Deposit Guarantee Schemes (DGS)** and **Investor Compensation Schemes (ICS)** are enhanced and an **Insurance Guarantee Schemes (IGS)** is being created.
- The **fair practices relating to mortgage credits to consumers** are improved, ensuring that all mortgage lenders and intermediaries are properly regulated and

supervized and that there is appropriate information on risk and costs for **packaged retail investment products (PRIPs)** and new governance rules on those selling the products.

- The **Units for Collective Investment in Transferable Securities Directive** are being strengthened to protect against fraud or loss of assets.
- The Bank Recovery and Resolution Directive.
- The Deposit Guarantee Directive.
- "Six-Pack" – The 2011 reforms of the Stability and Growth Pact addressed gaps and weaknesses in the framework identified during the financial crisis. These reforms significantly strengthened both the fiscal surveillance and enforcement provisions by adding an expenditure benchmark to review countries' fiscal positions, operationalizing the Treaty's debt criterion, introducing an early and gradual system of financial sanctions for euro area Member States, and requiring new minimum standards for national budgetary frameworks. The 2011 reforms also brought the surveillance of both budgetary and economic policies together under the so-called European Semester, to ensure the consistency of the policy advice given.
- "Two-Pack" – The two-pack regulations will further enhance fiscal surveillance, coordination and enforcement of budgetary discipline.
- "Fiscal Compact" – The Fiscal Compact complements and enhances key provisions of the Stability and Growth Pact to strengthen fiscal discipline and introduce stricter surveillance within the euro area, in particular by establishing a balanced budget rule (the "Fiscal Compact"). This also includes a correction mechanism at national level including the nature, size and timeframe of any corrective actions needed. The Stability Treaty entered into force on January 1, 2013.

13.12 Appendix: Data collection from the insurance sector

- Number of enterprises – national enterprises, branches of non-EU/EEA countries, branches of EU/EEA countries, branches in non-EU/EEA countries, branches in EU/EEA countries – for life, non-life, composites and reinsurance enterprises.
- Gross premium written – national enterprises, branches of non-EU/EEA countries, branches of EU/EEA countries, branches in non-EU/EEA countries, branches in EU/EEA countries – for life, non-life, composites and reinsurance enterprises.
- Main items of gross and net technical provisions in non-life – gross/net direct premiums, gross/net claims, gross/net operating expenses, gross/net technical provisions – for accident and health; third party liabilities; marine aviation and transport; fire and other damage to property; general liability; credit and suretyship, and other non-life.

- Main items of gross and net technical provisions in life – gross/net direct premiums, gross/net claims, gross/net operating expenses, gross/net technical provisions – for non-linked life assurance being pension arrangements and non-linked life assurance not being pension arrangements, linked life assurance being pension arrangements, linked life assurance not being pension arrangements, capital redemption assurance, tontine assurance, group pension assurance.
- Main items of the gross/net technical account in reinsurance – gross/net direct premiums, gross/net claims, gross/net operating expenses, gross/net technical provisions – for life, non-life, composites and reinsurance companies.
- Investments of insurance enterprises – lands and buildings; investments in affiliated enterprises and participating interests (of which shares in affiliated undertakings, of which debt securities issued by, and loans to, affiliated undertakings, of which participating interests, of which debt securities issued by, and loans to, undertakings with which an insurance undertaking is linked by virtue of a participating interest); shares and other variable-yield securities and units in unit trusts; debt securities and other fixed income securities; participation in investment pools; loans guaranteed by mortgages; other loans; deposits with credit institutions and other financial investments; deposits with ceding enterprises; investments for the benefit of life-assurance policyholders who bear the investment risk, subscribed capital unpaid, intangible assets, debtors, prepayments and accrued income, other assets – for life, non-life, composites and reinsurance companies.
- Capital, reserves and liabilities – reserve; profit or loss brought forward; profit or loss for the financial year; total of capital and reserves; subordinated liabilities; total of capital, reserves and subordinated liabilities; other provisions; deposits received from reinsurers; creditors; accruals and deferred income; subscribed capital or equivalent funds; share premium account; revaluation reserve – for life, non-life, composites and reinsurance companies
- Result and solvency – profit or loss for the financial year, total balance sheet assets (%), available solvency margin, required solvency margin – for life, non-life, composites and reinsurance companies.
- Concentration ratio by premiums written in the domestic market for the three, five and ten largest companies – for life, non-life, composites and reinsurance companies.

13.12.1 Insurance data collection for the "fast track report"

- Gross premiums written (of which premiums related to investment contracts); profit or loss (after tax, before dividends); acquisition and administration costs related to insurance business; net claims incurred – differentiated by life and non-life.

- Asset composition – shares and other variable-yield securities and units in unit trusts (of which investments for the benefit of life-assurance policyholders who bear the investment risk); debt securities and other fixed income securities (of which investments for the benefit of life-assurance policyholders who bear the investment risk); real estate; capital and reserves – differentiated by life and non-life.
- Gross technical provisions (of which technical provisions for life-assurance policyholders who bear the investment risk); total balance sheet assets (of which investments for the benefit of life-assurance policyholders who bear the investment risk); available solvency margin; required solvency margin – differentiated by life and non-life.
- Information where premiums are written (non-EEA and EEA countries).

13.13 Appendix: Data collection from occupational pension funds

- Contributions and benefits of OPF – total gross contributions receivable (of which to: defined contribution schemes, of which to: defined benefit schemes, of which to: hybrid schemes); reinsurance contributions ceded; total net contributions receivable; total gross benefits payable; reinsurance benefits received; total net benefits payable; transfer of pension rights from other pension funds or insurance company (entrants); transfer of pension rights to other pension funds or insurance company (leavers); net cash flow.
- Profit and loss OPF – cash flow; cash flow in % of total assets; total assets.
- Technical provisions for pensions' liabilities – technical provision for pensions; market value of total assets; other liabilities; net assets covering the technical provision; average cover ratio (%).
- Investment assets of OPF – debt and other fixed income securities; equity and other variable-yield securities (excluding UCITS); UCITS (of which in debt securities, of which in equity securities, of which in real estate, of which in other assets); derivative instruments; loans; real estate investments; other investments; reinsured technical provisions; other assets; total assets; return on assets – for defined benefit schemes; defined contribution schemes; hybrid schemes.
- Size of the OPF sector – number of Institutions for occupational retirement provision; NUMBER of occupational retirement schemes; NUMBER of members – for defined contribution scheme; defined benefit scheme; hybrid scheme; and for active members; deferred members; retired persons.
- Concentration ratio – for the three, five and ten largest companies.

13.3.1 Additional data collection

- To complement the qualitative report, statistics are provided on a best effort basis using partial data, estimates and/or projections, relating to "rate of return on assets," "funding ratio" and "equity exposures" over time.

Acknowledgements

The views expressed in this article are those of the author(s) and do not necessarily reflect those of the ECB, EBA, ESRB, EIOPA, ESMA. The authors would like to thank Aurel Schubert, Werner Bier and Carsten Detken (ECB), Laura Vajanne, Antonio Sanchez, (ESRB), Christoph Kuhn, Meri Rimmanen, Oleg Shmeljov, and Wolfgang Strohbach (EBA), Frederico Alcântara de Melo, Alberto Garcia, David Nadry, Olga Petrenko, Fabrizio Planta, Sarah Raisin, Thierry Sessin-Caracci, and Eva-Christina Smeets (ESMA) for the useful comments provided.

Bibliography

Commission White Paper on Insurance Guarantee Schemes, COM, 2010, 370 The White Paper is part of a broader package on compensation and guarantee schemes that also includes two proposals for amendment of the Directive on Investor-Compensation Schemes and of the Directive on Deposit Guarantee Schemes.

CEBS, 2009, Guidelines on the implementation of the revised large exposures regime, Committee of European Banking Supervisors, December 11, 2009, `http://eba.europa.eu/documents/Publications/Standards---Guidelines/2009/Large-exposures_all/Guidelines-on-Large-exposures_connected-clients-an.aspx`

De Larosière Report, 2009, Report of the High-level Group on Financial Supervision in the EU, chaired by Jacques de Larosière, February 25, 2009.

Directive 94/19/EC of the European Parliament and of the Council of 30 May 1994 on deposit-guarantee schemes, OJ L 135, 31.5.1994. On 12 July 2010, the Commission adopted a legislative proposal for a thorough revision of the Directive on Deposit Guarantee Schemes: COM(2010)368 final.

Directive 97/9/EC of the European Parliament and of the Council of 3 March 1997 on investor-compensation schemes, OJ L 84, 26.3.1997, p. 22. On 12.7.2010, the Commission adopted a proposal to amend Directive 97/9/EC: COM(2010) 371.

Directive 2003/6/EC of the European Parliament and of the Council of 28 January 2003 on insider dealing and market manipulation (market abuse), as amended, OJ L 96, 12.4.2003, p. 16.

Directive 2004/39/EC of the European Parliament and of the Council of 21 April 2004 on markets in financial instruments amending Council Directives 85/611/EEC and 93/6/EEC and Directive 2000/12/EC of the European Parliament and of the Council and repealing Council Directive 93/22/EEC, as amended, OJ L 145, 30.4.2004, p. 1.

Directive 2004/109/EC of the European Parliament and of the Council of 15 December 2004 on the harmonisation of transparency requirements in relation to information about issuers whose securities are admitted to trading on a regulated market and amending Directive 2001/34/EC, as amended, OJ L 390, 31.12.2004, p. 38.

Directive 2006/48/EC of the European Parliament and of the Council of 14 June 2006 relating to the taking up and pursuit of the business of credit institutions (recast), as amended, OJ L 177, 30.6.2006, p. 1, and Directive 2066/49/EC of the European Parliament and of the Council of 14 June 2006 on the capital adequacy of investment firms and credit institutions (recast), as amended, OJ L 177, 30.6.2006, p. 201.

Directive 2006/49/EC of the European Parliament and of the Council of 14 June 2006 on the capital adequacy of investment firms and credit institutions (recast), as amended, OJ L 177, 30.6.2006, p. 201.

Directive 2009/65/EC of the European Parliament and of the Council of 13 July 2009 on the coordination of laws, regulations and administrative provisions relating to undertakings for collective investment in transferable securities (UCITS), OJ L 302, 17.11.2009, p. 32.

Directive 2010/73/EU of the European Parliament and of the Council of 24 November 2010 amending Directives 2003/71/EC on the prospectus to be published when securities are offered to the public or admitted to trading and 2004/109/EC on the harmonisation of transparency requirements in relation to information about issuers whose securities are admitted to trading on a regulated market, OJ L 327, 11.12.2010, p. 1.

Directive 2011/61/EU of the European Parliament and of the Council of 8 June 2011 on Alternative Investment Fund Managers and amending Directives 2003/41/EC and 2009/65/EC and Regulations (EC) No 1060/2009 and (EU) No 1095/2010, OJ L 174, 1.7.2011, p. 1.

Dodd–Frank Wall Street Reform and Consumer Protection Act, 2009, Dodd–Frank Wall Street Reform and Consumer Protection Act. H.R. 4173, December 2, 2009.

ECB Statistics Quality Framework, ECB, April 2008, `www.ecb.europa.eu/pub/pdf/other/ecbstatisticsqualityframework200804en.pdf?1de2c0f94c0dceafa571af23d3ff4dea`

EIOPA, Second European Insurance Stress Test, 2011, EIOPA European-wide Stress Test Report, July 4, 2011.

EIOPA, Financial Stability Report, 2011, First half-year report, June 14, 2011. `https://eiopa.europa.eu/publications/financial-stability/index.html`

European Banking Authority, 2011,. European Banking Authority 2011 EU-wide Stress Test Aggregate Report, July 15, 2011.

Eurostat, 2005,. The European Statistics Code of Practice for the National and COMMUNITY statistical Authorities, February 24, 2005, `http://epp.eurostat.ec.europa.eu/portal/page/portal/quality/documents/code_practice_EN_0.pdf`

Financial Crisis and Information Gaps. Report to the G-20 Finance Ministers and Central Bank Governors. Prepared by the IMF Staff and the FSB Secretariat, October 29, 2009 [endorsed at the G20 meeting November 2009].

Financial Stability Board (FSB), `www.financialstabilityboard.org`

Group of twenty (G20), 2010, Seoul summit, Leaders' Declaration, November 11–12, 2010, `www.G20.org`

IMF, 2006, International Monetary Fund, Data Quality Assessment Framework, August 2006, www.dsbb.imf.org/Pages/DQRS/DQAF.aspx.

International Organisation for Standardisation, 2006, Securities and Related Financial Instruments – International Securities Identification Numbering System (ISIN), ISO6166, July 27, 2006.

Legislation on Legal Certainty of Securities and Dispositions, Consultation Document of the Directorate-General Internal Market and Services, European Commission, 5 November 2010, DG Markt G2 MET/OT/acg D(2010) 768690.

Legislation on Central Securities Depositories: not yet finalized; on 13.1.11, the Commission launched a Public Consultation on Central Securities Depositories (CSDs) and on the harmonization of certain aspects of securities settlement in the European Union, DG Markt G2 D(201)8641.

Potential Financial Stability Issues Arising from Recent Trends in Exchange-Traded Funds (ETFs), Financial Stability Board, April 12, 2011.

Proposal for a Regulation of the European Parliament and of the Council on OTC Derivative Transactions, Central Counterparties and Trade Repositories, 2010/0250 (COD).

Proposal for a Council Regulation conferring specific tasks on the European Central Bank concerning policies relating to the prudential supervision of credit institutions, com/2012/OSII final.

Proposal for a Directive of the European Parliament and of the Council amending Directives 2003/71/EC and 2009/138/EC in respect of the powers of the European Insurance and Occupational Pensions Authority and the European Securities and Markets Authority, COM(2011) 0008 final.

Proposal for a Directive of the European Parliament and of the Council on the access to the activity of credit institutions and the prudential supervision of credit institutions and investment firms and amending Directive 2002/87/EC of the European Parliament and of the Council on the supplementary supervision of credit institutions, insurance undertakings and investment firms in a financial conglomerate, COM(2011) 453 final.

Proposal for a Regulation of the European Parliament and of the Council on Short Selling and certain aspects of Credit Default Swaps, 15.9.2010, COM(2010) 482 final.

Proposal for a Regulation of the European Parliament and of the Council amending Regulation (EU) No. 1093/2010 establishing a European Supervisory Authority (European Banking Authority) as regards its interaction with Council Regulation (EU) No . . . / . . . conferring specific tasks on the European Central Bank concerning policies relating to the prudential supervision of credit institutions. Dated 12.9.2012 (COM 2012) 512 final.

Regulation (EU) No 1092/2010 of the European Parliament and of the Council of 24 November 2010 on European Union macro-prudential oversight of the financial system and establishing a European Systemic Risk Board, OJ L 331, 15.12.2010, p. 1.

Regulation (EU) No 1093/2010 of the European Parliament and of the Council of 24 November 2010 establishing a European Supervisory Authority (European Banking Authority), amending Decision No 716/2009/EC and repealing Commission Decision 2009/78/EC, OJ L 331, 15.12.2010, p. 12.

Regulation (EU) No 1094/2010 of the European Parliament and of the Council of 24 November 2010 establishing a European Supervisory Authority (European Insurance and Occupational Pensions Authority), amending Decision No 716/2009/EC and repealing Commission Decision 2009/79/EC, OJ L 331, 15.12.2010, p.48.

Regulation (EU) No 1095/2010 of the European Parliament and of the Council of 24 November 2010 establishing a European Supervisory Authority (European Securities and Markets Authority), amending Decision No 716/2009/EC and repealing Commission Decision 2009/77/EC, as amended, L 331, 15.12.2010, p. 84.

Council Regulation (EU) No 1096/2010 of 17 November 2010 conferring specific tasks upon the European Central Bank concerning the functioning of the European Systemic Risk Board, OJ 331, 15.12.2010, p. 162.

Regulation (EU) No 513/2011 of the European Parliament and of the Council of 11 May 2011 amending Regulation (EC) No 1060/2009 on credit rating agencies, OJ L 145, 31.5.2011, p. 30.

Schubert, A. and Nymand-Andersen, P. 2013, The role of statistics in generating trust within societies – necessary preconditions, forthcoming article in the *Journal of the International Association of Official Statistics*, IOS Press.

Shadow Banking, 2011, Scoping the Issues. A Background Note of the Financial Stability Board April 12, 2011.

Trichet, J.-C., 2010, Opening speech at the fifth ECB conference on statistics, Central bank statistics: what did the financial crisis change? 19 October, Frankfurt, `www.ecb.int/press/key/date/2010/html/sp101019.en.html`

United Nations Statistics Division, 2005, Fundamental Principles of Official Statistics of the United Nations Statistical Commission, `http://unstats.un.org/unsd/methods/statorg/FP-English.htm`

United Nations Statistics Division, 2005, Principles Governing International Statistical Activities, September 14, 2005, `http://unstats.un.org/unsd/methods/statorg/Principles_stat_activities/principles_stat_activities.htm`

14

Data needed for macroprudential policymaking

Laura Kodres

14.1 Introduction and motivation

The recent crisis underscored that ensuring the safety and soundness of individual institutions is insufficient to guarantee that the financial system as a whole is stable and well functioning. This implies the need for a "macroprudential" approach to ensuring financial system stability. That is, by its nature, a view of the aggregate risks and vulnerabilities posed by financial institutions, through markets and through their bilateral interactions, is necessary to adequately understand and mitigate systemic risk.[1]

Although financial crises have occurred throughout history, there are several developments in recent decades, and specifically just prior to the latest crisis, that require a different – a broader – focus for regulation.[2]

- Interactions among individual financial institutions through markets have become faster (due to technological advances) and involve more and different types of institutions than ever before.
- Many over-the-counter markets and bilateral trades make the size and interconnections across institutions difficult to see and monitor, amplified by the growing length of complex intermediation chains.
- Repackaging of exposures in complex products and creation of other tailor-made instruments make risks difficult to assess and monitor.
- Lack of consistent criteria and homogenous reporting standards imply data sharing, even when it occurs, may be suboptimal, making it hard to "connect the dots" to see vulnerabilities to the financial system or spillovers to the real economy.

[1] The type of externality envisioned in a macroprudential approach is not a "pecuniary externality" that transfers the costs of an externality from one set of individuals to others, but a "real" externality in that individual institutions fail to appropriately internalize the costs of their actions on the system as a whole – the total amount of risk is larger than that transferred from one party to another as in the pecuniary case.

[2] Many of these issues are discussed in greater detail in Chapter 1 of this Handbook.

Handbook of Financial Data and Risk Information, ed. Margarita S. Brose, Mark D. Flood, Dilip Krishna and Bill Nichols. Published by Cambridge University Press. © Cambridge University Press 2014.

Even though these new developments have captured the attention of regulators and supervisors, as well as central bankers and treasuries, it is important to recognize that some features of crises remain ever present. A buildup of leverage – in its myriad and ever-changing forms – continues to amplify business cycles, possibly sowing future crises. Much of the leverage is of the old-fashioned variety – too much debt relative to equity or too many loans relative to loss-absorbing capital. Another variation is the degree to which overvalued collateral is used to underpin credit extension – overvalued real estate underlying mortgages is a prime example that repeats itself across countries and across time.[3]

Macroprudential regulation needs to address both the new and the old features of financial crises.[4] To mitigate risks arising from interconnectedness, complexity, and cross-institutional relationships, more information about the size, direction, and types of risk exposures across institutions will be essential – a "cross-sectional" approach to systemic risk. Being able to identify how risks are transmitted through markets and what types of market structures act to minimize such spillovers and transmission is important.[5] To dampen the procyclicality in its various forms, information is needed about how amplification and buildups in risks occur over time – a "time-series" approach to systemic risk.[6] These two main ways of looking at systemic risk will help identify the right regulatory tools.[7] But to see which macroprudential tools to employ, the correct information will need to be collected and analyzed.

A number of practical considerations arise when attempting to collect, analyze and use information for the purposes of macroprudential oversight and regulation.

- What types of information are needed (quantitative "hard" data or qualitative "soft" data)?
- How does one obtain it, and from whom (which types of institutions or entities)?
- Who bears the costs of collection and storage (private or public sectors, a mixed model)?
- Who would get to see it and in what form (data sharing, transparency, confidentiality, public disclosure)?

These practical issues are taken up below.

[3] See Kiyotaki and Moore (1997) and Bernanke et al. (1999) for the basic models that incorporate collateral to generate procyclical credit dynamics.

[4] In this chapter, the terms *macroprudential risk* and *systemic risk* are used interchangeably.

[5] Spillovers in this context occur when a financial event in one country or institution (or group of institutions) alters the value or risks in another country or institution (or group of institutions). In most cases this is viewed as sequential in nature as opposed to simultaneous.

[6] Procyclicality refers to the positive reinforcement of an existing cyclical time series variable. For instance, procyclical credit cycles would imply that there is some force making the amplitude of a credit boom higher and the credit bust deeper than a cycle without this force applied.

[7] See IMF (2011a) for a more detailed description of these two dimensions to systemic risk.

14.2 Gaps in data and information

In the recent crisis (and for that matter in previous ones), policymakers realized that they did not have enough of the right kind of information to: (1) gauge the buildup in risks; (2) understand the transmission and spillover mechanisms and their outcomes; or (3) judge the potential effectiveness of various policy or market-based solutions.

A first step to determining the types of information needed to formulate effective macroprudential regulation is to distinguish the types of issues to be addressed. Using the two main approaches to systemic risk presented above, the following discussion outlines for each type of issue the type of data needed to enable the official sector to detect a problem, potentially create a tool to mitigate it, and see if the tool is effective.

14.2.1 Issue 1: procyclicality

The accumulation of risks over a business cycle is relatively slow moving. Time-series data at a monthly frequency are probably sufficient to detect trends in credit, asset prices, and most measures of leverage.[8] There are two difficulties to be overcome: the quality of the raw data and deciding whether other supporting data are needed to determine the magnitude or amplitude of the procyclicality.[9] For instance, how much of the cyclical development in, say, credit, poses little threat to financial stability and how much reflects a dangerous acceleration likely to end in a crisis? And do housing prices reflect "fundamentals" or is housing "overvalued"?

Looking first at credit data, most central banks collect adequate information, usually quarterly, but sometimes monthly, on banks' extension of credit to various sectors – the non-bank financial, the non-financial corporate, and the household sectors are common levels of granularity. Many central banks collect data even more finely, with different industrial sectors or sizes of enterprises, different types of financial institutions, and different types of loans to households.[10] While knowing which sectors or which types of units are borrowing is important, aggregated credit variables can help reveal impending crises.

[8] Asset prices here can refer to any contract or instrument that has value to the holder. Often such assets are accompanied by payments of cash flows from one party to another. Houses are assets; mortgages are assets; stocks (equity) of firms are assets; and bonds are assets. Note that often what may be an asset to one party is a liability (an amount owed) to another. A typical mortgage based on a single family home is an asset for banks but a liability for the household.

[9] See Andritzky et al. (2009).

[10] For instance, the German Bundesbank Monthly Report contains lending by maturity and by sector, `www.bundesbank.de/download/volkswirtschaft/monatsberichte/2012/201202mb_bbk.pdf`. The European Central Bank Monthly Bulletin also contains disaggregated lending information for banks in the euro area, `www.ecb.int/pub/pdf/mobu/mb201202en.pdf`. Many countries now provide such information.

Recent empirical work by economists at the International Monetary Fund (IMF) and the Bank of International Settlements (BIS) has found that bank credit-to-GDP ratios and a bank credit-to-GDP gap (that is, deviations from a trend determined by a Hodrick–Prescott filter[11]), can signal upcoming financial crises fairly well.[12] Panel regression techniques showed that both credit measures are almost equally good at predicting crises at a one to two year horizon, even though the predictive power for either measure is moderate. The gap performs better one year ahead, whereas the growth rate of the credit-to-GDP measure does better two years in advance. A measure of credit that combines bank credit and cross-border credit does even better, but most countries do not have such data readily available.[13] Hence, information about aggregate bank credit is a good starting point for examining when business cycles may be exaggerated to a degree that they may pose systemic risks.

When combined with other variables, particularly other types of asset price data, the information content of aggregate bank credit for the prediction of crises becomes even better. For instance, the IMF study shows that growth in equity prices and real exchange rate appreciations (for emerging economies) are particularly helpful. While housing price appreciation combined with excessive credit growth is quite indicative of future crises, equity price-related indicators have the highest predictive power. For instance, if credit-to-GDP growth is at 5 percentage points, equity price growth of 10% increases the probability of a systemic financial crisis to more than 15% within the next two years.

All this suggests that credit information alongside asset price data, particularly for housing and equity markets, is a key set of building blocks for detecting procyclicality – the time series dimension of systemic risk.

Most countries still need to improve data collection. In particular, data tend to be lacking on cross-border bank credit, preferably granular enough to detect which countries' banks are lending to domestic borrowers. The currency of denomination of the loans can also be important, as lending in foreign currency usually contains more risks unless borrowers have a natural source of foreign currency with which to make loan payments (e.g., exporters or households with members who earn foreign currency overseas and send it home).

Information about where credit is being distributed can also be relevant. If much of the credit is to households for real estate purchases, then a close look at the

[11] The Hodrick–Prescott filter technique is used in a variety of economic time-series applications as it is able to flexibly determine a dynamic trend; see Hodrick and Prescott (1997).

[12] See IMF (2011b) and Drehmann et al. (2011). Many other studies have also concluded that credit measures are useful signals of excessive lending and subsequent crises, but usually for individual countries and without comparing different measures or using thresholds as these studies do.

[13] Only eight countries out of a sample of 40 advanced, emerging, and developing countries had long enough time-series data on cross-border credit to perform rigorous econometric tests.

mortgage contracts and housing prices is in order. Loosened lending standards, high loan-to-value ratios, and high debt service-to-income measures are early indications that mortgages are being granted too easily.[14] If credit is flowing to particular industrial sectors (for instance, minerals or oil), then other types of prices should be monitored more closely, particularly if these industrial sectors are pledging collateral based on its current market price.

Real estate cycles are often paired with credit cycles and deserve separate scrutiny. The affect of a real estate bust on the real economy tends to be larger than for an equity price bust, in part owing to the higher leverage taken in real estate.[15,16] Better (disaggregate) real estate price and quantity/sales data can allow policymakers to gauge where and how damaging a real estate bust might be. Aggregated data do not allow pockets of vulnerability to be seen and addressed early enough to prevent greater problems later. Rapid increases in housing prices in large metropolitan areas, a large number of purchases of second homes and areas with vacation properties can be early clues of an overdone housing market. Commercial real estate has historically been very volatile and damaging to the economy as well. Hence, disaggregated data by geographical area and by type of real estate transaction are relevant pieces of information.

Attention to turnover and other measures of the housing market (affordability indices, numbers of homes in foreclosure, new starts) are also good peripheral information to judge the sustainability of house price increases. Ideally, a fundamentals-based model of house prices, combining information about supply-side and demand-side variables influencing home ownership can also help gauge whether house price growth may present a financial stability concern.[17]

Bond markets can also harbor risks. Bonds, especially bank and non-financial corporate bonds, which are often traded OTC, are difficult to track, while sovereign bonds are typically more homogeneous and therefore easier to follow. Mortgage bonds are somewhere in the middle in terms of ease of data collection, but not every country has these types of bonds, or if they do, they may be infrequently traded.

[14] A loan-to-value ratio refers the amount of the loan relative to the value of the collateral underpinning the loan. In a mortgage it is typically the loan amount relative to the appraised value of the real estate. A debt service to income ratio refers to the amount of interest paid on a loan relative to the amount income the household receives.

[15] See Claessens, Kose, and Terrones (2011a). Their paper analyzes the interactions between business and financial cycles using an extensive database of over 200 business and 700 financial cycles in 44 countries for the period January 1960–April 2007. Their results suggest that there are strong linkages between different phases of business and financial cycles. In particular, recessions associated with financial disruption episodes, notably house price busts, tend to be longer and deeper than other recessions.

[16] See Claessens et al. (2011b). Looking at only 21 advanced countries over the period 1960–2007, this paper provides a comprehensive analysis of financial cycles. Looking at three types of cycles – credit, house prices, and equity prices – they find that financial cycles tend to be long and severe, especially those in housing and equity markets. In addition, they are highly synchronized within countries, particularly credit and house price cycles. Moreover, financial cycles accentuate each other and become magnified, especially during coincident downturns in credit and housing markets.

[17] See Hilbers et al. (2001).

The development of basic bond indices, where their construction is flexible enough that they are representative of the current bond market pricing, is useful to track developments. For instance, during the latest crisis, the ABX is being followed intensively by private market participants as well as by central banks to determine what is occurring in the structured credit market.[18] In previous crises that involved the corporate sector, high-yield (low credit rated), "junk" bond spreads were useful indicators – with spread compression signaling a buildup of risks.[19]

Gathering data on specific markets that represent more risky, leveraged transactions is also useful, but frequently these markets have short histories or are too new to have anyone routinely collect relevant data. For instance, the number of leveraged buyouts and their associated terms of financing are typically not recorded by the official sector, although private sector entities sometimes collect such information on an ad hoc basis. The increased use of leveraged loans and the lenient loan conditions are another early indicator of out-of-control credit. These are typical transactions that peak during an upswing and cause detrimental busts, especially in the non-financial corporate sector.

Generally, any leveraged transaction whose growth rate is exponential (numbers of transactions or amounts) or whose spread across various types of counterparties (some of whom may be unfamiliar with the terms and conditions of the transaction) is rapid deserves scrutiny. The purpose of the transaction, the leverage involved, the financial soundness of the counterparties, and the disclosure of the risks should be well understood by the private parties involved and the macroprudential overseers or regulators.

14.2.2 Issue 2: abrupt amplifiers and cliff effects

Credit extension is not, by itself, a measure of systemic risk or a necessary precursor to systemic financial risk. It is the leverage involved in that credit extension – and the deleveraging that occurs if the underlying collateral or loss-absorbing capital or equity is insufficient – that causes abrupt and detrimental financial difficulties.[20] Smooth and regular business cycles (perhaps a figment of economists' imagination) do not threaten financial stability. It is excessive leverage and other magnifiers

[18] The ABX, launched in January 2007, serves as a benchmark of the market for securities backed by home loans issued to borrowers with weak credit (so-called "subprime" borrowers); see Greenlaw et al. (2008).

[19] The difference in yields between "junk" (or high yield) bonds and high grade (or low yield) bonds is considered a useful benchmark. Spread compression refers to the difference in yields becoming relatively small, suggesting the risks in the high-yield market are undervalued.

[20] Deleveraging refers to a reduction in the asset (or risk exposure) relative to the capital (or equity) related to asset (or risk exposure). Deleveraging can thus occur either by selling the asset (or lowering the risk exposure) or by increasing the capital (or equity) underpinning it. Typically, the effect of deleveraging is thought to be more damaging (to the growth prospects of the economy) when it entails a sale of assets (or lower risk exposures) rather than an increase in capital (or equity).

or amplifiers that create the abrupt downturns that wreak havoc on the financial system.[21] The mechanisms described below relate to the time-series component of systemic risk.

Excessive leverage is one amplifier, but it is not the only one. Traditional measures of leverage – debt-to-equity ratios, loan-to-deposit ratios, maturity mismatches of assets and liabilities – are the usual measures of leverage and they can be constructed from data collected from the corporate and financial sectors of most advanced countries. Leverage through excessive lending activities (say, through large maturity mismatches) is most often what amplifies financial and/or credit cycles and so is not separable from Issue 1. But even for some of the concepts like maturity mismatches, the usual measures are coarse – durations of cash flows are measured in years[22] whereas most of the recent problems for bank funding were in the one-week or less maturity bucket in the interbank market. Some measures, such as debt service-to-income ratios or other "ability to pay debt" measures are less commonly collected for the household sector. While these are not leverage measures, per se, they indicate vulnerabilities to an unwinding of a credit boom where households are exposed.

For some types of enterprises it is sufficient to collect the information at fairly infrequent intervals – semi-annually or quarterly – for instance, for non-financial corporations, since the risks they pose are slow moving. But for financial institutions, measures of leverage need to be collected at finer intervals – monthly for some and even weekly or daily for others. Some entities, like hedge funds or bank proprietary trading desks, depending on their trading strategies, would need to maintain leverage information at even higher frequencies (say daily or weekly).

The physical concept of leverage – a small amount balanced against a larger amount by moving the fulcrum of the balance closer to the smaller amount – needs to be adapted to other types of financial markets. Newer non-traditional measures of leverage need to be developed and routinely collected for the detection of systemic risks arising out of less common financial transactions. For instance, some types of derivatives require little "down payment" (called initial margin when dealing with futures contracts) to maintain a position. This amount is meant to indicate an interest in maintaining a position, and daily or intraday margin calls based on the daily price changes (called variation margin) ensure risks are kept low. However, if daily mark-to-market margining is not maintained and prices move in one direction for some time, the embedded leverage (the initial margin compared to the losses on one

[21] See Danielsson and Shin (2002) for an example of methods of amplification that are endogenous to the financial system – ones that they believe are the most damaging to financial stability.

[22] Duration is a concept of a weighted average net present value of future cash flows. The weights are the years (or other time dimension) used in the average. Hence, duration is shorter than the actual maturity of a fixed income security with interim cash flows. A security that makes one cash payment at maturity has a duration that is equal to its maturity.

side of the position) could become an issue if the underlying positions need to be unwound under distressed conditions.[23]

This type of "leverage" was present in some types of structured credit derivatives in which the collateral (the structured credit product) underpinned other positions, such as the asset-backed commercial paper. When the borrowed funds needed to be rolled over and the collateral underpinning the commercial paper was worth significantly less than initially (or was perceived to be worth less), the commercial paper investors decided not to reinvest. This set in train a string of "forced sale" transactions in which the collateral (the structured credit product) had to be sold at depressed, "fire sale" prices. This led to other similar products being marked-to-market at these distressed prices, leading other investors to decide to withdraw their funding for these products, putting more downward pressure on prices and a "funding liquidity spiral" ensued.[24] The amounts outstanding of such types of leveraged transactions (the true "risk exposure") would need to be quantified (and standards agreed upon) so that, after collection, such data can be aggregated and used to see the "true" extent of leverage.

Leverage taken in other off-balance-sheet instruments may have more abrupt effects when an unwinding is initiated.[25] Hence instead of collecting information on traditional measures of leverage, it may be useful to collect information about the market conventions and structures within which an abrupt deleveraging event could occur.

- Margin and haircuts. Data on margin and haircuts (that is, the degree to which a small fraction of the outstanding position needs to be collected from one or the other counterparty) on various instruments (e.g., repurchase agreements or "repos," collateral held for OTC derivatives position taking, futures variation margin) are not uniformly collected. Often when interest rates are low and lending conditions are loose, the margin and haircuts collected against such outstanding positions are small and when interest rates rise and lending conditions are tightened, margin and haircuts are high – that is, margin and haircuts are lower in upturns and higher in downturns. This implies that there is more money available for lending during upturns and less during downturns – margin and haircuts are typically procyclical. Hence, monitoring changes in these amounts can be used to monitor excessive risk-taking. The lower the amount held to buffer against

[23] Marked-to-market means that financial positions are valued ("marked") to the most current market price so as to reflect the most current valuations. In many accounting contexts, positions are valued at "book" value, or the value at which the position was established.

[24] See Brunnermeier and Pedersen (2009) and IMF (2010a).

[25] Off-balance-sheet instruments can be a variety of instruments – contingent liabilities (such as lines of credit that have not been drawn upon) and derivatives contracts are probably the most common ones for financial institutions.

short-term price chances, the more danger there is of a forced unwinding when a large price move occurs.

- Stop-loss orders, short-sale constraints, barrier options. Stop-loss orders entail the sale of a position when a specific loss threshold has been exceeded. Short-sale constraints are mechanisms that do not allow investors to sell an asset or instrument unless they already own it. This is a method of assuming the risk of (profiting from) falling prices. Barrier options are options that can be executed if a specific price (barrier) has been reached. They represent one-way bets. If the barrier is the same for many of the options purchased, then when the price moves through the barrier the action could precipitate a further large and abrupt change in prices. The use of these accelerators varies widely and some (like stop-loss orders) are easily executable on electronic trading systems. Some are less easily automated (e.g., how many barrier options are written at a given strike price). Some are institution specific (e.g., short-sale constraints might be imposed on certain types of clients by a broker or by their own risk management systems).
- Over-reliance on credit ratings. Another amplifier is the number of investors that make investment decisions on the basis of various credit ratings agencies' judgment about credit risks. Often sales of assets occur if the underlying obligor (the company, country, or issuer) drops from a rating associated with "investment grade" to "non-investment grade." A similar situation can occur when prudential regulations require financial institutions to rely on credit ratings for the purposes of holding regulatory capital. Capital need not be held against some highly rated securities (corporate or sovereign) and if the obligor credit rating falls, more capital may be needed to be held against those assets. Hence, there may be "cliff effects" whereby many types of investors sell assets when credit ratings fall below various thresholds. The largest effects are between investment and non-investment grade, but there are still large effects when credit ratings decline at other levels (such as AA to A, as determined by Moody's credit rating agency).[26]

These mechanisms can be detrimental to systemic stability because they can cause highly concentrated selling pressure (or purchases) within a short period of time. Or in some cases the mechanism can remove automatic stabilizers from the system (e.g., the removal of short-sales allows purchases to become excessive). Unless information is routinely collected about the extent of the use of these mechanisms in markets, public authorities will have difficulty determining the buildup of vulnerabilities and the likely downward pressure when some events occur.

[26] See IMF (2010b).

14.2.3 Issue 3: data to assess interconnectedness/spillovers

Collecting information about interconnectedness relates to the cross-sectional characteristic of system risk. Hence, the first task is to develop an understanding of what types of interconnectedness can be detrimental to systemic stability. Individual institutions that are "connected" to others through bilateral or multilateral contracts do not necessarily represent a downside risk or vulnerability. Indeed, during normal times, such connections can be a benefit to the financial system. Take, for example, a central counterparty that nets down contracts between two counterparties that are redundant. The fact that there are multiple trades between two counterparties does not necessarily mean the failure of one counterparty implies the failure of the next – it depends on the nature of the transactions and their legal commitments to net the exposures. Even so, it is important for overseers of systemic stability to know what kinds of transactions bind counterparties together and the legal repercussions if one of them fails to make good on the resulting cash flow transfers.

To assess the risks appropriately, bilateral gross and net risk exposures of the largest counterparties trading an instrument (or set of instruments) or those that are major intermediaries in a given market are needed. Note that a risk exposure is not the same as a position, since a risk exposure is a forward-looking concept and takes into account the potential change in the value of a position over time. Since no one knows with certainty what the value of a position will be at any given future time, there is a probabilistic or estimation process involved.[27]

For counterparty risk – the risk stemming from the failure of a single counterparty – to be assessed, gross exposures are needed. They allow overseers (and the counterparties themselves) to judge what would happen if a counterparty failed to pay its obligations on time (with no netting available). If netting is legally valid and upheld, net risk exposures are needed to judge whether the amount of remaining exposure poses a risk. Information on netted exposures is typically more difficult to attain since various risks may be offsetting even if they are in different types of contracts, different terms to maturity, or other closely related contingent cash flows. For instance, the complete "offset" of derivative contracts often requires that the two contracts have exactly the same nominal (or base) cash flows and maturity dates. Whereas net risk exposures is a concept which would allow, say, the hedging of a commodity exposure with a closely related, but not identical, commodity index.

A somewhat different aspect of measuring interconnectedness is to determine the level of aggregation useful for an overseer of systemic stability. Should the risk exposures be collected at the level of a conglomerate (e.g., all cash flows in and out of the conglomerate to outside parties), a specific type of institution (e.g., a

[27] Note that the same is true of concepts of leverage, where the future value of the collateral or capital (or equity) underlying the leverage measure matters for the extent of risk it poses to the financial system.

commercial bank in an interbank market), within an institution (especially cross-border subsidiaries) or across a group of institutions (e.g., a group of small money market mutual funds, MMMFs, connected to a given bank or set of banks)? For each area of concern, the level of collection of the data would need to be determined based on the size and risk posed by the grouping (which, unfortunately, in some cases cannot be determined without first collecting a lot of data).

The above set of questions also bears on whether a public overseer of systemic risk should collect information about relationships, not just financial exposures. Under what conditions would an entity support another one based on their tight relationship – even if their financial contracts appear minimal? For instance, prior to the crisis some institutions decided to support other institutions which were not linked to them through actual exposures. A highly celebrated case is the decision that BNP Paribas (Banque Nationale de Paris Paribas) made to financially support its two off-balance-sheet structured investment vehicles (SIVs) even though they were self-financing subsidiaries. The issue was not of actual, legal exposure, but of reputational risk.

Along these lines is also a notion of interconnectedness that is "guilty by association." The idea is that firms with similar business models or lines of activities, even if managed differently, may appear the same to outsiders who are not aware of such nuances. For instance, when several monoline insurers were implicated in the subprime losses, nearly all monoline insurers suffered drops in their share prices or in their business activity, even though some had not insured subprime-related mortgages or the structured credit products in which they were embedded. Just being in the same business line as a competitor that runs into difficulty can be a risk when a crisis strikes. This implies that knowing which institutions (especially non-banks) are doing what can be helpful in seeing these "invisible" connections and potential contagious effects.

So far, the discussion has focused on the collection of actual risk exposure data. What about inferring such interconnections through market-based information, rather than the reporting of actual exposures? This idea gets at the notion of "guilt by association" since the market may price the securities of a group of institutions similarly even though their actual exposures (if known) would imply very different prices. Given the collection and sharing difficulties (see Section 14.5), it is very appealing to use what the market already produces (price and quantity trade data) to infer or surmise how risk exposures of various important counterparties are spread across counterparties in their respective markets. The benefits are that even if the resulting proxies are not accurate reflections of their real risk exposures, if enough counterparties *believe* that they are accurate and a problem occurs they will act as if they are the "true" exposures – causing systemic instability through say withdrawal of trading or funding a particular counterparty. This sort of "pariah" phenomenon can be deeply harmful to solvent, but now illiquid financial institutions. Indeed, market equity prices may "overshoot" in some markets, perhaps penalizing some institutions unfairly. This can occur when information available to counterparties is imprecise or non-existent and rumors lead the market, rather than facts. That said, even if not accurate, knowing what "the market" reveals about various institutions can be useful for overseers to be able to surmise which institutions are likely to suffer from rumors or other shocks. Hence, market-based information can be used productively by both market participants themselves and overseers of systemic stability.

While promising in many ways, one needs also to be cognizant of the limitations of this approach. Usually "the market" is right about vulnerabilities, but may not provide much forewarning. Recent IMF work tests the ability of various systemic risk measures to signal an upcoming financial crisis.[28] Some measures do better than others, but most market-based, high-frequency indicators are only able to signal an upcoming problem within a few weeks. Those measures that contain some aspect of interconnectedness do poorly in detecting the failure of an institution (tested by observing whether they signaled the demise of Bear Stearns or bankruptcy of Lehman Brothers). In short, most information used to provide early warning systems of systemic risk based on market data does not provide early enough warnings for policymakers to respond.

Most recently, techniques have been developed that use firm-specific information *and* integrate market-price information. This has been relatively successful in the area of stress testing where position information alongside market assessments of future default probabilities on various financial contracts can be looked at side-by-side to examine, say, solvency conditions or liquidity risks.[29] In some cases, one can use market data alongside individual position data to condense the position data into a smaller set of individual specific risks and then use this smaller set (by institution) to aggregate across institutions. These techniques are still quite nascent and have not been widely used, but can help ameliorate the large collection of potentially confidential data, lowering the costs of monitoring.

14.3 Gaps in "soft" data

There are several types of "soft" or non-quantitative data that could usefully be monitored to better anticipate issues leading to financial instability. Some set of this "soft" data can be renamed "market intelligence" that is picked up by public

[28] See IMF (2011b).

[29] Barnhill and Schumacher (2011). The default probability refers to the probability that an institution defaults on their debts (fails to pay either interest or principle) and are calculated using various data and techniques depending on the final use of the information and data availability.

authorities' discussions with various market participants, often by asking detailed questions, performing surveys, and "behind the scenes" types of investigations.

New product developments. Typically, volumes of new products that potentially cause financial instability start out small, but grow exponentially. Often they are to take advantage of some market failure or regulatory loophole (e.g., difference in regulatory treatment across products or jurisdictions).[30] Understanding who is attracted to the new product and why is important. What do buyers (and sellers) hope to gain from the product? Is there enough understanding about its true risks? This information usually starts out as anecdotal information culled from conversations with market participants or from business periodicals (when the developer is willing to talk about it). When the activity or product is growing rapidly and has reached an appreciable size, quantitative data (usually prices and quantities) are typically requested by regulators or supervisors. Unfortunately, by the time the hard data arrive, the risks may already be systemic. Hence it is important to spend some resources watching the newly sprouted financial products, ascertain which are growing rapidly, and then look carefully for unintended consequences, insufficient transparency of risks and returns, and buyers who are unable (or not incentivized) to understand the risks they are undertaking.

An open question is whether innovative new products need to undergo some screening process by a regulatory body. This would undoubtedly slow the process of getting new products to market, many of which may be valuable for customers and pose little risk to the system. However, a few of them may be ill conceived, marketed to the wrong clients, or have financial stability consequences and thus society may benefit from a pre-issuance testing procedure.[31]

Legal, governance, and risk management structures. These data are difficult to collect in the usual form but are critical to be able to anticipate the firm's reactions to negative events and any subsequent spillovers. The legal obligations and the decision-making apparatus within a financial entity will often determine the counterparty risk, the reaction to trading violations, transfer of liquidity from one legal entity to another, and so on. Relationships between legal entities within a given institution may be quite important for the institution to avoid insolvency or a liquidity shortage. This is especially true in cross-border contexts where subsidiaries or branches of banks entail different access to funding or capital from the parent, differential

[30] For instance, some tranched securitizations were constructed to have an AAA-rated tranche to make them attractive to banks because no regulatory capital was required to be held against AAA securities.

[31] The US Commodity Futures Trading Commission (CFTC) used to require that new futures contracts, proposed by US futures exchanges, be approved by the CFTC prior to trading. The criteria included (among others) a demonstration that the contract would have bona fide hedgers and that its price could not be manipulated. Current rules allow trading to take place alongside a self-certification process in which it is assumed the CFTC will approve the contract within 90 days unless it violates the Commodity Exchange Act, www.cftc.gov/IndustryOversight/ContractsProducts/ListingProcedures/listingprocedures.

oversight by home and host supervisors (and different information sharing arrangements). When adversity hits a financial institution, knowledge of which legal entities are connected to other parts of the institution can help determine how various parties may position themselves (e.g., whether the assets of foreign subsidiaries can be "ring-fenced" by the authorities in other countries).

The recent move by the international regulatory community to create a single "legal entity identifier" or LEI for each legal entity active in financial markets is an important step in the right direction. Such a system, when in place, could be particularly helpful in determining relationships and likely spillovers both within institutions and across institutions based on legal entity. In March 2011, the Financial Stability Board (FSB) provided clarity on two technical points in order to give early guidance to the financial industry to help with expected forthcoming proof of concepts and to provide initial direction on the work of the Expert Group.[32] First, the Expert Group has agreed that a 20-character alpha-numeric code is a good basis for the global LEI code. Second, the Expert Group identified six data elements that will all form part of the minimum set of reference data regulators will require on the launch of the LEI. The FSB is planning to fulfill the G20 mandate "to prepare recommendations for the appropriate LEI governance framework, representing the public interest," by the G20 Summit in June 2012.

In addition to formal legal structures, the overall corporate governance structure and risk management governance structure within institutions are of prime importance. Compensation schemes, risk limitations, and economic capital allocations all contribute to the expected response of an institution to trouble and to its risk-taking actions during normal times. Certain steps to curb aberrant behavior by individuals ("rogue traders") and overzealous trading in certain activities can be taken by regulators and supervisors if they understand the internal incentives of the firm's management and employees.

Efforts in this direction are ongoing, though supervisors are often reluctant to consider compensation schemes as part of their supervisory ambit, viewing this area as outside their expertise. In October 2011, the results of an FSB-conducted peer review were released. The review surveyed supervisors and regulators of FSB member jurisdictions and 70 banks and broker-dealers that FSB members deemed as significant for the purpose of the review. The review examined compliance with the FSB Principles for Sound Compensation Practices and their Implementation Standards (Principles and Standards, P&S). These practices have been developed to align compensation with prudent risk-taking, particularly at significant financial

[32] The Expert Group is made of up FSB member country representatives and is supported by an Industry Advisory Panel composed of 34 representatives from different sectors and regions to help provide important industry input into the global public-private LEI initiative. See also Chapter 11 of this Handbook.

institutions, although importantly, they are not intended to prescribe particular practices or amounts of compensation. The review found there was good progress toward the P&S but that more progress was needed to reach implementation across the FSB members.

Customer and intermediary types, especially non-banks. It is important to know how financial institutions interact with each other and with "end users," comprising other financial institutions, non-financial corporations, or households (some of whom may be considered "sophisticated" investors). Examining types of entities that are closely related (in "clusters") or are in linearly connected relationships (e.g., through a mortgage securitization chain) will help detect the type of data (by types of institutions or location) to be collected. This constitutes part of the cross-sectional approach to systemic risk measurement.

In addition, keeping track of new types of entities or the concentrated use of a particular product or instrument by one type of customer (e.g., subprime borrowers) can be helpful to see how the financial system may be developing vulnerabilities. This is a situation that can quickly arise in less developed financial systems where new ways of intermediation are constantly emerging as the economy and financial sector outgrows more traditional institutions. Informal mechanisms of transacting, often through non-bank financial institutions, are notoriously underregulated during their early existence and often are the location of vulnerabilities and unanticipated risks.

Regulatory gaps (e.g., unregulated mortgage brokers in the USA) can also allow the growth of new entities that do not have appropriate oversight and often have incentives that lead them to grow too quickly without adequate risk management, sometimes threatening financial stability. Diversity of types of positions and types of institutions can be a stabilizing factor. In addition, diversity may reduce existing inefficiencies in intermediation as a particular type of institution may be particularly effective at certain types of transactions or the production of certain types of market infrastructures. Hence, the idea is not to limit automatically new or different entities, but to watch for concentration or other risks (either stemming from their own activities or the types of risk they distribute to customers).

14.4 Gaps of reporting entities – from whom to collect the information?

Much of the information for effective macroprudential policies is already available. Some information that is collected in monetary surveys or other official central bank macroeconomic databases is readily available to re-examine with financial stability in mind, rather than for its for original purpose. Examples are private sector bank credit data, money supply (with various definitions), deposits, and a host of bank balance sheet and off-balance-sheet information. Often the issue with these data is that either they are not recorded frequently enough or they do not have enough granularity (for instance, they may lack the type of credit provided, the type of borrower, the sector for which the credit is provided, the currency in which the credit is granted, or the maturity of the instrument or loan). For instance, knowing that a large proportion of adjustable rate mortgages that carry market "teaser" rates (initial rates well below market rates) will have an interest rate adjustment in the next six months may have an important bearing on the solvency of the mortgage borrowers (as it did in the USA). So the first task is to review the data already collected to see what adjustments should be made. In some cases, the only adjustment may be not to aggregate, but to provide the data instead in raw form to those evaluating systemic stability.

New data initiatives may well be aimed at entities that have not officially reported data to the official sector (or anyone else) before. Here, the need to collect information from various entities should be determined by what they do – their activities, not their corporate or legal identity. For instance, some non-financial corporations provide bank-like services but are not overseen by bank supervisors. Moreover, it is important to weigh the costs and benefits of collection before requesting it. One problem is that without looking at the data first, it is hard to know what is worth collecting. In this case a small prototype sample survey or other small-scale effort (with a request for an estimate of the costs) could be a useful starting point. In addition, finding out what data the new entities already produce in house for their own purposes would be relevant.

Just because the entity does not find it in their interest to collect or report information internally should not be used as an excuse by authorities not to request it – even if it is costly to produce – as it depends on its importance at detecting systemic risks. The nature of systemic risk is that it is not necessarily the sum of individual risks and hence it may not be in the incentive of individual institutions to collect relevant information themselves. Also conceivable is a situation in which the entity does not aggregate the data (or net them) in the same way a macroprudential regulator would. For instance, a large bank may not find it useful to know that it has sold certain products to a large number of smaller banks as long as it does not have an undue concentration risk to any one bank. A macroprudential regulator, by contrast, would want to know if (across those smaller banks) there was an aggregate risk that exceeded some threshold that could be dangerous to the system. Credit registers in some countries (e.g., Germany) can go some way to capturing high exposures that many banks may have to the same counterparties, but are typically only related to direct credit exposures.

In October 2011, the FSB released a set of recommendations to strengthen oversight and regulation of the "shadow banking system," which the FSB broadly describes as "credit intermediation involving entities and activities outside the regular banking system."[33] The FSB advised authorities to narrow their focus to those types of non-bank credit intermediation that have the potential to pose systemic risks, by focusing in particular on those involving the four key risk factors: (i) maturity transformation; (ii) liquidity transformation; (iii) imperfect credit risk transfer; and/or (iv) leverage. Then, by looking at other factors, such as the interconnectedness between the shadow banking system and the regular banking system, authorities could assess the impact of certain shadow banking entities/activities on the overall financial system. It is notable that the FSB focused on the functions that entities perform and their activities rather than their legal form. Frequently, regulatory oversight is only based on the legal definitions of an entity and fails to take into account how that entity functions within the financial system.

Aside from the criteria underpinning the shadow banking system, other potential types of entities that this crisis suggests should be better understood include the following.

- Fast-moving entities. The ability to move vast sums of money in short periods of time should influence which entities need to report various types of information (e.g., so-called high-frequency traders or algorithmic trading systems). Any entity that can influence the price of an asset or instrument single-handedly needs to be monitored. Large trade reporting systems are a good start, but it is not just size, but speed, that can matter.
- Entities whose activities are vital and not substitutable to ensure system-wide stability. This set of entities would include those whose withdrawal from a market may be disruptive. Obvious examples are repositories, central counterparties, or clearing houses for equities, bonds, foreign exchange, and derivative transactions (e.g., Bank of New York Mellon, BoNY, or Intercontinental Exchange, ICE). Trading venues can also be important to understand well enough to keep functioning.

It is important also to define entities that can be left outside the perimeter of information collection – those that are too small or too inconsequential. There should be some reasonable cut-off points so that the burden of reporting information is not prohibitively high. Nearly every reporting system has a "de minimus" category to exclude small participants. That said, sometimes groups of smaller institutions *do* make a difference for financial stability, especially if they have similar positions or activities, or are regarded as exemplary of a particular category of risk – a set of relatively small hedge funds all using a similar strategy may qualify. As noted above, a watch list should be established for the activities of institutions deemed to be highly innovative and that are growing exponentially (see Section 14.2 above).

[33] See FSB (2011a).

14.5 Costs and benefits of data collection

The benefits of better macroprudential data have been extolled above, but have to be weighed against the costs of collection. Data collection is not free – not for those submitting the information, not for those storing the information, and not for those processing and interpreting the information. Fortunately, the costs of data collection are mostly fixed costs, with the variable costs lower. That is, once a data template is constructed and linked with the electronic information system in an institution, the cost of submitting the same information in the same form decreases over time. One-off requests at the institutional level, however, are very expensive, so it makes sense to have a plan about how to use the information in advance.

Even though fixed costs are probably the bulk of the total costs, the variable costs cannot be ignored either. The costs of validation, storage, versioning, auditing, security, and archiving are not insignificant. One issue is whether the entity from which the data are collected should bear these costs or whether the recipient should do so. Some sharing of this burden seems appropriate since the cost of a systemic crisis is borne by all in the system (including the institutions that would provide the data) although the onus for observing its probability of occurrence and magnitude are on the overseers of the system.

The costs are also related to the degree of privacy or confidentiality that needs to be maintained. If the official sector collects publicly available information, for example, some types of trade data from a feed maintained by a data vendor (say, from an exchange), the amount of effort needed to maintain confidentiality is non-existent. If one collects private information, but information that is distributed to, say, contributors to a trade repository or a private data vendor, then there may be privacy or security concerns. If one collects position data from individual institutions, even if aggregated by instrument or counterparty, there would be need for solid protections so only intended recipients have access to the information (see below).

As a first step, it is useful for the official sector to inventory what is already collected from different sources and processors (both private and public) and consider how it might be repurposed. By doing this, it will likely lower the costs to the institutions that would normally provide information and data.

The following could be used to see what is already collected at more granular levels, but is not reported or aggregated in a useful way (perhaps even within an institution).

- What can bc composed from individual financial institutions' risk management systems.
- What can be composed from existing warehouses/repositories for trades.
- What can be collected in the same way in multiple cross-border jurisdictions.
- What can be collected from electronic trading sources without disturbing participants' activities.

As emphasized above, the data that are most useful have some forward-looking risk component, and so much of the raw information needs to be transformed to judge what it implies for risk buildups along various dimensions. For this reason, individual institutions' risk management systems may be the farthest along in this quest for risk exposure information that can be aggregated across institutions or markets. Even so, various concentrations can be seen relatively easily from trade warehouses and sometimes from clearing houses for some types of instruments (notably, exchange traded futures and, increasingly, over-the-counter derivatives).

As a final point, before asking for more data, one needs to analyze how the existing data could be used to enhance financial stability assessments. There is one fundamental problem with this approach of "knowing how to use it in advance of receiving it" – there are new ways of looking at risks that are discovered by looking at the raw data, data that had previously not been examined in that way. Hence, there may need to be an "overage" or "excess" data collection – i.e., more data than thought necessary to ensure there is scope for new measures of systemic risk.

14.6 Who should be able to see the data and why

14.6.1 Public disclosure for market discipline

The normal supposition is that the information collected for detecting and mitigating systemic risks will not be made public. That is, it will be collected by a macroprudential overseer (or by a set of regulators or supervisors) and kept within the confines of the public sector (even if shared across borders between macroprudential units or other regulatory or supervisory bodies). However, which data should and should not be available to the public is a decision to be made based on the circumstances.

One of the reasons the crisis was so deep and long-lasting is that market discipline – owners of equities, bonds, and other securities and counterparties of financial intermediaries – did not operate effectively. For market discipline to work, in the first instance, those doing the disciplining have to have enough information about the risks they have in these institutions to be able to price their respective exposures to them. These exposures could be in the form of equity prices, bond prices, credit limits, haircuts or margin, and so on. In general, if the firm is becoming riskier, providers of capital and other funds would require higher returns, thereby making it more expensive for the firm to take on additional risks.

However, enough information is only one prerequisite to getting market discipline to work – the other is that these discipliners need to have the incentive to do so.[34] For instance, there are several well-known cases where "herding" occurs that

[34] See Rajan (2005).

circumvents the incentives to use the information at hand. For instance, some studies have shown that mutual funds managers' performance can be linked to specific momentum strategies and not to using information that would show their ability to "stock pick."[35] That said, if enough varied types of holders of securities have access to information, they can, at least in principle, discipline excessively risky behavior of an individual firm.[36]

One thorny issue is whether individual investors and counterparties can appropriately reward good (or bad) systemic risk behavior if they see it in an institution. In general, the view is that systemic risk is greater than the sum of the individual risks and so is an externality that only a regulator or the official sector (and even then, perhaps only a systemic risk overseer) can address. This may be taking the argument too far. Leaving aside for the moment the possibility of official support for systemically important institutions (and the associated moral hazard),[37] if investors had information that suggested that firm XYZ is contributing significantly to systemic risk, and if such risk materialized, the firm (including the firm's investors) would suffer greatly, investors could increase the required rate of return on their assets to ensure the systemic risk event is covered. This argues that even information on the contribution of individual firms to systemic risk could provide market discipliners with a means to appropriately price in some of the systemic risk (at least to the extent it affects an individual firm).[38]

Hence, in an ideal world, much of the data to be collected for the purposes of financial stability should be released to the public and, in theory, investors could use it to price appropriately the risks undertaken by individual institutions and even those that relate to systemic risk. However, there are certain cases in which more disclosure is welfare-reducing, even abstracting from the costs of collection, so these cases need to be considered more carefully. Such cases include providing information that may cause the insolvency or a liquidity shortfall for an institution, which is more likely to occur during unsettled, uncertain, or volatile economic or market conditions. Hence, ideally, public releases of information should be made

[35] See Grinblatt et al. (1995) and Daniel et al. (1997) for evidence on mutual fund performance.

[36] However, in markets where arbitrage is undertaken by a small set of highly specialized investors, Shleifer and Vishny (1997) show that if deviations from fundamentals-based information are too large, arbitrage will become ineffective.

[37] Moral hazard refers to a tendency to take undue risks because the costs are not borne by the party taking the risk. A party makes a decision about how much *risk* to take, while another party bears the costs if things go badly, and the party insulated from risk behaves differently from how it would if it were fully exposed to the risk.

[38] Acharya et al. (2010b) argue that their expected shortfall measure could be used to the rank the systemic risk generated by financial institutions. While not expressly touted for its ability to influence investors, the authors of the systemic risk rankings suggest that "by making these rankings available over the web, we hope they enhance the dialog among academics, practitioners, and regulators alike, generating new ideas for measuring systemic risk," Professor Engle explained. "Because they use available market data, the ultimate goal of our work is to provide a new tool to evaluate global systemic risk that is more efficient, replicable and transparent than individual scrutiny of confidential company financial and accounting data."

routine in "good times" so that the fact the information itself is being released does not cause the problem it is trying to prevent during "bad times."

The release of stress-test results on individual financial institutions is a case in point. If stress tests were performed routinely and the information commonly released, the fall of an institution into a category suggesting it needed more capital would be less problematic for the system than if only one stress test is performed in the midst of a crisis. In cases where it is necessary to release information inimical to the health of individual institutions, public authorities should be prepared to mitigate its knock-on effects. The results should be released alongside a solution for these near-insolvent institutions – such as guaranteed recapitalization, restructuring plans, sales of identified business activities, or a plan to sell assets in an effort to deleverage. This was done in the case of the US bank stress tests, the Supervisory Capital Assessment Program or SCAP, in 2009.[39] Though not without risks, the idea of concurrently indicating a "plan B" solution alongside revelation of the problem is much better than no solution at all. Without a "plan B" the official sector could run real risks of widespread problems if the stress tests show weak institutions.

14.6.2 Public disclosure for self-discipline

In addition to reasons for disclosure to ensure market discipline, some forms of disclosure could aid in "self-discipline" – that is, disclosure aimed at informing wholesale market participants about the risks of certain products or strategies so that they can avoid over-investing in these themselves. If they are more cognizant of the risks, they may take smaller positions, commensurate with the risks. Traditionally such disclosure has been targeted at retail investors under the term "consumer protection," but disclosure should be extended to larger, so-called "sophisticated" financial entities that interact in wholesale financial markets. Indeed, it was these types of institutions (regional banks and other relatively large institutions) that held the securities (e.g., credit derivatives such as collateralized debt obligations (CDOs)) that suffered dramatic declines in value – declines that came as a surprise to some of them that did not truly understand their risk characteristics. Hence, in addition to disclosure of information to aid presumably unsophisticated retail consumers, disclosure may be advisable to ensure even some large financial institutions can perform the due diligence to make well-founded decisions (e.g., data about the pools of mortgages held in a SIV that underpin a structured credit product).[40]

[39] See Federal Reserve Board (2009).

[40] There are a number of recommendations for more information about the underlying assets and risks in securitizations that have followed the crisis. See IMF (2009) for initial proposals and Altman et al. (2010) for a discussion about regulations now in force in the USA regarding the transparency credit rating agencies must provide to the public about their ratings methodology and information underpinning their ratings.

14.6.3 Non-disclosure of proprietary information

Certain types of data could be misused by market participants. If position data are released at too granular a level other market participants could use them to alter their trading strategies to their own benefit.[41] For instance, a trader may use the information to position themselves in anticipation of another party's need to augment or reduce a position, thereby taking advantage of the resulting price impact. This, and other types of behavior based on specific position information, could lead to price volatility unrelated to the underlying value of the instrument. Situations could develop in which possibly destabilizing declines in market liquidity occur if market participants decide to avoid trading with parties thought to have excessively large exposures. In other words, trying to uncloak the "invisible hand" of classical economics, whereby neither buyers nor sellers know the other sides' inventory or holdings, may prevent markets from performing their job of finding prices to equilibrate buyers' and sellers' desires. Admittedly, it is rare for markets to operate with no inkling of participants' positions, but to disclose information that would guarantee that specific positions are transparent in real time would be going too far.

Even though very detailed position data should not be made public, supervisors and regulators should collect such information and use it to examine the concentration of positions across counterparties and their interconnections. The purposes of such information would be to improve their ability to anticipate the spillovers or knock-on effects across institutions that may cause adverse effects on the system as a whole – aiding the ability to see the cross-sectional component of systemic risk. If such information were known far enough in advance, supervisors and regulators (or macroprudential overseers) could impose some types of controls or other regulatory restraints on certain activities when they are deemed to threaten the system-wide stability. For instance, limits on foreign currency exposures are a fairly common tool used by bank supervisors in emerging market economies. One could envisage a similar type of concentration limit that would take into account the interconnectedness of institutions. Such preventive types of tools could be imposed based on a certain percentage of total asset or liability holdings, or OTC derivative exposures, with one or more interconnected entities. Overseers of financial stability should be able to anticipate cases of potential distress using proprietary information and, possibly, prevent the buildup. Barring this, they could use it to prepare themselves for the aftermath.

A case could be made that even some types of aggregate information, if disclosed during a period of high uncertainty, could be disruptive to markets. In these cases,

[41] See Admati and Pfleiderer (1991) who show that self-revelation of one's own trading quantity prior to trading can reduce trading costs for the initiator, but has an ambiguous affect on other traders and overall welfare.

either the overseer should refrain from releasing the information (though the rumors about what it may show may still precipitate the event that the overseer is attempting to avoid) or the information needs to be accompanied with other policy solutions aimed at combating the anticipated disruption.

When calm returns to financial markets, it may be possible to release a regular set of disclosures so that the next time a problem arises, the information is already in the public domain. A regular, anticipated, set of aggregated disclosures could desensitize markets to the information in good times and hence be less likely to aggravate a situation in bad times. The buildup of financial strength indicators (FSIs) and their now widespread dissemination is an example where a fixed set of information (after the Asian Crisis) was viewed as necessary and a large set of countries decided to produce and distribute it.[42]

14.7 Conclusions and challenges

There are a number of ancillary challenges posed by the collection and analysis of macroprudential data.

Domestic and international consistency. While a regulator or supervisor will be inherently interested in their own country's financial stability and will likely use the tools legally available to them, they will still need to pay attention to information from abroad. Some of this information may be publicly available and easily collected, other data may be able to be shared under certain circumstances, and still other types of data may be unable to be shared at all.[43] With the rising potential for external shocks and spillovers to affect a domestic economy, the international regulatory community should strive, wherever feasible, to collect data that can be aggregated and combined in meaningful ways to observe cross-border vulnerabilities. In a world in which much of the trading takes place on global platforms, it makes sense to coordinate the data collection and reporting requirements.

To the extent that jurisdictions are not allowed to share relevant information with regulators in other jurisdictions (even on a confidential basis) because of legal constraints, efforts should be undertaken to remove these constraints, including by redrafting and passing new legislation. The excuse that changing or eliminating restrictive laws is too difficult is no longer acceptable in a world where the effects of financial crises are large losses of economic growth, high rates of unemployment,

[42] See the IMF website for Financial Strength Indicators database at `http://fsi.imf.org`.

[43] Bank supervisors are often legally bound to not disclose proprietary data collected from banks. While cross-border sharing of information is sometimes facilitated by Memoranda of Understanding (usually for institutions that are connected across borders through subsidiaries or branches), ability to see second or third round connections is obviated.

and huge fiscal costs.[44] Although altering long-standing laws will take time, it is important for regulators and supervisors to begin the process of urging legislative bodies to take action as soon as practicable.

International fora such as the G20, IMF, FSB, BIS, BCBS (Basel Committee on Banking Supervision), International Organization of Securities Commissions (IOSCO), International Association of Insurance Supervisors (IAIS), International Accounting Standards Board (IASB), and others can be useful in establishing best practices in a number of data collection areas.[45] By ensuring consistency across these institutions on the basics, these institutions can reduce costs by providing guidance, templates, and consistent concepts for data collection. As noted above, an internationally agreed common Legal Entity Identifier is an important step to provide a lowest common denominator from which to aggregate along various types of business lines, institutions, and instruments. Consistency and continuity with already-existing work in the area of data collection is a worthy goal.

Potential need for new institutions. To the degree that new types of data from new sources need to be collected and there are no natural, extant institutions to do so, new institutions may need to be created. The typical domestic statistical office, for instance, would not have the expertise or means to collect high-frequency data on trading activities. Similarly, banking supervisors may not have the mandate to collect information from non-bank entities, even if their activities are bank-like.

A stand-alone institution to collect data may not be the most efficient if there is an existing set of institutions already collecting data from various sources. However, to ensure the data are deployed to examine and analyze systemic risks – to "connect the dots" – there may be a need to house the data on a common platform or location. Indeed a new institution for information collection may also be relevant if that same institution is also overseeing macroprudential policies (such as the Office of Financial Research in the USA that has been given extraordinary legal powers to collect information relevant to the monitoring of systemic risk on behalf of the Financial Stability Oversight Council).

A final note of caution is relevant. Too much data and too many coordinating institutions can lead to a false sense of security among the public, financial institutions, and politicians. The impression will be that armed with a wealth of data a supercomputer is able to compute and examine all the potential indicators, producing flashing yellow and red lights to alert authorities in real time about imminent

[44] Changing laws to share information across jurisdictions carries with it the possible stigma that a jurisdiction may appear to have weak supervisory institutions if the information shared showed that risks had built up but few preventive actions were taken.

[45] See the jointly produced implementation progress report by the FSB Secretariat and IMF staff, The Financial Crisis and Information Gaps, in June 2011 and by the BIS staff, IMF staff, and FSB Secretariat, Macroprudential Policy Tools and Frameworks Progress Report to G20, in October 2011, for recent information about progress on these fronts, FSB (2011b) and BIS (2011), respectively.

problems. By contrast, the key will be to distill from pertinent information the buildup of system-wide risks in a way that provides clues about the answers to the "who, what, where and why" questions. Moreover, designed correctly, the appropriate collection of macroprudential data can contribute to sensible and balanced implementation of policy measures to mitigate systemic risk.

Acknowledgements

I thank Srobona Mitra and Eli Canetti for comments on an earlier draft of this chapter.

Bibliography

Acharya, Viral V., Thomas F. Cooley, Matthew P. Richardson and Ingo Walter (eds.), 2010a, *Regulating Wall Street: The Dodd-Frank Act and the New Architecture of Global Finance*, Wiley.

Acharya, Viral, Lasse Pedersen, Thomas Philippon and Matt Richardson, 2010b, Measuring Systemic Risk, NYU Stern School of Business Working Paper, New York, http://vlab.stern.nyu.edu/public/static/SR-v3.pdf

Admati, Anat and Paul Pfleiderer, 1991, Sunshine trading and financial market equilibrium, *Review of Financial Studies*, **4**, 443–482.

Altman, Edward I., Sabri Oncu, Matthew Richardson, Anjolein Schmeits and Lawrence J. White, 2010, Regulation of ratings agencies, in: V. Acharya, T. Cooley, M. Richardson and I. Walter, (eds.), *Regulating Wall Street: The Dodd-Frank Act and the New Architecture of Global Finance*, Wiley, ch. 15, 443–468.

Andritzky, Jochen, John Kiff, Laura Kodres, Pamela Madrid, Andrea Maechler, Aditya Narain, Noel Sacasa and Jodi Scarlata, 2009, Policies to Mitigate Procyclicality, Staff Position Note No. 2009/09, International Monetary Fund, Washington, DC, www.imf.org/external/pubs/ft/spn/2009/spn0909.pdf

Bank of International Settlements, Financial Stability Board, and International Monetary Fund, 2011, Macroprudential Policy Tools and Frameworks: Progress Report to G20, October, www.imf.org/external/np/g20/pdf/102711.pdf

Barnhill, Theodore and Schumacher, Liliana, 2011, Modeling Correlated Systemic Liquidity and Solvency Risks in a Financial Environment with Incomplete Information, Working Paper, No. 11/263, International Monetary Fund, Washington, DC, www.imf.org/external/pubs/ft/wp/2011/wp11263.pdf

Bernanke, Ben S., Mark Gertler and Simon Gilchrist, 1999, The financial accelerator in a quantitative business cycle framework, in: J. B. Taylor and M. Woodford (eds.), *Handbook of Macroeconomics*, First edition, Elsevier Press, Volume 1, ch. 21, 1341–1393, www.faculty.wcas.northwestern.edu/~lchrist/course/Czech/BGG%201999%20Handbook%20chapter.pdf

Brunnermeier, Markus and Lasse H. Pedersen, 2009, Market liquidity and funding liquidity, *Review of Financial Studies*, **22**(6), 2201–2238.

Claessens, Stijn, Ayhan Kose and Marco Terrones, 2011a, How Do Business and Financial Cycles Interact? Working paper, No. 11/88, International Monetary Fund, Washington, DC, www.imf.org/external/pubs/cat/longres.aspx?sk=24805.0

Claessens, Stijn, Ayhan Kose and Marco Terrones, 2011b, Financial Cycles: What? How? When? Working Paper No. 11/76, International Monetary Fund, Washington, DC, www.imf.org/external/pubs/cat/longres.aspx?sk=24775.0

Daniel, Kent, Mark Grinblatt, Sheridan Titman and Russ Wermers, 1997, Measuring mutual fund performance with characteristic-based benchmarks, *Journal of Finance*, **LII**(3), 1035–1058.

Danielsson, Jon and Hyun Song Shin, 2002, Endogenous Risk, London School of Economics, Working Paper, http://hyunsongshin.org/www/risk1.pdf

Drehmann, Mathias, Claudio Borio and Kostas Tsatsaronis, 2011, Anchoring countercyclical capital buffers: the role of credit aggregates, *International Journal of Central Banking*, **7**(4), 189–240, www.ijcb.org/journal/ijcb11q4a8.pdf.

Federal Reserve Board, 2009, "The Supervisory Capital Assessment Program: Overview of Results, May, The Board of Governors of the Federal Reserve System, Washington, DC, www.federalreserve.gov/newsevents/press/bcreg/bcreg20090507a1.pdf

Financial Stability Board, 2011a, Financial Stability Board Publishes Recommendations to Strengthen Oversight and Regulation of Shadow Banking, Press Release, No. 55/2011, www.financialstabilityboard.org/press/pr_111027.pdf

Financial Stability Board Secretariat and the International Monetary Fund Staff, 2011b, The Financial Crisis and Information Gaps, June, www.financialstabilityboard.org/publications/r_110715.pdf

Greenlaw, David, Jan Hatzius, Anil K. Kashyap and Hyun Song Shin, 2008, Leveraged Losses: Lessons from the Mortgage Market Meltdown, U.S. Monetary Policy Forum Report No. 2, Rosenberg Institute, Brandeis International Business School and Initiative on Global Markets, University of Chicago Graduate School of Business.

Grinblatt, Mark, Sheridan Titman and Russ Wermers, 1995, Momentum investment strategies, portfolio performance and herding: a study of mutual fund behavior, *American Economic Review*, **85**, 1088–1105.

Hilbers, Paul Louis Ceriel, Qin Lei and Lisbeth Zacho, 2001, Real Estate Market Developments and Financial Sector Soundness, Working Paper, No. 01/129, International Monetary Fund, Washington, DC, www.imf.org/external/pubs/ft/wp/2001/wp01129.pdf

Hodrick, Robert and Edward C. Prescott,1997, Postwar U.S. business cycles: an empirical investigation, *Journal of Money, Credit, and Banking*, **29**(1), 1–16.

International Monetary Fund, 2009, Restarting Securitization Markets: Policy Proposals and Pitfalls, ch. 2, October, Global Financial Stability Report, International Monetary Fund, Washington, DC, www.imf.org/External/Pubs/FT/GFSR/2009/02/pdf/chap2.pdf

International Monetary Fund, 2010a, Systemic Liquidity Risk: Improving the Resilience of Financial Institutions and Markets, ch. 2, October, Global Financial Stability Report, International Monetary Fund, Washington, DC, www.imf.org/External/Pubs/FT/GFSR/2010/02/pdf/chap2.pdf

International Monetary Fund, 2010b, Uses and Abuses of Sovereign Credit Ratings, ch. 3, October, Global Financial Stability Report, International Monetary Fund, Washington, DC, www.imf.org/External/Pubs/FT/GFSR/2010/02/pdf/chap3.pdf

International Monetary Fund, 2011a, Macroprudential Policy: An Organizing Framework, International Monetary Fund, Washington, DC, www.imf.org/external/np/pp/eng/2011/031411.pdf

International Monetary Fund, 2011b, Towards Operationalizing Macroprudential Policies: When to Act, ch. 3, October, Global Financial Stability Report, International Monetary Fund, Washington, DC, www.imf.org/External/Pubs/FT/GFSR/2011/02/pdf/ch3.pdf

Kiyotaki, Nobuhiro and John Moore, 1997, Credit cycles, *Journal of Political Economy*, **105**(2), 211–248.

Rajan, Raghuram, 2005, Has Financial Development Made the World Riskier, NBER Working Paper, 11728, National Bureau of Economic Research, Cambridge, MA, www.nber.org/papers/w11728

Shleifer, Andrei and Robert W. Vishny, 1997, Limits to arbitrage, *Journal of Finance*, **52**(1), 35–55.

Index